M	NGC	Description	1980[a] Right Ascension (α) h m	Declination (δ) ° '	m_v
56	6779	Globular cluster in Lyra	19 15.8	+30 08	8.3
57	6720	Ring Nebula; planetary nebula in Lyra	18 52.9	+33 01	9.0
58	4579	Spiral galaxy (SBb) in Virgo	12 36.7	+11 56	9.9
59	4621	Elliptical galaxy in Virgo	12 41.0	+11 47	10.3
60	4649	Elliptical galaxy in Virgo	12 42.6	+11 41	9.3
61	4303	Spiral galaxy (Sc) in Virgo	12 20.8	+04 36	9.7
62	6266	Globular cluster in Scorpio	16 59.9	−30 05	7.2
63	5055	Spiral galaxy (Sb) in Canes Venatici	13 14.8	+42 08	8.8
64	4826	Spiral galaxy (Sb) in Coma Berenices	12 55.7	+21 48	8.7
65	3623	Spiral galaxy (Sa) in Leo	11 17.8	+13 13	9.6
66	3627	Spiral galaxy (Sb) in Leo; companion to M65	11 19.1	+13 07	9.2
67	2682	Open cluster in Cancer	8 50.0	+11 54	7
68	4590	Globular cluster in Hydra	12 38.3	−26 38	8
69	6637	Globular cluster in Sagittarius	18 30.1	−32 23	7.7
70	6681	Globular cluster in Sagittarius	18 42.0	−32 18	8.2
71	6838	Globular cluster in Sagitta	19 52.8	+18 44	6.9
72	6981	Globular cluster in Aquarius	20 52.3	−12 39	9.2
73	6994	Open cluster in Aquarius	20 57.8	−12 44	
74	628	Spiral galaxy (Sc) in Pisces	1 35.6	+15 41	9.5
75	6864	Globular cluster in Sagittarius	20 04.9	−21 59	8.3
76	650	Planetary nebula in Perseus	1 40.9	+51 28	11.4
77	1068	Spiral galaxy (Sb) in Cetus	2 41.6	−00 04	9.1
78	2068	Small emission nebula in Orion	5 45.8	+00 02	
79	1904	Globular cluster in Lepus	5 23.3	−24 32	7.3
80	6093	Globular cluster in Scorpio	16 15.8	−22 56	7.2
81	3031	Spiral galaxy (Sb) in Ursa Major	9 54.2	+69 09	6.9
82	3034	Irregular galaxy (Irr) in Ursa Major	9 54.4	+69 47	8.7
83	5236	Spiral galaxy (Sc) in Hydra	13 35.9	−29 46	7.5
84	4374	Elliptical galaxy in Virgo	12 24.1	+13 00	9.8
85	4382	Elliptical galaxy (SO) in Coma Berenices	12 24.3	+18 18	9.5
86	4406	Elliptical galaxy in Virgo	12 25.1	+13 03	9.8
87	4486	Elliptical galaxy (Ep) in Virgo	12 29.7	+12 30	9.3
88	4501	Spiral galaxy (Sb) in Coma Berenices	12 30.9	+14 32	9.7
89	4552	Elliptical galaxy in Virgo	12 34.6	+12 40	10.3
90	4569	Spiral galaxy (Sb) in Virgo	12 35.8	+13 16	9.7
91	—	M58?	—	—	
92	6341	Globular cluster in Hercules	17 16.5	+43 10	6.3
93	2447	Open cluster in Puppis	7 43.6	−23 49	6
94	4736	Spiral galaxy (Sb) in Canes Venatici	12 50.1	+41 14	8.1
95	3351	Barred spiral galaxy (SBb) in Leo	10 42.8	+11 49	9.9
96	3368	Spiral galaxy (Sa) in Leo	10 45.6	+11 56	9.4
97	3587	Owl Nebula; planetary nebula in Ursa Major	11 13.7	+55 08	11.1
98	4192	Spiral galaxy (Sb) in Coma Berenices	12 12.7	+15 01	10.4
99	4254	Spiral galaxy (Sc) in Coma Berenices	12 17.8	+14 32	9.9
100	4321	Spiral galaxy (Sc) in Coma Berenices	12 21.9	+15 56	9.6
101	5457	Spiral galaxy (Sc) in Ursa Major	14 02.5	+54 27	8.1
102	—	M101?	—	—	
103	581	Open cluster in Cassiopeia	1 31.9	+60 35	7
104	4594	Sombrero Nebula; spiral galaxy (Sa) in Virgo	12 39.0	−11 35	8
105	3379	Elliptical galaxy in Leo	10 46.8	+12 51	9.5
106	4258	Spiral galaxy in (Sb) Canes Venatici	12 18.0	+47 25	9
107	6171	Globular cluster in Ophiuchus	16 31.8	−13 01	9
108	3556	Spiral galaxy (Sb) in Ursa Major	11 10.5	+55 47	10.5
109	3992	Barred spiral galaxy (SBc) in Ursa Major	11 56.6	+53 29	10.6

[a]Positions and magnitudes based on a table in the *Observer's Handbook 1980* of the Royal Astronomical Society of Canada.

I leave to children the long, long days to be merry in, in a thousand ways, and the Night and the Moon and the train of the Milky Way to wonder at; and I give to each child the right to choose a star that shall be his, and I direct that the child's father shall tell him the name of it, in order that the child shall always remember the name of that star after he has learned and forgotten his astronomy.

Excerpt from last will and testament by unknown author

EXPLORING THE COSMOS

Fourth Edition

LOUIS BERMAN
University of San Francisco

J. C. EVANS
George Mason University

Little, Brown and Company
Boston   Toronto

Library of Congress Cataloging in Publication Data

Berman, Louis
  Exploring the Cosmos.

  Includes index.
  1. Astronomy I. Evans, J. C. (John C.), 1938–
II. Title
QB45.B47 1983   523   83-1014
ISBN 0-316-09184-7

Library of Congress Catalog Card No. 83-1014

ISBN 0-316-09184-7

9 8 7 6 5 4 3

MU

Published simultaneously in Canada by Little, Brown & Company (Canada) Limited

Printed in the United States of America

ACKNOWLEDGMENTS

We gratefully acknowledge permission to use the following material.

Page ii: background photo from Mount Wilson and Las Campanas Observatories, Carnegie Institution of Washington.

Chapter 1

Page 1: The Bettmann Archive; background photo by permission of Harvard College Observatory, Cambridge, MA. *Fig. 1.1:* both NASA. *Fig. 1.2:* Mount Wilson and Las Campanas Observatories, Carnegie Institution of Washington. *Fig. 1.3:* photo from Mount Wilson and Las Campanas Observatories, Carnegie Institution of Washington. *Fig. 1.4:* left, Lick Observatory; right, Palomar Observatory, California Institute of Technology. *Fig. 1.5:* Palomar Observatory, California Institute of Technology. *Page 9:* from Robinson Jeffers, "Margrave." Copyright, 1932 and renewed 1960 by Robinson Jeffers. Reprinted from *The Selected Poetry of Robinson Jeffers*, by Robinson Jeffers, by permission of Random House, Inc. *Page 11:* Culver Pictures. *Page 12:* Brown Brothers. *Fig. 1.7:* Mary Lea Shane Archives of the Lick Observatory, University of California, Santa Cruz.

Chapter 2

Page 17: Le Opere di Galileo Galilei, Florence, 1892. *Fig. 2.5:* Lick Observatory. *Fig. 2.7:* photos from Lick Observatory. *Page 32:* Brown Brothers.

Chapter 3

Page 41: Cambridge University Library; background photo by permission, Harvard College Observatory, Cambridge, MA. *Page 47:* courtesy Yerkes Observatory. *Page 58:* The Bettmann Archive.

Chapter 4

Page 68: American Institute of Physics Library, Margaret Russell Edmonson Collection. *Fig. 4.1:* The Bettmann Archive. *Fig. 4.6:* MIT Department of Physics, Professor A. P. French and Douglas Ely. *Fig. 4.11:* Lick Observatory. *Fig. 4.17:* Mount Wilson and Las Campanas Observatories, Carnegie Institution of Washington.

Chapter 5

Page 90: Palomar Observatory, California Institute of Technology. *Fig. 5.1:* top left to right, Harvard-Smithsonian Center for Astrophysics; Palomar Observatory, California Institute of Technology; Palomar Observatory, California Institute of Technology; Lick Observatory; middle left to right, NASA; Lick Observatory; Lick Observatory; bottom left to right, NASA; Lick Observatory; NASA. *Fig. 5.3:* Leiden Observatory. *Fig. 5.4:* Sacramento Peak, Association of Universities for Research in Astronomy. *Fig. 5.6:* Kitt

(Credits continue on page 528.)

Preface

Public awareness and knowledge of astronomy are probably as great today as at anytime in the past. From the general sources, such as newspapers, magazines, and television, to the highly visual big-production books, educational television series, and movies, to the beaming of *Voyager* pictures into our living rooms as they are received across billions of miles of interplanetary space, there has been an explosion in astronomical information available to the public. With all this information available, is it possible that college courses in astronomy are redundant or provide little enhancement of the general knowledge of astronomy? Obviously we do not believe so; otherwise we would not have written a textbook for such a course and continue to revise it through four editions.

As much as any of the sciences, in our estimation more so, a study of astronomy reveals to the student the interplay between collective observations of nature and the imagination of the human mind to conceptualize nature. Those efforts that attempt to capture and rivet the attention of the audience like an adventure novel, or those that are no more than a loose knit collection of facts, fail to reveal the beauty inherent in astronomy's conceptual framework. As important as what we know is how we know we know. Jacob Bronowski said it best when he observed, "Science is a very human form of knowledge. We are always at the brink of the known, we always feel forward for what is to be hoped. . . . Science is a tribute to what we can know although we are fallible. . . . We cannot reach certainty because it is not there to be reached. . . . The certain answers ironically are the wrong answers. . . . There is no God's eye view of nature . . . only a man's eye view." It is difficult, if not impossible in a textbook, both to explain what we know and to do justice to the development of our concepts of nature and the larger universe of which we are a part. What we strive to do in this book is to give the reader a mixture of facts to provide the context in which to see the progress of scientific thought.

Long after factual astronomy has faded from the reader's mind, if he or she still senses the awe and wonder of having explored one of the great quests in intellectual history — that of exploring the cosmos — then this book and the college course using it will have been successful.

In this fourth edition we have tried to continue the philosophy and spirit that guided the organization and written exposition in the third edition. It is our hope that this book can be the foundation for an introductory course yet possess sufficient flexibility to allow the instructor to bring out his or her own unique teaching abilities and areas of interest. In this edition we have made about a twenty percent change in organization, updated where necessary, revised about a third of the illustration program, and enhanced the learning aids at the end of the chapters. These changes, we believe, will provide instructors who were users of earlier editions with the currency of material in a familiar presentation and yet add a freshness so necessary for them to maintain their enthusiasm.

Organization plays a vital role in conveying a topic like astronomy to the uninitiated. In that regard we continue to find that the traditional approach with the ever-widening view from earth through the solar system to stars, galaxies, and the universe is the most intuitive and comprehensible approach to teaching and learning astronomy. We have made some major changes but have preserved the general division into four major units each containing five chapters. The first such unit comprises Chapters 1–5 and is on the history, conceptual foundations, and tools of astronomy. New in this unit is the introduction of atomic structure in Chapter 1 instead of Chapter 4 as in the third edition, while Chapter 3 has been restructured to emphasize the study of motion in Newton's laws, in atomic motion, and in relativity. A number of new figures have been introduced to strengthen the discussion.

As in the third edition, Chapters 6–10 constitute the unit on the solar system. Major organizational changes have been made here in addition to the extensive updating of the material to include the latest *Voyager* results and illustrations. Chapter 6 now provides the overview of the solar system including general features of the planets as viewed from earth, while the earth and moon are discussed in Chapter 7 as an introduction to the terrestrial planets. The discussion of the terrestrial planets in Chapter 8 and the Jovian in Chapter 9 are integrated discussions of the same aspects of each group, such as their atmospheres, surfaces, and interiors. Thus we can better highlight similarities and differences. Chapter 10 covers the origin and evolution of the solar system through considerations of life in the solar system.

Chapters 11–15 are devoted to the nature and evolution of stars and stellar systems, the third unit of the book. In those chapters we have interspersed observational material with theoretical discussion in order to help readers understand the interplay between these two as different sides of the same coin. Some organizational changes have been made, as suggested by users, along with a thorough updating of the text and illustrations.

The last unit of the book, Chapters 16–20, is devoted to galaxies, cosmology, and exobiology. Major organizational changes, expansions of the discussion, and updating have been made in this unit. A new chapter, Chapter 18, has been introduced to cover the many new developments in astronomers' knowledge of the large-scale structure of the known universe. Since the writing of the third edition, a number of important developments have occurred regarding superclusters of galaxies and the apparent absence of galaxies in great volumes of space — the cosmic voids. Although there is still a great deal of uncertainty in those areas, a revolution is brewing in our understanding of the large-scale structure as revealed by observations all across the electromagnetic spectrum, especially by X rays and gamma rays. Chapter 18 should also provide a better introduction to Chapter 19 on cosmology.

In addition to organizational changes, this edition includes an updating of all information, in both the text and the illustrations, so that the book will reflect the current state of astronomy as much as possible. For example, *Voyager 1* and *2* results for Saturn are included; new ideas are present on the heating of the sun's corona, stellar activity cycles, giant molecular clouds, death of stars, galactic structure and evolution, and life in the universe. The use of mathematics in the text remains the same as in the third edition; mathematical applications are placed in boxes as supplementary material. Thus the mathematical applications, as well as some nonmathematical topics and biographies in the boxed material, are available if desired, or they can easily be skipped. As in the third edition, we have tried as much as possible to confine factual material to tables or figure captions in order to avoid burdening the exposition in the text and to make it easier for readers to find while reviewing.

In this edition, we have completely revised the end-of-chapter material. All the chapter summaries have been rewritten so that they match the organization of the chapter. Following the summaries is a list of most of the key terms in the chapter that appear in italics when defined for the reader (successive uses of key terms are, in general, not italicized). Two types of questions now exist: One is a broad topical question that can be used as the basis for class discussions, while the second is a reading review question for which the answer is generally one word, phrase, or sentence. New to this edition are numerical problems in the appropriate chapters. They support the mathematical material in boxes so that the readers can test their grasp of the mathematical concepts. Finally a list of four or so readings on particular chapter topics are given. Most of them are at the level of difficulty of the book or are slightly above it.

In our preparation of the fourth edition we have been encouraged and helped by many users of previous editions who have generously offered suggestions and their time and expertise. We would like to acknowledge our great indebtedness to them as well as to a number of individuals who have reviewed all or parts of the manuscript. Our special thanks go to Richard G. Teske (University of Michigan), Richard L. Sears (University of Michigan), Thomas Gehrels (University of Arizona), Tobias C. Owen (SUNY at Stony Brook), Robert F. Garrison (University of Toronto), Edward R. Harrison (University of Massachusetts), Minas Kafatos (George Mason University), Dimitri M. Mihalas (National Center for Atmospheric Research), Herbert J. Rood (Princeton, New Jersey), Peter A. Strittmatter (University of Arizona), Wayne A. Christiansen (University of North Carolina), Stephen J.

Shawl (University of Kansas), Michael G. Emsley (George Mason Uniersity), and Averett S. Tombes (George Mason University).

Finally, we would like once again to express our appreciation to the very capable and diligent professionals at Little, Brown and Company, whose hard work, perseverence, and courage under fire have helped render a readable and visually exciting book. I especially want to thank Ian Irvine, who has helped greatly in the preparation of this edition as well as the two preceding ones. Elizabeth Schaaf, book editor, and Barbara Breese, editorial assistant, have been unfailingly supportive and have contributed generously of their professional skills. I thank them and Richard Maurer, art editor, and Adriana Wechsler, editorial assistant, for their contributions to this new edition.

This is the first edition in which Louis Berman has not directly participated, and I cannot let the opportunity pass without expressing my deepest appreciation to him. He conceived and did the first edition by himself, which was probably the most difficult part. Through the second and third editions, it has been a very rewarding experience working with him in spite of the awkwardness of being separated by a whole continent. I know few individuals who love or are able to maintain a youthful enthusiasm for astronomy more than Lou and who sincerely want succeeding generations to share the fascination of exploring the cosmos — qualities for which many of us are grateful that he possesses.

J. C. Evans
Fairfax, Virginia

Contents

8 THE INNER SOLAR SYSTEM: THE TERRESTRIAL PLANETS *169*

9 THE OUTER SOLAR SYSTEM: THE JOVIAN PLANETS *197*

10 EVOLUTION OF THE SOLAR SYSTEM AND LIFE *225*

11 THE SUN: OUR BRIDGE TO THE STARS *247*

12 STARS: "THE PALE POPULACE OF HEAVEN" *277*

13 THE H-R DIAGRAM: LIFE OF STARS *305*

14 INTERSTELLAR MATTER: BIRTHPLACE OF STARS *330*

EXPLORING THE COSMOS

1

Introduction to the Cosmos

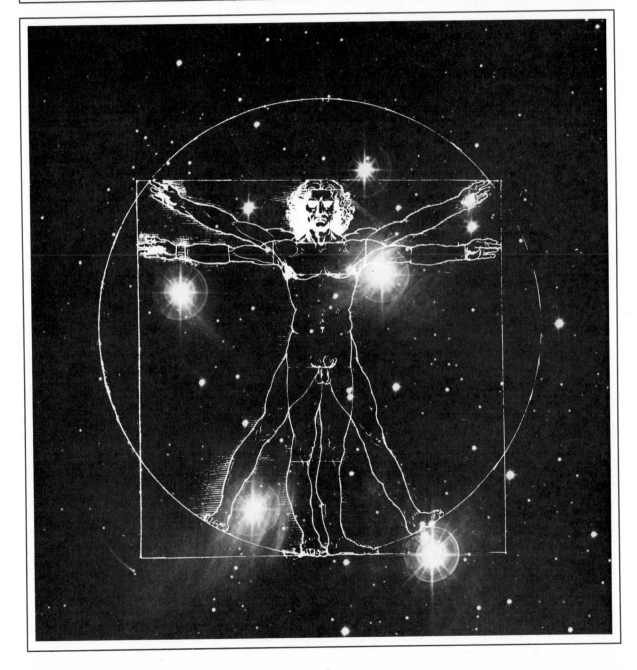

1.1
The Cosmic Design:
The World of the Large

Our contemporary view of the universe is far different from the view held thousands or even hundreds of years ago. By the word *universe* we mean not only everything that we know exists but also whatever will be discovered in the future; the universe is everything. The ancient Greeks used the word *kosmos* (in English *cosmos*) for the universe, but it implied more than just physical existence: It was the conception of a whole system that has order or dynamic arrangement (our "cosmetic" is derived from that signification). Our exploration of the cosmos will, we hope, help you to see the order, the arrangement, and the harmony of the universe. Let us begin *Exploring the Cosmos* with a brief overview of the universe as we understand it today.

◀ From the Stars Has Descended Man. The elements of our existence were forged in stars.

COSMIC OBJECTS: FROM THE EARTH TO THE GALAXIES

Only with the last 400 years have we been able to confirm the fact that the earth is not the center of the universe. It is only one of four comparatively small rocklike planets (see Figure 1.1) circling the sun; beyond these four there are four large planets made up of fluidlike matter; and beyond them is small and mysterious Pluto. Thousands of even smaller solid objects — asteroids and meteoroids — also orbit the sun; all these together would make a body only a fraction of the earth's size. Moving in highly elongated orbits about the sun are countless bodies of ice and rock called *comets*. Between all the objects that we have mentioned is a near void, containing only a few atoms and molecules and some tiny solid particles. That is the picture of our solar system that astronomers have pieced together over the last few centuries.

The central object of our solar system is the sun, a sphere of glowing gases so large that it could contain 1.3 million earths (see Figure 1.2). Yet the sun is only

FIGURE 1.1
Portion of Mercury's sun-illuminated side (left) and the giant planet Jupiter (right). Mercury is heavily cratered like the moon, with craters ranging up to nearly 200 kilometers in diameter. Like the other rocky, or terrestrial, planets — earth, Mars, and Venus — Mercury probably has a metallic iron-rich core and an outer silicate mantle. The banded appearance of Jupiter's atmosphere results from atmospheric currents. Unlike the terrestrial planets, the large outer, or Jovian, planets are composed mainly of hydrogen and helium; they may have a small rocky core.

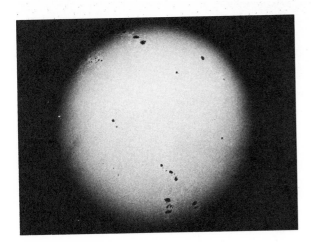

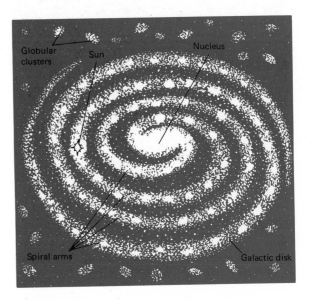

FIGURE 1.2
The sun, photographed in white light under excellent conditions, showing a large number of sunspots. Presumably other stars, if as close to us as the sun, would have a similar appearance.

one rather ordinary star among 400 billion or so stars in our Galaxy—a system of stars, gas, and dust shaped by gravity. From the earth we see our Galaxy as the luminous band of stars familiar to us as the Milky Way, shown in the photographic part of Figure 1.3. From outside the Milky Way we would see that its true shape is a spiral disk (as in the sketch of Figure 1.3), a disk so large that a ray of light takes 100,000 years to cross from one side to the other. We would also see that our sun lies in the disk away from the center of the Galaxy, toward its outer edge. The Galaxy rotates like a giant pinwheel, taking better than 200 million years to complete one rotation. In addition

FIGURE 1.3
Artist's conception (above) of the Milky Way Galaxy as seen from north of the Galactic plane. Below, a photographic mosaic of the Milky Way from Cassiopeia to Sagittarius. Shown here is a composite mixture of stars, bright discrete patches called gaseous nebulae, and dark, interstellar clouds of gas and dust.

to its billions of individual stars the Galaxy has bright patches of hot gas called gaseous nebulae, immense clouds of cold, dark gas and dust, and clusters of stars like raisins stuck in and around the disk.

Astronomers estimate that billions of galaxies can be photographed with the largest telescopes; our Galaxy is thought to be among the larger ones (see Figure 1.4). Within each are millions to hundreds of billions of stars of assorted size, brightness, and age, arranged singly or in groups. Galaxies themselves appear to come in a variety of sizes and shapes: Some are spirals, others have barred or elliptical shapes, and some have no classifiable shape at all.

A cluster of galaxies held together by mutual gravitational attraction is a frequent pattern throughout the universe. One example is shown in Figure 1.5. Such clusters are not the largest collections of matter that we are able to discern. There is convincing evidence that clusters of clusters of galaxies—superclusters—also exist. The superclusters appear to be separated by great gulfs containing few, if any, galaxies. Galaxies are the building blocks of the universe, and we can think of the clusters of galaxies and superclusters as houses and buildings in a city.

A CHANGING, EXPANDING UNIVERSE

The universe not only contains objects that were beyond our comprehension even a few decades ago but also behaves more exotically than dreamed of by science fiction. Early telescopic observations seemed to reveal a universe that was quiet, orderly, predictable, and populated with unchanging stars and galaxies. Occasionally stars would explode with the brilliance of a million or a hundred million suns, but these events appeared to be rare exceptions in the otherwise placid universe.

With the advent of radio astronomy in the late 1940s our picture of the universe began to change dramatically. Evidence mounted that the universe is not static; change, even violent change, is an integral part of it: Some galaxies are exploding, colliding, or spewing out great quantities of matter; quasars, objects like galaxies, emit enormous amounts of energy and are among the most distant objects in our visible universe.

Furthermore, not until this century did astronomers learn that the universe is expanding; galaxies appear to be separating from each other at speeds approaching the speed of light, and thus all of space is growing. The most widely accepted explanation is that about 15 to 20 billion years ago a superhot and superdense fireball exploded, beginning the universe's expansion. As this primeval fireball cooled, galaxies formed within it. "Cold" radiation detected from everywhere throughout the universe, astronomers now believe, is an echo of that initial "big bang." How long will the expansion continue? How much is it slowing down? Will the universe eventually cease to expand and begin to contract? There is still much to learn before we have answers to these questions.

OF TIME AND SPACE: ASTRONOMICAL SCALE

Astronomy not only has shown us that the earth is not at the center of things but has also given us a glimpse of time and distances otherwise beyond our everyday experience. In the chapters that follow we shall talk of objects trillions of trillions of miles away. In some cases light from the object that we now see left billions of years ago, even before the formation of the earth, and is only now arriving. That light tells about the object as it existed in the past, not what it is like today. Such times and distances are beyond our intuitions; so to help grasp what they mean, we can resort to some simple comparisons.

Imagine that the solar system could fit on the top of a typical dining-room table. Let the size of the solar system (the distance from the sun to Pluto) be about 40 inches in this model. Then the distance from the earth to the sun (which in reality is approximately 100

FIGURE 1.4
A spiral-shaped galaxy, M81, in Ursa Major (left). The nebulous knots scattered along the spiral arms are regions containing luminous stars and glowing gases similar to the Orion Nebula in our Galaxy. At right, an enlarged photograph of the small elliptical galaxy NGC 205, which is a satellite of the nearest spiral galaxy, known as the Andromeda Galaxy. M81 is about five times farther away than is NGC 205, which is so far that it takes light over 2 million years to reach us traveling at the rate of 300,000 kilometers per second (or 186,000 miles per second).

FIGURE 1.5
A cluster of distant galaxies in Corona Borealis, photographed with the 5.1-meter Hale reflector. The small, fuzzy, lens-shaped objects are the galaxies.

million miles) would be about 1 inch, and the sun itself would be smaller than a period on this page. The nearest star would be a little over 4 miles away from our table, and the radius of the Milky Way would be about 50,000 miles. In this model of the universe the great spiral galaxy in the constellation Andromeda, which is visible with the naked eye on a clear dark night, would be about ten times the actual distance to the moon. Finally, the most distant galaxies would be about sixty times the actual distance to the sun. In other words, if our solar-system model fitted on a large table, then the model of the universe would be one and a half times the *actual* size of our solar system.

Another series of comparisons may help to convey a sense of our place in the astronomical scale of time. As we have noted, scientists estimate that the big bang from which the universe evolved took place some 15 to 20 billion years ago. If we shrink this time span to 1 year, we come up with the following "universal" calendar: The earth would have formed in mid-September; the oldest known signs of life would have appeared in October; and the first mammals would not have appeared until December 26. All the events of human prehistory and history—from before the first known stone tools were made millions of years ago until *Voyager 2* beamed photos of Jupiter back to earth—would occur in the last hour of New Year's Eve.

1.2
The Atomic Design:
The World of the Small

THE ATOM

Some of the earliest philosophical speculations were concerned with what the material world is made of. Is each substance, such as rock or wood, infinitely divisible so that its subdivisions always yield the same properties as the whole substance? Or is there some level of structure below which the subdivisions will show new properties and forms? The Greek philosopher Democritus (460?–?370 B.C.) suggested that the material objects of our experience are actually made up of fundamental units. He called these units atoms, and he visualized them as indestructible, indivisible, and capable of being assembled into various forms and shapes. Therefore matter is never born

> Is it not astonishing that nearly all the inhabitants of our Planet, down to our times, have lived and died without knowing where they are, and without having the slightest conception of the marvels of our Universe?
>
> Camille Flammarion

from nothing but springs only from new combinations of atoms. The concept of the atom was not widely accepted until the English scientist John Dalton (1766–1844) developed his atomic theory about 150 years ago.

Modern science has shown that the atom too can be subdivided into more fundamental units. In 1897 the English physicist J. J. Thomson (1856–1940) identified the electron, carrying a unit of negative electrical charge, as a constituent of the atom. Also in England early in this century Ernest Rutherford (1871–1937) demonstrated that most of the atom is empty space and that nearly all its mass is concentrated in the nucleus. The principal constituent of the nucleus is the positively charged proton, which was identified by Rutherford about 1919. By 1932 the British physicist James Chadwick (1891–1974) was able to show that a second particle, having no electrical charge and thus called a neutron, resided in the nucleus.

The current picture of the atom is that of a nucleus made of protons and neutrons, with electrons in orbits surrounding the nucleus. The chemical identity of each atom of an element is determined by the number of protons in its nucleus, which in turn establishes the element's *atomic number*. The simplest nucleus is that of hydrogen, with one proton and atomic number 1; the atomic number of helium is 2, that of lithium is 3, and so on. As we have indicated, the proton has a unit of positive electrical charge equal and opposite to the negative charge of the electron; the mutual attraction between positive and negative charges holds the atom together. The neutral atom has as many electrons as protons (see Figure 1.6); with fewer electrons than protons, the atom is a *positive ion*. Atoms can combine to form molecules, with the number of atoms varying from two (as in the oxygen molecule we breathe) to many millions in the complex hydrocarbons that compose biological life. This conceptual

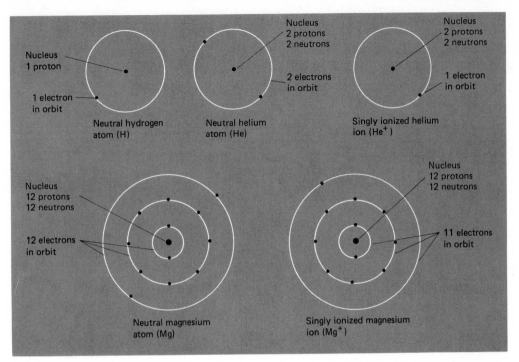

FIGURE 1.6
A simplified model of atoms and ions.

model, with later refinements, accounts satisfactorily for the periodicities in the physical and chemical properties displayed in the periodic table of the elements (see Appendix 4).

THE NUCLEUS

Atomic nuclei are made up of from 1 to about 260 protons and neutrons. The atom's mass is concentrated in the nucleus because the mass of the proton or neutron is almost 2,000 times greater than that of the electron.

Although the nucleus of any one element contains a fixed number of protons, it may have somewhat different numbers of neutrons. The resulting different nuclei of an element are called *isotopes.* All the elements in the periodic table, except 20, possess two or more isotopes so that their natural *atomic weights* depend on the relative abundance of each isotope. Among all elements there are about 300 stable isotopes and 1600 unstable, or radioactive, isotopes. *Radioactive nuclei* are ones that are undergoing spontaneous changes that make them generally into the nuclei of elements with smaller atomic numbers. Although there are a large number of unstable isotopes, most nuclei are stable.

The search for the ultimate indivisible constituents of matter that began before Democritus continues today. Scientists have discovered over 100 subatomic particles. Evidence grows that many of the subatomic particles may in turn be composed of more basic particles, called quarks.

SCALE SIZE OF UNITS OF MATTER

We can compare the sizes of objects in the universe in a way similar to our earlier comparison of distances. The universe is approximately a million billion billion billion billion (1 followed by 42 zeros) times larger than the nucleus of an atom. A whole host of objects lies between the size of a nucleus and that of the universe:

- An atom is approximately 10,000 to 100,000 times larger than its nucleus.

- Complex molecules can be 1000 times larger than an atom.

- A human being is 10 million times larger than the complex molecule.

- Humanity's home, the earth, is 10 million times larger than a human being.

- The radius of the sun is about 100 times that of the earth.

- The Milky Way Galaxy has a radius 1 trillion (1000 billion) times that of the sun.

- Clusters of galaxies are hundreds of times larger than the individual galaxies within them.

- Finally, the visible universe is on the order of thousands of times larger than a cluster of galaxies.

If you have difficulty comprehending such immense sizes and distances, take heart; it is an overwhelming scale even for astronomers. For convenience astronomers measure distances in units such as light years (defined on pages 13–14) and express large numbers in powers of 10. For example, 100 is written as 10^2, and 1000 as 10^3 (that is, $10 \times 10 \times 10$). A hundred times a billion then becomes $10^2 \times 10^9 = 10^{11}$, which is 100 billion. (For a full explanation, see Appendix 1.)

1.3
Philosophy of
Science and Astronomy

PHILOSOPHICAL ORIENTATIONS

Astronomy may well be the oldest of the sciences. The human pursuit of understanding about the world has been strongly shaped and guided by the wonder of, and awe for, the sun, moon, and starry sky. Intellectual thought has been inspired by the beauty and immensity of the night sky, but it also contains elements of fear, doubt, and uncertainty because mankind is dwarfed by the immensity of the cosmos. Astronomy has not only stimulated the other sciences of inquiry, such as physics and mathematics, but has also been an influence for creative efforts in art and literature.

The scientist's goal is to describe rationally and coherently *how* nature operates but not necessarily to explain *why* it behaves as it does—which can be a problem more in theology than in science. In this quest the philosophical orientations of scientists are far less uniform than one might believe. The practicing scientist, as Albert Einstein (1879–1955) has pointed out, "appears as realist insofar as he seeks to describe the world independent of the act of perception; as idealist insofar as he looks upon the concepts and theories as free invention of the human spirit; as positivist insofar as he considers his concepts and theories justified only to the extent to which they furnish a logical representation of relations among sense experiences."

Historically, there is much that we could say about the three philosophical orientations mentioned by Einstein and their many variations. However, space does not permit us more than simple definitions in order to give a sense of the possible debate among them. The first, known as *realism,* is the belief that experiences that come by way of our senses must reveal a "real" world that exists regardless of our interpretation of the acts of perception. The second is *idealism,* which says that there is no objective or absolute real world apart from the products of our imagination or mental constructs. From these mental constructs mankind can fashion imperfect copies of the ultimate reality, such as chairs, meter sticks, or laboratories. Finally, *positivism* does not speak of realities, does not speculate on origins and ultimate causes, but regards nothing as meaningful or understandable beyond that which is measurable or concepts that are derivable from measurements. Most scientists are in fact positivists when actually conducting experiments, but few scientists actively consider or participate in philosophical controversies on the orientation of scientific thought. That is an exercise they leave to the historians and philosophers of science.

SCIENTISTS: THE MODEL BUILDERS

In order to appreciate fully what astronomers know about the cosmos, one should understand how scientists work. Their methods can often be summarized by the following description: Phenomena are carefully observed, and information about them is recorded systematically. As part of this process, scientists first make a very fundamental assumption that

I believe most simply in the nobility of this great effort to understand nature, and what we can of ourselves, that is science.

J. Robert Oppenheimer

And the earth is a particle of dust by sand-grain sun,
  lost in a nameless cove of shores of a continent.
Galaxy on Galaxy, innumerable swirls of unnumerable stars,
  endured as it were forever and humanity
Came into being, its two or three million years are a moment,
  in a moment it will certainly cease out from being
And galaxy on galaxy endure after that as it were forever....

Robinson Jeffers

what occurs in nature is the result of cause-and-effect relationships. Then from the observed relationships scientists may generalize to form an *empirical law* or *hypothesis*—a general statement or explanation of the various observations.

How do scientists confirm the hypothesis? They make predictions on the basis of the hypothesis and apply them to new situations. The tests may show that the hypothesis is valid, or that it needs to be modified, or that the best we can do is to discard it and try again. Out of this process of trial and error emerges the general explanation, fitting many phenomena, that we call a *theory*. The theory is of no value, however, unless it agrees with what we already know and is supported by the future experiences to which we apply it.

As part of a theorizing process, scientists resort to a model, the framework of most modern theories. A *scientific model* is a mental picture, or idealization, based on physical concepts and aesthetic notions, that accounts for what scientists see regarding particular phenomena and allows them to predict a future course of events for the phenomena. The requirement of explanation and prediction as constraints on the scientific model dates only from the sixteenth and seventeenth centuries. Before then concepts of nature often had to satisfy only aesthetic or theological constraints. Although old and new scientific theories may not have been constrained by the same objectives, it is not unfair to compare their validity by using the models upon which they are based to explain and predict.

Models to deal either with something so encompassing as the entire universe or with limited phenomena such as lunar eclipses are widespread in all the sciences. But even though widely used, models must be carefully distinguished from the real world: Like metaphors in a poem, their early versions may be very figurative; that is, only vague approximations to the world of our experiences. Yet even in such forms scientific models can be beginnings that with time and more data may grow closer to accurate representations of the real world. Thus the astronomical model for planetary motion is probably closer to reality than models of the universe are.

1.4
A Brief History of Cosmological Thought

ANCIENT COSMOLOGICAL THOUGHT

The picture we sketched earlier of the universe is a fairly new one. Its roots go back thousands of years earlier when people first observed and wondered about the heavens and their origins. Later they developed concepts for the origin, structure, and evolution of the universe; these topics are the subject of *cosmology*.

Out of a great dark void the universe was created: That is the story of the beginning in most ancient cosmologies. The Bible's version of the beginning of the universe appears in Genesis: "In the beginning God created the heaven and the earth. And the earth was without form, and void; and darkness was upon the face of the deep." And then six days of divine labor fashioned the earth as we know it. Other cultures had their descriptions of the beginning of the world.

We can see a common thread in most ancient cosmologies: an inclination for particular groups to place themselves at the center of the universe. This egocentric aspect evolved later into a geocentric one; that is, the whole earth was put at the center of the universe. Primitive cosmologies were only idealized sketches against which the activities of nature took place; but few, if any, aspects of natural phenomena were actually incorporated into the cosmological picture. Not until centuries later were explanations of the details of nature considered to be part of cosmology.

ASTROLOGY

To many ancient peoples the heavens were more than the source of cosmological speculation; they held the keys to their earthly existence in that the celestial gods were believed to control human destiny. *Astrology* became the way to understand the mysterious divine activities and comprehend the course the gods dictated for human events.

Astrology can legitimately lay claim to one constructive act: It spurred the study of astronomy. As astrologers observed the sky wanderers in an attempt to improve their horoscopes, they paved the way for the large advances in planetary astronomy made in later centuries. Indeed right up to the seventeenth century some of the leading astronomers were practicing astrologers (including Johannes Kepler, whose astronomical contributions we shall examine in Chapter 2).

Today the gap between astrological predictions and scientific reality is unbridgeable. To think that stars and planets control human affairs by their configurations is a preposterous delusion, but one that still manages to make its way into most of our daily newspapers.

GEOCENTRIC AND HELIOCENTRIC COSMOLOGIES

The Greeks inherited much of their astronomy (and astrology) from the Egyptians and the Babylonians. After the seventh century B.C. astronomy and astrology clearly were separating, the former leading to the science of astronomy we know today. For the Greeks to see the earth, even though they were aware of the almost equally difficult concept that it was spherical by at least 400 B.C. or so, as the center of the universe was still not unreasonable. A geocentric cosmology certainly seems more intuitively obvious than does, say, a heliocentric one. With the geocentric concept, the Greeks tried a variety of conceptual models to account for the movement of the sun, moon, and five naked-eye planets—Mercury, Venus, Mars, Jupiter, and Saturn. The ultimate product of the geocentric cosmology was the Ptolemaic system, which we shall discuss in Chapter 2. That cosmology lasted until Co-

pernicus challenged it in the sixteenth century, and even afterward it did not totally disappear until the time of Newton in the late seventeenth century.

The heliocentric cosmology did not originate with Copernicus. It is known that in the third century B.C. the Greek natural philosopher Aristarchus proposed that the sun is at the center of planetary motion. His argument for the sun at the center was that it is the only self-luminous body in what we now recognize as the solar system. Even prior to Aristarchus the Greeks believed that the moon and planets shine by reflected sunlight, a notion that one might come to through studying the geometry of a lunar eclipse. It is probable that Aristarchus recognized that the stars were also self-luminous and conceivably like the sun but far away. What seems to have been one of the strongest arguments against the heliocentric system in the period immediately after Aristarchus was the failure to see a shift (which would be due to the earth's orbital motion) in the apparent position of nearby stars relative to more distant stars. That this phenomenon (known as the parallactic phenomenon) could be adduced as a criticism means that the Greeks were conceptually aware that the stars are not all at the same distance from the earth—a point that the ancients may have arrived at by arguing that all the stars have about the same brightness and that therefore the fainter ones are the more distant ones. But although they may have suspected that the stars are not all at the same distance from us, they apparently did not realize that even the nearest stars are too distant to reveal parallactic shifts to the naked eye. Within a couple of centuries after Aristarchus the heliocentric cosmology had lost out to the geocentric until its revival by Copernicus.

From the third century B.C. up to the latter part of the seventeenth century cosmological speculation was pretty much along the lines established by the Greeks. Although in that long interval many things were happening in the western world, for cosmology the most important were ultimately Newton's concepts of motion, gravity, space, and time. Within Newton's conceptual framework a whole new era in cosmological thought unfolded, made possible by the invention and refinement of the telescope.

THE UNITY OF THE WORLD

In the decades following Newton's work astronomers expanded Newtonian theory and developed some

CLAUDIUS PTOLEMY (A.D. 100?–?170)

Ptolemy was the last and perhaps the greatest of the astronomers of antiquity. He lived most of his life in Alexandria, Egypt, which at the time was the cultural capital of the Mediterranean world. Here scholars from many lands gathered to browse through its magnificent library. Ptolemy was both an observer and theoretician. He was the first to point out the effect of atmospheric refraction on the position of a heavenly body. And he asserted that natural phenomena require the simplest hypothesis that will coordinate the facts.

To Ptolemy must be given credit for collecting and handing down for posterity the scattered observations and principal discoveries of his Greek predecessors. These were incorporated in his famous work the *Great Syntaxes of Astronomy*. Translated later into Arabic about A.D. 827, it became known as the *Almagest* ("greatest"). The *Almagest* is a monumental encyclopedia consisting of 13 books. The work includes Ptolemy's refinement of the geocentric epicyclic theory into its most elegant and final form; the Hipparchus (190?–120 B.C.) catalog of some 1000 stars, plus some additions by Ptolemy; mathematical definitions and formulas; and tables by which the positions of the sun, moon, and planets can be predicted, derived from epicyclic theory. The predictions proved satisfactory within the limits of the naked-eye observations of the planetary motions. (It should be emphasized that the Ptolemaic model was a geometric representation of the planetary motions and not a physical model.) The *Almagest* became known in Western Europe through a Latin translation from the Arabic in 1175. So highly prized was the *Almagest* that it served as the standard treatise of astronomy for some 14 centuries prior to the time of Copernicus.

The veracity of Ptolemy's observational data has been questioned by Robert R. Newton in his book *The Crime of Claudius Ptolemy* (Johns Hopkins University Press, 1977). Ptolemy there stands accused of being "the most successful fraud in history" and is charged with faking the observations of his predecessors to support his theory of planetary motion. The figures that Ptolemy quoted, according to historians of science, were the ones that agreed best with theory, out of a larger body of observations. Their consensus, however, is that the case against Ptolemy is not proved.

Ptolemy also wrote a four-volume textbook on astrology known as the *Tetrabiblos*, which deals with various astrological influences and the casting of horoscopes. It still serves as the main reference for modern astrologers. Other works by Ptolemy include a compilation of geographical maps and a discussion of optics, including light, color, reflection, and refraction.

---

powerful techniques for analyzing planetary and stellar motions. They were able to predict the complex interactions between bodies in the solar system and to compare these results with observations, revealing a unity between the motions of common experience and the cosmic world.

During the last half of the eighteenth century and into the next century, astronomers gathered data on all sorts of astronomical phenomena. Foremost among the observational astronomers of the eighteenth century was William Herschel. With his skill as an observer and telescope-maker, he made many remarkable discoveries including his discovery of the planet Uranus in 1781. Herschel surveyed the Milky Way, extensively cataloging many stars that orbit each other; thus he found the first positive evidence that Newton's law of gravitation is valid far beyond the solar system. And from his analysis of the motions of nearby stars Herschel correctly deduced that the sun was moving relative to the nearby stars.

COSMOLOGICAL SPECULATIONS AFTER GREEK COSMOLOGY

Whether the universe "have his boundes or bee in deed infinite and without boundes" was the profound question English astronomer Thomas Digges (1546?–1595) asked himself. It appeared in Digges's 1576 book *Perfit Description of the Celestial Orbes*. A few decades later Galileo too wrote of an infinite and unbounded universe in his *Dialogues*, which was probably prompted by the fact he had seen with his

WILLIAM HERSCHEL (1738–1822)

William Herschel, destined to become one of the world's great observational astronomers, was born in Germany but lived most of his life in England. By the age of 34 he had established himself as a musician of note: teaching music; playing the organ at Bath; giving violin concerts; and composing military music, symphonies, and choral works. But his leisure hours were devoted to studying foreign languages, philosophy, and mathematics. At age 35 his simmering interest in astronomy was fired into action after reading Robert Smith's *Compleat System of Opticks* and Ferguson's *Astronomy*. Before long he was constructing his own refecting telescopes.

By 1773 Herschel had built a telescope whose focal length was 5.5 feet, and he was now ready to begin observing. Between 1774 and 1781 he recorded observations of numerous individual objects. His study of the sky led him to several important discoveries. For example, Herschel was searching for double stars in the expectation that a parallactic shift of the brighter component might be detected relative to the fainter and presumably more distant component. Instead, Herschel discovered that one star actually orbited around the other, the first tangible proof that gravity extended to the stars. Out of this work came the first *Catalogue of Double Stars.*

Another program Herschel initiated was the examination of every star in the standard star charts of his day. On the night of March 13, 1781, he made the historic discovery of a new planet in the constellation of Gemini, the planet Uranus. This discovery won him international fame and the royal patronage of King George III. After being knighted and awarded an annual stipend of £200 by the king, Herschel could devote all his time to astronomy, unhindered by the necessity of earning a living as a professional musician.

One of Herschel's important discoveries was of the sun's motion in space. From the proper motions of only thirteen stars he found that the sun is moving in space relative to its stellar neighbors toward a point in the constellation of Hercules, not far from the bright star Vega.

His most ambitious undertaking was an attempt to determine the structure of the Milky Way system. This involved "star gauging": making sample counts of the stars in the field of view of his telescope. By the time he finished nearly twenty years later in 1802, he had counted over 90,000 stars in 2400 sample areas. Along the way Herschel noted for future astronomers many objects of interest: variable stars, binary stars, dark areas in the Milky Way that looked like holes, irregular bright nebulosities, clusters of stars, and several thousand small nebulous objects or nebulae. The last three categories of objects were the subject of his 1802 *Catalogue of Star Clusters and Nebulae.* The Milky Way system, he concluded from his star counts, had the shape of a disk, like a grindstone, having a thickness of about one-sixth of its diameter. It was marked by many irregularities, and the sun was located near its center. Later studies confirmed Herschel's deduction that our Galaxy is disk-shaped, but the sun is not near the center and the system is considerably larger than Herschel supposed.

---

telescope that the Milky Way consists of myriads of stars.

What is the Milky Way? Swedish philosopher Emanuel Swedenborg (1688–1772) speculated in 1734 that the stars formed one vast collection, of which the solar system was but one constituent. Thomas Wright (1711–1786) of England theorized in his 1750 work, *An Original Theory of the Universe,* that the Milky Way seems to be a bright band of stars because the sun lies inside a flattened slab of stars. He also suggested that there were other Milky Ways in the universe. Immanuel Kant (1724–1804), the noted German philisopher, went beyond Wright's idea, suggesting in 1755 that the small oval nebulous objects seen with telescopes were other systems of stars or "island universes"; the phrase captured popular fancy a century and a half later. A modern photograph of Kant's island universes is shown in Figure 1.5.

Besides these speculations observational evidence was also accumulating that could be used to reveal the structure of the Milky Way. In 1785 Herschel gave the first quantitative proof that the Milky Way was a stellar

FIGURE 1.7
The Earl of Rosse's 1.8-meter telescope. This 6-foot reflector stood on his estate in Ireland. It was erected in the 1840s and was referred to by local residents as the Leviathan of Parsonstown.

structure shaped like a flat disk, a grindstone. Since Herschel and others did not suspect that starlight might be dimmed by obscuring material between the stars, he deduced that the sun was near the center of the system.

PREPARATION FOR MODERN COSMOLOGY
William Herschel and his son John (1792–1871), who carried his father's surveys to the southern skies, found over 5000 nebulous objects, cataloged in 1864 by John. Shortly thereafter the third Earl of Rosse in Ireland examined many of these objects with a 1.8-meter reflecting telescope, whose nineteenth-century appearance is shown in Figure 1.7. Several of the larger nebulosities appeared to consist of gaseous clouds, often with filamentary or ring-shaped forms; still others were resolvable into clusters of stars. However, many were not resolvable into either stars or gas clouds and had a spirallike structure.

As more nebulosities were found, the conviction grew that they belonged to one grand system, the Milky Way. If these objects were outside the Milky Way system, they could logically be expected to be spread more or less uniformly in all directions. But the nebulae were not distributed over the sky; their number increased in either direction away from the plane of the Milky Way. No one knew that this effect was illusory: Light was simply dimmed by the cosmic haze in our Galaxy's plane. The nebulae's seeming distribution and the appearance of stars and gas mixed in the

same nebula seemed to outweigh the possibility that the spiral nebulae might be unresolved stellar systems, separate from the Milky Way. The true nature of the Milky Way, the nebulae, and what they had to do with each other and the Milky Way was not to be demonstrated until after 1920.

After William Herschel's pioneering work, the next comprehensive study of the distribution of stars was done by Dutch astronomer Jacobus Kapteyn (1851–1922) toward the end of the nineteenth century. He then published in 1922 his final version of the shape, structure, and dimensions of our stellar system. Known as the Kapteyn universe, his was a sun-centered model of the Galaxy, 50,000 light years in diameter and 6000 light years thick. (A *light year* is the distance light travels, at the rate of approximately

> The most beautiful thing we can experience is the mysterious. It is the source of all true art and science. He to whom this emotion is a stranger, who can no longer pause to wonder and stand rapt in awe, is as good as dead: his eyes are closed.
>
> Albert Einstein

300,000 kilometers per second, in 1 year, or about 9.5×10^{12} kilometers.)

Five years before Kapteyn published his version of the stellar system, the American astronomer Harlow Shapley (1885–1972) had arrived at a markedly different conclusion about the sun's location and the dimensions of our system of stars. He found 69 globular star clusters that formed a spheroidal system centered on our flattened system of stars (see Figure 1.8), and the overall system, he thought, was 300,000 light years across and 30,000 light years thick, with the sun lying some 65,000 light years out from the center of the system. Shapley concluded that the spiral nebulae would be engulfed by our giant Galaxy and therefore were not island universes as some astronomers believed.

Astronomers accepted Shapley's finding that the sun was not the center of our Galaxy but were disturbed by the enormous scale he claimed for the Galaxy. He was challenged by Lick Observatory astronomer Heber Curtis (1872–1942), who insisted that the Milky Way's size had been considerably overestimated. In a great debate before the National Academy of Science on April 26, 1920, Shapley advocated the one-galaxy idea, and Curtis the multigalaxy, or island universe, concept. Their debate was inconclusive. Better observational data were needed, and they were not long in coming.

In 1924 Edwin Hubble (1889–1953), a Mount Wilson astronomer, derived distances to the spiral nebulae by analyzing photographs of the Great Nebula in Andromeda and another great spiral in Triangulum. Using the new 2.5-meter reflector, he was able to obtain photographs showing greater detail: The outer portions of these nebulae could be resolved into swarms of stars, and Hubble calculated a distance of 900,000 light years for the two spirals (an underestimate of about 100 percent). Doubt was at last gone that the spiral nebulae were the island universes Kant had long ago envisioned. Today we are aware of billions upon

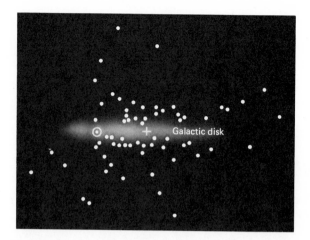

FIGURE 1.8
Shapley's globular-cluster distribution, proposed in 1917. The globular clusters are represented as white dots, the sun is shown as a circle with a central dot, and the cross marks the center of the Milky Way. Shapley pictured the Galactic disk as highly mottled so that the large concentration of stars around the sun could be the Kapteyn universe engulfed in our Galaxy.

billions of galaxies, often in giant clusters, spread over vast reaches of space.

In addition to the recognition that the universe is a universe of galaxies in which neither the earth nor the sun is central in location or significance, the conceptual framework of an absolute space and time provided by Newton has been replaced by new concepts of space and time known as relativity. In the early years of this century Einstein argued that—unlike Newton's belief in a perception of the universe independent of observers—our presence as observers is crucial. We cannot withdraw from the universe to watch it function; observation of the universe cannot be divorced from us as observers. This opened an entirely new chapter for cosmological thought. Throughout the next 17 chapters we shall consider the foundations upon which rests the edifice of modern cosmology to be discussed in Chapter 19.

SUMMARY

Cosmic objects. Earth is only one of thousands of objects in the solar system. The center of the solar system is the sun, one of about 400 billion stars in our Galaxy. And our Galaxy is one of the billions of galaxies that comprise the universe. Until recently, only the

most obvious details of the universe could be studied, first with the unaided eye and then with optical telescopes. Modern instruments have permitted astronomers to observe objects in the universe never before detected. As a result, our conception of the universe has changed from a static assemblage of matter to a dynamic organization of matter and energy.

Atomic design. The smallest unit of elemental matter is the atom. Even the atom is made of smaller parts — the nucleus of positively charged protons and electrically neutral neutrons and, orbiting the nucleus, a number of negatively charged electrons equal to the number of protons in the nucleus. Most of the mass of an atom is concentrated in the nucleus. Radioactive nuclei may spontaneously change, giving off mass and energy and, in the process, transforming from one element to another, lighter one. Most nuclei, however, are stable.

Philosophy of science. Astronomy is an ancient science dating back to the earliest human attempts to understand nature. The study of astronomy also stimulated progress in other branches of the sciences and arts. Scientists work by building models on careful observation, framing hypotheses based on those models, and testing those hypotheses by further observation. New tools and techniques of astronomy have permitted astronomers to build more comprehensive and detailed models and to test them more rigorously than in the past.

Cosmological thought. Cosmology is the study of the origin, structure, and evolution of the universe. The ancients could only frame stories about how the observed universe might have come into being from a time of darkness and chaos. Astrologers, especially those of Chaldea but also in Egypt and Babylonia, made detailed records of the movements of celestial bodies. From the time of the ancient Greeks to Copernicus, astronomers debated whether the earth or sun was the center of the universe. Geocentric and heliocentric models of the universe were devised and used to predict astronomical phenomena. By the eighteenth century, however, the heliocentric universe was the dominant model. Newton and Herschel prepared the way for modern astronomy by (1) developing universal laws governing phenomena throughout the universe and (2) properly placing earth and the solar system among the countless objects of the universe.

KEY TERMS

astronomy	hypothesis
atom	light year
atomic number	molecule
atomic weight	nuclei
cosmology	positive ion
empirical law	science
expanding universe	scientific model
galaxy	technology
geocentric	theory
heliocentric	universe

CLASS DISCUSSION

1. What does it mean to say, "The universe is a universe of galaxies"? Is it not equally accurate to call it "a universe of stars"?

2. Do you agree or disagree (and why) with the following statements?

    a. Jacob Bronowski: "Science leads us progressively to an understanding of man's own place in the natural order of things."

    b. Albert Einstein: "Why does this magnificent applied science which saves work and makes life easier bring so little happiness?"

    c. Hendrik van Loon: "I have come to have very profound and deep-rooted doubts whether science, as practiced at present by the human race, will ever do anything to make the world a better and happier place to live in or will ever stop contributing to our misery."

d. Duane Roller: "No significant scientific knowledge, no law, no principle can be established as true by appeal to observation."

e. George Bernard Shaw: "Science is always wrong. It never solves a problem without creating ten more."

f. Mark Twain: "There is something fascinating about science. One gets such wholesale returns of conjecture out of such a trifling investment of fact."

3. What is reality to the scientist? How does the scientist proceed to reveal scientific knowledge?

4. Without a telescope how could the Greeks have known that the sun was self-luminous, while the moon and planets were not?

5. Why could astronomers not immediately recognize the validity of the multigalaxy nature of the universe?

READING REVIEW

1. Name the major levels of structure in the universe and arrange them in order of relative size from smallest to largest.

2. How many stars are estimated to be in our Galaxy? How many galaxies do astronomers estimate to exist in the universe?

3. Is the universe static and unchanging, or does some change occur? If change does occur, how would you describe it?

4. Describe the structure of the atom, of molecules, and of larger pieces of matter.

5. How much larger is the visible universe than a nucleus? How much larger is a galaxy than a human being?

6. Does astronomy owe any debt to astrology?

7. What is the difference between a geocentric and a heliocentric cosmology? Which body proved to be the center of the universe?

8. Approximately when did astronomers accept Kant's island universe hypothesis?

SELECTED READINGS

Blacker, C., and M. Loewe, eds. *Ancient Cosmologies.* Rowman & Littlefield, 1975.

Culver, R. B., and P. A. Ianna. *The Gemini Syndrome: An Analysis of Astrology.* Pachart Publishing House, 1979.

Davidson, M. *The Stars and the Mind: A Study of the Impact of Astronomical Development on Human Thought.* Gordon Press, 1976.

Pedersen, O., and M. Pihl. *Early Physics and Astronomy.* Neale Watson Academic Publications, 1974.

2
The Cosmos from the Earth

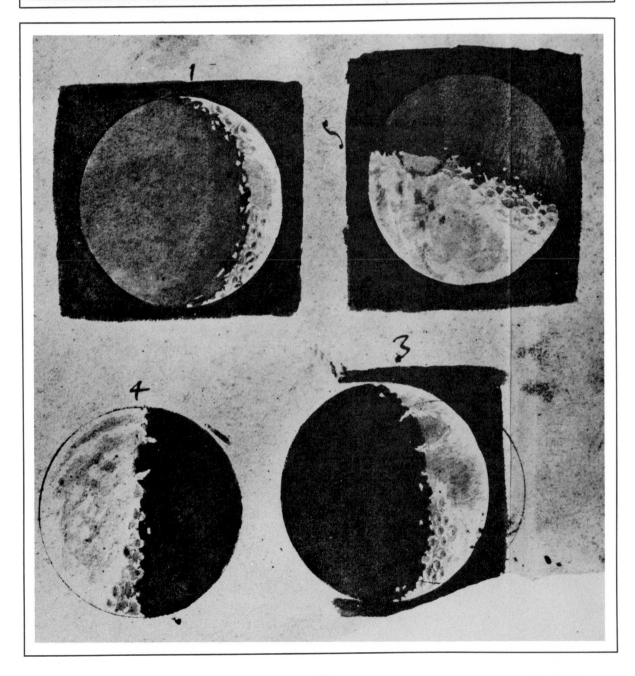

In the historical development of conceptual ideas about the universe an understanding of the common cyclic phenomena of the sky played an important part. Many of these phenomena were not discovered in a literal sense but have been known since long before the ability to write about them. A good example is the daily rising and setting of the sun. For this phenomenon what is new and has changed with time is an "explanation" of why the sun rises and sets. Space does not permit us to sketch the evolution of all the explanations of these various phenomena. Therefore, in the first sections of this chapter we shall discuss most of the common sky phenomena, which were influential in channeling the development of astronomy, although the explanations will reflect our present understanding.

This discussion will make it easier to understand the historical events leading to the development of geocentric and heliocentric cosmologies. Beginning with Kepler's laws of planetary motion and Galileo's concepts of motion, we can say that astronomy was truly on the path to our present-day understanding of the cosmos.

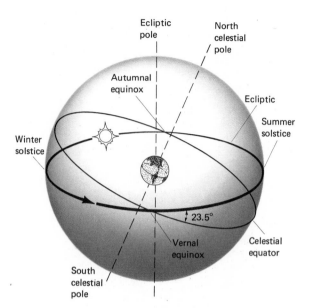

FIGURE 2.1
The celestial sphere. As envisioned by the ancients, the celestial sphere had the earth at the center with the stars emblazoned on the sphere of the sky. The ancients thought that the stars rose and set because the celestial sphere rotated, carrying the stars from east to west around two points on the sphere, the north and south celestial poles — projections of the earth's axis of rotation. The earth's equator projected on the celestial sphere becomes the celestial equator. It intersects the ecliptic, the yearly apparant path of the sun, at the vernal and autumnal equinoxes (March 21 and September 22) at an angle of 23.5° since that is how much the earth's axis is tilted from the perpendicular to the plane of the ecliptic. The solstices (December 22 and June 21) are the points on the ecliptic 90° from the equinoxes.

2.1
Ancient Astronomy: The Sky

CELESTIAL SPHERE

To our ancestors the sky must have looked like the inside of a huge dome covering the earth as far as they could see. Little wonder that ancient peoples thought of the earth (and themselves) as the center of a sphere, with the stars emblazoned on that sphere. This imaginary sphere is still used and known to astronomers as the *celestial sphere* (see Figure 2.1).

Watching the sky, early peoples saw the stars rise above the eastern *horizon* — the circle that divides the visible celestial hemisphere from the invisible one — cross the sky, and later set below the western horizon. Observing this daily behavior of the stars was one of the first steps in human understanding of the sky. The

ancients apparently thought that the stars rose and set because the celestial sphere actually rotated, carrying the stars from east to west. Today we know that the rising and setting of the stars is caused by the earth's rotation from west to east.

Ancient cultures also noticed that the stars appear to move around two points on the celestial sphere: the *north* and *south celestial poles* (Figure 2.1). To them these were the ends of the axis around which the celestial sphere rotated. To us they are projections of the earth's axis of rotation, through the north and south celestial poles, onto the celestial sphere. (Polaris, the *North Star*, now lies within 1° of the north celestial pole and is a relatively bright marker of its position.)

◄ Galileo's drawings of the phases of the moon. With his telescope Galileo made these drawings in about 1610, and they were published in his book *The Starry Messenger*.

The *zenith* is the point directly overhead for any observer. The imaginary arc on the celestial sphere running from the north point of the horizon through the celestial pole and the zenith to the south point of the horizon is the *celestial meridian.* This line is the dividing line between rising and setting because the highest position above the horizon that each star reaches in its daily motion is on the celestial meridian. For this reason it is useful for developing a conceptual frame for timekeeping. (Further descriptions of the systems of measurement used to locate objects on the celestial sphere are in Appendix 2; timekeeping is covered in Appendix 3.)

Not all stars daily rise above and set below the horizon. Some stars remain either above or below the horizon, depending on the observer's latitude. As ancient peoples traveled to different latitudes, they noticed that a different pattern of stars could be seen. For example, as observers traveled northward, they could see stars near the northern horizon that previously had not been visible. And stars previously visible near the southern horizon were now below it. Such an effect could be, and probably was, interpreted as evidence that the earth is a sphere.

CONSTELLATIONS

In the clear skies of the Tigris, Euphrates, and Nile valleys, where the earliest civilizations flourished more than 5000 years ago, watchers of the heavens observed groupings of stars, the *constellations.* They saw in these groupings mythological beings, animals, and monsters who reigned over heaven and earth, and they attached names to the constellations accordingly. The eminent British astronomer John Herschel once complained: "Innumerable snakes twine through long and contorted areas of the heavens where no memory can follow them; bears, lions, and fishes, large and small, confuse all nomenclature."

The names and shapes of our constellations are part of our heritage from the Greeks of antiquity although they did not originate in Greek culture but appear to have stemmed from earlier civilizations in Mesopotamia and from other peoples of the Near East. (Some constellation figures are shown in Figure 2.2.) Greek astronomers observed and named forty-eight constellations. Forty more were added, most of them in southern skies, by European mapmakers and astronomers in the seventeenth and eighteenth centuries. Note that the "catch figure," or *asterism,* often

> But who shall dwell in these worlds if they are inhabited? Are we or they Lords of the World? And how are all things made for man?
>
> Johannes Kepler

associated with a constellation should not be mistaken for the entire constellation. An example is the asterism of the Big Dipper, which is the recognizable figure for the constellation Ursa Major, the Great Bear. Appendix 2 contains star maps of the constellations and a discussion of their locations on the celestial sphere.

Standing in midlatitude of either hemisphere, with practice you could see, as a year goes by, four-fifths of the constellations. Of the eighty-eight constellations, about half lie in the Milky Way or near its borders. As you learn the constellations, hearing the name of one will bring to mind both its shape and its place in the sky, just as earthbound place names and their locations become familiar. Keep in mind that the stars of a constellation form an apparent grouping as seen from the earth and are not necessarily in proximity to each other in space.

NAMING THE STARS

To go with a constellation's name, its brighter stars are designated by lowercase Greek letters, assigned approximately in descending order of brightness. When a letter is added to the constellation's Latin name, the case ending changes to possessive or genitive forms, such as α Orionis. This scheme was invented by Johann Bayer for his 1603 map of the heavens. Almost 900 of the brightest stars also have Greek, Latin, or Arabic names. Brightest of the stars in the constellation Leo is α Leonis, whose proper name is Regulus. Second brightest is β Leonis (Denebola). Table 12.4 lists the 30 brightest stars in the night sky with their proper names.

But the Bayer scheme quickly ran out of Greek letters (the limit was 24) for the 5400 naked-eye stars that are dotted over the entire sky. John Flamsteed, first Astronomer Royal of England, cataloged the stars in 1725 and simply numbered them west to east across the constellation. A particular star in Cygnus is identified as 61 Cygni.

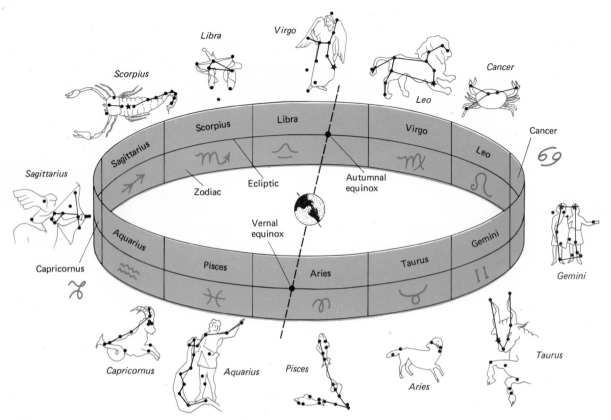

FIGURE 2.2
Signs of the zodiac and the constellations, with the earth at the center of the celestial sphere. Viewed from the earth, the moon and the planets move almost entirely along the band of sky known as the zodiac. Ancient priest-astronomers divided the zodiac into twelve constellational signs, but today these signs and their constellations are out of alignment because of the earth's precession (see Figure 2.4).

Fainter stars may be identified by their number in any one of several star catalogs. (For example, Lalande 21185 is entry 21, 185 in Lalande's catalog, which has numbers for the brighter stars as well. Other examples can be seen in Table 12.1 for the nearest stars.) There are still other systems for naming the stars, and enough different designations exist today to confuse even the astronomer.

2.2 Cyclic Phenomena of the Heavens

THE SUN AND THE SEASONS

The different seasons are caused by the tilt of the earth's axis of rotation relative to its orbital plane and by the earth's revolution about the sun once a year (see Figure 2.3). Relative to the stars, the sun appears to move eastward about 1° each day. Because of this, a particular star rises about 4 minutes earlier each night than it does the night before. And, at the end of 1 month the star rises approximately 2 hours earlier than it did in the previous month. By the end of 1 year the nightly change adds up to 24 hours, and the annual cycle of the heavens begins again.

Over the centuries people tracked the sun's yearly path across the sky, a path called the *ecliptic* (Figure 2.1). The earth's geographic equator projected onto the celestial sphere is called the *celestial equator*. The celestial equator intersects the ecliptic at two points: the *vernal equinox*, which the sun reaches about March 21, and the *autumnal equinox*, which the sun reaches about September 22. It was found that the moon and the planets moved almost entirely along a narrow band of sky, the *zodiac*, which was 16° wide and centered on the ecliptic. They divided the zodiac into twelve constellational divisions or *signs*, through which the sun passed in successive months (Figure 2.2).

The earth's axis of rotation is tilted 23.5° so that in the Northern Hemisphere we incline away from the sun in December and toward it in June (Figure 2.3). Consequently, the amount of sunlight falling on the surface of either hemisphere varies depending upon whether the hemisphere is inclined toward or away from the sun as shown in the figure.

In the Northern Hemisphere spring begins on or about March 21, when the sun crosses the celestial equator from south to north at the vernal (or spring) equinox. On this day all places on earth experience 12 hours of daylight and 12 hours of darkness. Summer starts on or about June 21, when the sun reaches its maximum distance of 23.5° north of the celestial equator at the *summer solstice*. During our summer the hours of daylight are longest and the sun is highest in the sky. Autumn begins on or about September 22, when the sun crosses the celestial equator from north to south at the autumnal equinox and the days and nights are again equal over the earth. Winter begins on or about December 22, when the sun is farthest south of the celestial equator by 23.5° at the *winter solstice*. The length of daylight is now the shortest and the sun is lowest in the sky. In the Southern Hemisphere the seasons are reversed; for example, Christmas there occurs during the warm summer months.

FIGURE 2.3
(a) Seasonal change in the location of the sun. The change in the apparent position of the sun as seen from the surface of the earth is caused by an inclined earth that orbits the sun. Shown are the earth's inclinations relative to its orbital plan at the times of vernal equinox (March 21); summer solstice (June 21); autumnal equinox (September 22); and winter solstice (December 22). Orion and Scorpius are winter and summer constellations, respectively. (b), (c) Enlargements of the earth at the time of the summer solstice and winter solstice, respectively. For two observers located at 30°N and 30°S the sun's rays strike the earth in the Northern Hemisphere more nearly vertically at the time of the summer solstice than they do in the Southern Hemisphere. The reverse is true at the winter solstice.

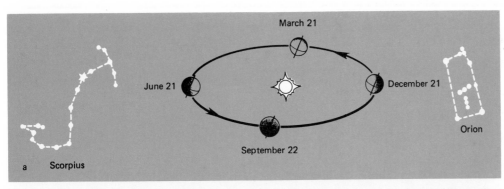

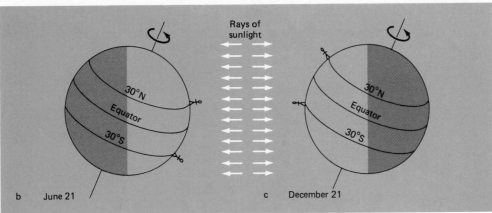

PHASES OF THE MOON

The rising and setting of the sun and the period of the phases of the moon were important cyclic events for ancient man. Both of the repetitive cycles of the sun and the moon were important in establishing the concept of time.

Our present-day understanding of the reasons for the phases of the moon predates Aristotle, who was

PRECESSION OF THE EQUINOXES

In the second century B.C. the Greek astronomer Hipparchus compared the positions of the principal stars in the zodiac with those astronomers had noted over 100 years earlier. He found that the vernal equinox had shifted westward about 2° along the ecliptic. What he was observing is called the *precession of the equinoxes.*

The earth is not a perfect sphere. It tends to bulge out in the equatorial region. Both the sun and the moon try to pull the earth's equatorial bulge into the terrestrial and lunar orbital planes. Acting like a spinning top, the rotating earth resists this pull. The effect of the sun's and the moon's attraction and the earth's resistance causes the earth's axis of rotation to move slowly westward around the pole of the ecliptic (see Figure 2.4). Because of the precession of the axis of rotation, the points of intersection between the celestial equator and the ecliptic shift westward along the ecliptic at a rate of about 50 seconds of arc per year. This causes the equinoxes, which are the points of intersection, to precess completely around the ecliptic in 26,000 years.

Two thousand years ago the vernal equinox (Figure 2.2) lay in the constellation of Aries. Three thousand years prior to that the vernal equinox was located in the constellation of Taurus. Today the vernal equinox occupies a position in the constellation of Pisces. The "age of Aquarius" dawns when the vernal equinox moves into the constellation of Aquarius, about 1000 years from now.

The axis of the earth retains its tilt of 23.5° throughout the cycle of precession. Today the axis is directed toward a point on the sky less than 1° from Polaris. In ancient Egypt the axis pointed in the direction of the star Thuban (α Draconis), which was then only a few degrees from the pole around which the heavens rotated. About A.D. 14,000 the very bright star Vega (α Lyrae) will be the "North Star" and mark the approximate position of the north celestial pole for our descendants.

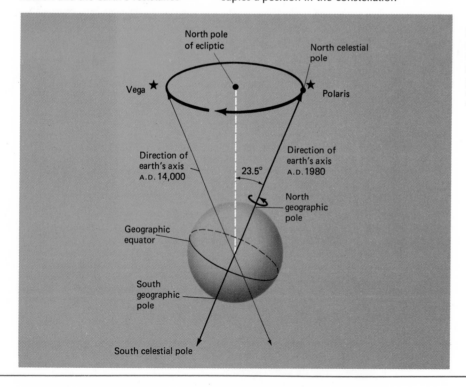

FIGURE 2.4
Precession. Resisting the pull from the sun and the moon, the earth's axis of rotation describes a cone about the perpendicular to the plane of the ecliptic in a period of 26,000 years.

aware that the moon "shines" by reflecting sunlight. The parallel rays of the distant sun always illuminate one-half the moon's surface as well as one hemisphere of the earth; we see the rays of the sun reflected off the moon.

When the moon is between the earth and the sun—the time of a *new moon*—its dark side faces us (see Figure 2.5), and we do not see it at all. Because the moon moves eastward relative to the sun within a few days a thin crescent appears low in the western sky after sunset, setting shortly after the sun sets. In the next few days the waxing (growing) crescent appears higher in the sky after sunset and therefore sets later on consecutive nights. One week after new moon, the moon is at *first quarter* and will be on the observer's celestial meridian at sunset; it will set about 6 hours after the sun. In the following week the gibbous moon (more than half a crescent but not yet full) waxes toward full as it continues its easterly movement around the earth. Two weeks after new

FIGURE 2.5
Lunar phases. The portion between the dashed lines represents the part of the moon we can see from the earth. The moon's age is counted from the time of the new moon. Since the moon keeps basically the same side toward the earth, brightening and darkening during the cycle of phases occur on the face toward earth.

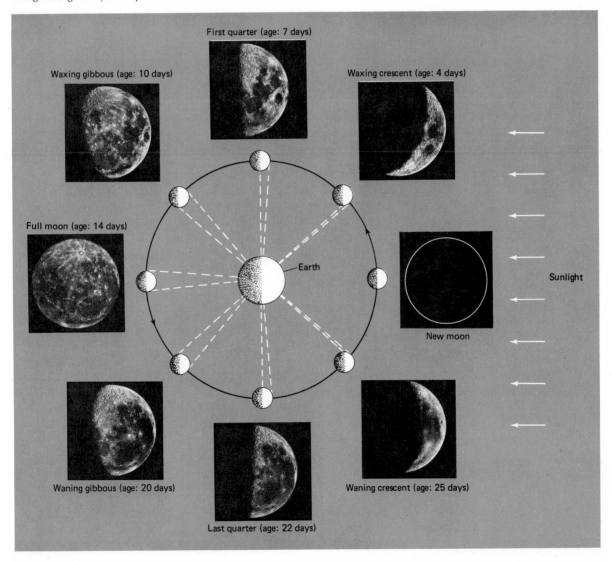

moon, the moon is *full,* and it lies on the opposite side of the earth from the sun. We see it rise at approximately 6 P.M. and set at about 6 A.M. One week later we see the moon at *last quarter* rise at about midnight and set at about noon. Finally we see the waning crescent moon rise shortly before sunup as it is about to overtake the sun 1 month after the previous new moon.

If we track the moon's movement against the stars, we find that it takes around 27.3 days to complete its orbit; this is the *sidereal month* (Figure 2.6). But because the earth is also moving around the sun, the time between each cycle of lunar phases is longer than the sidereal month. Although the moon has completed its revolution around the earth at the end of

FIGURE 2.6
Distinction between sidereal and synodic months for the moon. The moon requires 27.3 days to complete a 360° revolution about the earth, starting from the full moon phase. To return to the next full moon requires 29.5 days, or more than a 360° revolution. The first period is a sidereal month; the second is a synodic month.

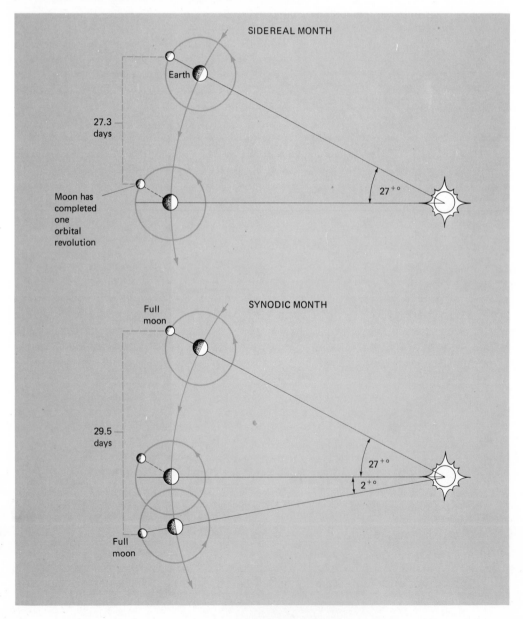

Hereafter, when they come to model heaven
And calculate the stars, how will they wield
The mighty frame, how build, unbuild, contrive
To save appearances, how gird the sphere
When centric and eccentric scribbled o'er,
Cycle and epicycle, orb in orb.

John Milton

27.3 days, it takes about 2 more days to bring the moon back to the earth-sun line so that it again appears as a new moon. Thus the period of lunar phases, the *synodic month,* is 29.5 days.

LUNAR AND SOLAR ECLIPSES

When the moon comes directly between the earth and the sun (see Figure 2.7), the moon's shadow or part of it falls on the earth and the sun is eclipsed. The moon's orbit is inclined to the earth's orbital plane (the ecliptic plane) by about 5°. Thus a solar eclipse is possible only when the moon is near a new phase and, in addition, is at or near one of the two points in its orbit known as *nodes* (where its orbit intersects the plane of the ecliptic). This lineup occurs at least twice each year; at most and rarely, five times a year.

The *umbra* of the lunar shadow is the totally dark portion; the *penumbra,* or partial shadow, is the semi-dark portion. If you stand in the umbra, you see a *total solar eclipse;* if you stand in the penumbra, you see a *partial solar eclipse.* The penumbra cast on the earth covers a much larger area than the umbra does, making a partial solar eclipse visible over a wider region than is a total solar eclipse.

The average length of the moon's conical shadow is not quite equal to the moon's mean distance from the earth. An *annular solar eclipse* takes place when the new moon is at or near the point farthest from the earth (apogee), at which point its umbral shadow is too short to reach the earth. Under these conditions the slightly smaller, dark disk of the moon is surrounded by a brilliant ring of the still-exposed sun.

Under the most favorable conditions the width of the moon's shadow on the earth's surface is about 270 kilometers. Totality then lasts longest (maximum length is about $7\frac{1}{2}$ minutes) in the equatorial zone, where the velocity of the moon's eastward-traveling

shadow in relation to earth's eastward rotation is smallest. The last such eclipse took place on June 30, 1973, over the northern part of South America, the eastern part of the Atlantic Ocean, and northern Africa. The next 7-minute total solar eclipse will occur on July 11, 1991, and will be visible in Hawaii, Central America, and Brazil.

Usually, if an eclipse of the sun occurs, an eclipse of the moon precedes or follows it by 2 weeks. Earth, moon, and sun then are sufficiently in line for the earth's shadow to fall on the full moon. The earth has nearly four times the moon's diameter, so its conical shadow is about four times wider at the base and four times longer than the moon's shadow. *Lunar eclipses* may be *partial* or *total,* everyone on the dark side of the earth seeing the lunar eclipse at the same time.

A year may bring as many as three lunar eclipses—or none at all. More often we have two eclipses of the sun and two of the moon in each calendar year. Centuries of observing eclipses taught the ancients that eclipses recur at regular intervals. After 18 years and 10 or 11 days, an interval named the *saros,* the circumstances of an eclipse are repeated approximately.

CONFIGURATIONS OF THE PLANETS

Ancient astronomers devised systems to describe the configurations of the planets. These early descriptions are basis for our present-day definitions of the configurations (see Figure 2.8). Between the earth and the sun revolve Mercury and Venus, the two *inferior planets.* Because their orbital periods are shorter than that of the slower-moving earth, they overtake and pass it. From the earth, relative to the earth-sun line, Mercury and Venus appear to move counterclockwise around the sun while swinging from one side of it to the other, as shown in Figure 6.6. The angular distance

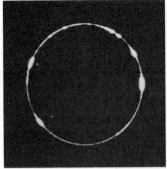

Annular eclipse of the sun.

Total solar eclipse.

Partial lunar eclipse.

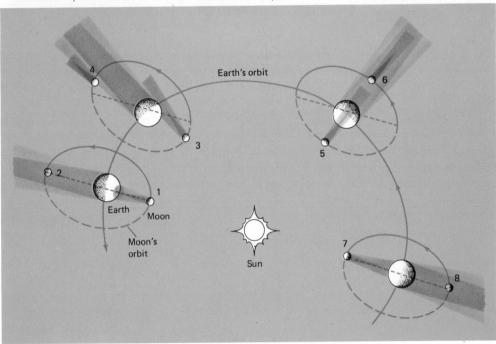

FIGURE 2.7

Solar and lunar eclipses. The lines with short dashes indicate the line of the nodes — where the moon's orbit crosses the plane of the earth's orbit. The lines with longer dashes show that portion of the moon's orbit that lies below the earth's orbital plane. The diagram shows eight positions of the earth and the moon relative to the sun as seen from north of the plane of the ecliptic:

1. Total eclipse of the sun: The new moon is at or near the line of nodes.
2. Total eclipse of the moon: The full moon is at or near the line of nodes.
3. Partial eclipse of the sun: The moon has moved far enough from the line of nodes so that the earth intercepts only the penumbra portion of its shadow.
4. Partial eclipse of the moon: The full moon intercepts part of the penumbra of the earth's shadow; this penumbral lunar eclipse is not visible to the naked eye.
falls below the earth.
5. No eclipse: The new moon is too far removed from the line of the nodes; its shadow falls below the earth.
6. No eclipse: The full moon is too far removed from the line of nodes; the moon passes above the earth's shadow.
7. Annular eclipse of the sun: The new moon's umbral shadow is too short to strike the earth even though the moon is at or near the line of nodes; however, a partial solar eclipse is visible outside the extended umbral shadow.
8. Total eclipse of the moon: The full moon is at or near the line of nodes.

(in degrees) either planet appears east or west of the sun is its *elongation*.

From its position closest to earth—called *inferior conjunction*—when it is in line with the sun, the planet appears to move rapidly west of the sun as its phase changes from new to crescent (Figure 2.8). When an inferior planet reaches its greatest angular distance west of the sun, known as *maximum western elongation,* it is conspicuous as a morning star; its phase is quarter. Thereafter the inferior planet appears to reverse its course and move toward the sun until its elongation is a minimum at *superior conjunction*. It is now on the other side of the sun from the earth, and its phase is full. Past superior conjunction, the inferior planet swings east on its way toward *maximum eastern elongation,* when it becomes the evening star in a quarter phase. It moves back toward inferior conjunction to complete the circuit and the cycle of moonlike phases.

Planets with orbits outside earth's are called *superior planets* (Figure 2.8). Their orbital periods are longer than that of the earth; so their motion relative to the earth-sun line appears to be clockwise around the sun. When a superior planet is nearest to us and also brightest, it is in *opposition*—opposite the sun in the sky and visible throughout the night. As the superior planet moves 90° east of the sun, it passes through *eastern quadrature,* when it rises at noon and is an

evening star. Continuing on, the superior planet moves westward, or its elongation is decreasing. It reaches *conjunction* when it is farthest from the earth, at which time it rises and sets with the sun. Next the planet passes through *western quadrature,* 90° west of the sun. At which point the superior planet rises at midnight; it is a morning star. Finally, the superior planet returns to opposition, its cycle of configurations complete. As seen from the earth, superior planets do not exhibit a cycle of phases.

The length of time for one orbit of a planet around the sun is known as its *sidereal period.* It is the time taken to complete a 360° circuit around the sky relative to the stars. From the earth we actually observe the *synodic period*—the time it takes a planet to return to a particular configuration with respect to the sun as viewed from the earth (such as from opposition to opposition). The synodic and sidereal periods differ because the earth is advancing in its own orbit as a given planet revolves around the sun.

Since the earth's orbital period is shorter than that of a superior planet, the earth must overtake a superior planet like Mars and pass it. This occurs while the planet's configuration changes from western quadrature to opposition to eastern quadrature. During this period the planet's normal west-to-east motion is temporarily interrupted, and it moves from east to west. This countermotion is known as *retrograde mo-*

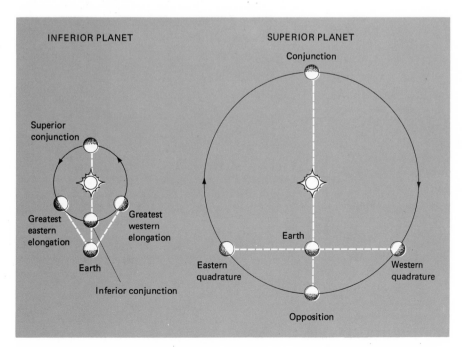

FIGURE 2.8
Configurations of an inferior planet and a superior planet. The planet's motions are shown as they are seen from the earth; viewed from the sun, the planets move in the same direction — eastward. Relative to the earth-sun line, the angular distance of the planet from the sun is called its *elongation*. For both the inferior and the superior planet there are four values of elongation that carry special names; these are shown beside the planet.

tion, in which the superior planet executes a closed or open loop (⌢ or ⌣) and then continues its usual path eastward. The geometrical reason for apparent retrograde motion is illustrated in Figure 2.9.

Now that we have surveyed some of the common earth-sky relationships and the regularity of the heavens, we shall discuss the historical development of astronomy, leading up to the modern conception of the dynamics of planetary motion.

2.3
Geocentric Solar System: Greek Cosmology

SIZE AND SHAPE OF THE EARTH

Some natural philosophers of the ancient world envisioned the earth as a flat disk afloat in water or riding in midair. Later it seemed impossible to doubt that the earth was spherical: How else, argued Aristotle, could it project a circular shadow when it eclipsed the moon?

Eratosthenes (273–193 B.C.), geographer and librarian of the museum at Alexandria in Egypt convinced that the earth was a sphere and that the sun was far enough from the earth for its rays to be parallel when reaching the earth, was able to measure its circumference, as shown in Figure 2.10. He chose observing stations at Alexandria and at Syene to the south (where the Aswan Dam is now located on the Nile River). For the experiment he fixed on local noon on the day of the summer solstice, which comes at the same moment at both sites because they are very nearly on the same meridian of longitude. He selected that day as a matter of convenience because the sun was directly overhead at local noon at Syene. (Legend says that he looked down a well and observed that no shadow was cast on its sides, clearly indicating when the midday sun was at its zenith.)

At noon an observer in Syene observed that the sun was directly overhead, while an observer in Alexandria found the sun to be 7° south of the zenith. Measurers paced off the distance between the two cities as about 4900 stadia (1 stadium ≃ 0.16 kilometer). Because a straight line cuts two parallel lines at equal angles, the angle at the center of the earth is equal to the zenith angle, 7°. Working a simple proportion, he could find the earth's circumference:

$C/4900$ stadia = $360°/7°$, or $C = 252,000$ stadia, or about 40,320 kilometers. In principle, Eratosthenes' experiment was correct. Although his measuring technique was inaccurate by modern standards, his results were extremely close to today's mean value of 40,030 kilometers.

HISTORICAL DEVELOPMENT OF THE GEOCENTRIC CONCEPT

Even though the spherical shape of the earth was, in general, accepted by the early natural philosophers,

FIGURE 2.9
The heliocentric model, showing retrograde motion of a planet. The diagram shows 10 positions of the earth and a planet as they orbit the sun and the corresponding apparent positions of the planet in the sky. Because the earth is moving more swiftly, it overtakes the planet. As the earth passes, the planet appears to backtrack (between points 4 and 6) before it seems to resume its forward motion.

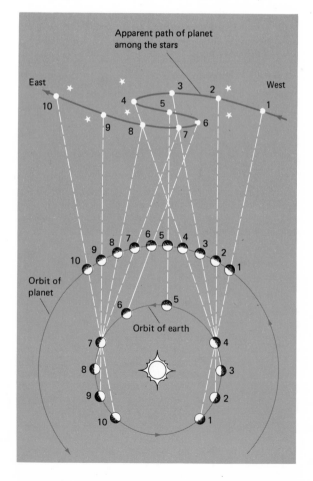

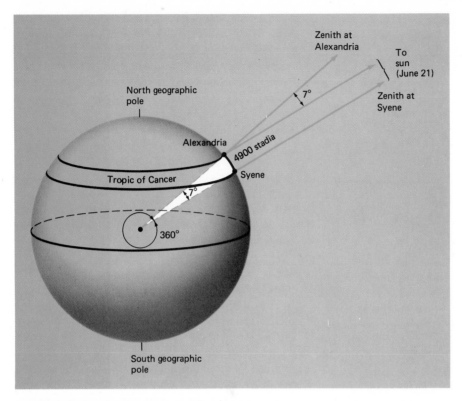

FIGURE 2.10
Eratosthenes' determination of the earth's circumference. Observations of the sun's position, together with the measured distance between Alexandria and Syene, enabled Eratosthenes to calculate the earth's circumference, as described in the text.

they still thought the earth was an immovable and solid globe surrounded by the fixed sphere of the heavens. Certainly the most immediately obvious and simplest geometrical model was one in which the earth was located at the center of the universe, while between the earth and the heavens was a second sphere, a rotating one, that would account for the stars' daily rising and setting.

However, with this model the problem was to explain the other movements in the solar system — those of the sun, moon, and five planets,[1] all of which seemed to wander among the stars. Never changing their course, the sun and moon move eastward relative to the stars on the celestial sphere, with the planets generally following the same path from west to east. The planets confuse matters, though, because sometimes they temporarily reverse their direction in retrograde motion. They further compound the problem by displaying irregularities in their motions (as do the sun and moon) and variations in brightness.

Explaining the universe with a logical system was a challenge for the natural philosophers of ancient

Greece. For centuries various schools of philosophy proposed, debated over, and elaborated on several *geocentric* (earth-centered) theories. With their taste for balance and symmetry, the Greeks reasoned that nature arrayed her celestial bodies in the perfect geometric figure, the sphere, and moved them on flawless circles. Their philosophers, beginning with Plato (427?–347 B.C.) or earlier, thought that planetary movements were accounted for by combinations of uniform circular motions with the earth at the center.

Hellenistic culture spread throughout the eastern Mediterranean world and after about 300 B.C. centered in Alexandria, where a new establishment for science was started. Alexandrian astronomers set themselves to removing discrepancies between the geocentric theory and the motions they observed. Thus geocentric theory maintained its dominance in philosophical thought — except for a brief departure in the third century B.C. — until the sixteenth century.

During the third century B.C. a *heliocentric* (sun-centered) scheme was proposed by Aristarchus (320?–?250 B.C.). He thought it natural to put the largest and only self-luminous body, the sun, at the center of the system. According to Aristarchus, the heavens

[1]The word *planet* is of Greek origin, signifying "wanderer," in contrast with the fixed stars.

> Finally we shall place the Sun himself at the center of the Universe. All this is suggested by the systematic procession of events and the harmony of the whole Universe, if we only face the facts, as they say, with both eyes open.
>
> Copernicus

moved each day because the earth rotated on its axis. Annual changes in the sky and the planets' irregular motions could be explained if they and the earth then revolved about the sun.

Contemporaries objected: Why suggest that the earth moves when no one has evidence for that movement? If the earth did rotate, unattached objects should fly off, and orbital motion should show up in an apparent shift in position of the nearby stars as seen against more distant ones. No such parallactic shift was ever observed. As mentioned in Chapter 1, we know today that these stars do appear to change place, but they are so far away from us that the apparent angular displacement, or parallactic shift, is extremely small, even for the closest stars, and can be measured only with a large telescope. Alexandrian astronomers, therefore, rejected the heliocentric theory and looked for some way of improving the geocentric picture. To work, the theory would have to represent more accurately the many small cyclic changes and the large general motions of the world system.

They found the system they wanted in the work of several astronomers, especially Hipparchus (190?–?120 B.C.), Appollonius (261?–?190 B.C.), and Ptolemy (A.D. 100?–?170). This system had a number of combinations of circles and off-center motions. The final version, which retained the circular motions and uniform rates from Aristotle's (384–322 B.C.) system, was published in the *Great Syntaxis of Astronomy*, an astronomical encyclopedia compiled by the last of the great Alexandrian astronomers, Claudius Ptolemy.

PTOLEMY'S GEOCENTRIC MODEL

Ptolemy's geocentric system had each planet (as in Figure 2.11) moving uniformly around a small circle called an *epicycle*. The center of the epicycle in turn revolved uniformly around the earth on the circumference of a large circle called a *deferent*. By means of the proper combination of rates in the epicycle and deferent, the planetary motions could be mostly direct and occasionally retrograde. The sun and moon had no epicycles; they completed their turns around the earth in 365.25 days and 27.3 days, respectively.

Each of the two inferior planets, Mercury and Venus, moved on their individual epicycles in their particular period of phases. The center of their epicycle, always located on the earth-sun line, described a complete revolution around the earth in 1 year. Mercury and Venus, therefore, swung from one side of the sun to the other as "morning" and "evening" stars.

Each of the three outer planets, Mars, Jupiter, and Saturn, moved once around their individual epicycles in their synodic periods. The period of the motion of the center of the epicycle on the deferent was thought to correspond to the sidereal period of the planet.

Later refinements using off-centered motions were made to better account for variations among the planets and the nonuniform motions of sun and moon as described in Figure 2.11. Ptolemy and his contemporaries devised the geocentric system as a mathematical conception of the planet's motions, not a physical model. Today's astronomy is descended from ancient Greek astronomical thought, through Arab culture, from which it spread across medieval Europe.

2.4
Heliocentric Solar System: The Copernican Revolution

Halfway through the thirteenth century, knowledge of astronomy had spread throughout Europe as Greek manuscripts were translated into Latin in the newly founded European universities. The Renaissance blossomed in the next two centuries, ending the dominance of ecclesiastical concerns in medieval thought and beginning the development of a broader range of intellectual considerations, including cosmology. Renaissance scientists proved to be creative in picturing the physical world, not letting themselves be dominated by past dogmas. They prepared the way for a deep change in scientific thought and viewpoint.

FIGURE 2.11
Simplified schematic of the Ptolemaic system. (a) Each planet moves along a small circle called an *epicycle*, which in turn revolves on the circumference of a large circle called a *deferent*. On the side of the epicycle farthest from the earth the planet appears to move in the same direction as the deferent — eastward, or direct. As it approaches the half of the epicycle closest to the earth, the planet appears to slow down, temporarily halt, then quickly reverse its direction, and halt once more before resuming its direct motion, when it returns to the other half of the epicycle. (b) The planet has apparently executed a complete loop. (c) Eccentric motion is a refinement of epicyclic motion. In this model the earth is not located at the center of the deferent. A second refinement is to allow the center of the epicycle to move uniformly about a point called the *equant*.

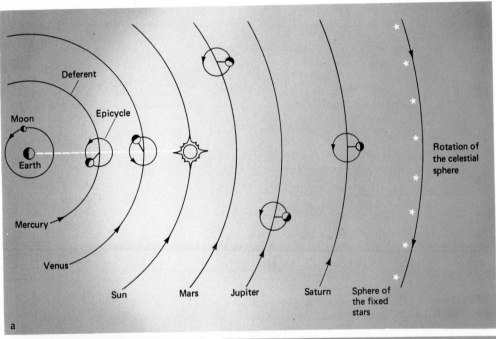

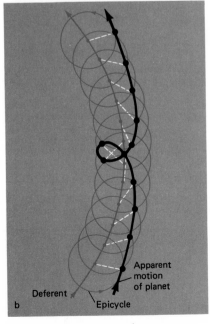

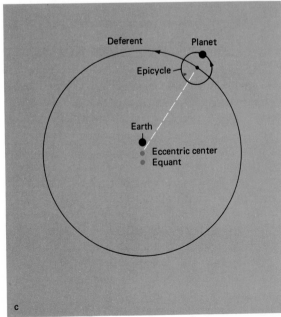

NICOLAUS COPERNICUS (1473–1543)

The five-hundredth anniversary of Copernicus's birth was celebrated throughout the world in 1973. Various governments, including ours, issued commemorative stamps in his honor; the last of the two orbiting astronomical observatories, launched in 1972, was named *OAO-Copernicus;* and historians of science met to eulogize the accomplishments of the man who revolutionized astronomy with his heliocentric system.

After the death of his father when Copernicus was 10 years old, an uncle, who was also a bishop, raised him and saw to it that he had an excellent education. Copernicus studied mathematics, philosophy, astronomy, and astrology at the University of Cracow; he studied law and medicine at the universities of Bologna and Padua. When he returned to Poland, he lived for a while in his uncle's castle. There he spent time as a physician, engaged in diplomatic activities, and undertook various administrative duties. After he was elected a canon through his uncle's influence, he had sufficient income to devote more of his time to astronomy, his first love.

Beginning in 1512, Copernicus set himself the task of examining critically the various systems of the world that had been proposed in the past. After several decades of study he became dissatisfied with the complexity and improbability he found in the Ptolemaic system. Placing the sun at the center of the solar system simplified matters. ". . . In the center of everything the sun must reside; . . . there is the place which awaits him where he can give light to all the planets." So wrote Copernicus. In his development of the heliocentric system he retained the notion of uniform circular planetary motion. He was thereby compelled to introduce a number of epicycles and eccentrics in order to account for the variable movements of the planets.

In 1530 Copernicus circulated a summary of his ideas among his friends. Knowledge of it spread to others. Eventually the fruits of his full labor appeared in print in 1543. To a cardinal friend who had inquired about his theory Copernicus wrote: "Although I know the thoughts of a philosopher do not depend on the judgment of the many, yet when I considered how absurd my doctrine would appear, I long hesitated whether I should publish my book." His book, divided into six volumes, includes discussions of the heliocentric concept, the geometry of the spheres of the heavens, and the earth's motions, including precession, lunar theory, and planetary motions.

THE COPERNICAN SYSTEM

Cultural ideas, including astronomy, proliferated after the 1430s, when the printing press was invented. Although the Ptolemaic system had been immensely successful in describing general aspects of planetary motion for over 13 centuries, by the fifteenth century easily recognizable discrepancies had arisen in the observed and predicted positions of some planets.

About the time the New World was being discovered, Nicolaus Copernicus (1473–1543), a Polish canon of ecclesiastical law and astronomer, began wondering whether any other arrangement of the planetary system might not be simpler, more reasonable, and more aesthetically pleasing than the Ptolemaic one. He resurrected Aristarchus's heliocentric idea and built a new cosmology based on it. After nearly four decades of study, Copernicus's monumental book *On the Revolutions of the Heavenly Orbs* was published in the year of his death, 1543. Dedicating the work to Pope Paul III, he died without seeing his theory accepted, except by a few friends to whom he had given his manuscript years before its publication. The public reception given at the time to what was much later a revolution in the concept of the universe varied between indifference and open hostility.

Because Copernicus still believed in the Greek idea that heavenly bodies must move in perfect circles, he had to explain the deviations from uniform motion. To clear those up he postulated a number of epicycles and other mathematical structures. His system, then, was not much more accurate or simpler than Ptolemy's, but the Copernican system was a tremendous step in cosmological thought for its time. The heliocentric model was as capable of explaining retrograde motion and all the observed motions as was the geocentric model. In the next century this change led to acceptance of the concept that celestial physics was not a supernatural matter but only an extension of terrestrial physics; Isaac Newton was later to make that clear.

TYCHO BRAHE (1546–1601)

Appearing at an opportune time, the right man for the next advance in astronomy was Danish nobleman-astronomer Tycho Brahe. With financial help from King Frederick II he constructed in 1582 a superbly equipped observatory on the island of Hveen, about 32 kilometers northeast of Copenhagen. There, with the most accurate pretelescopic observing instruments ever designed, Brahe determined positions with a precision of one minute of arc, far surpassing any previous measurements.

Brahe observed the sun, moon, and stars regularly instead of haphazardly as others had in the past. An uninterrupted record of their movements over many years was thus available for study and analysis. Brahe had reservations about adopting the entire heliocentric theory. He accepted the idea that the five planets revolved around the sun but not the idea that the heavy and sluggish earth moved. Earth's motion would be felt, he argued—and besides, a moving earth was contrary to scriptural belief. Neither could he detect the earth's orbital motion by parallactic shifts in the positions of the brighter stars. Consequently Brahe's cosmological system was a compromise: The planets orbited the sun; the sun and moon, in turn, orbited a fixed earth. There was relatively little interest in Brahe's cosmology, and it never really won a place in cosmological thought.

JOHANNES KEPLER (1571–1630)

In the years just prior to 1600 the Renaissance and Reformation were coming to an end. Copernicus's works were read by a few astronomers who recognized the computational advantages of the Copernican system but were not willing to take seriously its philosophical and physical implications. But a devoted Copernican, Johannes Kepler (1571–1630), the German assistant and successor to Tycho Brahe, was destined to change its acceptance. Brahe's observations of the celestial bodies bore fruit in the hands of Kepler when he pulled from Brahe's records history-making discoveries.

The question that concerned Kepler was what the clockwork was that governed the celestial machinery. After 17 years of labor, during which he rejected many ideas because they did not fit Brahe's observations, Kepler explained in two books, published in 1609 and in 1619, how the planets moved. His science was uncompromising and contributed to making today's standard: A scientific model must satisfy all observational facts or fail. "By the study of the orbit of Mars," he said, "we must either arrive at the secrets of astronomy or forever remain in ignorance of them."

2.5
Kepler's Laws of Planetary Motion

Kepler solved the problem over which so many astronomers had labored for centuries. The orbits of the planets are ellipses, not circles, and their variable motion is due to their varying distances from the sun. These are his first two laws, briefly stated (see Figure 2.12). The third contains the relationship between the planets' orbital periods and their distance from the sun.

THE FIRST LAW: ELLIPTICAL ORBITS

Kepler's first law of planetary motion can be stated as follows:

KEPLER'S FIRST LAW (LAW OF ELLIPTIC ORBITS): *Each planet moves in an elliptic orbit around the sun, with the sun occupying one of the two foci of the ellipse.*

The ellipse, a "family" of mathematical curves, is important in discussing the orbits of any bodies about each other, not just planets. Other families of curves related to the ellipse will be discussed in Chapter 3. Roughly speaking, an *ellipse* is a circle with the opposite ends of a diameter pulled outward, thus distorted into an oval figure. The long axis of the ellipse (Figure 2.12) is called the *major axis*, and perpendicular to it through the center of the figure is the *minor axis*. There are two points on the major axis, called the

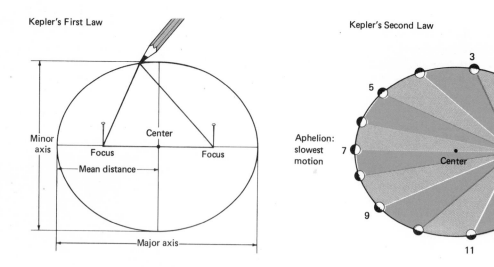

Kepler's First Law

Minor axis

Center

Focus

Focus

Mean distance

Major axis

Kepler's Second Law

3

5

Aphelion: slowest motion

7

Center

Perihelion: fastest motion

1

9

11

Kepler's Third Law

Planet	Sidereal Period P (yr)	Mean Distance a (AU)	P^2	$\simeq$	a^3
Mercury	0.24	0.39	0.06		0.06
Venus	0.62	0.72	0.38		0.37
Earth	1.00	1.00	1.00		1.00
Mars	1.88	1.52	3.53		3.51
Jupiter	11.9	5.20	142		141
Saturn	29.5	9.54	870		868
Uranus	84.0	19.2	7,060		7,080
Neptune	165	30.1	27,200		27,300
Pluto	248	39.4	61,500		61,200

FIGURE 2.12
Kepler's three laws of planetary motion. Kepler found that the planets move in elliptical orbits. (To draw an ellipse, loop a string taut around two tacks — the foci — and a pencil as shown.) For Kepler's second law the elliptical orbit is shown with the sun off center. The mean distance of the planet is equal to one-half the length of the major axis, or the semimajor axis, whose end points are 1 and 7. Because the areas of all sectors, gray and colored are equal, the planet passes through the numbered positions in equal intervals of time. The planet moves fastest when near the sun and slowest when farthest away. Kepler's third law is illustrated by the numbers in the fourth and fifth columns of the table. If the sidereal period is measured in years (the earth's sidereal period) and the mean distance in astronomical units (AU, the earth's mean distance from the sun), the two columns are approximately, but not precisely, equal. The slight discrepancy is removed by Newton's modification of this law.

foci (the singular form is *focus*), about which the figure is roughly symmetrical. The sum of the distances from each of the foci to every point of the ellipse is a constant. This immediately suggests a means of drawing an ellipse: Loop a piece of string around two tacks (the foci), and wield a pencil as shown in Figure 2.12. In a planet's orbit the sun occupies one focus; the other one is empty.

The farther the foci are from each other, the more elongated the ellipse; the closer together they are, the more nearly circular the ellipse. Thus the ratio of the distance of a focus from the center to the length of the semimajor axis (half the major axis), known as the *eccentricity*, determines the shape of the elliptic orbit. When the ratio is zero, the foci and center coincide, and the elliptic orbit degenerates into a circular

By denying scientific principles, one may maintain any paradox.

Galileo Galilei

orbit. The more elongated the elliptical orbit is, the nearer the eccentricity is to 1. The eccentricities of the planets' orbits are given in Table 6.2. They vary from near zero for Venus to around 0.2 for Mercury and Pluto. Thus the planetary orbits are not very elongated but very nearly circular, which is why it was not obvious to Kepler or to his predecessors that the orbits of the planets were not indeed perfect circles as Plato had envisioned. Finally the size of the elliptic orbit is set by the length of the major axis.

THE SECOND LAW: SPEED IN ORBITS

If the planets were orbiting the earth in circular orbits, they would traverse equal angles on the celestial sphere in equal intervals of time anywhere in their orbits. The fact that they do not actually do this but traverse variable angles in equal intervals of time (depending on where they are located in their orbits) was well known to the Greeks. This irregular motion was what necessitated epicycles in Ptolemy's geocentric system and in Copernicus's heliocentric system. In the elliptical orbits determined by Kepler the planets' distance from the sun, which occupies one focus, varies with position in the orbit. In addition the speeds in the orbits vary from one position to another such that planets move fastest when closest to the sun and slowest when farthest away. This is a consequence of Kepler's second law of planetary motion (illustrated in Figure 2.12), which may be stated as follows:

KEPLER'S SECOND LAW (LAW OF AREAS): *The imaginary line connecting any planet to the sun sweeps over equal areas of the ellipse in equal intervals of time.*

The mean distance of a planet from the sun is the average of the distance between the point of closest approach, called *perihelion*, which is located at one end of the major axis (point 1 in Figure 2.12), and the most distant point of the orbit, *aphelion*, which is located at the other end of the major axis (point 7 in Figure 2.12). The average is one-half the length of the major axis, or the semimajor axis, as shown in the figure. The alternate gray and colored sectors in the figure are of equal area. Therefore, according to Kepler's second law, the planet passes through the numbered positions in equal intervals of time.

THE THIRD LAW: YARDSTICK FOR THE SOLAR SYSTEM

Kepler's third law is extremely important in the sense that it provides a means of determining the relative size of the solar system in units of the mean earth-sun distance, the astronomical unit (AU). The law may be stated as follows:

KEPLER'S THIRD LAW (HARMONIC LAW): *The square of any planet's period of orbital revolution is proportional to the cube of its mean distance from the sun.*

The period of orbital revolution is the planet's sidereal period; that is, the time to move through 360° in its orbit. The sidereal period cannot be measured directly since there is no marker along the orbit to tell us when the planet has come back to its starting point 360° later. But astronomers can measure the synodic period directly, say, the time from opposition until the planet returns to opposition again. And from a simple mathematical relation the sidereal period can be computed from the synodic period. If we know the sidereal period, from Kepler's third law in units of years and astronomical units we can compute the mean distance (semimajor axis) of a planet's orbit. The table in the caption of Figure 2.12 illustrates this computation.

Newton later modified Kepler's third law (see page 53) to show that the mass of the sun and the planet under consideration enter into the equation. This modification is all important; for it allows astronomers to determine masses not only for the planets of the solar system but also for many stars and even galaxies. These topics will be taken up in Chapters 12 and 17.

UNIVERSAL NATURE OF KEPLER'S LAWS

Kepler's laws are universal. They apply to any two bodies gravitationally bound to each other, whether in the solar system or elsewhere in the universe.

Kepler's work put to rest any notion that the planets moved in perfectly circular orbits because nature had decreed that the heavenly bodies must show perfection in their movements. Of course, bodies may move in circular orbits when the two focal points coincide with the center, in which instance a circle is a special case of an ellipse.

Kepler did not know why the planets move by these empirical relationships, which he had established from Brahe's observations. He sought a cause of which his three laws were the effect. As he stated, "I am much occupied with the investigation of physical causes. My aim in this is to show that the celestial machine is to be likened not to a divine organism, but rather a clockwork. . . ." Kepler vaguely sensed that bodies have a natural "magnetic" affinity for each other and guessed that the sun has an attractive force. It remained for Newton, half a century later, to formulate a unified theory of motion, which includes planetary motion with gravity as its cause.

The substitution of a kinematic planetary model for a purely geometric one was Kepler's primary achievement. He prepared the way for the modern theory of force and the mathematical analysis of motion in terms of forces. A chief proponent of this new idea was the Italian physicist-astronomer Galileo Galilei. From his experiments with bodies in motion and the forces controlling them came the foundations of modern mechanics.

2.6
Early Concepts of Motion: Galileo

GALILEO'S CONCEPTS OF MOTION

Galileo Galilei (1564–1642) was a contemporary of Johannes Kepler. Yet for some reason he seems not to have been significantly influenced in his work by Kepler or Kepler's three laws of planetary motion. Galileo approached an understanding of motions in the cosmic world by studying terrestrial motion, especially falling bodies.

The dominant concepts of motion during the Renaissance were still those of Aristotle, who had defined motion as either natural or forced. A rock falling toward the ground was an instance of natural motion, or the tendency of earthly materials to return to their natural place — the ground. And no cause was needed to assist the motion; it was a natural tendency. On the other hand a thrown rock required a force or cause both to set it into motion and to continue it in motion, and thus this was an unnatural tendency — referred to by Aristotle as forced motion.

Galileo did not formulate the principle of gravity that we recognize today; that was for Newton to do later. However, he did conceptualize a force as something that brings about a change in the motion of bodies, with the earth exerting an attractive force (that is, gravity) that influences falling bodies. He also recognized the tendency of bodies either at rest or in motion to resist a change in their nonmotion or motion; we now call the resistance the *inertia* of the body. Thus to Galileo and to his scientific contemporary in France, René Descartes (1596–1650), rest and uniform motion were natural states of affairs. To change such states, it was necessary to have a force act on the body regardless of whether it was falling straight down or moving across the surface of the earth. The departure from uniformity in motion is now referred to as *accelerated motion*.

THE TELESCOPE: OBSERVING MOTION IN THE HEAVENS

Although mechanics was possibly his most significant accomplishment, Galileo also revolutionized astronomy in 1609 by designing and building a telescope. As the first telescopic explorer of the heavens, he established his place in history through such discoveries as Jupiter's four large satellites, craters and mountains on the moon, the phases of Venus, and individual stars in the Milky Way. (The frontispiece to this chapter reproduces Galileo's drawings of the moon.) Kepler had been the first to demonstrate that the heliocentric system was valid, and Galileo gave the Copernican theory observational support: For example, he observed that on a smaller scale Jupiter's satellites moving around the planet were analogous to the planets orbiting the sun. They were obviously heavenly bodies not in orbit about the earth, and here also was evidence disputing Aristotle's contention that a moving earth would leave the moon behind. Jupiter retains its satellites; logically, then, the earth can move around the sun without losing its satellite.

Theological hostility loomed over Galileo for supporting Copernican cosmology. Pope Paul V instructed his emissary, Cardinal Bellarmine, to warn Galileo against teaching or upholding Copernican doctrine, and from the Holy Office in February 1616 came this stern decree:

The following propositions are to be censured: (1) that the Sun is at the center of the world and the universe. . . . Unanimously, this proposition has been declared stupid and absurd as a philosophy, and formally heretic because it contradicts in express manner sentences in the Holy Scripture. . . . (2) that the Earth is not the center of the world and motionless, but changes its place entirely according to its diurnal movement. Unanimously, this proposition is declared false as a philosophy. . . .

But more liberal Pope Urban VIII took office, and Galileo obtained permission to teach both the Ptolemaic and Copernican systems; he was, though, to present the latter as an unproved alternative. Encouraged, Galileo began work on a masterly astronomical commentary, which passed censorship and was published in 1632 as *The Dialogues of Galileo Galilei on the Two Principal Systems of the World: The Ptolemaic and Copernican*. Powerful enemies soon convinced the Pope that Galileo had cast the Ptolemaic system in an unfavorable light, the book was officially banned, and in the year 1633 the great scientist was publicly humiliated before a papal tribunal in which he recanted his Copernican views.

The last 9 years before his death in 1642 Galileo spent in his villa in Arcetri, some distance from Florence, under strict house arrest. He was forbidden to publish, to discuss the forbidden philosophy, and even to speak to Protestants although he was able to finish *Two New Sciences* and have it published in Leyden in the Netherlands in 1638. By then he was 74 and totally blind—about which he writes, ". . . this universe, which by my remarkable observations and clear demonstrations I have enlarged a hundred, nay a thousand fold beyond the limits universally accepted by the learned men of all previous ages, are now shrivelled up for me into such a narrow compass as is filled by my own bodily sensations." The silencing of Galileo acted to silence Catholic scientists in the south of Europe, and from there the scientific revolution moved to northern Europe. Galileo died in the same year, 1642, that Isaac Newton was born in England.

GEOMETRICAL AND MATHEMATICAL FORM OF KEPLER'S LAWS

In mathematical shorthand we write Kepler's third law as (see Appendix 1 on measurements and computations)

$$P^2 = Ka^3,$$

where P is the sidereal period of revolution, a is the mean distance of the planet from the sun, and K is a constant of proportionality depending upon the units for P and a. Even though Kepler did not know the numerical value of the constant K, he could eliminate the constant by assuming that K was the same for each planet. He then considered the ratio of the equation for any planet to that for the earth:

$$\frac{P^2}{P_\oplus^2} = \frac{a^3}{a_\oplus^3},$$

where the subscript $\oplus$ stands for the earth. If we express the planet's sidereal period in years and its mean distance in units of the mean earth-sun distance (the *astronomical unit*, now known to be equal to 149,597,871 kilometers), then $P_\oplus = 1$ and $a_\oplus = 1$, and Kepler's third law becomes

$$P^2 = a^3.$$

Example: Let us calculate the mean distance of Jupiter from the sun, given that Jupiter's sidereal period is about 11.9 years. From Kepler's third law we obtain an expression for a:

$$a = \sqrt[3]{P^2} = \sqrt[3]{(11.9)^2} = 5.2 \text{ AU}.$$

SUMMARY

Ancient astronomy: The sky. Astronomers call the apparent dome of the sky the celestial sphere. The ancients, unlike modern astronomers, conceived of the earth as being at the center of the sphere with various celestial objects moving from east to west because of the rotation of the celestial sphere. Ancient astronomers also noted the rotation of celestial objects about two fixed points, the north and south celestial poles. In order to facilitate the observation of the motions of celestial objects, ancient astronomers named constellations of stars and the more prominent individual stars.

Cyclic phenomena. Since ancient times, people have observed and tried to predict cyclic phenomena of the heavens. Ancient astronomers tracked the path of the sun across the sky and recognized seasonal events on the basis of the sun's position. Phases of the moon, lunar and solar eclipses, and configurations of the planets are common earth-sky relationships carefully observed by ancient astronomers.

Geocentric solar system. By the time of Aristotle and Eratosthenes, natural philosophers seemed to have recognized that the earth was a sphere. Nevertheless, they also believed that the earth was an immovable globe surrounded by the sphere of the heavens. Movements of celestial bodies such as planets could be explained by attaching them to a movable sphere placed between earth and the celestial sphere. However, to make Ptolemy's geocentric model of the solar system work — especially to explain the apparent retrograde motion of certain planets — required elaborate combinations of uniform circular motions. Despite the complexity of the Ptolemaic model, the system dominated astronomical thought in Arab culture and medieval Europe. .

Heliocentric solar system. Aristarchus proposed a heliocentric cosmology, accounting for motions of celestial bodies on the basis of earth's movements. Ancient Greek astronomers were unable to make the observations and measurements that would have confirmed Aristarchus's contention. Copernicus revived interest in the heliocentric concept and had sufficient data to use a heliocentric model. But the model proposed by Copernicus was not immediately accepted. Later, Kepler's study of the orbit of Mars established that the heliocentric concept was a basis for understanding planetary motion.

Kepler's laws. By analyzing data collected by Brahe, Kepler determined that planetary orbits are ellipses with the sun at one focus and that their variable motion is due to their varying distances from the sun. Kepler's third law (relating the planet's period of orbital revolution to the planet's mean distance from the sun) is especially important because it provides a means of determining the relative size of the solar system in units of the mean earth-sun distance. Over time Kepler's three laws were recognized as universal kinematic laws, and thus they prepared the way for modern mathematical models of the solar system. Newton modified Kepler's third law as a result of his universal law of gravity and laws of motion.

Motion. Galileo made major contributions to understanding motion. Contrary to Aristotelian views, Galileo (and Descartes) maintained that uniform motion was the natural state of affairs in the universe. Any change from uniform motion was caused by a force acting on the body in motion. Thus Galileo introduced the concepts on which celestial mechanics would be based.

KEY TERMS

aphelion
astronomical unit
celestial equator
celestial meridian
celestial poles
celestial sphere
conjunction

eccentricity
ecliptic
ellipse
foci
major axis
opposition
perihelion

phases
retrograde motion
sidereal period
solar eclipse

synodic period
vernal equinox
zenith

CLASS DISCUSSION

1. If the celestial sphere is a hypothetical concept, of what value is it to the practice of astronomy today?

2. What arguments might ancient Greeks have used to convince themselves that the earth is spherical? Are there others besides the ones mentioned in the text? Can one actually see the curvature of the earth while on its surface?

3. Why do we on earth not witness an eclipse of the sun and one of the moon every month? If you were to speculate, is the earth the only planet on which a solar eclipse (total, partial, or annular) can be viewed?

4. How did the Ptolemaic system account for the retrograde motions of the planets and the irregular motions of the sun, moon, and planets? How did the Copernican system account for them?

5. Did Copernicus have strong observational evidence confirming his model of planetary motions? If so, what was it? If not, why did he propose a change?

6. If you had to choose the work of Copernicus, Kepler, or Galileo as the beginning point for the modern practice of science, which would you choose, and why? Is it entirely arbitrary?

READING REVIEW

1. How did the ancients differentiate between a star and a planet?

2. What is the zodiac?

3. How much does the rotation period of the moon have to do with the phases of the moon?

4. Should the synodic periods be longest or shortest for Venus and Mars compared with other planets?

5. Was the heliocentric concept original with Copernicus? Explain.

6. How did Brahe's cosmology differ from that of Ptolemy and Copernicus?

7. State in your own words Kepler's three laws of planetary motion.

8. What determines the size of an elliptic orbit? What determines its shape?

9. At what point in its orbit is a planet moving the slowest? The fastest? Does the variation in speed in the orbit depend upon the shape of the orbit?

10. What were Galileo's most significant telescopic discoveries?

PROBLEMS

1. If Jupiter is at opposition and the time of night is midnight, what is Jupiter's elongation, and where is it located relative to the celestial meridian?

2. If Venus is located at superior conjunction and the time is sunset, what is Venus's elongation, and where is it located relative to the astronomical horizon?

3. If a body were found orbiting the sun with a sidereal period of 292 days, what is the semi-major axis of its orbit? Which planet would pass nearest to it?

4. Measuring the sidereal and synodic periods for the planets in years (the earth's sidereal period), the relation between them is given by

$$\frac{1}{\text{sidereal period}} \pm \frac{1}{\text{synodic period}} = 1$$

where the minus is for an inferior planet and the plus is for a superior planet. If Mars's synodic period is 2.14 years, what is its sidereal period?

SELECTED READINGS

Gingerich, O. "Copernicus and Tycho," *Scientific American*, December 1973.

Koestler, A. *Watershed: A Biography of Johannes Kepler*. Doubleday, 1960.

Kuhn, T. S. *The Copernican Revolution*. Vintage Books, 1959.

Shapere, D. *Galileo*. University of Chicago Press, 1974.

Wilson, C. "How Did Kepler Discover His First Two Laws?" *Scientific American*, March 1972.

3

A Universe in Motion

PHILOSOPHIÆ
NATURALIS
PRINCIPIA
MATHEMATICA.

Autore *IS. NEWTON,* Trin. Coll. Cantab. Soc. Matheseos Professore Lucasiano, & Societatis Regalis Sodali.

IMPRIMATUR
S. PEPYS, *Reg. Soc.* PRÆSES.
Julii 5. 1686.

LONDINI,
Jussu *Societatis Regiæ* ac Typis *Josephi Streater.* Prostant Venales apud *Sam. Smith* ad insignia Principis Walliæ in Coemiterio D. *Pauli,* aliosq; nonnullos bibliopolas. *Anno* MDCLXXXVII.

In our everyday experience and throughout the universe the motion of objects is the rule rather than the exception: Even though you may be at rest on the surface of the earth, the surface is moving because of the earth's rotation on its axis and revolution about the sun. In the seventeenth century Isaac Newton (1642–1727) unified various concepts by means of his laws of motion and through the idea of gravity included terrestrial and cosmic motion as parts of a universal motion. For Newton gravity was the spring that ran the cosmic clockwork. The clockwork universe persisted as a philosophical concept until Albert Einstein (1879–1955) led science to see that motion does not occur in a framework of absolute space and time but is relative and cannot be divorced from us as observers. But we shall examine Newtonian mechanics before turning to Einstein's relativity.

3.1
Newton's Laws of Motion

The various aspects of motion and the unity of terrestrial and cosmic motion are realized in the work of Isaac Newton. He developed a unifying concept of gravity and the physical laws relating to matter, force, and motion. These Newton published in 1687 in his *Principia*, whose complete translated title is *The Mathematical Principles of Natural Philosophy*. The concept of gravity as underlying cause was an abundant source of solutions to old problems. With this one theory Newton accounted for the rise and fall of tides, the motions of the moon and the planets, and the earth's flattening at its poles. His theory also explained the precession of the equinoxes (see page 22) and opened the door for the later discovery of the planet Neptune.

FORCE: THE CAUSE OF MOTION
Before discussing Newton's laws, let us consider in more depth the fundamental concepts of motion. The first, the concept of force, is sometimes used as if its

◀ The title page of Newton's famous publication *The Mathematical Principles of Natural Philosophy*. The noted diarist Samuel Pepys was president of the Royal Society at the time.

meaning were obvious. A *force* is a push or a pull that makes a body change its state of motion (a definition for "state of motion" is given below). We should note that rest (or no apparent motion) is one possibility for a state of motion. The forces most familiar to us are *mechanical forces;* these are exerted by one body in contact with another, such as a bat and a baseball.

Fields of force, which are exerted without contact between bodies, make up the other class of forces. There are four types (see page 86): the *strong* and *weak nuclear forces,* which act on the subatomic scale; the *electromagnetic forces,* which can be either attractive or repulsive; and the *gravitational force,* which exerts a pull, or attraction, but never a push, or repulsion. Empty space between matter is no barrier to fields of force, as experience shows in the case of the gravitational force.

MASS: A MEASURE OF INERTIA
Every material body possesses a property called *inertia,* the resistance it offers to any change in its state of motion. The more matter a body has, the greater is its inertia. *Mass* is a measure of the amount of matter a body contains; therefore, mass measures its inertia. Massive bodies are more resistant to a change in their state of motion than less massive ones are, as we know from common experience.

A material body has the same mass regardless of where in the universe it is located, but its weight depends on its position relative to various attracting masses. A body's *weight* is a measure of the gravitational force that an attracting object exerts on the body. For example, a person weighing 90 kilograms on the earth's surface would weigh 15 kilograms on the moon's surface because the moon's gravitational pull is one-sixth that of the earth's. However, the person's mass is the same whether on the moon or on the earth (see Figure 3.1).

An important concept related to mass is that of compactness, or density. *Density* is defined as the amount of matter (mass or weight) in each unit of volume. Water has a density of 1 gram per cubic centimeter, while that of lead is 11.3 grams per cubic centimeter. Bodies may have the same mass but quite different densities, such as those of a feather pillow and a book (see Figure 3.1). If matter is distributed unevenly throughout a body—as it is in the earth— then the mass divided by the volume gives a *mean density.*

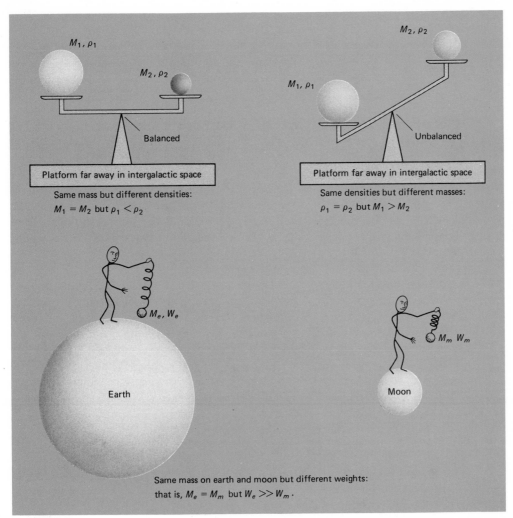

FIGURE 3.1
Distinctions among mass, weight, and density. Mass is a measure of the inertia of a body or the amount of matter it contains, while weight is the force exerted on a body by the gravitational field of a massive body, like the earth or moon. In the lower figure a mass hung on a spring extends the spring fully on the earth, but it extends it only partially on the moon since the moon's gravitational attraction is one-sixth that of the earth's. The illustration at the top left shows two bodies of the same mass but different densities on a balance in intergalactic space. They are the same mass since they balance, and they must have different densities since one is much smaller (that is, has smaller volume) than the other. At the top right the two bodies have the same densities, with the one on the left proportionately larger, but that makes it more massive; so they do not balance.

SPACE AND TIME FOR NEWTON

Distance is a familiar idea, but for our use in describing the motion of a body we need the notions of the *origin* from which and the *direction* in which it is being measured. The place where we observe and measure motion is called a *frame of reference*. It has an origin from which distances can be measured relative to some reference direction. As an example, let's consider the distances to the planets: To measure these distances, we must first select a reference frame, such as that with the sun as the origin (see Figure 3.2); and as our reference direction we might choose the direction toward a given star lying in the ecliptic plane. This is only one of several possible reference frames that could be used to measure the distance to the planets.

Time is equally familiar from our everyday experience. Our intuitive concept of *time* is based on changing patterns and events in our lives—which is quite different from our intuitive concept of distance; the separation between tables and chairs, for example. We can measure from an arbitrarily chosen origin, such as a historical event. We can move forward and backward over a distance, but in our common experience we move only forward in time, never backward.

Newton and most scientists after him believed in absolute time and absolute space as unchanging qualities of the universe. ("Absolute" here means that we can get outside the clockwork universe and define locations in space and events in time unequivocally.) It was not until the late 1800s that Ernst Mach (1838–1916), an Austrian physicist-philosopher, questioned the concept of absolute time. During his life, few took Mach's criticism seriously. Later Einstein, in

developing his theory of relativity, would be influenced by Mach's thinking. However, in the period between Newton and Einstein the concepts of absolute time and space were the pervasive influences in the development of science.

DESCRIBING MOTION:
VELOCITY AND ACCELERATION

For a moving body the distance traversed divided by the elapsed time is the *speed* of the body. If we take account of the direction of motion as well as the speed, we define the *velocity* of the body. A change in the *state of motion* or velocity, which is known as *acceleration,* can be due to a change in the speed or the direction of motion or both. Thus acceleration is measured as the rate of increase or decrease of a body's speed or as the rate at which its direction of motion changes. Since distance depends upon a frame of reference, then so also will velocity and acceleration. Figure 3.3 is a space-time diagram in which we couple spatial and temporal measurements to show the graphical meaning of velocity.

FIGURE 3.2
Frame of reference for measuring planetary distances. Shown is a very simple reference frame, with the sun as the origin and the reference direction as the direction toward the bright star Regulus (α Leonis), which is almost in the plane of the ecliptic. Two directions at right angles to the reference direction, one in the plane of the ecliptic, complete the three axes of the three-dimensional reference frame. Units of distance, in this case arbitrary ones, are used to the planets at any one moment relative to the three axes. Thus the earth is located by going six units along the reference direction, five units perpendicular to it, and zero units perpendicular to the orbital plane of the planets. This frame of reference is only for illustration and is not one actually used by astronomers.

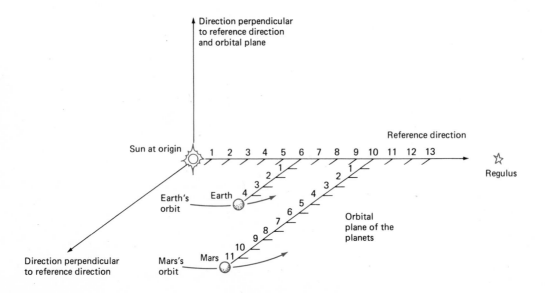

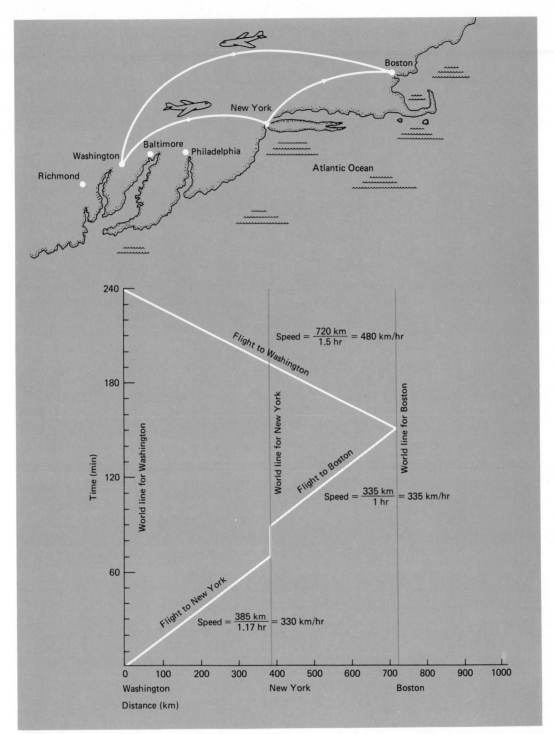

FIGURE 3.3
Space-time diagram of a trip by airplane between Washington and Boston with a stop in New York. The horizontal axis is the distance between the three cities; the vertical axis is the time elapsed since leaving Washington. We fly first to New York, stop over, and then continue to Boston. In Boston we realize that our appointment is not until next week, so we board another plane and fly nonstop back to Washington. Our spatial position goes forward and backward while time moves only forward. The vertical axis and the two other vertical lines represent what happens to the three cities: Time advances, but the cities don't change position. The three lines are called *world lines*. The speed (actually the average speed) can be found for each leg of the trip by dividing the distance traversed by the elapsed time, as shown.

An airplane's velocity, for example, is usually measured relative to the surface of the earth, say, 600 miles per hour. We could measure it relative to the sun, in which case the earth's rotational and orbital velocities would have to be added to that of the airplane. Any fixed point on the surface of the earth is continuously changing velocity—that is, accelerating—because of the spin of the earth on its axis of rotation and the revolution of the earth about the sun. What we really want to talk about in the case of the orbital motion of planets is the velocity at one single moment of time since an instant later the velocity will be different owing to acceleration. This concept of velocity is called *instantaneous velocity*.

QUANTITY OF MOTION: MOMENTUM

As you might guess, there is more to defining the motion of a body than simply finding a value for its velocity. For example, in the collision of two billiard balls moving with different velocities, it may appear that they simply exchange their states of motion so that the total quantity of motion is conserved and just redistributed between them. But in the case of two billiard balls of different masses moving with the same speed, the more massive one is able to transfer a greater quantity of motion in a collision than the small-mass ball can. Thus the concept of quantity of motion, or *momentum*, involves both the velocity and the mass of the body, as we illustrate in Figure 3.4. To find momentum, we multiply the body's mass by its velocity.

It is not difficult to visualize philosophically the possibility that the total quantity of motion, or momentum, in the universe is a constant—that is, the total momentum of the universe is conserved although the interaction of various bodies in the universe with each other redistributes the momentum of individual bodies. This concept of *conservation of momentum* was first annunciated by the French natural philosopher René Descartes in his book, *Principles of Philosophy*, published in 1644.

FIGURE 3.4
Meaning of quantity of motion, or momentum. The two bodies whose momenta are being compared are a cannonball and a musket ball. Both are given velocities of 100 meters per second, but the cannonball has 1,000 times the mass of the musket ball (10 kilograms and 0.01 kilogram, respectively). Upon striking the stone wall, the cannonball, having greater momentum, can transfer some to the stones of the wall, where they move in the original direction of the cannonball. On the other hand the musket ball has insufficient momentum to move any of the stones a significant amount.

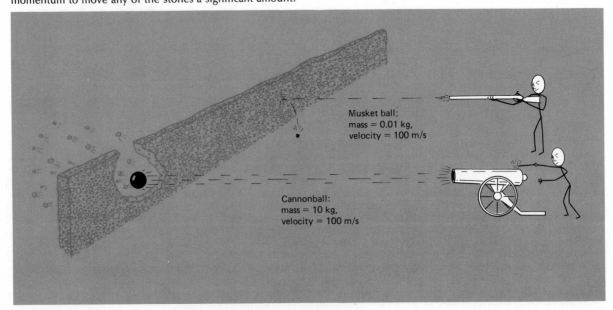

Musket ball:
mass = 0.01 kg,
velocity = 100 m/s

Cannonball:
mass = 10 kg,
velocity = 100 m/s

ISAAC NEWTON   (1642–1727)

At his birth on Christmas day, 1642, in Woolsthorpe, Lincolnshire, Newton was so tiny and frail that he was not expected to live. Yet despite his boyhood frailty, he lived to the age of 85. As a delicate child, he was a loner, interested more in reading, solving mathematical problems, and mechanical tinkering than in taking part in the usual boyish activities.

Up to the time Newton entered Cambridge University in 1661, there was little inkling of his mental prowess. His shyness kept him from making friends easily, and he did not mix with his more boisterous fellow students. At the university he took courses in Latin, Greek, Hebrew, logic, geometry, and trigonometry, and he attended lectures in astronomy, natural philosophy, and optics. His leisure time was spent reading works by Kepler and by Descartes, the inventor of analytic geometry, and filling his notebooks with remarks on the refraction of light, the grinding of lenses, and the extraction of roots of algebraic equations.

Newton received his BA degree—without any great distinction—and then returned to his home because of the Great Plague that was sweeping Europe. Practically all his time in his early twenties was spent at his home in Woolsthorpe. These were the most productive years of his life. During this period he discovered the expansion of the general binomial $(a + b)^n$; he invented the "fluxions" (differential calculus); with prisms he demonstrated the composite nature of white light; he discovered the law of gravitation; and he laid the foundations of celestial mechanics.

In 1668 Newton constructed the world's first reflecting telescope. It had an aperture of 1 inch and a tube length of 6 inches, which led Newton to say of it: "This small instrument, though in itself contemptible, may yet be looked upon as the epitome of what may be done this way." Not satisfied with his first effort, he completed an improved and somewhat larger reflector with an aperture of nearly 2 inches.

The publication of his *Principia* (1687), embodying his mathematical principles and his ideas on gravitation and the system of the world, marked the peak of Newton's creative career in science. The *Principia* represents the thought and study of more than 20 years, and it ranks in importance with Ptolemy's *Almagest* and Copernicus's *De Revolutionibus*. His treatise *Opticks* appeared in 1704, but most of it was written many years earlier.

Many tributes followed Newton's death in 1727. One that stands out was made by the great French mathematical astronomer Lagrange, who said: "Newton was the greatest genius who ever lived, and the most fortunate; for we cannot find more than once a system of the world to establish." Newton himself acknowledged: "If I have seen further than other men, it is because I have stood upon the shoulders of giants." In poet Alexander Pope's *Epitaph for Newton* are these lines:

Nature and Nature's laws lay hid in night;
God said, "Let Newton be!" and all was light.

NATURAL STATES OF MOTION: NEWTON'S FIRST LAW OF MOTION

In his *Principia* Newton formulated three laws that describe and predict the behavior of bodies in motion. As discussed in the last chapter, both Galileo and Descartes recognized that rest and uniform motion were natural states of motion. Newton's first law, often referred to as the principle of inertia, is about uniform motion:

NEWTON'S FIRST LAW: *A body remains at rest or moves along a straight line with constant velocity as long as no external force acts upon it.*

If a body is in motion in a straight line at constant velocity (that is, its state of motion is a uniform state of motion), it continues to move along that line without changing its speed or direction as long as no force

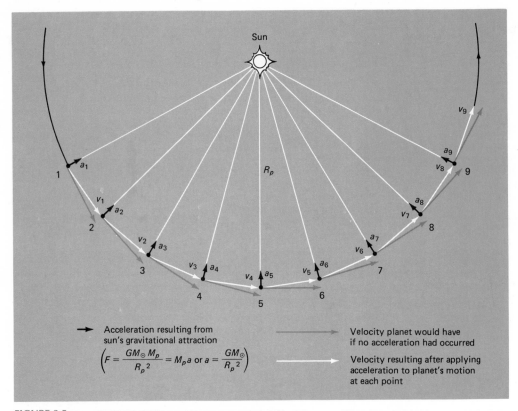

FIGURE 3.5

Newton's second law of motion. In an idealized situation suppose a planet is orbiting the sun in a circular orbit (Venus's orbit is very close to circular). With arrows whose direction represents the direction of the velocity of the planet at one instant and whose length represents the speed of the planet, we consider nine points that show the effect of the sun's acceleration on the motion of the planet. At each instant denoted by the numbers 1 through 9 the velocity is perpendicular to the radial direction to the sun. If there were no gravitational attraction, the planet would move off in a straight line shown by the arrow tangent to the circular orbit. But the gravitational attraction produces an acceleration toward the sun that changes the velocity such that the planet moves from one number to the next. In reality the acceleration produced by the sun's gravitational attraction happens all the time. Therefore, if we let the number of points increase indefinitely and come closer together, we have an illustration that more closely approximates the real situation.

acts on the body. Another way of saying this is that its quantity of motion or its momentum remains the same. Thus the natural state of affairs in motion is for the body to continue whatever motion it has until an external force changes its motion.

CHANGE OF STATE OF MOTION: NEWTON'S SECOND LAW OF MOTION

Newton's second law deals with nonuniform motion, or velocity change, and why it occurs:

NEWTON'S SECOND LAW: *A body acted upon by a force will accelerate in the direction of the applied force. The greater the force, the greater the acceleration will be.*

If a body's state of motion is basically defined by its velocity (or, more correctly, its momentum), then acceleration is a measure of the change. Acceleration is proportional to the magnitude of the acting force—as one increases, the other does too—and inversely proportional to the body's mass—as one increases, the other decreases.

If a moving body is subjected momentarily to an external force, it will momentarily accelerate in the direction given it by the applied force. Its velocity changes to a new value but not necessarily in the same direction as the force. But if the force is applied continuously, then a continuous change of velocity, or acceleration, takes place. For example, as a planet orbits the sun, there is a continuous change of speed and direction, or a continuous acceleration, as shown in Figure 3.5. The change must be due to some continuously applied external force. In this case we know that that force is the gravitational attraction of the sun.

ACTION AND REACTION: NEWTON'S THIRD LAW OF MOTION

Newton's third law deals with the relation between forces:

NEWTON'S THIRD LAW: *A body subjected to a force reacts with an equal counterforce to the applied force; that is, action and reaction are equal and oppositely directed.*

Two forces are involved here: action and reaction. One force never acts alone. We can see equal and opposite reactive forces every day. A bird taking off from an overhead power line produces a reaction that moves the line backward; water forced out of a lawn sprinkler produces a backward reaction that rotates the sprinkler. The next section is for astronomy a classic example of action and reaction.

3.2 Gravity: Newton's Universal Law

NEWTON DERIVES THE LAW OF GRAVITATION

The recognition that the earth exerts a pull on a body was not original with Newton, for Galileo recognized that the earth exerts a force on a falling body. It was Newton, however, who recognized that the pull of the earth can extend all the way to infinity. Such a pull is a universal phenomenon to which all material bodies are subject.

By analyzing Kepler's second law mathematically, Newton showed that the force acting on a planet must be one directed toward the sun. Only one kind of force would satisfy Kepler's requirement that the sun

MATHEMATICAL FORM OF NEWTON'S THIRD LAW

In mathematical form Newton's third law can be written as

$$F_1 = F_2,$$

where F_1 is the force acting on body 1 and F_2 is the reactive force acting on body 2. From a combination of Newton's second and third laws, it follows that

$$m_1a_1 = m_2a_2,$$

where the subscripts refer to body 1 and body 2.

Example: The force of attraction the earth has for you (F_1), standing on its surface, is *equal* and *oppositely directed* to the force of attraction you have for the earth (F_2). Both the earth and you are accelerated. But because the earth's mass is very large, the acceleration it gets is infinitesimal compared with what you get since your body's mass is very small. That is why we can see the acceleration of a body falling at the earth's surface (980 centimeters per second squared) but not the acceleration of the earth, which is about 1.5×10^{-21} centimeter per second squared for a 90-kilogram person. Only when two masses are more nearly the same can we easily observe the accelerations of both bodies, such as occurs in some binary-star systems.

be at the focus of the ellipse and still be consistent with Kepler's third law, relating the planets' periods to their distances from the sun. The force between the planets and the sun must then be an inverse-square force; that is, the intensity of the force must weaken as the square of the distance between a planet and the sun increases. Several contemporaries of Newton had suspected this relationship, but they could not prove it.

Newton then took into account his third law of motion and assembled his results in one comprehensive statement, the law of gravitation. He showed that it is universal by applying it to a falling apple and the earth, to the moon's motion around the earth, and to the planets revolving around the sun. He even imagined gravitation at work beyond the solar system, a thought that was verified later.

NEWTON'S LAW OF GRAVITATION: *Objects in the universe attract each other with a force that varies directly as the product of their masses and inversely as the square of their distances from each other.*

Newton proved that spherical bodies act as if their *gravitational mass* is concentrated at their centers.

THE LAW OF GRAVITATION AND THE MASS OF THE EARTH

If m_1 stands for the mass of one body, m_2 for the mass of a second body, d for the distance between their centers, F for the mutual force of gravity between them, and G for the constant of gravitation, the law of gravitation can be stated mathematically as

$$F = \frac{Gm_1m_2}{d^2}.$$

The constant of gravitation, G, was first measured in the eighteenth century by Henry Cavendish (1731–1810). Its numerical value in the metric system is 0.0000000667 (6.67×10^{-8}) centimeter cubed per gram second squared.

Although he apparently did not, Galileo is credited with having demonstrated that balls of different weights dropped from the Leaning Tower of Pisa fell with constant acceleration. Why? From the second law of motion we know that a body of mass m, subjected to the earth's gravitational force of attraction F, undergoes an acceleration at the earth's surface of $g = F/m$. From the law of gravitation this force is $F = GmM_\oplus/R_\oplus$, where $M_\oplus$ is the mass of the earth and $R_\oplus$ is the separation between the centers of the two bodies, or the earth's radius. Assuming that we know G, we have

$$mg = \frac{GmM_\oplus}{R_\oplus^2} \text{ or } g = \frac{GM_\oplus}{R_\oplus^2},$$

where the mass of the attracted body has canceled out. Thus the acceleration of the attracted body does not depend on its own mass; it depends on the mass of the attracting body, which in this case is the earth.

Example: The observed acceleration g at earth's surface is 980 centimeters per second squared. From the known values of the gravitational constant G and the earth's radius (6.38×10^8 centimeters), we can find the mass of the earth by rearranging the equation above to obtain

$$M_\oplus = \frac{gR_\oplus^2}{G} = \frac{(980 \text{ cm/s}^2)(6.38 \times 10^8 \text{ cm})^2}{(6.67 \times 10^{-8} \text{ cm}^3/\text{g}\cdot\text{s}^2)} = 5.98 \times 10^{27} \text{ g,}$$

and from this the earth's mean density is

$$\rho_\oplus = \frac{M_\oplus}{(4/3)\pi R_\oplus^3} = 5.5 \text{ g/cm}^3.$$

This simplifies the mathematical treatment of such bodies; the distance between their centers is ordinarily used in calculating their mutual gravitational attractions. Figure 3.6 shows examples of how the force of gravity varies under different circumstances.

ORBITS UNDER THE LAW OF GRAVITATION

Newton used his laws of motion and gravitation to show that Kepler's third law was only an approximation to the actual relation. The actual form of the law is much more useful, for it allows us to determine the masses of other celestial bodies (see Chapters 12 and 17). Newton also showed that the orbit of a body revolving around a central force always matches in shape one of the class of curves called *conic sections,* illustrated in Figure 3.7.

These curves are called conic sections because they are formed when we pass a plane (like a knife blade) through a cone at different angles. For the *ellipse,* of which a planetary orbit is one example, the cutting plane intersects opposite sides of the cone's slant edge. For a *circle* the plane cuts the cone at right angles to the vertical axis. The other two conic sections are open at one end: The *parabola* is formed

when the plane passes through the cone parallel to its slant edge; and the *hyperbola* is formed when the cone is intersected at an angle between that for the parabola and parallel to the vertical axis (as shown in Figure 3.7).

In a parabolic or hyperbolic orbit the body will pass by the attractive central force only once, approaching from and receding toward infinity, never to return.

FIGURE 3.6
Comparison of gravitational forces. The greater the mass of the body, the darker the shading in the drawing. (a) If the gravitational force between the two masses m_1 and m_2 at distance d is equivalent to 1 gram centimeter per second squared, or 1 dyne, then (b) at half the original distance but with the same masses the force is quadrupled. (c) If we replace m_1 with a smaller but denser body, having doubled the mass, we double the original force. (d) Doubling both masses but halving the original distance results in a gravitational force 16 times as strong as the original force. Note that the mass of the body is the important element, not the size. The relationship can be expressed as $F \propto m_1 \cdot m_2/d^2$.

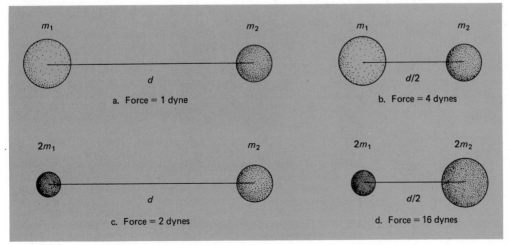

a. Force = 1 dyne

b. Force = 4 dynes

c. Force = 2 dynes

d. Force = 16 dynes

The different members of the sun's family move around the sun in closed paths, either ellipses or something approaching a circle. An object approaching the sun from outside the solar system would, when attracted by the sun, travel by it in a parabolic or hyperbolic orbit. If its motion is significantly influenced by the gravitational attraction of a planet during a near encounter with one such as Jupiter, it might be forced into an elliptical orbit around the sun, in which case we say it has been "gravitationally captured."

3.3
Energy:
The Most Important Concept in Science

The so-called fundamental quantities in the study of motion (mechanics) are usually taken to be distance, mass, and time, from which the other concepts can be derived or intuitively rationalized: velocity, acceleration, force, and momentum. Each of these helps to describe motion and to understand the cause-and-effect relationships in motion. However, these quantities still do not provide a complete understanding of motion.

Recall the arguments about colliding bodies that led us to the concept of momentum, or quantity of motion. The cannonball striking the rock wall in Figure 3.4 redistributes its momentum to the rocks of the wall, and they are set into motion. But momentum does not account for the destructive capabilities of the cannonball. For example, if we double the velocity of the cannonball, we find that it doubles the impulsive force necessary to set the stones into motion but has four times the destructive capability. So the concept yet needed is that of energy. No concept in science is more important than that of energy and its conservation principle.

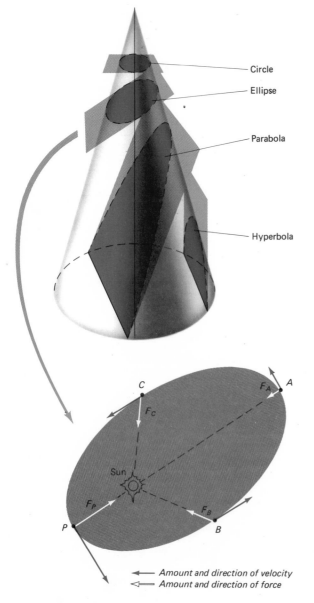

FIGURE 3.7
Permissible orbits under the law of gravitation. The orbital eccentricity e, which is a measure of how stretched out the orbit is, for each shape is as follows: circle, e = 0; ellipse, 0 < e < 1; parabola, e = 1; hyperbola, e > 1. The factors involved in shaping these orbits are clarified by the lower drawing, which shows the case of the ellipse. At A the body is moving at right angles to the direction of the attracting force F, which tends to pull it toward C, causing it to change direction toward C and to speed up. Closer in at C the tangential velocity increases because of the stronger force. As the body approaches closer to P, the increasingly stronger attractive force causes the body to curve around it, attaining maximum velocity at P. As the body swings outward past P, its motion is retarded by the attractive force; after passing B the body decelerates until it finally reaches A, and then the action repeats.

Labels in figure: Circle, Ellipse, Parabola, Hyperbola, C, F_A, A, F_C, Sun, F_P, F_B, P, B

← Amount and direction of velocity
⇐ Amount and direction of force

DEFINING ENERGY

The historical development of the concept of energy was long and laborious. It took more than 150 years from the first attempts at quantitative formulation, by the Dutch contemporary of Galileo, Descartes, and Newton, Christian Huygens (1629–1695), to the point at which the appropriate terminology was established. In general, energy does not have properties like those of matter, such as size, shape, and color. Also unlike matter, it cannot in general be said to occupy space. Nevertheless, as we shall develop in Chapter 13, matter is but one more manifestation of energy so that we must qualify our statements by saying "in general."

How then do we define this somewhat abstract concept? We can say that *energy* is a measure of the ability of a physical system to perform work when the system undergoes a change. ("Change" implies that we should be able to describe the system accurately before and after in order to be able to say that it has changed.) Yet change alone is not a sufficient means of defining energy: As human beings, we are physical systems, but changing our feelings for other human beings does not perform work. A genuine example of energy is what happens when a stream turns a waterwheel as it flows over a dam; the stream performs work on the waterwheel, which rotates a grindstone, which grinds grain. The energy of the stream is a measure of the ability of the stream to perform useful work.

Although energy does not in general have the properties of matter, it can be measured and quantified. One unit used by astronomers is the *erg,* the amount of energy needed to accelerate a mass of 1 gram at a rate of 1 centimeter per second squared as it moves a distance of 1 centimeter.

CONSERVATION OF ENERGY

The most important of all physical laws is the law of conservation of energy:

LAW OF CONSERVATION OF ENERGY: *Energy may be neither created nor destroyed but only transformed from one form to another.*

NEWTON'S MODIFICATION OF KEPLER'S THIRD LAW

Recall Kepler's third law relating a planet's period of revolution to its mean distance from the sun:

$$P^2 = Ka^3.$$

Newton worked out a more precise version of Kepler's law, which is expressed mathematically as

$$P^2 = \frac{4\pi^2 a^3}{G(M_\odot + M_p)}.$$

Thus $K = 4\pi^2/[G(M_\odot + M_p)]$, $M_\odot$ being the mass of the sun, M_p the mass of the planet, P the orbital period, and a its mean distance from the sun. Notice that K is not quite a true constant, as Kepler had supposed, because the mass of each planet, though a small fraction of the solar mass, is not truly zero.

Example: Let us determine the mass of the sun. Applying Newton's modification of Kepler's third law to the earth and the sun, we have

$$M_\odot + M_\oplus = \frac{4\pi^2 a^3}{GP_\oplus^2} = \frac{(39.5)(1.496 \times 10^{13}\text{cm})^3}{(6.67 \times 10^{-8}\text{ cm}^3/\text{g}\cdot\text{s}^2)(3.156 \times 10^7\text{ s})^2} = 1.99 \times 10^{33}\text{ g,}$$

where $M_\odot$ is the mass of the sun; $M_\oplus$ is the mass of the earth; a is the earth's mean distance from the sun, 1.496×10^{13} centimeters; G is the gravitational constant, 6.67×10^{-8} centimeters per gram second squared; and $P_\oplus$ is earth's period of revolution around the sun, 3.156×10^7 seconds. Thus the mass of the sun is 1.99×10^{33} grams, since the earth's mass is negligible.

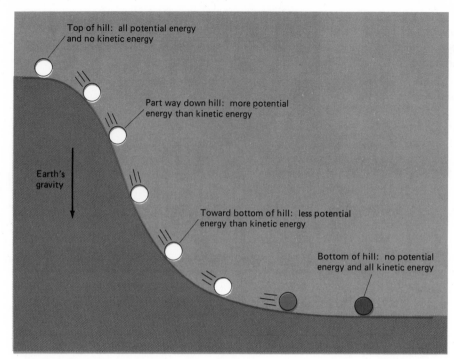

Top of hill: all potential energy and no kinetic energy

Part way down hill: more potential energy than kinetic energy

Earth's gravity

Toward bottom of hill: less potential energy than kinetic energy

Bottom of hill: no potential energy and all kinetic energy

FIGURE 3.8
Conservation of energy by changing potential energy to kinetic energy. Measuring potential energy relative to the bottom of the hill, the energy of the body is all potential at the top of the hill, and all kinetic at the bottom.

Motion involves mechanical energy, which has two forms: one is *kinetic energy*, and it is the energy a body has because of its state of motion; the other form is *potential energy*, and it is the energy a body has because of its position in a field of force. A stone at the top of a hill, for example, can be said to have energy by virtue of its position in the earth's gravitational field. If it is pushed, the stone will roll down the hill, converting potential energy to kinetic energy, as we illustrate in Figure 3.8.

We commonly refer to such other forms of energy as chemical energy and electrical energy. We can also understand these forms in terms of kinetic and potential energy; but in most of this book we shall be less specific about it and just say "energy." The form of energy with which we are most concerned in astronomy is radiant energy, the subject of Chapter 4. The important point to remember throughout is that the conversion of energy from one form to another does not create energy or destroy energy.

TRANSPORT OF ENERGY: WAVES

Energy can be moved from place to place, and thus it must be transported in some form. Waves are one way of transporting energy. What is a wave? It is the trans-

port of a disturbance; that is, a wave is a disturbance that moves. How does this account for the transport of energy? Imagine that two people several feet apart hold the ends of a rope; when one jiggles the rope, a wave will travel from one end to the other. No particles are moving from one end of the rope to the other, but a disturbance is traveling along it. We know that energy is transported by the disturbance since, when the disturbance arrives, the receiving hand is jiggled. That is, the wave in the rope does work on the hand, giving it kinetic energy. Another example of a wave is the disturbance that propagates across the surface after a stone is dropped into a pond. A wave can thus be defined as a disturbance capable of transferring energy from one place to another.

To understand waves better, consider how scientists describe them quantitatively. In Figure 3.9 the distance between successive crests or troughs is called the *wavelength*. The number of complete cycles of the disturbance passing a fixed point per second is called the *frequency*. The *speed* of the wave is the distance it travels per unit of time; this is just the length of each wave (its wavelength) multiplied by the number of waves passing a fixed point per unit of time (its frequency). Thus the *speed of the wave* is the product of its wavelength and frequency. The last quantity used to describe a wave is its *amplitude*. This is the greatest height the crests reach or the greatest depth to which the troughs fall. The greater the amplitude, the more energy the wave transports from one point to another. In fact, the *energy of the wave* is measured by the amplitude squared.

Waves are one means of transporting energy but, as we have mentioned, not the only one: The collision of bodies is another way to transfer energy and thus transport it from one place to another. The best common example of this process is the behavior of atoms and molecules in a gas. Therefore let us next focus the discussion of motion on the world of the atom.

3.4
Motion in the Atomic World

STATES OF MATTER

First let us consider the behavior of collections of atoms. Matter exists in three states: In a solid atomic particles are bound to permanent positions relative to each other; in a liquid the particle bonds are weak and temporary; by contrast in a gas there is no significant bonding between atomic particles, and the particles have no permanent positions relative to each other. Most of the matter in the universe is in the form of gas. Frequently matter is in the form of *plasma,* a gas composed of free electrons and positive ions, which are atoms from which one or more electrons have been stripped. Such ionization is generally the result of very high temperatures, as we shall discuss in the next section.

The particles of a gas can be molecules (which consist of two or more atoms), atoms themselves, or ions and electrons. In the molecule the atoms may be of the same element, as the two atoms of the oxygen molecule we breathe, or of different elements, such as the two hydrogen atoms and one oxygen atom in the water molecule. In the plasma there is a variety of possibilities. Each atom can be stripped of one electron—so that there are two independent particles per atom—or stripped of two electrons—three particles per atom—and so on. Only some of the atoms may lose an electron, or in the extreme case all atoms lose all their electrons.

FIGURE 3.9
Description of a wave. Two cycles of the wave are shown on a graph, with the wavelength and amplitude marked. The wavelength is the distance between either successive crests or successive troughs (8 centimeters), while the amplitude is the maximum height of the crests above the undisturbed position or the maximum depth of the troughs below the undisturbed position (3 centimeters). If the wave is moving to the right and the two cycles pass the origin (O) of the graph in 1 second (its frequency), then the speed of the wave is 8 centimeters times two per second, or 16 centimeters per second.

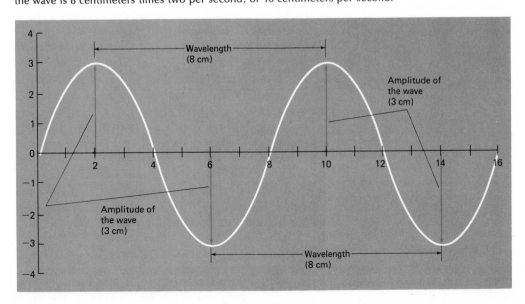

In the atomic world (as in a gas) gravity is not the cause for changes in motion as it is in the macroscopic world. The atomic world is dominated by electromagnetic forces. As with the force of gravity the intensity of electric and magnetic fields weakens as the inverse square of the distance from their source. At first glance this suggests that Newtonian mechanics ought to describe motion in the atomic domain, gravity as a cause simply being replaced by electromagnetic forces. Such is not the case in general, and the mechanics of the atom is called *quantum mechanics*. Its details go beyond our needs in this book, so we only point out that motion in the atomic world has a *discrete* nature rather than the *continuous* characteristics of our everyday experience.

RANDOM THERMAL MOTION

Within a gas particles dart about rapidly, colliding millions of times each second and changing their direction of motion just as frequently. Each gas particle has a kinetic energy proportional to the product of its mass and the square of its speed. After a collision the speed can be either greater or smaller than it was before the collision; the kinetic energy of each particle changes in its repeated collisions. Collectively, however, the gas particles will have some average kinetic energy, which changes only when energy is added to the gas or removed from it. Another way of saying this is that the average kinetic energy changes when the gas is heated or cooled.

Temperature is a measure of the average kinetic energy of gas particles. The motion of the particles composing a body, like ice or water or water vapor, is called *random thermal motion* and is illustrated in Figure 3.10 for a gas. It increases as the temperature goes up, and it decreases as the temperature goes down. Absolute zero is reached when the average kinetic energy is zero. Seen in terms of the motion of the particles in the gas, temperature is a measure of that motion: The greater the temperature of the gas, the greater the random thermal motion.

Temperatures in astronomy are usually measured on the absolute, Kelvin (K), scale. In this system there are 100 divisions (degrees) between the freezing point (273 K) and the boiling point (373 K) of water.

TEMPERATURE AND HEAT TRANSFER

When we heat the air in a vessel, we are increasing the kinetic energy of each particle, which means a higher average kinetic energy. As a result the gas particles move about faster, and they collide more frequently and more violently with their surroundings. If the density of the gas particles (the number per unit volume) is quite large, then the hot gas can transport a great deal of heat (or thermal energy) from one place to another. The flow of energy in nature is from regions in which the energy content is high to those in which it is low; natural physical processes tend to even out the energy content, whether on a small scale or a large one, and in general large energy differences and large volumes of space over which energy is to be transported require long periods of time to smooth out the differences. This fact will help us to understand many aspects of the evolution of astronomical objects.

3.5 Relativity

EINSTEIN'S CONCEPTS OF MOTION, SPACE, AND TIME

The Newtonian theory of gravitation has been very successful in explaining most gravitational problems involving the material objects of the macroscopic world. As we have pointed out, in another world, the submicroscopic universe of the atom, nuclear and electromagnetic forces dominate the very weak gravitational force, and they determine form and behavior in that realm. But it is the movement of bodies at high velocities or in the presence of very strong gravitational fields, where Newton's theory of gravitation breaks down, that we are interested in at this point. Here relativity has changed cosmology profoundly, helping to explain cosmic reality, as we shall discuss in Chapter 19.

The special theory of relativity of 1905 was not originally designed as a refinement of Newtonian gravitational theory but was addressed by Einstein to some aspects of electromagnetic radiation, that is, light. *Special relativity theory* deals with physical phenomena in which gravitational forces are not involved. By 1916 Einstein had worked out another, more comprehensive theory, which was an alternative to the gravitational theory of Newton. He incorporated in his *general theory of relativity* a description of nonuniform, or accelerated, motion between observers.

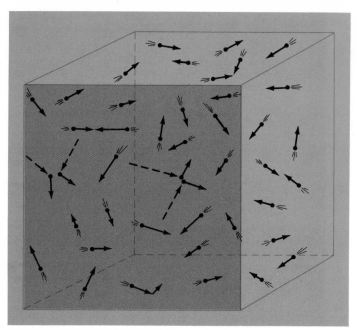

FIGURE 3.10
Random thermal motion of atoms in a gas. The atoms, which we are assuming are all of the same element, such as hydrogen, dart about at great speeds but travel only short distances before they collide with another atom. (If the gas in the box were typical of the air at the surface of the earth, then there would be about 10^{19} atoms per cubic centimeter; each particle would collide about 10^8 times per second.) All these collisions redistribute the kinetic energies of the individual particles, but the average remains constant as long as energy is neither added nor removed from the gas. If energy is added, the random thermal motions quickly distribute it among the atoms of the gas and raise the average kinetic energy, which is what one means by saying that the temperature has been raised.

The general theory of relativity has important implications for our ideas of motion, mass, and the geometry of space and time.

What is relative in relativity? Let us consider an example. If we walk down the aisle of a moving train car, the car can be thought of as the frame of reference for our motion. Looking out of the train window, we see our frame of reference moving relative to the ground as another frame of reference. We know that the ground moves relative to the center of the earth as another frame of reference. The center of the earth moves relative to the sun as still another frame of reference. And so it is on a larger scale—we cannot ascertain our possible absolute motion in the universe since we can find no frame of reference that is absolutely still.

Einstein coupled the three dimensions of space together with that of time and taught us that when we measure space and time, we find no absolute results—the answers are relative, depending on the observer. Two people who are moving relative to each other see the same events going on at different places and different times; each experiences the event in his own frame of reference. We need relativity in astronomy because we cannot drop anchor in the universe and passively watch what happens around us: Each of us is both observer and participant. The first truth in the relativistic universe is that *all motion is relative;* that is, *there is no preferred frame of reference in which space and time are defined absolutely.*

Newtonian space has three numbers, the spatial coordinates that describe where an object is: forward-backward, right-left, and up-down. Location has nothing to do with time. According to Newton, "Absolute space, in its own nature, without relation to anything external, remains always similar and immovable. . . . Absolute, true and mathematical time, of itself, and from its own nature, flows equably without relation to anything external. . . ."

Einstein's theory on the other hand coupled space and time, making time a fourth dimension. In his general theory of relativity Einstein described the geometry of four-dimensional *space-time,* in which we, planets, stars, and galaxies exist. The space-time continuum shows deformities (that is, space curvature) in the vicinity of material bodies; the more massive the body, the stronger is this curvature. From Newton's viewpoint an object moves in a curved path in response to the gravitational force of a massive body. For Einstein an object moves in a curved path as a natural consequence of the space curvature produced by the massive body. Where no massive body exists, the space-time geometry is flat, showing no curvature, and the object moves uniformly in a straight line. In the Newtonian world it moves that way because no force is acting upon it.

ALBERT EINSTEIN (1879–1955)

In 1979 many countries celebrated the centenary of Einstein's birth. To mark the occasion, a large bronze statue of Einstein was sculpted and placed on the grounds of the National Academy of Sciences. The second *High Energy Astronomy Observatory (HEAO-2)*, launched November, 1978, was re-named the *Einstein Observatory*.

As a child Einstein developed slowly. During his early years he showed no special aptitude in ele-mentary and secondary school, whose rigid methods of instruction he dis-liked. He was fascinated by mathe-matics and science, subjects that he studied on his own. He became a high-school dropout, and he left school to join his family in Milan.

Two years later he was able to en-roll at the Swiss Federal Institute of Technology in Zurich after making up a number of subject deficiencies. At the institute the academic fare did not suit him either; so he continued to follow his own inclinations. He man-aged, however, to pass the required examinations for his degree by cramming from the excellent lecture notes supplied by a close friend. In the 2 years following his graduation in 1900 he subsisted on odd teaching jobs. In 1902 he secured a position as patent examiner at the Swiss patent office in Bern.

During the next 7 years he worked at the patent office, and without any academic connections he found time to publish, at the age of 26, three trailblazing papers. One dealt with the random thermal motions of mole-cules in colloidal solutions, called Brownian motion. In 1827 Robert Brown, an English botanist, had ob-served through a microscope the zig-zag paths of tiny pollen particles in a drop of water being buffeted by the much smaller, invisible atoms and molecules in the fluid. Einstein worked out the correct mathematical expression for this action on the basis of the random thermal motions of the atomic and molecular constituents, and he did that at about the time the structure of the atom was first being

probed. His second paper reinforced the quantum theory of light devel-oped by Max Planck in 1900. Einstein established the photon nature of light in accounting for the recently discov-ered photoelectric phenomenon. For this contribution Einstein was awarded the Nobel Prize in physics in 1921. His third and most famous pa-per dealt with the special theory of relativity.

Einstein's last years were mostly spent in a vain search for a unified field theory, for a universal force that would link gravitation with the elec-tromagnetic and subatomic forces. Others who have tried since have not been successful either. Einstein was filled with reverence for the works of nature and said: "The most incom-prehensible thing about the world is that it is comprehensible." He thought of himself more as a philos-opher than as a scientist. In a way he followed the Greek philosophers, who tried to account for natural phe-nomena by logical deductions instead of experimentation. He succeeded where the ancients had failed because he could draw on the insight of his predecessors and the powerful analyt-ical tools of mathematics developed in the 2000 years since Plato and Aris-totle, combining them with his un-erring cosmic intuition.

SPECIAL THEORY OF RELATIVITY

In the nineteenth century space itself was thought to be an absolute frame of reference relative to which absolute motions could be measured. Space was also imagined to be filled with a stationary invisible me-dium, the *ether*, which carried electromagnetic waves (as air transports sound waves). If light must move through such a medium, it was argued that its speed should differ depending on its direction relative to a moving observer. Thus the earth's absolute velocity could be measured against this stationary medium by timing the speed of light in various directions.

In 1887 the American physicists Albert Michelson (1852–1931) and Edward Morley (1838–1923) sought to detect a difference in the speed of light beams propa-gated over the same distance, one parallel to the earth's motion around the sun and one simulta-neously transmitted perpendicular to it. The time light would spend moving *across* the earth's path, back and forth, was calculated to be $\sqrt{1 - (v^2/c^2)}$ shorter than the time it would take light to move *parallel* to the earth's path in one direction and then in the opposite direction; here v is the absolute velocity of the earth's motion and c is the velocity of light. It was reasoned that the anticipated minute difference in the relative speeds of the light beams would lead to an evaluation of the earth's absolute motion in space. But no matter in which direction measurements were made or at what time of year the experiment was performed, the result never changed: There was *no* measurable difference in the times of the light beams. This led to the conclusion that we cannot measure earth's absolute velocity in space.

The Michelson-Morley experiments were bewildering to the scientific community at the time Einstein began work on his special theory. Einstein rejected the notion of ether and concluded that the solution to the problem lay in recognizing the inadequacy of the Newtonian concepts of space and time. To develop the mathematical formulation of his theory, he laid down two postulates:

FIRST POSTULATE: *The laws of physics are the same for all observers in uniformly moving frames of reference.*

SECOND POSTULATE: *The velocity of light (299,792 kilometers per second) is the same for all observers in space regardless of the motion of the source or of the observer.*

Let us amplify these two postulates. Suppose two observers are moving uniformly relative to each other (not accelerating). They are said to occupy *inertial* frames of reference. Neither frame of reference is preferred by the laws of physics; there is no way to distinguish one from another. They have equal physical status; and for that matter *all* uniformly moving frames of reference have equal status. Neither observer is able to determine the uniform motion of his own reference frame by any experiments conducted within that frame of reference. For example, a person occupying an inside cabin (no portholes) of a ship cruising at uniform speed in calm seas could not conduct any kind of experiment to indicate whether the ship was moving uniformly at sea or still tied to the dock. That is, experiments like throwing a ball, flipping a coin, or playing a game of billiards would not distinguish between the two situations. If the person performed another experiment, using a frame of reference different from the reference frame provided by the cabin, say by looking out of a porthole at the ocean, he or she could determine the state of motion relative to that other frame of reference. But exclusive of that possibility, the only means of detecting motion would be under acceleration as when the ocean became stormy—then the deviation from uniform motion would be painfully evident.

Einstein's second postulate is obviously consistent

RELATIVE VELOCITIES

An aspect of relativistic and Newtonian motions that makes an informative comparison is velocity. If v_A and v_B are the velocities in space relative to some fixed frame of reference of two observers A and B, then their relative velocity v_{AB} is, according to Einstein,

$$v_{AB} = \frac{v_A + v_B}{1 + (v_A v_B / c^2)},$$

where c is the velocity of electromagnetic waves (about 300,000 kilometers per second). But if A's and B's velocities are insignificant compared with the velocity of light ($v_A/c \approx 0$ and $v_B/c \approx 0$), Einstein's formula reduces to the Newtonian expression

$$v_{AB} = v_A + v_B.$$

Example: If two light beams are transmitted simultaneously in opposite directions from the same point, they will recede from each other at the relative velocity of light (c = 300,000 kilometers per second) and not at 600,000 kilometers per second (or 2c). We verify this answer by substituting $v_A = v_B + c$ in Einstein's formula, and we get $v_{AB} = c$.

with the Michelson-Morley experiment: The speed of the light beam in the experiments did not depend on the direction of the light; it did not depend on the light source; it was not affected by the earth's motion. In other words the speed of light is independent of the relative velocity of the source and the observer. Regardless of whether we measure the velocity of light in our reference frame or in some other, we find that the velocity of light always has the same value. Einstein reasoned that what is different between frames of reference moving relative to each other is the measurement of space and time.

SPACE AND TIME IN SPECIAL RELATIVITY

Since light brings us the information about the universe and it travels at a finite rate (albeit a very large one), we are looking back in time when we look into our night sky. The nearer the object, the shorter the look back in time; but the objects we see farthest away are the objects we are seeing as they were longest ago.

Thus the night sky alone shows us the coupling of space and time.

To make this point somewhat more precise, let us consider in Figure 3.11 a space-time diagram like the one we introduced in Figure 3.3. In Figure 3.11 we have suppressed two spatial dimensions and show only one spatial dimension and the time dimension. If the origin represents where we are located, "Here and now," then two, A and B, of the three points labeled "There and then" are accessible to us by trips that are within our realm of experience. But to go to point C requires that we travel faster than the speed of light, which is contrary to Einstein's second postulate. Thus the first consequence of Einstein's two postulates is that there are realms of the space-time continuum from which we could not have come in our past and realms in the future to which we cannot go. Or in other words not all realms of space and time are readily accessible to us either in the past or in the future. Let us examine some other consequences of

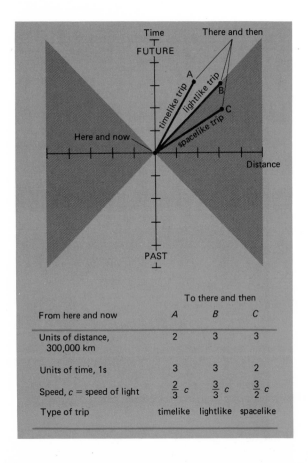

From here and now	To there and then		
	A	B	C
Units of distance, 300,000 km	2	3	3
Units of time, 1s	3	3	2
Speed, c = speed of light	$\frac{2}{3}c$	$\frac{3}{3}c$	$\frac{3}{2}c$
Type of trip	timelike	lightlike	spacelike

FIGURE 3.11
Space-time diagram. The origin is "Here and now;" that is, where we are located at this time. We choose units of distance that are each 300,000 kilometers in length and units of time that are 1 second in length so that a beam of light moves along the two 45° lines leading from the origin. If we choose three points, "There and then," in the future labeled A, B, and C, what does it take to go from here and now to any one of the three? As the table shows, point A can be reached by moving at $\frac{2}{3}c$, a trip known as a timelike one. Such a trip is within our realm of experience. To point B the necessary speed is c, the speed of light. Again, that is within our realm of experience for a beam of light, if not for us directly. But to reach point C requires a velocity in excess of the velocity of light ($\frac{3}{2}c$) and is clearly not possible according to Einstein's second postulate. The shaded areas of the diagram are not accessible, requiring what is called a spacelike trip to us here and now, either having come from there in our past or going there in our future. Thus there are realms of space-time with which we have no contact.

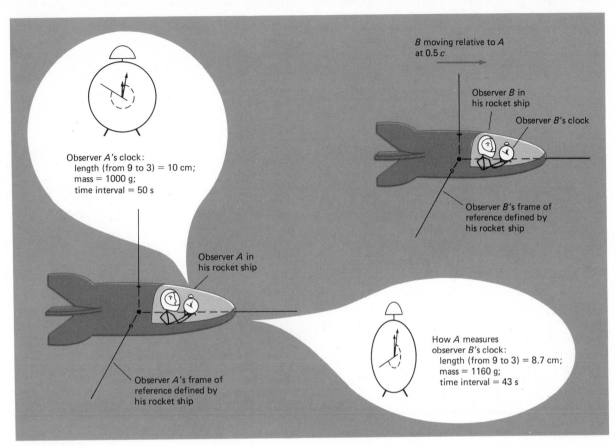

FIGURE 3.12
Uniformly moving frames of reference in which space, mass, and time can be measured. This illustration is drawn with you as observer *A*; that is, you are in this frame of reference, and you are watching observer *B* uniformly moving relative to you at a speed of 50 percent of the speed of light. As you make measurements on your clock, they are as shown; but when you measure *B*'s identical clock, it has contracted in the 9 to 3 direction (or the direction of motion), increased its mass, and is ticking slower than yours. Your interpretation is that in *B*'s realm of space-time, space has contracted, while time has slowed down compared to your realm of space-time. *B*'s interpretation for measurements of your realm of space-time would be exactly the same as yours is of his: Space has contracted, and time has slowed down.

these two postulates, having to do with the measurement of space, mass, and time.

Suppose we have two observers in different frames of reference that are at rest relative to each other. If both make measurements of lengths in space and intervals of time, they would agree on the measurements; so they must occupy the same realm of space-time. Now suppose the frame of reference containing observer *B* moves at a large fraction of the speed of light relative to the frame containing observer *A*. Do the two observers still occupy the same realm of space-time as before?

On the basis of his two postulates Einstein derived three important formulas for length, mass, and time in which the factor $\sqrt{1 - (v^2/c^2)}$, called the Lorentz *contraction factor*,[1] plays a crucial role. The length, mass, and time interval for an object, say, a clock (see Figure 3.12), in *B*'s reference frame are measured by observer *A*, who sees the object in motion. If *A* has an identical clock with him, then his measurements of *B*'s clock are related by the Lorentz contraction factor to his identical clock's length, mass, and time interval. The

[1]Named after the Dutch physicist Hendrik Lorentz, who derived it in 1904 from his mathematical analysis of electromagnetism.

effect of the Lorentz factor on length, mass, and time is shown graphically in Figure 3.13. These formulas are unimportant for all ordinary speeds ($v/c \approx 0$) so that A and B still occupy approximately the same realm of space-time. But this is not the case for speeds comparable to the velocity of light.

What do the formulas tell us? The observer A, who sees the object moving with B's reference frame, will measure a shorter length for the object in the direction in which it is moving than will observer B, to whom the object is stationary. For example observer A would notice that B's rocket ship is also contracted in length compared with his own rocket ship, which is, of course, his frame of reference. Thus there is a *contraction of space* in the direction of motion for the moving frame of reference. Observer A will also measure a greater mass for the object in B's moving rocket ship than observer B will; and observer A will also measure a longer time between two events, such as the tick of the clock taking place on B's rocket ship, than observer B will for the same two events. The latter effect is called *time dilation*, or spreading out of time. The faster observer B travels relative to observer A, the slower his clock appears to run as observer A sees it (observer A's own clock is seen by observer B also to be running slow). Neither observer sees any effect on his own clock. Thus measurements of length, mass, and time vary with the frame of reference. Clearly, observer A and observer B occupy different realms of space-time.

From the discussion above what can we infer about space and time for the universe as a whole? Newton's concept of an absolute space and time envisions a material universe inserted into a realm of space and time. But in the Einsteinian concept space and time

are in the universe; that is, the universe defines space and time. There is no space beyond the universe, and there is neither time before nor after the universe. Space and time and their local features are properties of the universe.

So contrary to ordinary experience are the concepts of relativity theory that they seem to violate common sense and to be too abstract to be of any consequence. This is far from the case; for relativity theory has amalgamated old ideas of space and time into a unified arrangement leading to new and unsuspected revelations that make it possible to test the consequences of the theory. At the appropriate points in the remaining chapters we have inserted these tests of relativity theory as supplemental reading.

GENERAL THEORY OF RELATIVITY

In 1916 Einstein advanced his theory of relativity greatly by making it apply to observers (reference

FIGURE 3.13
Change in length, mass, and time for a uniformly moving frame of reference as a fraction of the velocity of light. We denote our reference frame as A, in which we are at rest. Let B be a frame of reference moving uniformly relative to us. Then the three mathematical relations leading to the white and colored lines tell us what values length, mass, and a time interval have in the moving frame of reference (l_B, m_B, t_B) compared to what they would have in our frame of reference (l_A, m_A, t_A) for various fractions of the velocity of light.

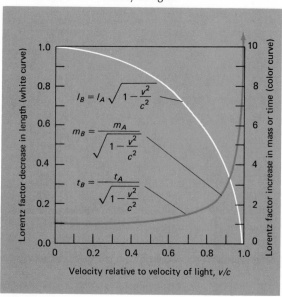

$$l_B = l_A \sqrt{1 - \frac{v^2}{c^2}}$$

$$m_B = \frac{m_A}{\sqrt{1 - \frac{v^2}{c^2}}}$$

$$t_B = \frac{t_A}{\sqrt{1 - \frac{v^2}{c^2}}}$$

Velocity relative to velocity of light, v/c

Lorentz factor decrease in length (white curve)

Lorentz factor increase in mass or time (color curve)

frames) moving nonuniformly relative to each other. Nature's fundamental laws, he reasoned, remain invariant throughout the universe in all frames of reference, whether the observers are accelerated or not.

In his second law of motion Newton had showed that the force it takes to accelerate a body is proportional to its inertial mass. (Inertia is the resistance a body offers to an applied force.) He was also aware that the gravitational force on a body is proportional to its gravitational mass. Otherwise, bodies of different masses would not fall to the ground at the same rate—that is, with constant acceleration—as we know they do (see page 50). In 1889 a Hungarian physicist, Baron von Eötvös, first proved experimentally and very precisely that inertial mass and gravitational mass are equivalent, an equality that had long been taken for granted. Modern experiments confirm that the inertial mass and the gravitational mass are the same to about one part in 10^{12}.

Einstein argued that the equality of inertial and gravitational mass must mean that "the same quality of a body manifests itself according to circumstances as 'inertia' or as 'weight.'" The consequence of this is that it is impossible to distinguish between the effect of an inertial force or a gravitational force on accelerated motion. He worked the idea into this principle:

PRINCIPLE OF EQUIVALENCE: *A gravitational force can be replaced by an inertial force that is due to accelerated motion without any change in the physical activity.*

By way of illustration imagine an observer in a rocket ship constantly accelerating to 1g (see Figure 3.14). Because 1g of acceleration is the acceleration we experience on the surface of the earth, the observer should experience within his reference frame (defined by the cabin of the rocket ship) what he would experience on the surface of the earth. To show

It is difficult to say what is impossible, for the dream of yesterday is the hope of today and the reality of tomorrow.

Robert H. Goddard

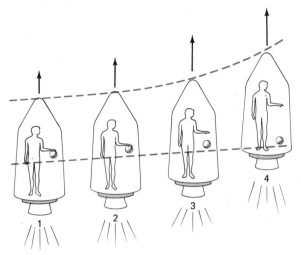

FIGURE 3.14
Experiment in an accelerating rocket ship, illustrating the principle of equivalence. To the observer in the rocket ship, accelerated motion mimics the action of gravity.

this, suppose he has some sort of ball in one hand. When the ship is in position 1, he releases the ball. It continues to move upward with the velocity the ship had at the moment of release (the successive positions along the lower broken line in Figure 3.14). If the ship were moving upward at constant velocity, the ball would remain suspended in the same place because ship and ball move the same amount. But the ship is accelerating; so the floor moves upward faster than the ball, colliding with the ball in position 4. The observer in the ship could attribute this to the force of gravity of some massive body if he does not know he is accelerating. From a vantage point outside the ship, however, another observer has been watching what is going on inside. The second observer's explanation of the sequence of events is simple: All those actions are explained by the ship's accelerated motion, and one need not postulate a gravitational force.

Einstein pointed out that each observer has a right to his or her own description of events. We can replace the force of gravity by an inertial force caused by an accelerated motion. An inertial force is not a real force but an effect of the nonuniform motion of the observer's frame of reference. You have already experienced such a fictitious force from accelerated motion while standing on the floor of a merry-go-round. You felt a force called a *centrifugal* force, that tended to move you toward the rim.

In general-relativity theory spatial curvature of local space-time is dictated by the presence of material bodies. If no mass is there, the curvature of nearby space is zero, and it is a flat space whose geometrical properties are described by ordinary Euclidean geometry (the kind you learned in high school). In the warped geometry of space-time that surrounds a large mass less massive objects move along curved paths. A planet's elliptical motion is an example: The planet moves in a curved path in the warped space surrounding the sun. The only way we have of illustrating curved space for you is in two dimensions rather than three. Figure 3.15 illustrates the deformation in two dimensions for bodies of two different masses.

The force we call "gravity," then, is nothing more than natural behavior by bodies moving within the geometrical framework of space-time. The Newtonian would say that the body moves according to action from a distance dictated by a force called gravity. The

> Nothing puzzles me more than time and space; and yet nothing troubles me less.
>
> Charles Lamb

Einsteinian says that a body moves naturally in response to the local structure of curved space-time.

Massive bodies, in addition to warping space in their vicinities, also alter time. If we could position ourselves where there is little if any warping of space-time, far from a massive body, and watch what happens as a clock approaches the massive body, we should see the clock ticking slower and slower, the closer it gets. Clearly the realm of space-time in the immediate vicinity of massive bodies is different from

FIGURE 3.15
Curvature of two-dimensional space in the vicinity of massive bodies. The two-dimensional space is so represented that the up-down direction is not a direction in space nor is it the time dimension. The bottom figure shows the warping of space near a massive body, while the upper one shows the greater space curvature caused by an even more massive body.

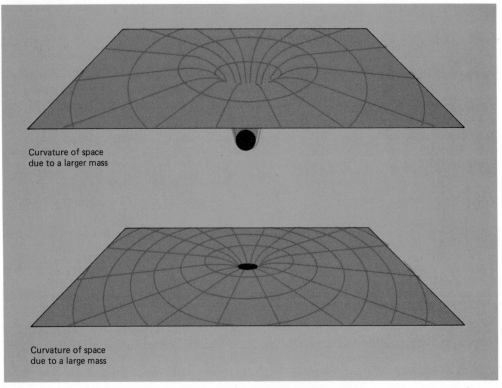

Curvature of space
due to a larger mass

Curvature of space
due to a large mass

that far from any mass. We may summarize motion in general relativity with the following principle:

PRINCIPLE OF GENERAL RELATIVITY: *Curved space-time tells matter how to move, and in turn matter tells local space-time how to curve.*

As the next twelve chapters describe with some exceptions, the realms of space-time in which we find various astronomical objects are approximately the same as ours. Thus Newton's laws of motion and law of gravitation describe fairly well what is happening in the cosmos. We shall again face the cosmic reality of different realms of space-time when we discuss black holes in Chapter 15 and cosmology in Chapter 19.

SUMMARY

Newton's laws of motion. Newton's laws of motion are a unifying theory explaining both terrestrial and cosmic motion. A body's state of motion, which is defined by its velocity as specified in some reference frame, can be changed only by a force with the change being an acceleration in the direction of the force. The natural state of motion is to continue at a constant speed in the same direction. The resistance or inertia a body shows to changing its state of motion is measured by its mass. Newton envisioned space and time as absolute properties of the universe against which absolute motion could be measured. The total quantity of motion or momenta of all bodies in the universe is constant with their interactions with each other redistributing it.

Gravity. Gravity is an intrinsic property of matter in which each body in the universe attracts every other body with a force proportional to the product of their masses and inversely to their separation. Gravity as the force holding bodies in mutual orbit about each other prescribes that the paths allowed will be conic sections: ellipses, circles, parabolas, hyperbolas. From his laws of motion and gravitational law, Newton showed that the constant of proportionality in Kepler's third law depends on the masses of the orbiting bodies, which provides astronomers with a means of determining masses for celestial bodies.

Energy. Energy is a measure of the ability of a physical system to perform useful work when changing in a describable way. Energy can be either or both the kinetic energy of motion or the potential energy in a field of force. Energy can neither be created nor destroyed, only transformed from one form to another; that is, energy is conserved. Waves, which are propagating disturbances, transport energy from one place to another.

Atomic motion. Motion on the atomic scale is governed by electromagnetic forces rather than gravity, and the Newtonian laws of motion have a restricted applicability there. The atoms and molecules composing matter are in constant motion, colliding and exchanging kinetic energy in a state known as random thermal motion. Energy added to a gas is distributed among all its particles by random thermal motion with temperature being a measure of the average kinetic energy of the gas particles. Hence energy will flow from high temperature (high energy) regions to ones of low temperature (low energy).

Relativity. All motion is relative in that there is no absolute reference frame; one's perception of space and time is dependent on one's frame of reference. In Einstein's special relativity, physical laws are the same in all reference frames moving uniformly relative to each other, and all observers measure the velocity of light as a constant regardless of their reference frame or relative motion. What differs between different frames of reference moving uniformly is the measurement of space and time, meaning that masses, lengths, and time are not measured as being the same in different reference frames. Contrary to Newton's vision, Einstein conceived of different realms of space and time in the universe being defined by the universe itself on a local basis. General relativity is an alternative to Newtonian gravity in which gravitation and acceleration are equivalent interpretations of the same action. Curved space-time tells matter how to move and in turn matter tells local space-time how to curve.

KEY TERMS

acceleration
amplitude
density
energy
field of force
force
frame of reference
frequency
inertia
kinetic energy

mass
momentum
potential energy
random thermal motion
space-time
temperature
time dilation
velocity
wave
wavelength

CLASS DISCUSSION

1. What is meant by a field of force? What are some examples of fields of force? How would you show that a field of force is present at some point in space? What is meant by the intensity of the field? Does intensity vary with the distance from the source of the field of force?

2. What is meant by an absolute space and time? How does Newton's absolute space and time differ from Einstein's concept of space-time? For the various scale sizes of physical existence described in Chapter 1, whose concept, Newton's or Einstein's is correct? Is it fair to use right and wrong in this context?

3. What is a scientist talking about when referring to a body's state of motion? Give some examples.

4. If the effect of gravity is not noticeable between two human beings, why is it so obvious in planetary motions? Why is gravity more important in planetary motions than electromagnetic forces?

5. Define energy and give some examples of its transformation from one form to another. Is the statement that the world is running out of energy correct? If not, why?

6. Does relativity have any practical meaning for your life? Does relativity change the way you perceive human existence and what we are capable of knowing?

READING REVIEW

1. What causes a change in a body's state of motion?

2. How does the force of gravity depend upon mass and distance?

3. Are mass and density synonymous? If not how do they differ?

4. How does momentum differ from the concept of energy?

5. Is the acceleration in Newton's second law of motion in the same direction as the force? Is the velocity also in the same direction?

6. Is the amount of energy in the universe finite or infinite?

7. Does temperature measure mass, momentum, energy, all of these, or none of these?

8. What is a frame of reference? Is it different in Newton's concept of motion from Einstein's concept?

9. What criteria does one use to tell which direction energy will be transported?

10. What is meant by time dilation?

11. What is the difference between inertial and gravitational mass?

12. Is the material world embedded in a space and time predefined before the existence of the universe?

PROBLEMS

1. If a rocketship is traveling at a constant velocity of 3000 km/s, how far in kilometers will it travel in 10 hrs? If at the end of the 10 hrs, it begins to increase its speed in the same direction from 3000 km/s to 3500 km/s in the first second, to 4000 km/s in the next second, to 4500 km/s in the third second, and so forth, what is its acceleration in km/s²? If the rocket ship's mass is 254,000 kilograms (about 250 tons), what is the magnitude of the force producing this acceleration in newtons?

2. Considering the superior planets at the time of opposition, which one exerts the strongest gravitational force on the earth? Which one is second? Third? How does the strongest compare with the gravitational attraction of Venus at inferior conjunction?

3. The total energy E of a planet at any velocity v and distance r from the sun is

$$E = -\frac{GM_\odot M_P}{2a} = (1/2)\, M_p v^2 - \frac{GM_\odot M_P}{r},$$

where $M_\odot$ = mass of the sun, M_P = mass of the planet, a = semimajor axis of the elliptical orbit, and G = constant of gravitation. The negative signs arise since potential energy (the second term on the right) is defined as negative, making the total energy also negative. For a planet moving in a circular orbit, how does the velocity v depend on the semimajor axis a? Derive the dependence of the sidereal period on the semimajor axis? Whose equation is this?

4. In Figure 3.16, observer B's rocket ship is moving at 0.5c relative to observer A's ship. If B were moving at 0.8c, what values of the length, mass, and time interval should A measure for B's clock?

SELECTED READINGS

Bronowski, J. *Ascent of Man*. Little, Brown, 1973.

Drake, S. "Newton's Apple and Galileo's Dialogue," *Scientific American*, August 1980.

Kaufmann, W. J. *The Cosmic Frontiers of General Relativity*. Little, Brown, 1977.

Manuel, F. E. *A Portrait of Newton*. Harvard University Press, 1968.

4
Light: Message System of the Universe

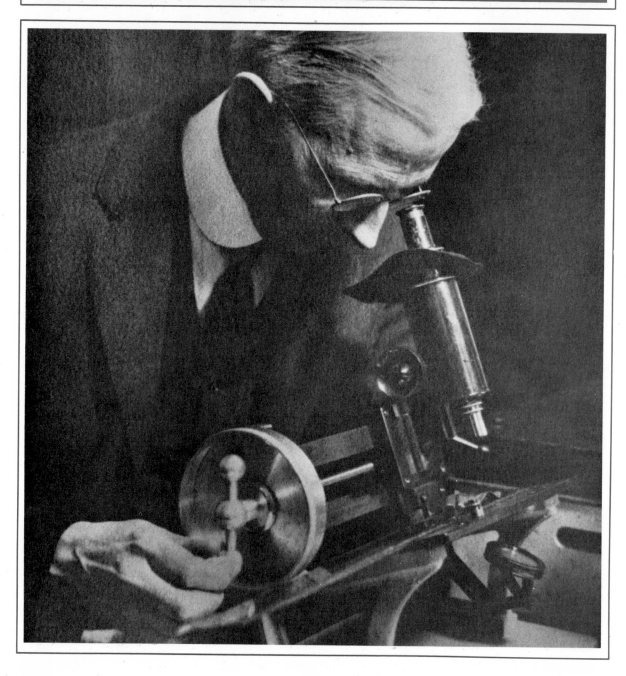

Until about a century ago astronomers were almost exclusively concerned with the positions and motions of celestial bodies. They knew little or nothing about the physical makeup or nature of these bodies and were really not able to find out. Yet today concern with the physical nature of celestial bodies, which is the field of *astrophysics,* is one of the most important topics in astronomy. This change in orientation is primarily the result of two developments in the growing knowledge of the nature of light.

The first was the invention of the *spectroscope,* a device capable of breaking down white light from a distant source into its component colors. In his *Opticks* Newton described how one saw a rainbow of colors when sunlight was passed through a prism, but it was actually William Wollaston (1766–1828) in England and Joseph Fraunhofer (1787–1826) in Bavaria who were primarily responsible for developing the spectroscope (see Figure 4.1).

The second development was the recognition that each different chemical element emits a specific set of colors that is peculiar to it, much like the fingerprints of an individual. As early as the 1830s this fact was suggested in connection with the presence, identity, and abundance of different elements in ores. The real beginning of the field of spectroscopy was made in the last half of the nineteenth century by chemist Robert Bunsen (1811–1899) and physicist Gustav Kirchhoff (1824–1887) at the University of Heidelberg.

From his experimental work in the laboratory Kirchhoff was able to formulate three empirical laws of spectroscopic analysis. These laws describe the physical conditions under which matter will produce light having one of three different spectra of colors. The most important astronomical application was the potential of determining the chemical composition of the sun and stars.

By 1864 the English astronomer Sir William Huggins (1824–1910) had identified nine elements in the bright star Aldebaran in the constellation of Taurus. Sir Norman Lockyer (1836–1920) in 1868 detected an element in the solar spectrum that was unknown on the earth. It was later found in natural gas, but it still carries its solar name—helium. In the years following, more elements were identified in stars, along with the discovery that the spectra of colors in the white light coming

◀ With a measuring engine Henry Norris Russell analyzes a spectrogram of a star's light.

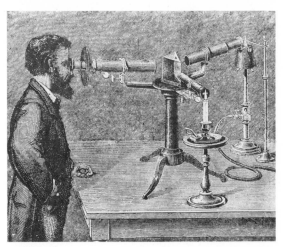

FIGURE 4.1
Spectroscopic analysis. A woodcut from about 1887 of a scientist looking through the eyepiece of a prism spectroscope that is analyzing the light from both a candle (foreground) and a gas flame (background).

from stars are not the same for all stars. However, in the spectra of stars sufficient similarities existed so that the stars could be arranged into broad spectral classes.

By the beginning of this century an important relationship had been established between the researcher in the laboratory and the astronomer in the observatory. From it new means of viewing the universe appeared, which have fundamentally altered our concepts of the universe. The extension of spectrum analysis to radiation in parts of the electromagnetic spectrum other than the visible and the ability to move above the earth's obscuring atmosphere to view radiation coming from the depths of the universe are the Rosetta stone of today's astronomy.

4.1
Electromagnetic Radiation

LIGHT AS WAVES

Astronomers have learned most of what we know about stars and galaxies by analyzing the electromagnetic radiation coming from them. Electromagnetic radiation, of which the light that our eyes respond to is one part, is a form of energy. Without any material aspects, it is energy that can move through the empty reaches of the universe.

Historically, this conceptual view began in 1862, when the Scottish physicist James Clerk Maxwell (1831–1879) showed that light is energy carried in the form of a traveling wave composed of electric and magnetic fields. The electric and magnetic fields vary in intensity, and they are at right angles to each other and to the direction in which the wave is propagating (see Figure 4.2). The electric and magnetic fields continually interact, forming an *electromagnetic wave*. These fields maintain themselves and continue to propagate until the energy of the wave is converted to some other form of energy. At that point the electromagnetic wave ceases to exist. Maxwell's proposal that light is an electromagnetic wave, as we shall soon see, was not the last word in explaining the physical nature of light. But visualizing light as waves spreading out from a radiating source helps us to understand many aspects of it.

The speed of light measured in empty space is 299,792 kilometers per second (3×10^5 kilometers per second in round figures, or 186,300 miles per second). This appears to be the speed limit for all energy transported in the universe. As we discussed in the last chapter, the speed of light is a fundamental constant of nature and apparently has the same value throughout the universe.

Electromagnetic waves possess a range of wavelengths, the distance between successive crests or

> Up to the twentieth century, "reality" was everything humans could touch, smell, see, and hear. Since the initial publication of the chart of the electromagnetic spectrum . . . humans have learned that what they can touch, smell, see, and hear is less than one-millionth of reality. Ninety-nine percent of all that is going to affect our tomorrows is being developed by humans using instruments and working in ranges of reality that are nonhumanly sensible.
>
> R. Buckminster Fuller

troughs (see Figure 4.2), and a range of frequencies. The product of the wavelength and frequency gives the speed at which the electromagnetic wave travels. The amount of energy the wave transports is proportional to the square of the wave's amplitude. (Also see the discussion in Section 3.3.)

ELECTROMAGNETIC SPECTRUM

Our eyes are sensitive to only a very limited portion of the entire range of wavelengths for electromagnetic radiation. If we order this range of wavelengths from the shortest on the left to the longest on the right, we produce an array of wavelengths called the *electromagnetic spectrum*, which is shown in Figure 4.4. Toward the short-wavelength end we find that portion to which our eyes are sensitive, called the *visible spectrum*. The physiological response of the eye to the various wavelengths of the visible spectrum is color.

Short wavelengths in the visible spectrum are violet, with progressively longer wavelengths producing the response we identify as the range of hues from blue, green, yellow, and orange to red in the color spectrum. Visible light is electromagnetic radiation with wavelengths between approximately 35×10^{-6} and 70×10^{-6} centimeter. These wavelengths correspond to frequencies between 8.5×10^{14} to 4.3×10^{14} hertz. One hertz equals one cycle, or oscillation, of the wave per second. The lowest frequencies of visible light appear red to our eyes; the highest fre-

FIGURE 4.2
Propagation of electromagnetic waves. We do not actually "see" these waves or know exactly what they look like. This is a schematic representation, depicting electromagnetic radiation as oscillations, or variations of electric and magnetic fields at right angles to each other. As shown, the electric-field intensity varies in the up-and-down direction, and the magnetic-field intensity varies in the back-and-forth direction. The electromagnetic wave is advancing in the direction shown.

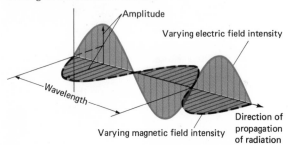

quencies appear violet; and between these are the rest of the color spectrum.

All types of electromagnetic radiation show the properties of a wave; all propagate in the same way with the same speed in empty space; and all transport energy. For convenience, however, we divide the nonvisible portions of the electromagnetic spectrum into regions according to the radiation's wavelength, such as the ultraviolet or the infrared and so on. We label these different regions not because of any intrinsic difference in the radiation but because we have different ways of detecting radiation depending on its wavelength. (These detection methods are discussed in Chapter 5.)

Gamma rays, X rays, and ultraviolet radiation are the wavelength regions shorter than visible light (see

VELOCITY OF AN ELECTROMAGNETIC WAVE

Consider an electromagnetic wave moving between two points 30 centimeters apart. In Figure 4.3a each wave moves the 30 centimeters in 1 billionth (10^{-9}) second at a constant velocity c, and 5 complete waves pass a fixed point in 10^{-9} second. The distance between the crests, the wavelength λ (λ is the Greek lowercase letter lambda), is $30/5 = 6$ centimeters. The number of waves passing a point in 1 second, the frequency f, is 5 billion. In Figure 4.3b each wave still moves the 30 centimeters in 10^{-9} second, but now 10 waves go by a fixed point in 10^{-9} second. Here the wavelength $\lambda = 30/10 = 3$ centimeters, and the frequency $f = 10$ billion waves per second. In both examples the velocity of the waves is the same, 3×10^{10} centimeters per second. (See Appendix 1 for abbreviations of units.) For light we can express this important relationship in any of the following forms:

$$\lambda \cdot f = c \quad \text{or} \quad \lambda = c/f \quad \text{or} \quad f = c/\lambda,$$

where λ is the wavelength, f is the frequency, and c is the velocity of light.

FIGURE 4.3
Electromagnetic waves of different wavelengths and frequencies moving at the same speed. Imagine that the star represents the source of the electromagnetic waves. The waves move in three dimensions but are shown in just two. The solid color lines represent the crests of the wave and the dashed lines represent the troughs.

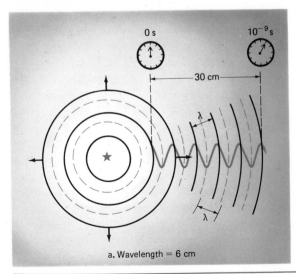

a. Wavelength = 6 cm

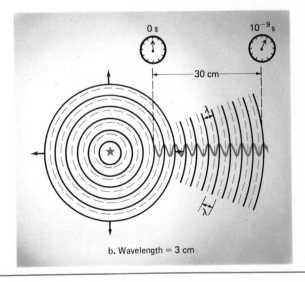

b. Wavelength = 3 cm

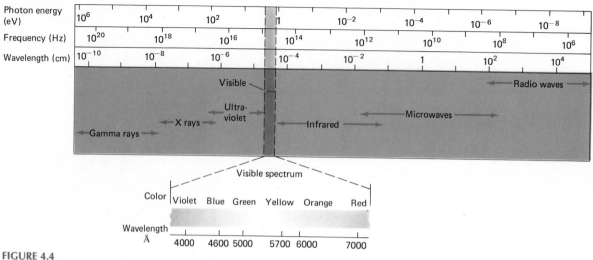

FIGURE 4.4

The electromagnetic spectrum, showing the various spectral, or wavelength, regions. The spectral regions are not divisions imposed by nature. They are based on the means by whch we detect radiation of different wavelengths.

Figure 4.4); most of this radiation that comes from outer space is absorbed high above the earth's surface by our atmosphere (see Figure 5.1). Infrared radiation is the first wavelength region beyond visible light, and it is also partially absorbed by the earth's atmosphere. The next wavelength regions are the microwave and radio regions, for which there is a broad electromagnetic window (Figure 5.1) through the earth's atmosphere.

Because of the wide range of wavelengths, some units of measurement are more convenient than others for describing different regions of the electromagnetic spectrum. For the visible spectrum angstroms are convenient. An *angstrom* is a hundred-millionth of a centimeter (1 Å = 10^{-8} cm). Visible radiation lies approximately between 3500 angstroms (the violet end of the spectrum) and 7000 angstroms (the red end). X rays are also measured in angstroms, but infrared wavelengths are generally expressed in microns (1 micron = 10^4 angstroms = 10^{-4} centimeter). Astronomers use the hertz as the unit for measuring frequency of all radiation.

WAVE PROPERTIES OF LIGHT

Light traveling through empty space moves in a straight line. In our everyday experience we encounter light not in empty space but passing through various media—light partially absorbed by the atmosphere, scattered by dust, transmitted through a window or a telescope. In these circumstances the speed of light may be slowed and the direction of the light wave may be changed. These changes are best understood through the wave properties of light.

Several properties illustrate the wave characteristics of light. One is *reflection,* which occurs when light strikes the boundary between two media of different materials, such as glass and air. When a light ray moving in air reaches the boundary, part of it may be reflected, as shown in Figure 4.5. The reflected ray lies in the plane formed by the incident ray and the perpendicular to the boundary. The ordinary mirror, or looking glass, illustrates reflection.

Also, part of the incident ray may be transmitted through the glass rather than being reflected. The transmitted ray does not, however, continue along the same straight line; it is bent toward the perpendicular shown as the dashed line in Figure 4.5. This change in direction is called *refraction*. If the medium into which the ray moves is more dense than that from which it comes, the angle of refraction will be less than the angle of incidence. If its density is less, then the angle of refraction is greater. A good example of refraction is a spoon sticking out of a glass of water. The handle looks bent at the point where the spoon enters the water because part of the handle is in the same medium (air) as you, while for the part under water light must pass through the water-air boundary, where it is refracted.

Light shows another wave property, *diffraction,*

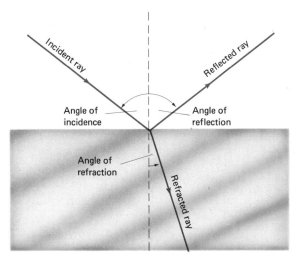

FIGURE 4.5
Reflection and refraction. The angle of incidence equals the angle of reflection. Also, the reflected ray is always in the plane formed by the incident ray and the perpendicular to the boundary (shown as a dashed line). Here the angle of refraction is less than the angle of incidence and also lies in the same plane as the incident ray and the perpendicular.

which is the spreading out of light past the edges of an opaque body (see Figure 4.6). Instead of being propagated in a straight line, light, like sound waves, bends around corners. The spread is greater for longer wavelengths. Because light's wavelength is very small, we do not normally observe diffraction in the everyday world. We can see diffraction, though, in the laboratory. Optical instruments, such as telescopes and microscopes, depend upon these wave properties of light for their operation.

Nearly all natural light sources, such as stars, emit electromagnetic waves composed of many wavelengths. How do waves of different wavelengths add to produce a composite wave? If waves of the same wavelength from two sources are superimposed so that their crests and troughs coincide, they are said to be *in phase* with each other, and their amplitudes add to produce a sum greater than the amplitudes of the individual waves; the light is said to "interfere constructively." If the crests of one set of waves fall on the troughs of the other, they are said to be *out of phase* with each other, and their amplitudes cancel each other; the light is said to "interfere destructively." Interference is common to all types of waves; in fact its occurrence was strong evidence that light is a wave phenomenon. Light waves of one or many different wavelengths may interfere constructively or destructively. Such waves are called *composite waves,* or *white light,* since that is the physiological response they evoke. If we can add waves together, then we must also be able to separate a composite wave into its constituent wavelengths. Indeed we can, as we shall describe in Section 4.2.

BRIGHTNESS OF ELECTROMAGNETIC WAVES

The surface area covered by an expanding sphere of light (or a portion of it) increases as the square of the radius of the sphere, that is, as the square of the distance from the light source. Since the total amount of energy leaving the light source in all directions is the same at any distance, the amount of radiation passing through each unit of area of the expanding sphere must diminish with the square of the distance (see Figure 4.7).

For example suppose at 1 meter from a light source that the apparent brightness of the radiation over 1 square meter of surface area is 1 unit. At twice the distance each square meter will receive one-fourth of a unit of illumination; at three times the distance,

FIGURE 4.6
Shadow-diffraction pattern of a razor blade and ball bearing on a microscope slide. Diffraction bands are seen both inside the hole in the razor blade and around the outer edges of the blade, slide, and ball bearing.

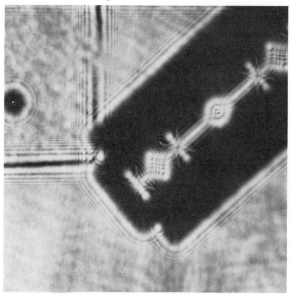

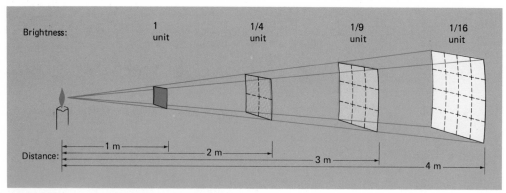

FIGURE 4.7
Inverse-square law of brightness. Light from a radiating source spreads out as the square of the distance from the source. The amount of illumination reaching surfaces that are equal in area but at different distances from the source decreases with the square of the distance from the source.

one-ninth of a unit; at four times, one-sixteenth of a unit; and so on. This relationship between apparent brightness and distance is known as the inverse-square law of light:

INVERSE-SQUARE LAW OF LIGHT: *The apparent brightness* b *varies inversely as the square of the distance* d *from the light source; that is,* $b \propto 1/d^2$.

This law is applied in many kinds of astronomical work, as we shall see in Chapters 11, 12, and 16.

DOPPLER EFFECT

If an observer is moving relative to a source of light or the source is moving relative to him, then the observer will see a change in the wavelength of the light:

DOPPLER PRINCIPLE: *Electromagnetic radiation received by an observer will have a shorter wavelength if source and observer approach each other and a longer wavelength if they recede from each other; the amount of change in wavelength is directly proportional to the velocity along the line between source and observer.*

To see why, suppose a stationary light source is radiating concentric waves of one wavelength in all directions. Then observers in any direction, if station-ary, would see successive crests of the wave passing them at the same rate at which they were emitted by the source. If, on the other hand, the light source begins to move at uniform speed toward the right (as in Figure 4.8), the two observers O and P along the line of motion would see crests passing them at rates different from that with which they were emitted. Observers Q and R, located at right angles to the moving source, would detect no change in the rate for crests passing them. Observers elsewhere would notice some change, the amount depending on the angle

When we step through the gateway of the atom, we are in a world which our senses cannot experience. There is a new architecture there, a way that things are put together which we cannot know; we only try to picture it by analogy, a new act of imagination. The architectural images come from the concrete world of our senses, because that is the only world that words describe. But all our ways of picturing the invisible are metaphors, likenesses that we snatch from the larger world of eye and ear and touch.

Jacob Bronowski

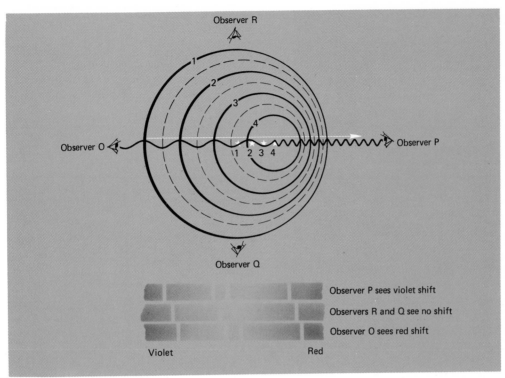

Observer R

1
2
3
4

Observer O

1 2 3 4

Observer P

Observer Q

Observer P sees violet shift

Observers R and Q see no shift

Observer O sees red shift

Violet   Red

FIGURE 4.8
Doppler line shifts. In the drawing, as the light source recedes (points 1 to 4) from observer *O*, he sees longer waves and fewer of them per second passing by, and the light's spectral lines show a Doppler shift toward the red. Observer *P* sees shorter waves rushing by at a higher rate, and the spectral lines show a Doppler shift toward the violet. Observers *Q* and *R* see no change in wavelength or frequency—and no Doppler shift.

between their radial direction to the source and the line of motion. This phenomenon is known as the *Doppler effect,* named for Christian Doppler (1803–1853), the Austrian physicist who first explained it.

Let us consider observers *O* and *P* in more detail. Wave 1 was produced when the light source was at position 1; wave 2, when it was at position 2; and so on. Because of the greater distance the wave travels in reaching observer *O*, each successive wave crest passes him at a slower rate (lower frequency) than when the source was stationary. Because the waves travel a shorter distance to reach *P*, the successive crests pass at a faster rate (higher frequency). The wavelength is shifted toward longer wavelengths (red shifted) as the source recedes from *O* and toward shorter wavelengths (blue shifted) as the source approaches *P*.

An example of the Doppler effect familiar to all of us is the change in frequency of sound waves in the rising and falling pitch of a train whistle as the train

approaches and then moves away. It is immaterial whether the light source is in motion, or the observer, or both: The size of the Doppler effect seen for light waves depends only on the net relative motion along the line of sight between the light source and the observer.

The amount of the wavelength shift due to the Doppler effect is directly proportional to the velocity of approach (blue shift) or recession (red shift) as long as the relative velocity is well below the velocity of light. (Later in this book we discuss the relationship when the relative velocity is a substantial fraction of the velocity of light.) The constant of proportionality is the ratio of undisplaced wavelength to the velocity of light. This means that the wavelength shift is greater the longer the wavelength of the radiation.

As an example, if we are approaching two stationary radiation sources, with one emitting electromagnetic radiation twice the wavelength of the other, then we should expect twice the wavelength shift from that one. In the nearby box we present the mathematical form of the Doppler effect. Because of the continual movement of all bodies in the cosmos, the Doppler effect is an important tool for detecting and measuring the amount of motion.

4.2 Information in the Spectrum of Light

SPECTROSCOPY

Just as sound waves of various wavelengths are transported simultaneously through the same region of air, electromagnetic waves of different wavelengths can move through the same point in space and superimpose to form composite waves, or white light. Stars, for example, are white-light sources though the color of the composite light from the stars may be white, red, yellow, or blue (see Table 12.5). As stated earlier, white, or composite, light can then be *dispersed,* or separated, into its component colors, or wavelengths,

to form a spectrum. The study of spectra is called *spectroscopy.* Let us briefly explain how we can accomplish the dispersion of composite light by using a triangular piece of glass called a *prism.*

In the refraction of a ray of light the angle through which light is refracted depends on wavelength; the angle of refraction is greater for shorter wavelengths than it is for longer ones. Consider white light passing through the slit of a narrow diaphragm and then through a glass prism, as in Figure 4.9. The light disperses into its component wavelengths, with short-wavelength light refracted through larger angles than long-wavelength light is. Thus waves of different wavelengths disperse in different directions. The result is a rainbow-colored sequence of images of the slit, containing an image for each wavelength present in the white light.

Another means of dispersing white light to produce a spectrum is the *diffraction grating.* Unlike the ordinary glass prism, which is transparent only to visible and infrared radiation, the grating is useful over a broad spectrum, from X-ray to infrared wavelengths. In its simplest form the diffraction grating is a plate containing a very large number of very narrow parallel slits, uniformly spaced at distances that are only a few

MEASURING THE DOPPLER EFFECT

The Doppler displacement can be measured if we know the wavelength when there is no relative motion between the source and the observer. Let $\Delta\lambda$ (Greek letters delta and lambda, where delta means "difference") be the shift in wavelength equal to the difference between the measured wavelength λ_m and the wavelength λ when there is no relative motion, or $\Delta\lambda = \lambda_m - \lambda$. Then the formula relating the Doppler shift for light to the relative velocity is

$$\frac{\Delta\lambda}{\lambda} = \frac{v}{c},$$

where c is the velocity of light (approximately 300,000 kilometers per second) and v is the relative line-of-sight velocity between the observer and the light source, or, as astronomers call it, the *radial velocity.* The wavelength shift is considered positive if the wavelength shifts to the red (distance increasing) and negative if it shifts to the blue (distance decreasing).

Example: The measured red shift of an absorption line at 5000 angstroms in the spectrum of a star was found to be +0.5 angstroms. From the Doppler formula we find the radial velocity of the star relative to the earth to be

$$v = \frac{\Delta\lambda}{\lambda}c = \frac{(+0.5\ \text{Å})(3 \times 10^5\ \text{km/s})}{(5 \times 10^3\ \text{Å})} = +30\ \text{km/s},$$

where the plus sign denotes a velocity of recession. If the algebraic sign of the velocity had been negative (that is, the wavelength shift had been negative), then the star would have been approaching the earth.

times the wavelength of light. (By "large number" we mean many thousands of slits per centimeter.) The spectrum is viewed in the direction of the light source, as in Figure 4.9. Since the amount of the bending, or diffraction, of electromagnetic waves at each slit depends on its wavelength, composite light is dispersed into its component colors. (For more details on these dispersing devices as they are used in an instrument called a spectroscope, see Section 5.2.)

KIRCHHOFF'S LAWS: THE NATURE OF LIGHT SOURCES

When we analyze light from various astronomical sources, we do not always find a continuous rainbow-colored sequence of wavelengths. Spectra can be classified and interpreted according to laws formulated by the German chemist Gustav Kirchhoff more than a century ago. The three basic types of spectra—continuous, emission, and absorption—and the physical conditions under which they are formed are given by Kirchhoff's laws and are illustrated in Figure 4.10:

KIRCHHOFF'S FIRST LAW—CONTINUOUS SPECTRUM: *The spectrum of a radiating solid, liquid, or highly pressurized gas is an uninterrupted sequence of wavelengths known as a continuous spectrum.*

KIRCHHOFF'S SECOND LAW—EMISSION, OR BRIGHT-LINE, SPECTRUM: *The spectrum of a radiating rarefied gas is a set of isolated or discrete wavelengths whose appearance is a series of bright-colored lines that form a pattern characteristic of the chemical composition of the gas.*

KIRCHHOFF'S THIRD LAW—ABSORPTION, OR DARK-LINE, SPECTRUM: *Light from a radiating source producing a continuous spectrum will, if it passes through a cooler gas, have certain specific wavelengths characteristic of the cooler gas removed from the spectrum. The spectrum appears continuous except where it is crossed by dark lines, which indicates that these wavelengths have been removed.*

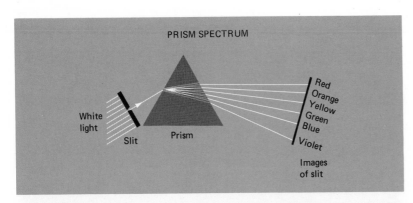

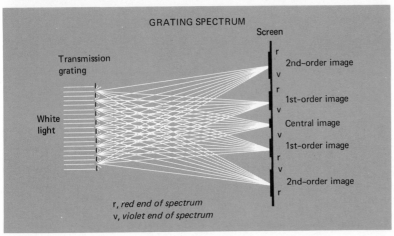

FIGURE 4.9
Two methods of producing a spectrum. The prism produces a single spectrum whose dispersion (spreading out of colors) is greater at the blue end than at the red end. The grating produces a bright central image flanked on either side by multiple fainter, higher-order spectra.

Illustrations of the different types of spectra are shown in Plate 1 as well as Figure 4.10. There are many common examples of light sources whose spectrum is one of the three basic types. As one example, the spectrum of the glowing filament of an electric light bulb is a continuous spectrum. The spectrum of a neon sign is an example of an emission spectrum. The spectrum of a gas composed of molecules (which consist of two or more atoms) is actually many sets of very closely spaced spectral lines known as emission bands. And, as a final example, the spectrum of the sun and most stars is an absorption spectrum.

We will see additional examples of each type of spectrum at many points in the remaining chapters. The important point to remember is that the type of spectrum for a light source tells us something about the conditions in and around that source.

IDENTIFYING THE ELEMENTS FROM EMISSION OR ABSORPTION SPECTRA

An astronomical light source, such as a star or a gaseous nebula, contains a mixture of chemical species, each either emitting or absorbing its own set of wavelengths of electromagnetic radiation. With the aid of laboratory spectral analysis of the different chemical elements, astronomers can identify individual elements in the light source from the measured wavelengths of its spectral lines, regardless of whether they are emission or absorption lines.

Identification is done in the following way: Light from a celestial body is collected by a telescope and then passed through a spectrograph in order to disperse the white light from the light source and form its spectrum. The photographic plate on which the spectrum is recorded is called a *spectrogram*. As a standard against which unknown wavelengths in the astronomical spectrum can be measured, an emission spectrum of a known gas, such as neon or vaporized iron or titanium, is placed above and below the astrono-

FIGURE 4.10
Kirchhoff's laws of spectrum analysis. The incandescent lamp produces a continuous spectrum (first law); the glowing rarefied gas produces a bright-line, or emission, spectrum (second law); the rarefied gas (either luminous and cooler than the continuous source or nonluminous) lying in the light path of the continuous source results in a dark-line, or absorption, spectrum.

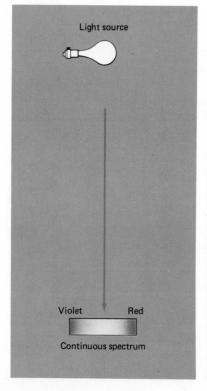

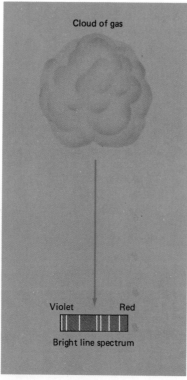

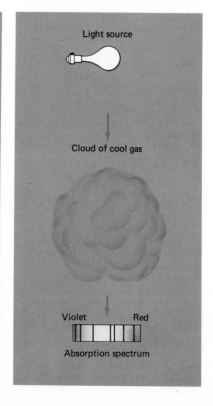

mical spectrum. (The mechanism for placing the laboratory spectrum on the astronomical spectrogram is a part of the telescope and spectrograph.) With these comparison lines of known wavelength the astronomer can determine the unknown wavelengths of the astronomical object's spectral lines. An example of a star's absorption spectrum is shown in Figure 4.11; the absorption spectrum is gray with black absorption lines and the comparison spectrum of neon shows white emission lines on a black background.

Kirchhoff's laws of spectrum analysis tell us about the general physical conditions of the light source. And if the spectrum of the light source contains absorption or emission lines, we can measure their wavelengths and identify the chemical elements that are present.

Can more detailed information about the light source be found? Suppose we want to know the temperature of the light source. Can this be done? Yes it can, for special types of light sources known as ideal radiators, or blackbodies. In Chapter 12 we shall outline the basis for learning more detailed information about stars, such as density and rotation.

INFORMATION IN CONTINUOUS SPECTRA

All objects radiate and absorb some form of electromagnetic radiation; the wavelength region and the amount of energy depend generally on the body's temperature and physical state. From laboratory experiments and from theory physicists in the nineteenth century analyzed how various bodies emit and absorb radiation as a function of temperature and wavelength. From this work they developed the concept of an idealized radiator called a *blackbody*.

BLACKBODY CONCEPT: *A blackbody is an imaginary body that, when cool, absorbs all the radiant energy falling on its surface so that it is black in color; when hot, the blackbody emits energy with 100 percent efficiency.* (Real matter is generally less than 100 percent efficient when it radiates.)

For our purposes the most important feature of the blackbody is the way in which emitted radiant energy is spread in wavelength, or the spectral energy distribution. Scientists have found that the distribution of energy depends only on the blackbody's temperature

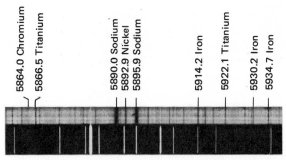

FIGURE 4.11
A portion of the yellow part of the spectrum of an orange-colored giant star. A number of absorption lines are marked, giving the wavelength in angstroms and the elements responsible for them. Below the star's spectrum is a comparison spectrum of neon used to establish the scale of wavelengths.

and not on its chemical composition. Figure 4.12 illustrates this point; note how the amount of radiant energy emitted by a blackbody varies with wavelength in a very recognizable way, even for different temperatures. The emission of radiant energy (or the brightness at each wavelength) covers a continuous range of wavelengths so that *the spectrum of a blackbody is a continuous spectrum;* that is, there are no color bands missing from its spectrum. At room temperature, lampblack (a finely powdered black soot) is very close to being a blackbody because it absorbs almost all the radiation incident upon it and reflects very little. Fortunately, the radiation emitted by stars tends to be much like that emitted by a blackbody. We shall use this fact later in our study of the sun and stars in Chapters 11 and 12.

In 1900 the German physicist Max Planck (1858–1947) derived a mathematical expression, now called *Planck's law*, that describes the distribution of brightness in the spectrum of a blackbody, which is shown in Figure 4.12. There are two other distinguishing characteristics of the spectrum of blackbody radiation in Figure 4.12: First, the energy emitted by the blackbody is greater at every wavelength as the temperature increases. Thus the total amount of radiant energy emitted increases with increasing temperature, which is known as the *Stefan-Boltzmann law*. Second, the greatest amount of radiation (the peak of the curves in Figure 4.12) is found toward shorter wavelengths (blue end of the spectrum) as the temperature increases. This is known as *Wien's displacement law*.

The significance of the blackbody-radiation laws—Planck's law, the Stefan-Boltzmann law, and Wien's law—is that bodies that emit electromagnetic radiation because they are hot, such as stars, do so much like a blackbody. Thus the blackbody-radiation laws are powerful diagnostic tools for measuring the temperature of these *thermal sources of radiation*. For the study of bodies that emit radiation not because they are hot (called *nonthermal sources of radiation*) but because of some selective physical processes, the blackbody-radiation laws are of no use. Fortunately, most of the celestial bodies—all the stars—are thermal sources of radiation and emit much like a blackbody. Some everyday examples of thermal sources of radiation are an incandescent light bulb, the burner on an electric stove, and the flame of a cutting torch. Examples of nonthermal sources are a fluorescent light, lightning, and a television screen.

4.3
Photons:
The Discrete Nature of Light

From his theoretical study of the emission of radiation by blackbodies Planck concluded that they do not emit or absorb energy in a continuous fashion but only discontinuously in discrete units, which later were called *photons*. This means that the energy transported by an electromagnetic wave is not continuously distributed over the wave front; it is located at discrete points (the photons) along the wave and moves with the wave. In 1905 Einstein used Planck's idea of a discrete nature for the emission of light to explain a phenomenon discovered in 1887, known as the photoelectric effect. This effect cannot be understood if light has only a wave nature. Since that time an extensive body of experimental and theoretical evidence has been collected to validate the photon concept of light.

What are some of the properties of photons? They move with the speed of light, have no inertia, are electrically neutral, and are massless. Picture a radiating body as emitting photons of differing discrete amounts of energy in all directions (see Figure 4.13). The photons retain their energy while traveling through space. Their arrival rate, or *flux,* at any point in space decreases with the square of their distance

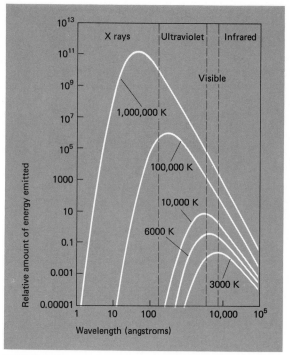

FIGURE 4.12
Blackbody energy curves according to Planck's radiation law. Note that, as the temperature increases, the peak energy shifts toward the shorter wavelengths and the energy emitted at every wavelength increases.

from the radiating source. Hence the inverse-square law of light can be understood in terms of numbers of photons (brightness of radiation) as well as in terms of electromagnetic waves.

The energy of each photon is inversely proportional to its wavelength. The shorter the wavelength, the more energetic is the photon; the longer the wavelength, the less energetic is the photon. That is why, for example, very-short-wavelength X-ray and gamma-ray photons can destroy molecular structures in living tissue while photons of visible light cannot. If you find that talking about the wavelength of a photon while saying that photons are the localizations of energy in an electromagnetic wave seems paradoxical, then you are perfectly normal. Just as in our discussion of relativity in Chapter 3, where we encountered a reality not evident in our human existence, so it is with our conceptual picture of light as both wave and particle. Wavelength is a characterization of the wavelike properties of light, while the energy content of a photon refers to its discrete nature. The fact that we can link wavelength and energy content in a mathematical equation is strong validation of our conceptual picture.

Photons may be absorbed by an atom, scattered by particles of matter, or converted into matter by interaction with other photons. They are created inside atoms and in violent collisions between subatomic particles. When they lose their identity, they transfer their energy to some other physical system; and when they are created, they obtain their energy from some other physical system. Their creation and destruction is a classic example of the conservation of energy. The concept of light as being simultaneously both discrete photons and continuous waves is not contradictory but is a reality not borne out in our everyday life. Yet experiments are designed to inquire about either light's wave nature or its corpuscular nature; no experiment will simultaneously yield the discrete and the wave properties of light.

The theory of the discrete nature of light began a conceptual revolution in twentieth-century physics and astrophysics. It was used by Niels Bohr to formulate a new model for the atom that can be used to understand how light is created and destroyed inside the atom.

FIGURE 4.13
Photon emission. The incandescent filament of the light bulb emits photons in all directions in space. Since the source is a continuous source, the photons have a continuous range of energies (wavelengths).

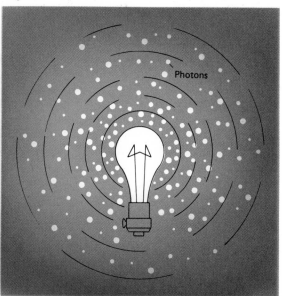

4.4 Creation and Destruction of Photons

THE BOHR ATOM

One of the most perplexing problems for early-twentieth-century physicists was why the atom emits a discrete pattern of spectral lines. By 1913, when the structure of atoms was reasonably well known, Niels Bohr (1885–1962), a Danish physicist, proposed a theory for the structure of the simplest atom, hydrogen, whose one electron orbits around a proton. He suggested that *the electron can occupy only a selected number of prescribed concentric orbits about the nucleus,* rather than having an unlimited and unspecified orbital distance. Also, the electron normally resides in the lowest energy orbit, which is the one closest to the nucleus. Orbits representing higher levels of energy are increasingly farther from the nucleus. The diameter of the first orbit corresponds to the normal size of the hydrogen atom, about 10^{-8} centimeter in diameter (see Figure 1.6).

When the atom absorbs energy, it is said to be *excited,* and the electron appears in one of the outer orbits, which have successively higher energies than the lowest orbit does. The electron's change (up or down) from one allowed orbit to another is called an *electron transition.* An atom in a gas may acquire the internal energy necessary to excite an electron by random thermal collisions with other gas atoms, collisions with subatomic particles such as free electrons, or absorption of a photon traveling through the gas.

Of all the photons striking the atom only those possessing an amount of energy equal to the energy difference between a higher energy orbit and the one in which the electron is located will be absorbed and excite the atom. For example, in the hydrogen atom it takes 10.2 electron volts, or 1.63×10^{-11} erg, of energy to raise the electron from its lowest energy orbit to the next higher energy orbit. Photons with energies below 10.2 electron volts will not be absorbed, and consequently the electron will not be excited. Photons with energies in excess of 10.2 electron volts cannot raise the electron to the second orbit, but they may, if they have the right amount of energy, excite the atom by lifting the electron to even higher energy orbits.

How long will an excited atom remain that way? If

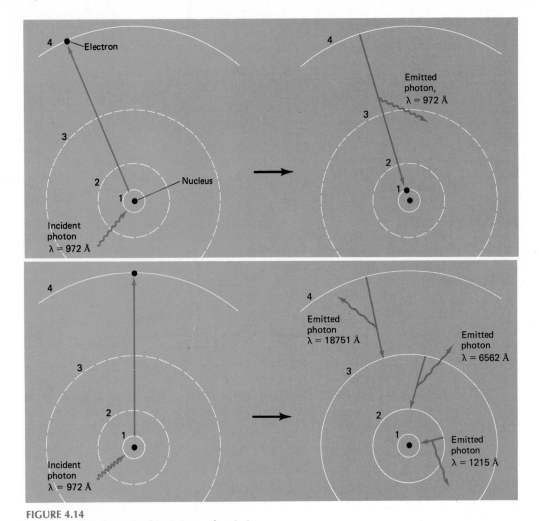

FIGURE 4.14
Two examples of atomic absorption and emission. When the hydrogen atom absorbs a photon (top, at left), it is excited, and an electron jumps, for example, from orbit 1 to orbit 4. Within 10^{-8} second the atom radiates a photon of equivalent energy as the electron drops from orbit 4 to orbit 1. Here (bottom), the hydrogen atom absorbs a photon, and an electron jumps from orbit 1 to orbit 4, as in the first example. But this time the energy is quickly released in three steps (at right) as the electron drops from orbit 4 to orbit 3 to orbit 2 to orbit 1, each time giving up a photon of different energy, that is, different wavelengths. The emitted energies of the three photons are equal to the absorbed energy of the incident photon.

a gas atom is excited, then in about a hundred-millionth of a second it will rid itself of any energy in excess of that of the lowest energy orbit by emitting the energy in the form of one or more photons. Somewhat like a ball bouncing down a staircase, the electron will drop in succession into one or maybe several lower energy orbits on its way to the lowest energy level, where it can reside indefinitely. With each

downward transition a photon of electromagnetic radiation is emitted. This photon represents the energy difference between the two orbits between which the electron makes the transition. The greater the energy difference, the greater is the amount of energy released in the form of a photon and, consequently, the shorter is the wavelength of the photon. Bohr was led to such a model for the atom as the most straightforward means of accounting for the discrete amounts of energy contained in photons. (Laboratory experiments had already shown that light really possessed the properties of a discrete phenomenon.)

Two examples from the countless electron transitions possible in the hydrogen atom are shown in Figure 4.14. Besides emitting energy spontaneously, an excited atom, before it can emit a photon, may collide with another atom in the gas and transfer energy to it as kinetic energy of motion. In this case no photon will be emitted. Such a radiationless transition is shown in Figure 4.15 along with the three means of exciting an atom discussed earlier.

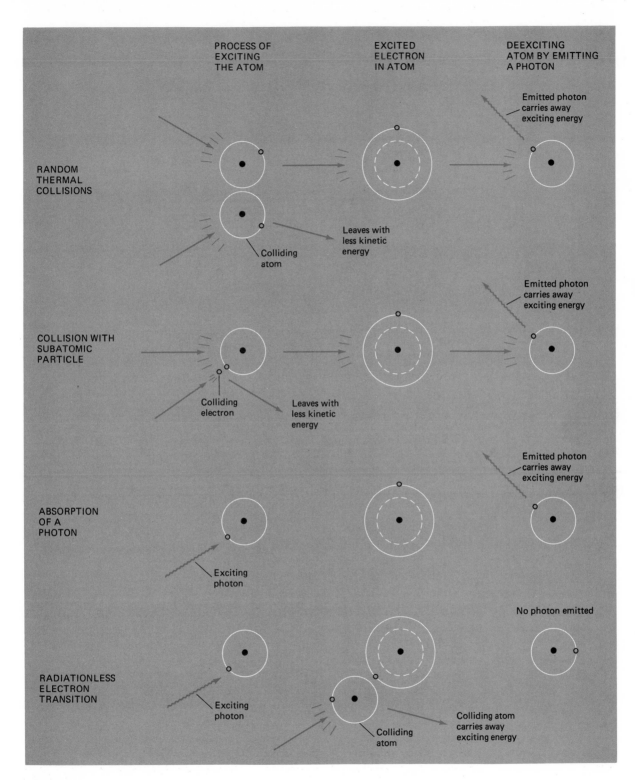

PROCESS OF
EXCITING
THE ATOM

EXCITED
ELECTRON
IN ATOM

DEEXCITING
ATOM BY EMITTING
A PHOTON

Emitted photon
carries away
exciting energy

RANDOM
THERMAL
COLLISIONS

Leaves with
less kinetic
energy

Colliding
atom

Emitted photon
carries away
exciting energy

COLLISION WITH
SUBATOMIC
PARTICLE

Colliding
electron

Leaves with
less kinetic
energy

Emitted photon
carries away
exciting energy

ABSORPTION
OF A
PHOTON

Exciting
photon

No photon emitted

RADIATIONLESS
ELECTRON
TRANSITION

Exciting
photon

Colliding
atom

Colliding atom
carries away
exciting energy

FIGURE 4.15
Means of exciting and de-exciting atoms. The first three rows illustrate the means of exciting an atom: random thermal collisions between atoms, where the energy to excite
comes from the kinetic energy of the colliding atom; collision with a subatomic
particle—in this case an electron; and absorption of a photon. In each case the de-
excitation of the atom occurs by emitting a photon. In the fourth row the de-excitation is
radiationless in that a colliding atom carries away as kinetic energy the excess energy
that otherwise would have gone into an emitted photon.

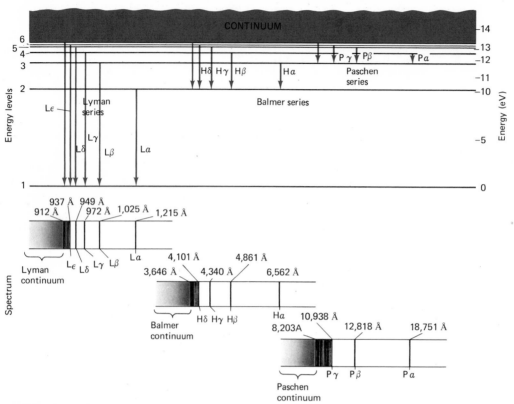

FIGURE 4.16
Energy levels and series in the hydrogen atom. Each series occupies a different portion
of the electromagnetic spectrum; each series comes to a limit, known as the series limit.
Downward transitions shown here give rise to emission lines, diagramed below. Upward
transitions give rise to absorption lines.

SPECTRUM OF HYDROGEN ATOM

In addition to the model of the atom that represents it
by its electron orbits, we can make a model using the
energy of each allowed electron orbit. Such an
energy-level model for hydrogen appears in Figure
4.16. The number of the energy levels corresponds in
a one-to-one fashion with the number of the electron
orbits. The distance between successive electron orbits increases with higher orbit numbers, but the differences in energy between successive orbits grows
smaller as the orbit number increases. Compare Figures 4.14 and 4.16, which illustrate this point. Notice
in Figure 4.14 that an electron can de-excite in many
ways.

Suppose a hydrogen atom is excited so that its electron is in the third energy level. Then it may de-excite
directly to the ground state, emitting one photon. The
photon would have a wavelength of 1026 angstroms,

which corresponds to the energy difference between
these two orbits in the hydrogen atom. Or it may
de-excite to the second energy level, emitting a photon with a wavelength of 6562 angstroms, and then
de-excite from the second energy level to the ground
state, emitting a photon with a wavelength of 1216
angstroms. The total energy emitted in both cases is
the same, but the wavelengths that result and hence
the spectral lines differ.

The hydrogen-line spectrum in the visible region,
known as the *Balmer series,* is prominent in the absorption spectra of most stars. It arises from electron
transitions originating on the second energy level of
the atom. In the same way all the possible transitions
from the ground level are known as the *Lyman series,*
which is in the ultraviolet part of the electromagnetic
spectrum. Those transitions from the third level up to
higher energy levels are the *Paschen series,* which is in

the infrared; and so on for the remaining series, whose lines appear in the far infrared on out to the microwave region of the spectrum.

Each series of spectral lines comes to a limit toward shorter wavelengths. The uppermost levels, representing the electron's highest energy orbits, crowd together toward a *series limit,* which represents the point beyond which the proton can no longer bind the electron to it. In this case the electron has been removed from the atom (it has been ionized) and is free to take on any energy.

If the electron is given enough energy, either by collision or absorption of a photon, it can escape the electrical attraction of the nucleus. The atom is then ionized and is in the form of a positive ion. The ionized hydrogen atom cannot absorb or reradiate energy in the form of discrete lines until it captures a free electron. It can execute the capture because of the electrical force of attraction between the positively charged nucleus (the proton) and the negatively charged electron. Figure 4.17 shows the Balmer series in the absorption spectrum of an actual star. Note that it converges toward its series limit at approximately 3646 angstroms.

SPECTRA OF OTHER ELEMENTS

In the Bohr atom, besides limits on the size of electron orbits, there is a limit to the number of electrons that may occupy a given orbit. These allowed orbits with a prescribed number of electrons in them are called *electron shells* (see Figure 1.6).

In general, as one goes through the periodic table, electrons are added to balance the number of protons in the nucleus by filling the shells from the one closest to the nucleus outward. In hydrogen there is one electron in the innermost shell, which has room for a maximum of 2 electrons. Helium's 2 electrons fill, or close, the shell so that for the element lithium the third electron must start a new shell, which is the next innermost. In the second shell there is room for only 8 electrons; in the third, 18 electrons; in the fourth, 32; and so on.

Each element has a unique set of energy levels. Consequently, the wavelength of the spectral lines originating from electron transitions between various energy levels is also unique for each element—a clear fingerprint of the element (see Plates 1 and 3).

The amount of energy needed to ionize an atom varies from one element to the next depending on the number and "position" of the electrons. For example, to remove the outermost electron from helium takes five times as much energy as it does to do the same for sodium. Also, for a given element each additional ion-

FIGURE 4.17
Balmer series in the spectrum of the star HD 193182, showing the hydrogen lines in absorption converging toward the series limit to the left of H40 at 3646 Å. Beyond that extends the Balmer continuum (a continuous spectrum of diminishing intensity), produced by the removal of the electron from the atoms. Balmer lines between H9 and Hα (the first Balmer line) are also present although they are not shown here. The comparison spectrum of iron, above and below the star's spectrum, establishes the scale of wavelengths.

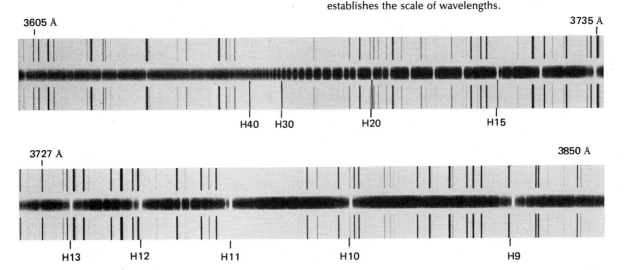

3605 Å — 3735 Å

H40 H30   H20   H15

3727 Å — 3850 Å

H13   H12   H11   H10   H9

ization takes more energy to free an electron from an inner orbit than from an outer one because the inner one is more tightly bound to the nucleus. Thus, with carbon as an example, it takes more than twice as much energy to remove the second electron than it does to remove the first; four times more for the third electron than the first; almost six times more for the fourth electron; thirty-five times more for the fifth electron; and a whopping forty-four times more for the innermost sixth electron than for the outermost first electron.

Multiple ionization of a carbon atom brings a corresponding readjustment of the energy levels because of the altered electrical attraction between the positive nucleus of the carbon atom and the reduced number of electrons. Altering the energy corresponding to each allowed orbit produces different spectral lines with each succeeding ionization of the carbon atom. So we see not only different wavelengths for absorption lines or emission lines between different unionized elements but for the same element a different spectrum after each ionization. That is, the spectrum

QUANTUM THEORY AND THE FORCES OF NATURE

Gravity was the first force in nature to be understood in at least a mathematical sense. Newton's theory shows that, even though separated by enormous distances, pieces of matter can influence each other's state of motion. Although less familiar in many respects, the electric and magnetic forces have been known since ancient times. Like gravity they both weaken as the square of the distance away from their source. In 1873 James Clerk Maxwell (1831–1879) showed that a relationship exists among electricity, magnetism, and light—an amazingly unifying step.

In 1924 the French physicist Louis de Broglie (1892–) pointed out that, like light, subatomic particles also have a wave nature, as well as a discrete nature. This has been verified experimentally many times. It is now an accepted fact that matter and radiant energy have dual natures in that they show both wave and discrete properties. Taking de Broglie's idea, Erwin Schrödinger (1887–1961), an Austrian physicist, and Werner Heisenberg (1901–1975), a German physicist, independently constructed mathematical theories for atomic structure at about the same time (1925). Their theories were consolidated by Paul Dirac, an English physicist, into the mathematical formulation called *quantum mechanics,* the most rational and logical approach so far for understanding a vast variety of

atomic phenomena. In reality there are no discrete electron orbits like those of planets in the solar system. Within the hydrogen atom, for example, are spherical regions surrounding the proton. In these regions the electron is spread into a pattern of standing waves, whose distribution corresponds to a discrete energy state of the atom.

All atomic properties are known to be the consequence of the electrical interaction between the nucleus and the electrons surrounding it. This electromagnetic interaction is responsible for the characteristic structure of each atomic species. These characteristic structures are responsible for the basic forms of matter, from simple rocks and crystals to flowers and even human beings. The electromagnetic force between the electron and the nucleus is 10^{39} times stronger than the gravitational force between them; no one has detected, nor is there any prospect of detecting, the effects of gravity within atoms or molecules.

By 1932 it was known that the nucleus was composed of protons and neutrons. This raised the problem of what force holds the nucleus together against the mutual electrical repulsion of the protons for each other. The solution of this question was the discovery of the strong *nuclear force* of attraction. It is about a hundred times more powerful than the electro-

magnetic force, but of very short range, and is capable of holding together nuclei with as many as a hundred or so protons.

Finally, a fourth force was discovered around 1935, the weak *nuclear force,* which is about 10^{-5} times as strong as the strong nuclear force, or about a thousandth as strong as the electromagnetic force. This force is responsible for some changes in the nature of the nucleus that occur in radioactive decay. It is also a very-short-range force. There is recent evidence that suggests that the electromagnetic force, the weak nuclear force, and possibly the strong nuclear force are actually different manifestations of the same force acting differently at different distances between particles. Linking all four forces into one universal expression, the so-called unified field theory, still eludes us.

The reason we are familiar with the gravitational and electromagnetic forces is that they operate on the scale of our experiences. The other two, the strong and weak nuclear forces, are confined to the nuclear scale of existence. The gravitational force increases its intensity with increasing mass, whereas the other forces are independent of mass. In the cosmos, as we shall see in later chapters, gravity dominates. Gravity is responsible for motion and form in the cosmic realm.

of singly ionized carbon differs from the spectrum of neutral carbon; doubly ionized carbon differs from singly ionized; triply ionized differs from doubly ionized; and so on.

Using the properties of electromagnetic radiation, the atom's structure, the interaction between matter and energy, and spectrum analysis, astronomers have gained much information about the universe from the radiation it emits. And as we develop an even greater understanding of the nature of radiation—its formation, propagation, interaction with matter, and destruction—we can explore more deeply the dim sources in the outer reaches of the cosmos, almost back to the beginning of time.

SUMMARY

Electromagnetic radiation. Electromagnetic radiation is made up of electric and magnetic fields of force continually interacting and propagating in the form of waves at the speed of light. Electromagnetic waves possess a continuous range of wavelengths from short-wavelength gamma rays through X rays, ultraviolet, visible, infrared and microwaves to long-wavelength radio waves. Light displays all the properties of wave phenomena. White (composite) light is composed of a range of wavelengths. The apparent brightness of light source varies inversely as the square of its distance.

Information in the spectrum of light. The wavelengths, or spectral composition, of composite waves contain information about the physical nature of the light source and its environment. The study of the spectra, produced by dispersing white light into its constituent wavelengths, is called spectroscopy. Kirchhoff's laws are diagnostic in that one can infer the physical conditions of the light source by the type of spectrum its light forms: continuous, emission, absorption. Spectral lines in emission and absorption spectra uniquely identify the chemical composition of the emitting or absorbing gas. Three radiation laws— Planck's, Stefan-Boltzmann, and Wien's—can be applied to the analysis of the continuous spectrum of a blackbody to determine its temperature. Thermal sources of radiation like the sun and stars emit radiant energy much like a blackbody.

Photons. Electromagnetic radiation has properties showing it to be discrete as well as wavelike. Photons are discrete units of electromagnetic energy that have no inertia (are massless), are electrically neutral, and move at the speed of light. The energy content of a photon is inversely proportional to its wavelength. Photons are created by taking energy from atoms and destroyed by transferring their energy to the internal energy of atoms.

Emission and absorption of radiation. The atomic processes responsible for the emission and absorption of photons are summarized in a model for the atom known as the Bohr atom. In hydrogen (and similarly for other atoms), the electron can occupy only a selected number of allowed orbits; that is, not all possible radius values are permitted for electron orbits; the electron normally resides in the lowest energy orbit, which is the one closest to the nucleus. Electrons make transitions to larger allowed orbits when atoms absorb photons and transitions to smaller allowed orbits when atoms emit photons. An electron can remain in a higher energy orbit for a very short time before it spontaneously drops to a lower energy state emitting a photon. All electron transitions beginning for absorption or ending for emission in an allowed energy state constitute a spectral series, such as the Balmer series in hydrogen.

KEY TERMS

absorption spectrum
angstrom
blackbody
composite waves
continuous spectrum

diffraction
dispersion
electromagnetic spectrum
electromagnetic wave
electron transition

emission spectrum
energy level
excitation
photon
reflection

refraction
spectrogram
spectroscopy
visible spectrum
white light

CLASS DISCUSSION

1. What properties of light characterize its wave nature? Its photon or discrete nature?

2. Can you give some everyday examples of the inverse-square law of light?

3. Under what circumstances can the chemical composition of a radiating source be determined? How would you determine the chemical composition of a star? Does this apply to the entire star?

4. How are photons created and destroyed? Is this a violation of conservation of energy? If not, how is energy conserved when an atom absorbs a photon?

5. What are the processes by which an atom is excited? Ionized?

6. How does the Bohr model of the atom differ from the mechanics of the sun and planets?

READING REVIEW

1. What are the various wavelength regions of the electromagnetic spectrum from short to long wavelength?

2. How does the apparent brightness of a light source vary with distance?

3. If a light source is moving away from an observer, is the wavelength of its radiation longer, shorter, or the same compared to when it is stationary?

4. What do Kirchhoff's laws reveal about the spectra of light sources?

5. What type of spectrum does blackbody radiation have?

6. What type of spectrum does radiation from the sun and stars have?

7. What is the Stefan-Boltzmann law about? Wien's law? Planck's law?

8. How does the energy content of a photon depend upon its wavelength?

9. When a hydrogen atom absorbs a photon, what happens to the electron? What will happen later to the electron?

10. How do we define a series of spectral lines like the Balmar series or Lyman series in hydrogen?

11. Why are the energy levels of the atoms of each chemical element different? Of each ion of that chemical species?

PROBLEMS

1. An electromagnetic wave traveling through space has a frequency of 6.0×10^{14} hertz. What is the wavelength of the radiation? To what color in the visible spectrum does this correspond?

2. If a light source at a distance d from us is

moved to a new distance such that the light source appears ⅕ as bright as before, then in terms of *d* how far away has the light been moved? If *d* is 5 meters, then what is the new distance in meters?

3. The Balmer delta line has a wavelength of 4101 Å. However, in a certain star its wavelength is 4101.82 Å. Assuming that this wavelength shift is produced by the Doppler effect, what is the velocity of the star in kilometers per second? Is the star approaching or receding?

4. How much more energy does a violet photon whose wavelength is 3500 Å have than a red photon with a wavelength of 7000 Å?

SELECTED READINGS

Adler, I. *The Story of Light.* Harvey House, 1971.

Connes, P. "How Light Is Analyzed," *Scientific American,* September 1968.

Hansch, T. W., A. L. Schawlow, and G. W. Series. "The Spectrum of Atomic Hydrogen," *Scientific American,* March 1979.

Malville, K. *A Feather for Daedalus.* Cummings, 1975.

Nassau, K. "The Causes of Color," *Scientific American,* October 1980.

5
Telescopes: Our Window on the Universe

The earth receives electromagnetic radiation of all wavelengths from various directions in outer space; yet wavelengths from only two regions of the electromagnetic spectrum are able to penetrate the earth's atmosphere freely. Most of the electromagnetic spectrum is screened out by the atmosphere well above the earth's surface, as we see in Figure 5.1. The two spectral windows in the atmosphere through which we can observe the universe from the earth's surface are called the *optical window*—from about 3,000 angstroms to about 10,000 angstroms, or roughly the visible-wavelength region—and the *radio window*—which includes the wavelength region from about 1 millimeter to 30 meters. The telescopes we build on the earth's surface to take advantage of these two windows are thus logically called *optical telescopes* and *radio telescopes*.

With the advent of the space age, astronomers have been able to take advantage of aircraft, balloons, rockets, and satellites to extend our vision of the universe by going above part or all of the earth's veiling atmosphere. Astronomers are aghast at what space telescopes carried by these vehicles have revealed through radiation in the ultraviolet, X-ray, gamma-ray, and infrared regions.

This chapter is devoted to a brief study of telescopes and their accessories, while what has been seen with these instruments is taken up in successive chapters.

5.1 Optical Telescopes

FORMATION OF AN IMAGE

In optical astronomy the object with which we work is the *image* of the light source formed by the principal image-forming part of the telescope, which is called an *objective*. The objective of the optical telescope is either a lens or a mirror, as shown in Figure 5.2. Light rays from the light source are refracted in passing through a lens and are reflected from a mirror. The image is produced where the light rays converge to a position known as the *focus*. The *focal length* of the

◄ This moonlight view on Mount Palomar shows the 5.1-meter Hale telescope dome with the shutter open.

objective is the distance behind the lens to the focus or the distance in front of the mirror to the focus. The image of a star is just a point of light, as shown in Figure 5.2, while that of an extended object, such as the moon, is inverted, as can also be seen in Figure 5.2.

In telescopes using either mirrors or lenses an eyepiece magnifies the image much as a reading glass magnifies small print. Or a photographic plate may be inserted into the focal plane of the objective instead of the eyepiece, transforming the telescope into a giant camera. In this case the objective lens or mirror serves as the camera lens. The advantage of photography over observing with the eye is that the photograph is available for later study and time exposures can record fainter sources than those the eye sees.

PROPERTIES OF AN IMAGE

The image formed by either a lens or a mirror has certain properties that depend upon the diameter, or *aperture*, of the objective and its focal length. One property is the *size of the image*. Since the image of a star is a point, size is not an important property for it. For an extended object the image size depends upon the angular size of the light source on the sky and upon the focal length of the objective.

Image brightness is important since it determines whether the image is above the threshold of visibility or how long it would take to photograph. The *brightness of an image* of a point source, such as a star, depends on how much light is intercepted by the objective. Hence its brightness is proportional to the area of the objective or to the square of the aperture. Doubling the aperture but leaving the focal length the same increases the area of the objective or its light-gathering power by four times, concentrating four times as much light into the same-size image.

When we photograph an extended object, the surface brightness of the image depends on the amount of radiant energy per unit area of the image. The objective's area (or the square of its aperture) still determines the total amount of energy collected, but the total energy is distributed over the entire image. Thus the larger the image's area, the smaller the energy per unit of area. The image size of an extended object increases in proportion to the focal length; so for a given telescope aperture the surface brightness of the image decreases as the focal length is made longer.

FIGURE 5.1

Observing the universe from below and from above the earth's atmosphere. *Top row:* X-ray image of Andromeda galaxy, Hercules galaxy cluster in visible light, galaxies M87 and (far right) Andromeda in visible light. *Center:* X-ray image of a globular cluster, the Horsehead Nebula, and the Crab Nebula in visible light. *Bottom:* the sun in X ray, visible image of a comet, visible image of Jupiter.

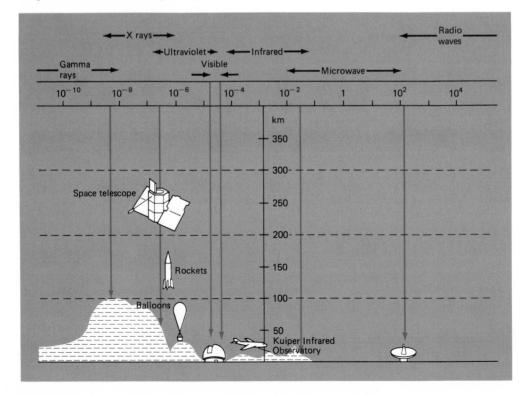

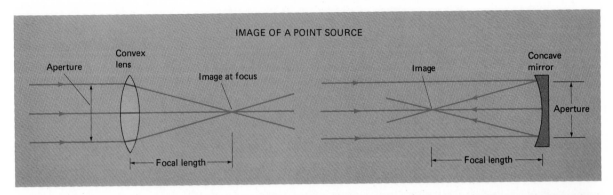

IMAGE OF A POINT SOURCE

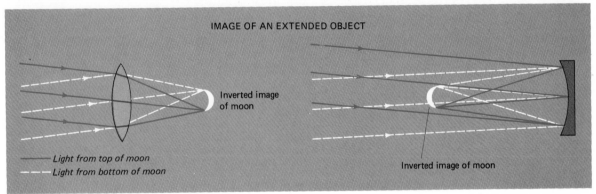

IMAGE OF AN EXTENDED OBJECT

FIGURE 5.2
Comparison of images formed by point sources and extended objects. The extended object's image is inverted because of the way light is refracted by the lens or reflected by the mirror.

How well a telescope discriminates between two adjacent objects or shows fine details is called its *resolving power*. Because of the wave nature of light, the image of a point source produces a *diffraction pattern*; it appears as a bright central spot, called a *diffraction disk*, surrounded by progressively fainter rings. When the diffraction patterns of two stars that are close together no longer overlap, we can see separate stellar images, as shown in Figure 5.3. The larger the telescope's aperture, the smaller is the diffraction disk of each image. A large aperture therefore improves the resolution of closely adjoining features by making the diffraction effect of adjacent objects overlap less. We define resolving power as the smallest angle between two close objects whose images can just be separated by a telescope. This critical angle is directly proportional to the wavelength of the observed radiation and inversely proportional to the aperture of the objective.

VIEWING PROBLEMS

The theoretical resolving power of any optical telescope is never fully realized. The lower layers in the earth's atmosphere are unsteady and turbulent; this turbulence blurs and distorts the star's image and makes it twinkle, or *scintillate*. The rapid scintillations break the starlight into many dancing specks of light, which in long exposures merge to form the fuzzy stellar images we see in photographs. When the atmospheric turbulence is low, the stars twinkle, or scintillate, less, and the so-called *seeing* is improved. A planet, on the other hand, shines with a steady light because each point on the tiny disk twinkles out of step with neighboring points; we see an average of all the twinkling points.

A technique called *speckle photography*, which can be used with large telescopes, can get around the smearing and wiggling of the image that comes from atmospheric turbulence. In the exposure of the photographic plate for an extremely short time (less than 0.01 second), each star image appears as a cluster of sharp specks of different brightness. Then the photograph is run through a high-speed light-sensing device which measures the variations in brightness

PROPERTIES OF AN IMAGE IN MATHEMATICAL FORM

The most important properties of an image formed by either a lens or a mirror are its size, brightness, and the resolution of points within it. As for the size of the image, it is a fundamental law of optics that, as seen from a lens or mirror, both the object being viewed and its image formed by the lens or mirror have the same angular size. Thus an objective that has a long focal length—that is, forms an image far away from itself—produces a larger image than a lens with a short focal length. If s is the size of the image, θ the angular size of the object in seconds of arc as seen from the objective, and l the focal length, then

$$s = \frac{\theta l}{206265} \propto l,$$

where the image size is measured in the same units as the focal length.

Example: The angular size of the moon is about 0.5°, or about 1800″ of arc. For a telescope with a focal length of 2.5 meters, or 250 centimeters (about 98 inches), the image size is

$$s = \frac{(1800)(250 \text{ cm})}{206265} = 2.2 \text{ centimeters},$$

which is almost an inch.

The brightness of an image is a measure of the amount of radiant energy concentrated into a unit area of the image. If you think of the light spreading out from the source in concentric spheres, then the amount of radiant energy in the image formed by the objective depends upon how much of the surface area of those concentric spheres the objective intercepts. That is to say, it is the area of the objective or the square of objective's aperture that determines the total energy collected, or the objective's light-gathering power. And the amount of energy per unit area of the image is set by the size of the image, which, as we learned above, depends directly on the focal length of the objective. If b represents the brightness, or the amount of radiant energy per unit area, then

$$b \propto \left(\frac{a}{l}\right)^2 = \frac{l}{f^2},$$

where a is the aperture in centimeters and l is the focal length in centimeters, and f is called the *focal ratio* (that is, the ratio of focal length to aperture). The smaller the focal ratio, the brighter is the image of an extended object like the moon.

Example: Let us compare the brightness of the image of a star as observed through two telescopes of the same focal length, one having twice the aperture of the other. The ratio of the image brightness is

$$\frac{b_1}{b_2} = \frac{(a_1/l)^2}{(a_2/l)^2} = \left(\frac{a_1}{a_2}\right)^2 = \left(\frac{2a_2}{a_2}\right)^2 = 4,$$

where $a_1 = 2a_2$. Thus the star image in a 2-meter (79-inch) telescope is four times brighter than the image in a 1-meter (39-inch) telescope if they have the same focal length.

Finally the resolution in the image can be stated in terms of the critical angle above which we can see adjacent points or details. This critical angle is inversely proportional to the objective's aperture and directly proportional to the wavelength of the radiation. The formula for the critical angle d in arc seconds is

$$d = \frac{(2.06 \times 10^{-3})\lambda}{a},$$

where λ is the wavelength in angstroms and a is the aperture in centimeters.

Example: For a binary (two stars in orbit about each other) whose angular separation is 1″ of arc (which is very close) we wish to find what aperture is needed to just separate the two stars in green light. Thus $d = 1″$ of arc and $\lambda = 5000$ angstroms, which yields

$$a = \frac{(2.06 \times 10^{-3})(5 \times 10^3)}{(1)} = 10.3 \text{ centimeters (or 4.1 inches)},$$

or a relatively small telescope. If we were trying to resolve them at a wavelength of 10 centimeters (10^9 angstroms), then

$$a = \frac{(2.06 \times 10^{-3})(10^9)}{(1)} = 2.06 \times 10^6 \text{ cm, or 21 kilometers},$$

which is an impossibly large radio telescope.

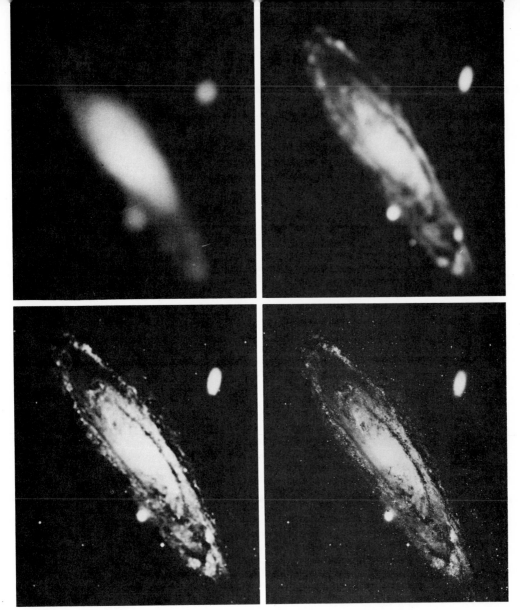

FIGURE 5.3
Resolution of an image. The spiral galaxy M31 as it would appear with resolutions of
(top, left) 12 arc minutes; (top, right) 3 arc minutes; and (bottom, left) 1 arc minute;
(bottom, right) M31 as seen with a large optical telescope.

across each speck. The information from the assemblage of specks in each of many photographed images is fed into a computer that is programmed to analyze and reassemble the information into the unsmeared image of the star. Figure 5.4 shows the reconstructed image of Betelgeuse (α Orionis), demonstrating that speckle photography can resolve minute sources of light, such as the disk of a large star.

Other nuisances hamper our observation of the heavens. The night sky's transparency varies as smog, dust, and atmospheric haze cloud it. The upper atmosphere is also suffused with a faint light called *airglow*. Atmospheric atoms and molecules absorb the ultraviolet photons in sunlight and reradiate the energy in a few wavelengths of the green, red, and infrared spectral regions. On long exposures airglow fogs a photograph and reduces the contrast between the faintest images and the sky background.

Another problem is that starlight entering the atmosphere is bent increasingly toward the vertical so that we see a star slightly closer to the zenith (the point directly above the observer) than it really is (see Figure 5.5). This *atmospheric-refraction effect* is greatest near the horizon (about 0.5°), for there the light's path through the air is the longest. When we observe the rising or setting sun, it is really below our horizon,

FIGURE 5.4

Speckle images (above) of three stars taken at the Kitt Peak National Observatory. The synthesis of such images produces the extended image of Betelgeuse shown below in an enhanced photograph from Kitt Peak. The slightly mottled surface and tenuous outer edge of the image are thought to be partly genuine features of the star and partly artifacts of the photographic technique.

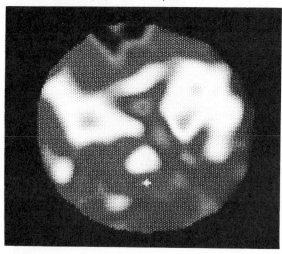

FIGURE 5.5

Bending of starlight or sunlight in earth's atmosphere due to refraction.

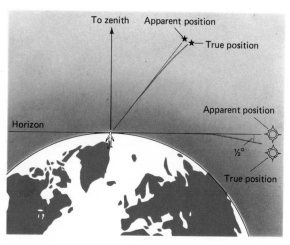

but refraction raises the sun's image above the horizon by an amount equal to its apparent diameter, which is 0.5°.

Other viewing problems are related to the geographical location of the observatory. An ideal site for an optical observatory is a mountaintop where the air is dry, transparent, and steady, and the sky is dark. An observatory also needs a minimum amount of wind and relatively easy access. The southwestern part of the United States satisfies most of these conditions and has many clear days and nights. Kitt Peak National Observatory is located there, about 65 miles southwest of Tucson, Arizona. A panoramic view of Kitt Peak is shown in Figure 5.6.

REFLECTING AND REFRACTING TELESCOPES

Telescopes that use lenses for the objective are known as *refracting telescopes,* while those that employ a mirror are called *reflecting telescopes.* The objectives of the early refracting telescopes could not form sharp images because of a condition known as *spherical aberration;* single lenses also failed to bring all colors to a common focus, a failure called *chromatic aberration.* A *compound lens,* or two lenses of different types of glass cemented together was invented to minimize these aberrations in refracting telescopes.

Spherical aberration also occurs in a reflecting telescope. If the surface of the mirror is parabolic rather than spherical, that aberration is eliminated although some minor deficiencies still remain.

Why are the big modern telescopes of the reflecting type? Reflecting telescopes have many advan-

FIGURE 5.6
Panorama of the Kitt Peak National Observatory.
The dome for the 4-meter Mayall reflector is in the
foreground (right), and the McMath solar telescope building can be seen in the background
(left).

tages over refractors: The reflecting telescope is free
from chromatic aberration, making it ideal for all-
purpose photography and spectroscopy. Also, since a
lens must be supported by its edges, there is a material limit to how large a lens system can be. But a
mirror, such as the one shown in Figure 5.7, can be
held both at its edges and from the back, the supports
allowing a wide range of sizes for mirror systems. The
largest refractor has an aperture slightly larger than 1
meter, but the largest reflector is 6 meters in diameter.

There are other advantages to reflectors: The glass
for the mirror in a reflecting telescope need not be so
optically pure and homogeneous as that required for
a large lens because the light reflects off the front
surface and does not pass through the mirror, as it
does through a lens. And the mirror has only one
surface that must be painstakingly ground—the ach-
romatic lens has four. To counter changes in temperature that would affect the focal length of the
reflector, large mirrors are constructed of fused
quartz or of a zero-expansion pyroceramic material.
The mirror's surface is coated with a thin layer of
highly reflecting aluminum that is replaced many
times during the life of the telescope.

FOCAL POSITION FOR
REFLECTING TELESCOPES

Reflecting telescopes can be designed for many kinds
of astronomical work through choice of the focal arrangement (see Figure 5.8) to suit the type of observation. For photography, photometry, and spectroscopy of faint objects the *prime focus* is best
because its small focal length lessens the exposure
time required. The *Newtonian focus*, most useful for
small telescopes, is now little used by professional
astronomers. In both these arrangements the observer works at a considerable distance above the observatory floor since both focal positions are near the
entrance of the telescope.

In the *Cassegrain focal arrangement* a convex secondary mirror positioned at the top in front of the
focus slows the rate at which light rays converge, effectively increasing the telescope's focal length. The
secondary mirror reflects the converging rays to the
bottom of the telescope and through a hole in the
objective mirror to focus behind the objective. This is
a much more convenient observing position since it is
near the floor and behind the telescope. Of all the
observations made with the 5.1-meter Hale telescope
on Palomar Mountain 75 percent are from the Cassegrain focus.

We might think that putting the secondary mirror
and its supports or the observer's cage for the prime
focus into the path of the light rays would obscure part

FIGURE 5.7
Modern large reflecting telescopes.

Top row:
Hale reflector (5.1 meters). The horseshoe collar on the right slowly rotates the great telescope about its polar axis. The observer's cage is the cylindrical centerpiece at the top of the tube.

Construction used to make the Multiple Mirror Telescope and Space Telescope. The mirror is of the so-called egg-crate design, a lattice sandwiched between two thin sheets of material.

Mayall reflector (3.8 meters), Kitt Peak National Observatory. The mounting is a horseshoe type with a mirror of fused quartz.

Center:
Multiple Mirror Telescope on Mount Hopkins in southern Arizona. It uses six small, independent mirrors to form an image rather than one large mirror and is the equivalent of a 4.5-meter telescope.

3.6-meter European Southern Observatory telescope, located in Chile.

The Soviet 6-meter reflector. It has an altazimuth mounting with a glass mirror.

Bottom row:
3.9-meter Anglo-Australian reflector at Siding Spring Mountain, Australia.

The 4-meter reflector at Cerro Tololo, Chile.

The mirror for the 5.1-meter Hale telescope as it appers before the surface is coated with a reflecting aluminum.

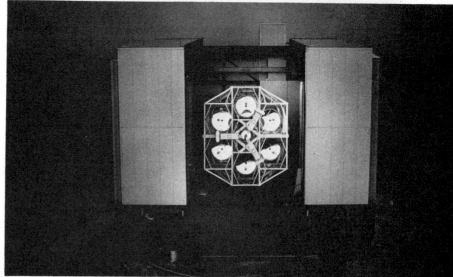

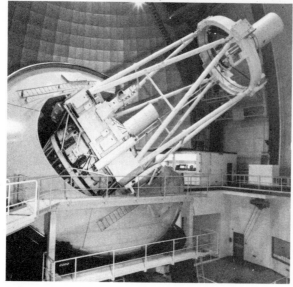

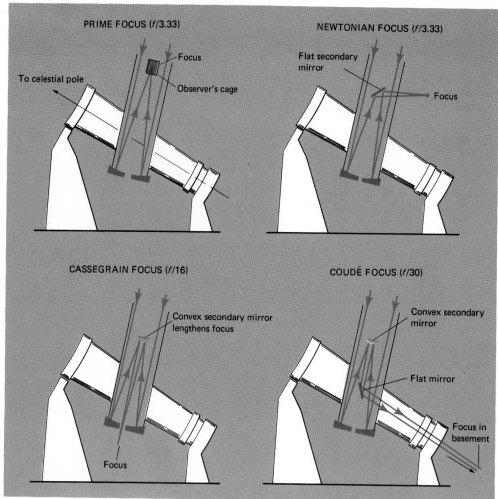

FIGURE 5.8
Focal arrangements of a large reflector. The prime focus, Cassegrain focus, and coudé focus are the primary ones used in astronomical research. The focal rations (ratios of the focal lengths to the apertures of the objectives) are those of the 5.1-meter Hale reflector.

of the image; but the only effect is to cut down the amount of light reaching the objective; the loss is small, and the quality of the image is not affected.

Equipment that is too heavy and bulky to be attached to the back of the primary mirror or is sensitive to changing gravitation as the telescope moves can be placed in a room below the observatory floor. An auxiliary flat mirror diverts the long converging beam down the hollow polar axis around which the telescope rotates, and with this *coudé focal arrangement* the focus can remain stationary no matter which way the telescope points.

TELESCOPE MOUNTINGS

An optical telescope, in order to follow an object as the earth's rotation carries it across the sky, must be free to move. To track stars accurately and to permit a telescope to be conveniently pointed in any direction, the *equatorial mounting system* is used for most telescopes (Figure 5.9). This system has two axes of rotation: The telescope can rotate in an east-west direction, called hour-angle, around its polar axis, which is aligned with the earth's axis of rotation; another allows the telescope to swing in a north-south direction about the declination axis, which is perpendicular to the polar axis. (For a discussion of astronomical coordinates, see Appendix 2.)

Large telescopes are usually positioned by a computer from an operating console and guided to the exact location with hand controls. Once a large telescope is properly set, the computer operating a clock drive slowly turns it westward around its polar axis at the same rate as the earth turns eastward, keeping the stellar images locked in position in the field of view. The great simplicity in equatorial mounting is that tracking requires continuous motion about only one of its two axes. The disadvantage, which obtains in the largest telescopes now in operation and planned for the future, lies in the stresses on the polar axis due to gravity. The polar axis is inclined in the earth's grav-

itational field and must rotate on one edge of its end. For a very large telescope that is a difficult engineering problem.

One means of removing some of the stress from the primary axis is to align it with gravity. Such a mounting is known as *altazimuth mounting;* with it a telescope rotates about a vertical axis and about a horizontal axis. This mounting's disadvantage is that, to track a star, it must turn continuously about both axes at the same time. When the telescope approaches the area of the sky directly overhead, continuous tracking becomes virtually impossible. Even with this disadvantage the altazimuth mounting will be the primary mounting for very large telescopes to be constructed in the future.

FIGURE 5.9
Schematics of equatorially mounted and altazimuth mounted refracting telescopes. The telescope turns about both the polar axis and the declination axis in the equatorial mounting. The polar axis moves the telescope east and west while the declination axis allows a north-and-south movement. In the altazimuth mounting the telescope turns about the azimuth axis to sweep the horizon and moves up and down from the horizon about the altitude axis.

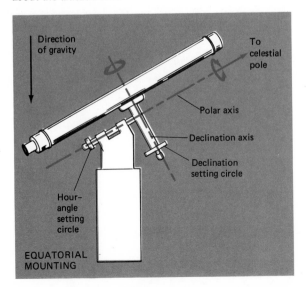

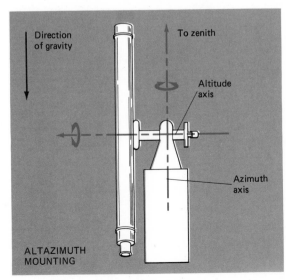

OTHER APPROACHES TO MAKING TELESCOPES

The principal problems in building very large telescopes on the earth's surface today are costs and construction time. A new 5-meter Hale telescope would now cost about 25 million dollars and take 10 years to build, while a 10-meter telescope would cost 200 million dollars and take 20 years to build, and a 25-meter telescope would require about 3 billion dollars and 50 years to construct. Clearly some dramatic changes in design are needed to lower cost and construction time.

A new telescope design, called the Multiple Mirror Telescope (MMT; see Figure 5.7), which is well suited for infrared observations, has been installed on Mount Hopkins in Arizona. It uses a mosaic of independent mirrors of small size to collect and focus light in order to simulate the collecting ability of a large-aperture single mirror. The MMT consists of a circular array of six identical 1.8-meter mirrors on an altazimuth mounting; the array has light-gathering power equivalent to that of a 4.5-meter single mirror.

The six mirrors are not thick solid ones but are of a new lightweight design, as shown in Figure 5.7. They are partially hollow, which requires a smaller mechanical structure to move them; thus they save money and construction time. The six images from the six mirrors may be either superimposed to form a single image or aligned along a spectrographic slit, one on top of the other, to take full advantage of slit geometry. The pointing directions of the six mirrors are locked together by laser beams. This instrument has been successful in demonstrating the practicality of the multiple-mirror concept, and it may be the forerunner of telescopes that are equivalent to a 25-meter (82-foot) telescope. Under consideration is an MMT consisting of eight 5-meter lightweight mirrors, having the light-gathering power of a 14-meter telescope, the angular resolution of a 22-meter telescope, and (it is hoped) the cost of a 4-meter telescope.

The MMT is not the only new design; others are displayed in Figure 5.10 which shows an artist conception of these new designs for future large telescopes,

FIGURE 5.10
Artist conceptions of experimental telescope designs. These unconventional designs are intended to reduce the cost and construction time for telescopes of the future although there are no present commitments to build any of them. It is likely that a very large new telescope for ground-based studies will be started before the turn of the century.

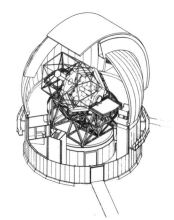

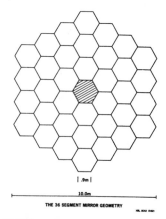

.9m

10.0m

THE 36 SEGMENT MIRROR GEOMETRY

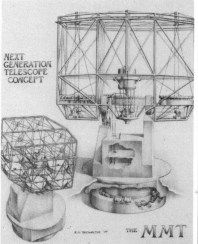

NEXT GENERATION TELESCOPE CONCEPT

NEXT GENERATION TELESCOPE CONCEPT

THE MMT

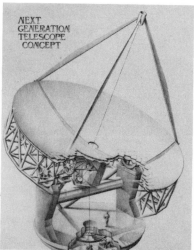

NEXT GENERATION TELESCOPE CONCEPT

NEXT GENERATION TELESCOPE CONCEPT

THE SINGLES ARRAY OF 2.4M TELESCOPES

but it is not now certain whether any of them will ever be built. The success of the Space Telescope, a 2.4-meter conventional reflector that is to be put into orbit in early 1985, will not lessen the need for a mammoth new telescope on the ground but will probably increase it. Since Space Telescope does not have to contend with light losses produced by the atmosphere, a very large telescope will be required on the ground to observe in visible wavelengths what Space Telescope is able to observe at shorter wavelengths, where it will primarily operate.

5.2 Accessory Instruments for Telescopes

RADIATION DETECTORS

Before discussing the accessory instruments used with optical telescopes, let us consider briefly the most important component of these instruments: the *radiation detector*. The telescope is capable of collecting light over a very wide range of wavelengths, but it is the radiation detector that determines what the telescope sees. One radiation detector with which we are all familiar is the human eye. It possesses most of the properties of radiation detectors in general and is thus illustrative of the points we wish to make about them.

The properties of interest are the wavelength regions to which the detector is sensitive, the differing response of the detector over that wavelength region, and the nature and range of detector response (see Figure 5.11). Using the human eye, we can briefly illustrate each of these properties.

PROPERTIES OF RADIATION DETECTORS

The eye is sensitive to the narrow wavelength region between about 3500 and 7000 angstroms, as shown schematically in Figure 5.11. However, the eye does not respond equally to all colors in the visible spectrum. It is most sensitive to the middle of the wavelength region, the green wavelengths, and the sensitivity drops to zero toward either the violet (short wavelengths) or the red (long wavelengths).

The nature and the range of detector response are expressed by the ways in which the eye responds to one photon and to a tremendous flood of photons.

Common experience tells us that the eye does not respond in the same way for both. There is some threshold number of photons, depending upon their wavelength, necessary to make the eye respond as shown in Figure 5.11. In other words, there is a limit to how faint a light source we can see, and that visibility limit depends upon whether we are looking at violet, green, or red light.

All of us have experienced the loss of response of the eye when we try to look at too bright a light. That is, the eye saturates—it no longer responds—and no scene is visible to us, just an intense and painful brilliance. To be useful, the radiation detector's dynamic range between threshold and saturation of visibility should be quite large, say, a factor of 100 or 1000. Now we may ask, "What is the response of the eye to doubling the number of photons in between the lower and upper limits of threshold and saturation?" If we double the number of photons, do we observe that the light is twice as bright? The answer in general is no. By and large, over the dynamic range of response of the eye, doubling the stimulus does not double the response; in other words, we say that the response is nonlinear. This concept of linearity is important because, in seeking the amount of radiant energy emitted by an astronomical source, astronomers usually compare the unknown light source against one of known energy output. Thus they have to know how their radiation detector responds to increasing or decreasing numbers of photons.

Now we look at two other radiation detectors, the photographic emulsion and the photoelectric device.

PHOTOGRAPHIC EMULSION

The *photographic emulsion* records photons by undergoing a chemical change (a photochemical effect) that will ultimately deposit silver on a glass plate or acetate film. The photographic plate can be made to respond to different wavelength regions within and beyond either end of the visible spectrum, which makes it much more versatile than the eye. Also, its response over a wavelength interval can be made much more uniform than that of the eye. The photographic plate, like the eye, is nonlinear in its response; it has a rather complicated response depending upon the position in its dynamic range. A simulated wavelength response is shown in Figure 5.11 for a fictitious photographic emulsion sensitive to infrared photons.

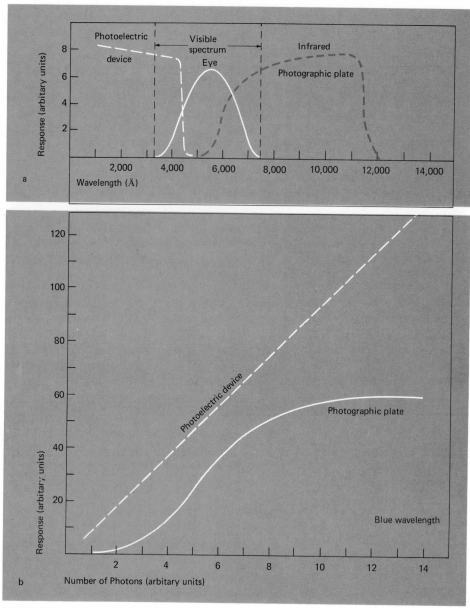

FIGURE 5.11

Illustration of the basic properties of radiation detectors, using a fictitious photographic emulsion and photoelectric device. (a) The wavelength region to which the detector is sensitive and the difference in response to different wavelengths in that wavelength region. (b) The nature of the response, for a photographic emulsion and for a photoelectric device for a blue wavelength, to increasing numbers of photons.

The photographic plate has a strong advantage over the eye since it will build up a response by storing the image. Thus time exposures allow the astronomer to collect information on a photographic plate about very faint light sources that cannot be detected by the eye through the same telescope. How faint a star can we photograph? The telescope's aperture sets the initial limit. Ultimately, however, the limit is set by the weak illumination in the night sky. This background interference comes from starlight scattered by the earth's atmosphere and from diffuse radiation in the atmosphere (airglow). Unfortunately, the photographic plate's photon-capturing efficiency is low. The emulsion can record only 1 or 2 percent of the incident photons (those that activate the light-sensitive coating). Facing this inefficiency, astronomers have found other types of radiation detectors to improve the telescope's performance.

PHOTOELECTRIC DEVICES

The *photoelectric device* is an application of the photoelectric effect. The basic principle is to liberate electrons from a metal surface by exposing it to photons in a light beam and then to measure the number of electrons with electronic circuitry. The photoelectric device, like the photographic emulsion, can be made to respond to different wavelength regions by varying the metals used in making the surface of a device. The biggest advantage of the photoelectric device is that it can be manufactured to have a very large dynamic range of response; in addition its response is linear to the number of incident photons, as shown in Figure 5.11 for a fictitious device. With modern electronics it is possible to adapt the photoelectric device to count individual photons or to use a mosaic of devices to form a picture much as a photographic plate does.

As an illustration of the photoelectric device's importance as a radiation detector, only about 15 percent of the nights of observing on the 5.1-meter Hale telescope are devoted to photographic work. On 85 percent of the nights some kind of photoelectric detecting device is being used.

IMAGE INTENSIFIERS

Electronic *image intensifiers* do as their name implies, they intensify, or amplify, the light from weak sources of radiation. In one such system photons from the telescope are focused onto a photocathode surface, which ejects electrons. The electrons are increased in number, accelerated, and focused by means of electric and magnetic fields onto a phosphorescent screen, which emits a spark of light for each electron that strikes it. Thus the faint light from the astronomical source is amplified by the device into light sufficient to record the image on a photographic plate. Alternatively a computer circuit can be used to count the electrons during the exposure. Still other image-intensifying techniques are in use or in developmental stages; these techniques can reduce exposure times by factors of 50 to 100 over those for photographic systems.

SPECTROGRAPHS

The photographic plate and the photoelectric device enhance our ability to detect light from different astronomical sources, but they are not basically analyzing instruments. We can equip an accessory instrument with either of these detectors and attach it to the telescope to analyze light. The two basic types of analyzing instrument are the spectrograph and the photometer.

The *spectrograph* disperses the composite light from the source into its component wavelengths so that we can, for example, determine the elements that compose the light source. Spectroscopy, which is the study of the spectra of light sources, is astronomy's fundamental interpretive tool.

A prism or grating spectrograph (see Figure 5.12) receives the concentrated light coming from the telescope's objective on an entrance slit. The light diverging past the slit enters a collimator, which delivers a beam of parallel rays to the dispersing device. Then these rays pass through either a prism or reflect off a grating, which separates the light into its constituent wavelengths. The dispersed light is focused by a camera system onto a radiation detector (a photographic plate or a photoelectric device) as individual color images of the entrance slit. Each wavelength forms a distinct image of the slit. The images of the slit in the different wavelengths are arrayed in an orderly progression of colors from red to violet to create the observed spectrum of the composite light falling on the entrance slit.

PHOTOMETERS

The *photometer* is an accessory device that the astronomer attaches to the telescope at the focal position of the objective to measure the amount of radiation coming from the astronomical object. Where the spectrograph is used to examine the spectral composition of radiation, the photometer can be made to scan the spectrum formed by the spectrograph. It measures the amount of radiant energy, on either a relative or an absolute scale, at one wavelength or in a band of wavelengths. The photometer is much like an exposure meter on a camera: Incident light is converted

> The giant Hale 200-inch telescope on Mount Palomar could easily record the images of a million galaxies in the bowl of the Big Dipper.
>
> Harlow Shapley

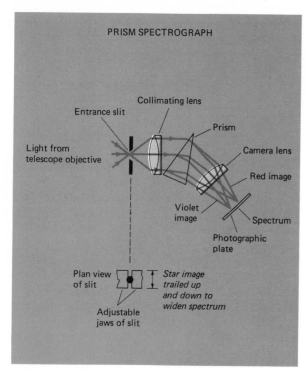

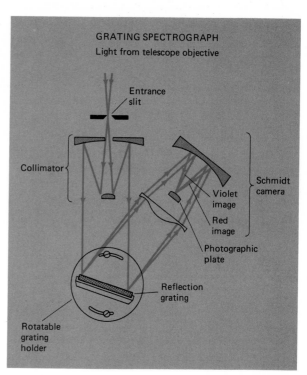

FIGURE 5.12
Schematic drawings of the optical arrangement in a prism spectrograph and a grating spectrograph. Composite light enters through the entrance slit to fall on the collimator, which renders the light rays parallel as they fall on the dispersing device. It is either a prism or a reflection grating that separates electromagnetic waves of different wavelengths and sends them in slightly different directions to the camera. The camera then focuses the dispersed light on the radiation detector—in this case a photographic plate. The spectrum formed is a sequence of images of the entrance slit in each wavelength present in the composite light.

into an electrical current. One can use a variety of techniques to define the wavelength region for the photometer, such as a spectrograph or color filters. And the radiation detector is generally today a photoelectric device.

The photoelectric photometer is usually limited to measuring only one light source, such as a star, at a time. But the limitation is compensated for by the photoelectric photometer's very great accuracy. Because of its quick response to changes in amounts of light, the photoelectric photometer is particularly useful in continually monitoring the change in brightness of an object whose emission of radiant energy varies with time (for example, a number of stars are known to be variable light sources).

5.3
Infrared Devices

In 1800 William Herschel detected the infrared component of solar radiation by positioning thermometers beyond the red end of the sun's visible spectrum and thus foreshadowed the astronomy of invisible spectral regions. What we have seen over the last 15 years in these regions has revolutionized our concept of the universe.

INFRARED TELESCOPES
We can subdivide the infrared spectrum into three segments. A large part of the infrared spectrum is not visible at ground level because of absorption by water vapor, carbon dioxide, and molecular oxygen, which lie between the ground and about 15 kilometers altitude. Consequently, airplanes, balloons, rockets, and satellites are extensively used to lift the infrared telescope above the veiling atmosphere. Astronomers can also locate infrared facilities on mountaintops, such as

FIGURE 5.13
A panorama of Mauna Kea summit in Hawaii, with the NASA infrared-telescope facility in the lower right and the United Kingdom 3.8-meter telescope in the upper center.

the one in the Hawaiian islands shown in Figure 5.13, to make ground-based infrared observations.

The liquid-helium-cooled infrared detector can be used with the appropriate analyzing instruments on an ordinary optical telescope to study the cosmos. But because of the longer wavelength of infrared radiation, the image-producing quality of the telescope objective need not be so fine as it must be for the visible region. Thus a number of new telescopes have been designed and built for infrared astronomy only. A national observatory for infrared astronomy is built high on the 4200-meter inactive Hawaiian volcano Mauna Kea (Figure 5.13). A 3.0-meter infrared telescope constructed by NASA and the University of Hawaii is in operation there along with a 3.8-meter infrared telescope belonging to the United Kingdom. Other major infrared telescope facilities are the MMT in Arizona, the University of Wyoming facility, and Mexico's 2.1-meter reflector.

Now that we have described the optical window, somewhat expanded to include available parts of the infrared, we should shift our focus to the radio spectral window in the atmosphere. Thus the next section is on radio telescopes.

5.4
Radio Telescopes

DISCOVERY OF CELESTIAL RADIO WAVES

In 1931 a Bell Telephone engineer, Karl Jansky (1905–1950), was trying to find where the interference disrupting transatlantic radiophone circuits came from. He discovered that some of the radio noise was not from the earth; it was extraterrestrial. The primary source was the center of the Milky Way, in the constellation of Sagittarius. In 1936 an Illinois radio engineer, Grote Reber (1911–), pursued the phenomenon farther. He built the first parabolic radio telescope, 9.5 meters in diameter, and made the first radio map of the sky. The strongest signals he found came from the star clouds in Sagittarius and from several discrete sources toward the center of our Galaxy. The next major discovery was in 1942, by British radar operators and scientists tracking down suspected radar jamming during World War II; they discovered that the interference was radio emission from the sun.

At first astronomers did not grasp just how significant Jansky's work was; they were preoccupied with their observations of the universe through the optical window of the earth's atmosphere. But after World War II radio astronomy came into its own when physicists, radio engineers, and astronomers joined forces to build larger and more efficient radio telescopes. Radio astronomy since then has led to startling discoveries, such as interstellar molecules, pulsars, and the enigmatic quasars. Today our concept of any cosmic body is based upon its appearance all across the electromagnetic spectrum, with the radio region an extremely important component.

RADIO-TELESCOPE DESIGN

Because the physical nature of a radio wave is exactly the same as that of a light wave, the problem of designing a radio telescope is similar to that of designing an optical telescope. There are some practical differences. Radio waves pass through most materials without any interaction; thus it is not feasible to design a "lens" for radio waves that will focus them in a refracting telescope. But any metal will reflect radio waves; so a dish-shaped metal mirror will focus radio waves, just as a glass mirror focuses light waves. The reflecting surface of the dish can be an open, fine-wire mesh or a solid metal with a parabolic shape. Radio waves are reflected from the surface and converge toward a focal point, where a small collector aerial absorbs the concentrated energy, as shown in Figure 5.14, turning it into an electrical current. From there the current or signal is carried by an electrical cable to the receiving equipment, which processes the signal just as in your home radio receiver.

After amplification the signal variations are recorded in one of several ways (Figure 5.14). The signal changes can then be fed into a computer for analysis. When the computer has done its job, a formerly invisible part of the universe is revealed as in the radio map in Figure 16.14.

Astronomers can increase the sensitivity of radio telescopes, with better accuracy in pointing and with higher resolution, by expanding the collecting area of the dish or by improving the capabilities of the receiver. With the largest radio telescopes we can obtain a resolution of about 1' of arc, comparable to that of the eye. That is like seeing a penny 65 meters away. The most powerful radio telescopes can detect energy from sources whose power is comparable to that of a terrestrial FM broadcast station many light years away.

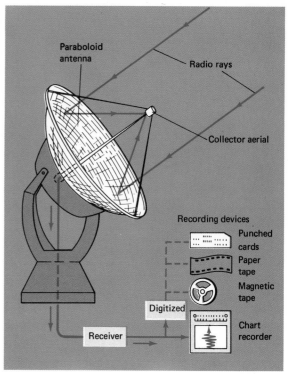

FIGURE 5.14
Simplified diagram of a radio-telescope system. Radio waves are reflected from the antenna and focused on the collector aerial. There the radio waves are converted to an electric current, which is carried by cables to the receiver. The output of the receiver is displayed on a chart recorder or recorded in digital form on punch cards, paper tape, or magnetic tape.

The radio telescope is remotely controlled by the astronomer from an electronic console. Moderate-sized radio dishes, up to about 100 meters or so in diameter, are steerable and have equatorial mountings that follow the rotating sky just as optical telescopes do. Larger and heavier dishes employ an *altazimuth mounting*—one that rotates about a vertical axis and a horizontal axis. This minimizes the distortion in the shape of the dish due to changing the orientation of the dish in the earth's gravitational field while tracking an object. A computer directs the rotation about both axes.

Even larger and more unwieldy antennas are fixed, pointing upward while the rotating earth sweeps much of the sky by the antenna's field of view. Arecibo, Puerto Rico, has the biggest fixed antenna, a metal dish 305 meters across contoured out of a natural bowl in the ground (see Figure 5.15). It can survey

FIGURE 5.15
Large fixed-position radio telescope. In a natural bowl in the ground in Arecibo, Puerto Rico, this 305-meter radio telescope is the largest single-dish telescope now in operation. The rotation of the earth sweeps the sky across its field of view, and its movable overhead collector aerial allows it to see within 20° of the zenith, a covering of about 40 percent of the whole sky.

the sky to within 20° of the zenith, allowing coverage of about 40 percent of the entire sky.

RADIO INTERFEROMETRY

Astronomers have searched endlessly for better resolving power. It can be achieved by building bigger telescopes or by observing at shorter wavelengths for a chosen aperture or by using the phenomenon of interference (see Figure 5.16). *Interferometry* is a technique involving two or more radio telescopes. Radio radiation from an astronomical source received at the individual telescopes is combined to obtain data that have a spatial resolving power equal to that of a single telescope as large as the distance between the individual receivers. With interferometry an astronomer can obtain details about the spatial structure of a given celestial object that a single radio telescope could never reveal.

For many years the separation between the individual telescopes of the interferometer was limited by the lengths of the cables connecting the antennas because the technique depends upon combining, at the same instant, the signals received by the separate telescopes. With the advent of the atomic clock (a clock governed by the vibrations of certain atoms) it became possible to record the signals received by the different

telescopes, along with the precise time, and to compare them later. This allowed the individual telescopes to be greatly separated, even on opposite sides of the earth. The technique is called *Very Long Baseline Interferometry* (VLBI). It has been used with a geosynchronous or geostationary satellite as the link in the communications channel between the telescopes.

In the Very Large Array (VLA) radio interferometer recently put into operation in New Mexico (see Figure 5.17) signals from each of 27 individual radio telescopes are combined by a computer. Each dish is 25 meters in diameter, and the 27 individual telescopes are moved along railroad tracks arranged in the shape of an enormous 21-kilometer Y. Nine dishes will be

FIGURE 5.16
Simplified schematic of a radio-telescope interferometer. Incoming wave crests from the radio source arrive at slightly different times at antennas A and B. They may be in phase, and, if so, the signal is reinforced. Or the crest of a wave may arrive at antenna A while the trough of the same wave or a succeeding wave arrives at B, so that they are out of phase and the signal is canceled. As the sky rotates, the path lengths to each antenna constantly change, resulting in variations in signal strength with time. The signals from each antenna are mixed in an electronic device called a correlator for proper analysis.

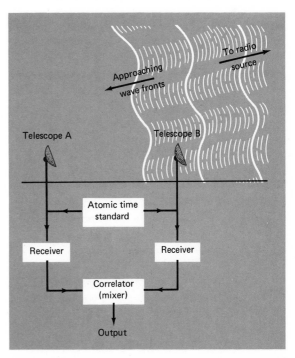

FIGURE 5.17
The Very Large Array (VLA) radio interferometer near Socorro, New Mexico. The telescope contains 27 dishes in a large Y-shaped configuration.

located on each branch of the Y, and the system can provide a total of 351 interferometer pairs of antennas. The energy-collecting power of the VLA is roughly equivalent to a single 122-meter telescope. The VLA will be able to achieve spatial resolution of about 1 second of arc in 10 hours of observing, or about that of the 5.1-meter Hale optical telescope. This ability makes it comparable to large optical telescopes in ferreting out the structure of various cosmic bodies.

5.5
Ultraviolet, X-Ray, and
Gamma-Ray Telescopes

Although much useful and important observational work remains to be done from ground-based observatories, an increasing portion of future astronomical research will be carried out from platforms outside most of the veil that is the earth's atmosphere. Up to the middle of this century nearly all our knowledge about the cosmos had come from studying the visible light of astronomical objects, and the visible and radio windows still constitute our most readily accessible

sources of information. Yet much of today's research is centered on the invisible regions of the electromagnetic spectrum, which do not penetrate the earth's atmosphere (as shown in Figure 5.1). To explore these regions—the infrared, the ultraviolet, the X-ray, and the gamma-ray wavelengths—new techniques and equipment are being developed, which must be flown above the atmosphere in some type of space vehicle.

SPACE VEHICLES
High-altitude aircraft and balloons are the least expensive way of investigating invisible extraterrestrial radiation. Jet aircraft can ascend to about 15 kilometers, while balloons are useful up to about 30 kilometers, above which only 5 percent of the atmosphere remains. Rockets, though their flights are short compared to balloon flights, lasting for minutes instead of hours, can climb five times higher than balloons can. Artificial satellites cost much more than rocket flights, but satellites can continuously monitor events over different regions of the electromagnetic spectrum for long periods of time, an advantage that outweighs their additional cost. Since 1958 the United States has launched hundreds of instrumented satellites that have either orbited the earth or been sent to search the solar system.

Up until about 1983 the means of getting satellites off the surface of the earth have been rockets. The advent of *Space Shuttle* has now provided another way to launch satellites. *Space Shuttle* is a true aerospace launch vehicle in that it takes off like a rocket, maneuvers in earth orbit as do other spacecraft, but lands like an airplane. This launch vehicle was designed to carry heavy loads into space and to be reusable. Satellites can be carried into orbit in the *Shuttle's* cargo bay; when the *Shuttle* is in orbit, the satellites can be lifted out by a retractable arm and placed in their orbits. This also means that satellites can be retrieved from orbit to be brought back to the earth's surface or serviced and returned to orbit. *Space Shuttle* gives us the capability of carrying pieces of immense spacecraft, including manned space stations, into orbit to be assembled there. Its versatility signals a new generation of space exploration.

An important group of space vehicles has been the observatory satellites placed in orbits several hundred kilometers above the earth. Two of the most sophisticated and costly observatory satellites have been the

Orbiting Astronomical Observatories; OAO-2 and *OAO-Copernicus.* (A drawing of *OAO-Copernicus* is shown in Figure 5.18.)

The *Skylab* program cost $6 billion. Three crews of three men each spent 171 days in *Skylab* between May and November, 1973. These astronauts carried out dozens of astronomical, biomedical, and technological experiments. The abandoned station was to remain in orbit for several years; it was hoped that with *Space Shuttle* astronauts would be able to reuse the station in the future. However, *Skylab* was dragged down by the atmosphere to a fiery demise over the Indian Ocean, scattering pieces over western Australia on July 11, 1979.

In early 1985 NASA expects to place a 2.4-meter unmanned reflecting telescope named *Space Telescope* in orbit at an altitude of 500 kilometers (Figure 5.18). Its optics and instrumentation will be enclosed in a cylindrical tube 13 meters long and 4.3 meters wide. Auxiliary apparatus includes two imaging cameras, faint-object and high-resolution spectrographs, a photometer, and other specialized devices. These analyzing instruments are designed to cover the wavelength range from about 1000 angstroms in the ultraviolet to 8000 angstroms in the near infrared. Data from the telescope will be radioed in digital (number) form through the Goddard Space Flight Center in Greenbelt, Maryland, to the Space Telescope Sciences Institute on the Johns Hopkins University campus for processing.

Out in space no atmospheric absorption or turbulence will distort the images produced by the telescope. Thus the telescope should see astronomical sources up to 50 times fainter than those visible from the earth's surface; in terms of distance a faint object can be seven times farther away than could be seen from the surface of the earth. *Space Telescope's* spatial resolution will be 10 times better than the best earth-based reflectors. With proper maintenance from *Space Shuttle,* the telescope could operate for at least a decade. Unlike its ground-based counterparts, *Space Telescope* will scan the electromagnetic spectrum from the deep ultraviolet to the infrared.

Other space observatories are the planetary probes (see Figure 6.13), which are literally the most exotic. Their role is to go to a planet to photograph and analyze from a close flyby, to orbit the planet, or in some cases to land. As examples, the *Viking 1* and *Viking 2* spacecraft landed on the surface of Mars (we shall discuss them later). Other examples are the *Voyager 1* and *Voyager 2* spacecraft launched to encounter Jupiter, Saturn, and perhaps Uranus and Neptune. Much of our understanding of the nature of the universe is changing—rapidly and dramatically—because of these space observatories.

ULTRAVIOLET TELESCOPES

The ultraviolet portion of the electromagnetic spectrum has been divided by astronomers into three segments, more or less derived from the time in which serious research into them began. First there is the ground-based ultraviolet, from 4000 angstroms to the atmospheric cutoff at 3000 angstroms; next the far ultraviolet from 3000 to 1000 angstroms; and last the extreme ultraviolet from 1000 to 100 angstroms.

Ultraviolet observations began after World War II, in October 1946, when a captured German V-2 rocket carried a small ultraviolet spectrograph to a height of 100 kilometers. During the ascent it recorded the ultraviolet portion of the solar spectrum down to 2200 angstroms. Telescopes, analyzing instruments, and radiation detectors for ultraviolet research are basically the same kinds of instrument used in visible and infrared observations. The principle difference is that a number of types of glass are not transparent to ultraviolet photons but are highly absorbing. Therefore, special materials must be used for lenses and entrance windows into the instrument. The principles of operation are the same as those for visible radiation. Since ultraviolet-sensitive film cannot be retrieved from an orbiting satellite, photoelectric devices have been the primary radiation detectors, so that data could be radioed back to ground stations.

Between 1962 and 1975 eight *Orbiting Solar Observatories* (*OSO-1* through *OSO-8*) were launched for study of the sun in ultraviolet wavelengths and X-ray and gamma-ray radiation. In December 1968 the first *Orbiting Astronomical Observatory* (*OAO-2*) began sampling the ultraviolet and far-ultraviolet radiation. By the time *OAO-2* ended its useful life in February 1973, it had carried out photometry on more than 1000 objects from planets to galaxies. Its successor, illustrated in Figure 5.18, *OAO-Copernicus*, launched in August 1972, carried an 0.8-meter ultraviolet telescope and three small X-ray telescopes, and was even more active than *OAO-2*.

In January 1978 the *International Ultraviolet Explorer*, an orbiting observatory, shown in Figure 5.18, was launched by NASA. This was a joint undertaking

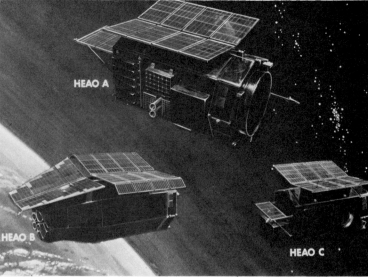

FIGURE 5.18
Earth-orbiting space telescopes.

Top row:
OAO-Copernicus satellite was placed into earth orbit in August 1972. The satellite was used for exploration in the ultraviolet wavelengths.

The High Energy Astronomy Observatory (*HEAO*) project of NASA includes three earth-orbiting satellites. The satellites launched between 1977 and 1979 were designed to collect data in the gamma-ray, X-ray, and ultraviolet regions as well as on cosmic-ray particles.

Center:
The *Advanced X-Ray Astrophysics Facility* is a second generation Einstein Observatory and is scheduled for launch in 1987.

The *International Ultraviolet Explorer*, launched in 1978, has provided an immense amount of ultraviolet data.

Bottom:
Space telescope in an artist's concept is a 2.4-meter reflector for ultraviolet, visible, and near-infrared investigations and will be launched in 1983.

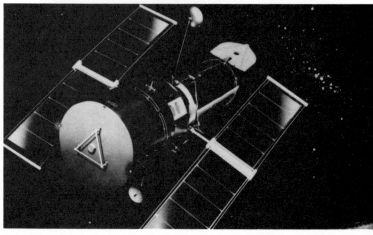

112

by NASA and several western European countries. Its facilities have been used for studies of planets, stars, galaxies, and the interstellar medium in the wavelength range from 1150 to 3200 angstroms. Astronomers conduct their experiments from an elaborate console of controls located at the Goddard Space Flight Center.

X-RAY DEVICES

X-ray astronomers divide "their" portion of the electromagnetic spectrum into two categories: *soft X rays,* from about 10 to 100 angstroms, and the more penetrating *hard X rays,* from approximately 1 to 10 angstroms. Both X rays and gamma rays are emitted by regions of space characterized by very high temperatures, low density, and high-speed subatomic particles—that is, wherever there are extreme conditions involving nuclear and atomic reactions. The observed radiation is in part thermal radiation but mostly nonthermal radiation (see page 80).

Most people are aware that X rays are more penetrating than visible light since they pass through the human body when making X-ray pictures for medical diagnosis. In this great penetrating power lies the difficulty in making telescopes to focus X rays, analyzing instruments, and radiation detectors for X rays; for glass lenses and mirrors do not refract or reflect X rays impinging directly on them. If X rays strike a smooth surface at a very shallow angle, less than a couple of degrees, they will reflect off the surface. This phenomenon has been used successfully to design an all-grazing-incidence reflector, which focuses X rays as an optical telescope focuses visible light. Such an X-ray telescope was flown as the heart of the second *High Energy Orbiting Observatory,* known as the *Einstein Observatory* and shown in Figure 5.18. We discuss the results from the *Einstein Observatory* in Chapters 13, 16, and 18.

At the focus of the X-ray telescope is the radiation detector, just as in the case of an optical telescope. Photographic emulsions can be made that are sensitive to X rays and can record an X-ray picture. For hard X rays special crystalline materials will absorb X rays, converting their energy into photons of visible wavelengths that can be detected with photoelectric devices. And there are solid silicon detectors whose ability to conduct electrical charges is influenced by their absorption of X-ray photons.

Astronomers first began using X-ray detectors in balloons and rockets and in a few unmanned satellites during the 1960s. By 1967 they had discovered some 30 discrete X-ray sources. Then in December 1970 NASA's *Explorer 42* satellite (*Uhuru*) was launched off the coast of Kenya, Africa. By the end of its useful life in 1973 it had scanned nearly the whole sky and had located nearly 200 X-ray sources. The newly discovered X-ray objects were named after the constellation in which they appeared, followed by X-1, X-2, and so on, in the order of discovery. For example, Taurus X-1, the first X-ray object discovered, is the Crab Nebula (see Section 14.3). Today, with a growing number of X-ray discoveries, it is convenient to designate the source by a catalog number.

A second generation of NASA satellites (the *High Energy Astronomy Observatories*), designated *HEAO-1, HEAO-2,* and *HEAO-3,* was launched in August 1977, November 1978, and September 1979, respectively. The three *HEAO* satellites are designed specifically to study X rays, gamma rays, and subatomic particles (called cosmic rays). The satellites shown in Figure 5.18 are each about 6 meters long and weigh approximately 3000 kilograms. Instruments aboard these satellites were designed to search the sky for discrete and diffuse background sources of X rays and gamma rays, to measure their total energy output and how that varies with wavelength, and to measure the ranges of energy, the composition, and the numbers of cosmic rays.

Proposed for launching in 1987 is a follow-up mission to *Einstein Observatory (HEAO-2),* the *Advanced X-ray Astrophysics Facility (AXAF)* shown in Figure 5.18. The satellite will be put into orbit from the *Space Shuttle,* which will also service it and eventually retrieve it for modernization. *AXAF*'s 1.2-meter grazing-incidence telescope will have 4 times the spatial resolution and at least 20 times the X-ray photon-collecting power of the X-ray telescope in the *Einstein Observatory.* This space vehicle represents as big an advance for the X-ray region of the electromagnetic spectrum as *Space Telescope* does for the visible and ultraviolet. Beyond *AXAF* an even larger telescope facility is in the planning process. To be known as the *Large Area Modular Array (LAMAR) X-ray* telescope, it will have 10 times the sensitivity to X rays that *ASAF* does.

GAMMA-RAY DEVICES

This new field of astronomy has burgeoned in the last decade or so from modest beginnings, employing balloons and rockets, to satellites (European and American) carrying highly sensitive gamma-ray detectors. Like that of X rays, the great penetrating power of gamma-ray photons makes observation and detection different from those for visible photons. Gamma-ray photons carry the highest energy of any photon. The primary gamma-ray detector used for space astronomy is a crystalline material that absorbs the gamma-ray photon, converting its energy to a flash of visible light. The visible photons can then be detected by a photoelectric device.

In 1978 came the first evidence of several gamma-ray spectral lines. The search for additional discrete lines will be greatly extended with the launch of the *Gamma Ray Observatory* in 1985. The *GRO* satellite will allow astronomers to move from survey activities, which we shall discuss in Chapter 16, to a detailed study stage.

The universe is highly transparent to gamma rays because of their great penetrating power. One would therefore expect them to be capable of carrying the imprint of their origin from far-distant places because they pass so easily through interstellar matter. Thus the unique penetration of gamma rays reveals directional and temporal information about their origin in regions that are too dense for visible photons and even X rays to penetrate. Gamma rays serve as a probe to provide us with new insights into the structure of the cosmos.

SUMMARY

Optical telescopes. Optical telescopes collect and focus electromagnetic radiation in the range of 3000 to 10,000 angstroms reaching the surface of the earth (optical window). In optical astronomy, astronomers work with an image of the celestial light source. The image may be formed by refracting lenses or by reflecting mirrors. In either case, important properties of the image are size, brightness, and resolution of adjacent points. The properties depend in different ways on the aperture and focal length of the objective. Turbulence in the earth's lower atmosphere blurs and distorts a star's image, causing it to scintillate. Other atmospheric conditions hamper our view of the heavens: weather conditions, atmospheric pollution, airglow, and the refraction of light passing through the atmosphere.

Reflecting and refracting telescopes. Telescopes that use lenses for the objective are known as refracting telescopes. Telescopes that use mirrors are known as reflecting telescopes. Modern reflectors are arranged so that images can be formed at three different focal positions: the prime, the Cassegrain, and the coudé foci. Each provides certain advantages depending on the brightness of the celestial object and the type of data to be obtained.

Large modern telescopes are all reflecting telescopes. They are free from chromatic aberration. They may be supported by their edges and from the back unlike lenses. Furthermore, reflection occurs from the front surface, decreasing the quality requirements on the mirror material.

Accessory instruments. Several types of radiation detectors, such as photographic emulsions and photoelectric devices, are attached to optical telescopes to analyze light in particular wavelengths. The most important properties of the radiation detector are its wavelength sensitivity, response over the wavelength interval, and the nature and range of response. Accessory instruments are used to analyze the wavelength composition of the radiation (i.e., type of spectrum) or the amount of energy in the radiation (i.e., photometry).

Radio telescopes. Radio astronomy developed rapidly after the late 1940s. Devices to detect and locate precisely sources of radio radiation disclosed that celestial objects radiate energy in the radio portion of the electromagnetic spectrum as well as in the visible light portion. The collecting element of a radio telescope is a parabolic metal dish. The "image" is a plot of the brightness of radio radiation at particular wavelengths as the telescope scans the source of radio radiation.

Ultraviolet, X-ray, and gamma-ray radiation. Telescopes, analyzing instruments, and radiation detectors are available to study cosmic radiation in the ultraviolet, X-ray, and gamma-ray portions of the electromagnetic spectrum. Because earth's atmosphere absorbs radiation in these wavelengths, such radiation studies must be carried outside the atmosphere by balloons, high flying aircraft, rockets, or satellites.

KEY TERMS

altazimuth mounting
aperture
atmospheric refraction
Cassegrain focus
coudé focus
equatorial mounting
focal length
image
interferometry
objective

photoelectric device
photographic emulsion
photometer
radiation detector
radio telescope
reflecting telescope
refracting telescope
resolving power
seeing
spectrograph

CLASS DISCUSSION

1. Which focal position on a large reflecting telescope should be used to obtain the largest picture of a planet? An image of a faint star? A picture of two stars close together in a binary system?

2. The night sky was clear and dark, but after developing the photographic plate, an astronomer noticed that the star images were unusually large and blurry. What is the probable explanation?

3. If space for storing information is an important consideration, which is the more economical, a photographic plate or a photoelectric device? Which is better for quick study? Which is more advantageous for detailed numerical study and why?

4. Why doesn't the parabolic shape of a large radio dish have to be so exact as the parabolic shape of the mirror in a large reflector?

5. Are astronomers able to do spectroscopic studies in the microwave and radio regions of the electromagnetic spectrum as they can in the visible? If so, how do you suppose they accomplish it?

6. What problems, if any, does the astronomer face when designing telescopes, accessory instruments, and radiation detectors for satellites to be placed in orbit about the earth?

READING REVIEW

1. What is meant by the "resolving power" of a telescope?

2. When astronomers refer to good seeing or bad seeing, what are they talking about?

3. What is chromatic aberration?

4. What are some of the principle advantages of the reflecting telescope over the refractor?

5. What is the primary advantage of a coudé focal position over any other focal arrangement?

6. What are the properties of importance when comparing different radiation detectors?

7. In a spectrograph what does the dispersing device do? What are the two primary types of dispersing devices?

8. How can astronomers improve the spatial resolution of a radio telescope? Is there more than one means?

9. Do radio telescopes produce a picture like that of an optical telescope? If not, what form do the results take?

10. Are ultraviolet, X-ray, and gamma-ray telescopes basically like telescopes designed to operate with visible light? If not, how do they differ? What about accessory instruments and radiation detectors?

PROBLEMS

1. If a new-generation reflector with an aperture of 25 meters were someday to be built and if its focal ratio for the prime focus were to be the same as that of the 5.1-meter Hale telescope on Mount Palomar, at the prime focus how much brighter an image of a galaxy than that of the Hale reflector will it produce? How much brighter if it has the same focal length at the prime focus as the Hale telescope? (See Figure 5.8.)

2. What aperture telescope is necessary to just resolve in green light two stars in a binary 0.2 arc seconds apart?

3. Using the focal ratios given in Figure 5.8 for the prime and Cassegrain focuses of the 5.1-meter Hale telescope, compare the size of an image of a galaxy formed at the Cassegrain focus to that formed at the prime focus. How do the image brightnesses compare?

4. For a small refracting telescope the eyepiece acts to magnify the image formed by the objective lens. Therefore, the magnifying power of the telescope is the ratio of the focal length of the objective to that of the eyepiece. What magnifications are possible for a 4-inch telescope with a 60-inch focal length, using eyepieces whose focal lengths are 1 inch, 0.5 inch, and 0.25 inch?

SELECTED READINGS

Asimov, I. *Eyes on the Universe: A History of the Telescope.* Houghton Mifflin, 1975.

Hey, J. S. *The Evolution of Radio Astronomy.* Neale Watson Academic Publications, 1975.

Kirby-Smith, H. T. *U.S. Observatories: A Directory and Travel Guide.* Van Nostrand-Reinhold, 1976.

Page, T., and L. W. Page. *Space Science and Astronomy.* Macmillan, 1976.

Robinson, L. J. "Monster Mirrors and Telescopes," *Sky and Telescope,* June 1980.

6

The Solar System

Throughout the first five chapters of this book, we have primarily been laying a foundation for our exploration of the cosmos. The first unit in the cosmos for us to study is logically the solar system, for it is the sun and its attendant planets among which our home, the earth, is located. Yet for purposes of study there is an awkward mixture of bodies in the solar system: For example, the sun is a star—a very different kind of body from planets, or satellites, or asteroids, or comets; therefore we shall delay a detailed discussion of the sun and instead use it to introduce our discussion of stars. In Chapters 6 to 10 we concentrate on the other occupants of the solar system, asking what they are like, reasons for their being that way, and what reasonable scenarios exist for their evolution.

This chapter will acquaint you with some of the facts and figures about the solar system. We say "figures" because words alone are inadequate to describe the full range of the cosmos from the smallest subatomic particle to the largest supercluster of galaxies: Words define common human experience, which

◄ Artist's illustration of the flight of the *Voyager 2* spacecraft to Jupiter and Saturn. The 1800-pound spacecraft, after leaving Saturn, is journeying on to Uranus and possibly Neptune for a complete survey of all the Jovian planets.

does not include direct contact with such entities as other planets or stars. So scientists use numbers to compare objects, and in this chapter we try to give you some sense of the scale of the solar system and the things in it.

6.1
Surveying the Solar System

SCALE OF THE SOLAR SYSTEM:
PLANETARY ORBITS

The primary, standard, measuring scale of the solar system is based on the earth's average distance from the sun, which, as we have seen in Chapter 3, is known as the astronomical unit (1 astronomical unit is about 150 million kilometers).[1] In deriving the scale of the solar system, astronomers have employed several independent techniques, of which the most accurate is that of timing the round trip of a pulsed radio signal reflected from a planet. Combining this information

[1]Earth's distance from the sun ranges between 147,097,000 and 152,086,000 kilometers as described in Kepler's laws with a mean distance (semimajor axis) equal to 149,598,000 kilometers.

TABLE 6.1
Bode's Law of Planetary Distances

Planet	Titius-Bode Rule	Actual Distance from Sun (AU)	Spacing Ratio
Mercury	$0 \times 0.3 + 0.4 = 0.4$	0.39	
			1.3
Venus	$1 \times 0.3 + 0.4 = 0.7$	0.72	
			1.4
Earth	$2 \times 0.3 + 0.4 = 1.0$	1.00	
			1.5
Mars	$4 \times 0.3 + 0.4 = 1.6$	1.52	
			1.8
Minor planets[a]	$8 \times 0.3 + 0.4 = 2.8$	2.81 (average)	
			1.9
Jupiter	$16 \times 0.3 + 0.4 = 5.2$	5.20	
			1.8
Saturn	$32 \times 0.3 + 0.4 = 10.0$	9.54	
			2.0
Uranus[a]	$64 \times 0.3 + 0.4 = 19.6$	19.18	
			1.6
Neptune[a]	$128 \times 0.3 + 0.4 = 38.8$	30.07	
			1.3
Pluto[a]	$256 \times 0.3 + 0.4 = 77.2$	39.44	

[a]These bodies had not yet been discovered when the rule was formulated.

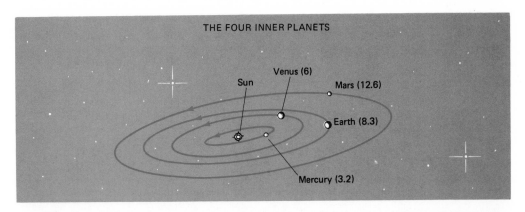

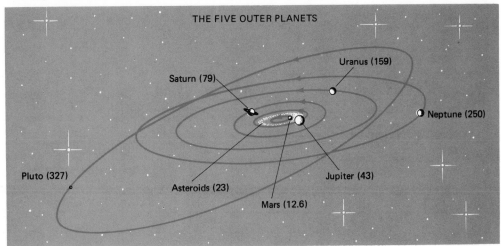

FIGURE 6.1

Orbits of the planets. The scale of the lower diagram is one-twentieth that of the upper diagram. The numbers give the planets' mean distances from the sun in light-minutes, the distance light travels in 1 minute of time, or about 1.8×10^7 kilometers.

with the planets' distances in astronomical units, from Kepler's third law, leads to the absolute size of the solar system in kilometers.

A German astronomer, Johann Bode (1747–1826), called attention in 1772 to a numerical scheme, or rule (originally discovered by Johann Titius (1729–1796) in 1766), that seemed to predict the mean distances of the then-known planets from the sun (Table 6.1). Although not a physical law in the same sense as Newton's laws, it is known as Bode's law; both Uranus, discovered in 1781, and the first asteroid, Ceres, found in 1801, adhered fairly well to this rule, but it broke down later when Neptune and Pluto were discovered. Despite the rule's having no unique physical basis, a similar rule relating the separations between planets seems to be characteristic of the formation of

bodies in the gravitational field of a star. We shall return to this point in Chapter 10.

All the planets are much alike in orbital characteristics. They revolve around the sun in the same direction in roughly circular orbits that lie nearly in the same plane (Figure 6.1). Mercury, the innermost planet, and Pluto, the outermost planet, depart most from this regularity. Between the terrestrial planets (Mercury, Venus, the earth, and Mars) the average spacing is much smaller than that separating the Jovian planets, Jupiter, Saturn, Uranus, and Neptune. The planets orbit at mean distances ranging from 40 percent of the earth's distance from the sun to 40 times earth's distance, with orbital periods between a quarter of a year and 248 years. Orbital data for the planets is given in Table 6.2.

PRECESSION OF MERCURY'S ORBIT: TESTING RELATIVITY THEORY

Einstein's relativity theory cleared up a long-standing problem that had plagued astronomers for decades. This was the moving of the perihelion point in Mercury's orbit in the direction of the planet's revolution around the sun, which is called the *precession* of Mercury's orbit. Although gravitational perturbations by the other planets are the primary cause (see Figure 6.2), astronomers observed that the orbit was precessing by 43 arc seconds per century more than Newtonian theory could explain. To account for the additional 43 seconds, they faced the unwanted choice of either increasing the mass of Venus by an inadmissible one-seventh or postulating the existence of a never-observed planet called "Vulcan" within Mercury's orbit. Einstein's theory of general relativity removed the difficulty in 1915.

The general theory of relativity predicts, as does the classical theory of Newton, that the orbit of a planet rotates in its own plane. In either theory the change in perihelion is most pronounced for Mercury's orbit since it is closest to the massive sun and has the most eccentric orbit. But Einstein's equations for the elliptic motion of a planet about the sun include a term not present in the Newtonian equations: Its contribution is a tiny fraction, the 43 arc seconds per century, of the total orbital precession. This additional 43 arc seconds adds up to one extra revolution in 3 million years. Looked at relativistically, the planet's eccentric orbital motion periodically moves it into stronger and weaker gravitational fields, where it encounters a different space-time structure.

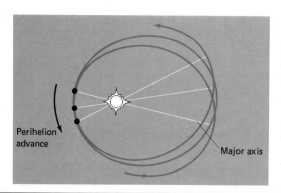

FIGURE 6.2
Precession of the major axis of Mercury's orbit. The observed advance in the perihelion of Mercury's orbit is 573 arc seconds per century. All but 43 can be accounted for by the perturbing influence of the other planets, as in Newtonian theory.

Perihelion advance

Major axis

THE SUN

What is it about the sun that forces planets, asteroids, comets, and other bodies of the solar system to orbit around it? It is its immense gravitational reach since the sun contains 99.86 percent of the solar system's mass. (The natural satellites of the planets, on the other hand, are gravitationally bound to their parent planets because of their closeness to them.)

Like other stars, the sun is a gas from center to surface and generates its radiant energy deep within its hot interior. This giant gaseous sphere has a radius 109 times greater and a mass 333,000 times greater than that of the earth. The sun's family of planets thus intercepts only a minute amount of the radiation that streams from the sun, flooding the solar system.

In addition to the steady emission of radiant energy there are numerous transient phenomena occurring in the sun's outer layers, such as sunspots, plages, flares, prominences, and coronal holes. Associated with these is a flow of protons, electrons, other atomic particles, and magnetic fields out through the orbital planes of the planets. These so-called solar-wind par-

ticles and magnetic fields impinge upon the planets and their magnetic fields, producing a variety of phenomena, such as the earth's aurora (northern and southern lights).

THE PLANETS

Outside the sun the nine planets contain the next important share of the mass of the solar system (see Figure 6.3). Although the planets had a common origin, they have significant chemical, physical, and geological differences. Such diversity stems mostly from their different masses and distances from the sun. At the time of the planets' formation these factors determined the ability to retain matter and defined the chemical composition of that matter.

The matter composing the planets, their satellites, and the minor members of the solar system can be roughly divided into three broad classes on the basis of the ease with which it will vaporize. Those materials that are solids for temperatures less than about 2000 K are the *rocky materials,* such as iron, magnesium, and their oxides and silicates. The second class is the *icy materials,* which can remain solid only up to temperatures of a few hundred degrees Kelvin. Examples are the ices of water, carbon dioxide, ammonia, and methane. The remaining class consists of those molecules that are gases down to almost absolute zero, such as hydrogen and helium.

On the other hand, we know of sufficient chemical and physical similarities among the planets to enable us to divide them into two well-defined categories, summarized in Table 6.3.

One category is an inner group composed mostly of rocky material, the *terrestrial planets*: Mercury, Venus, the earth, and Mars. Since the moon is not significantly smaller than Mercury, many astronomers include the moon among the terrestrial planets. Though like the moon in size, Pluto should have an icy composition more like that of comets. Thus it does not belong to the terrestrials, and it is not like the second group either.

The second group, the *Jovian planets*—Jupiter, Saturn, Uranus, and Neptune—are farther from the

sun; these planets are larger and consist mainly of the lighter elements, primarily hydrogen and helium, the most abundant elements in the universe. Jupiter and Saturn apparently have the same chemical composition as the sun. Uranus and Neptune seem to have less hydrogen and helium and presumably more of the icy materials (frozen gases such as water, ammonia, methane, and carbon dioxide). Table 6.4 contains a summary of the specific physical properties of the planets.

> All around me:
>   planet, moon, sun, riverbed, marsh:
>   grew out of cataclysms galore;
> nothing ever sprang whole, stays put.
> I feel the earth beneath my feet
>   suddenly shale away. . . .
>
> Diane Ackerman

TABLE 6.2
Planetary Orbital Data

Data	Mercury	Venus	Earth	Mars	Jupiter	Saturn	Uranus	Neptune	Pluto
Mean distance from sun (AU)	0.39	0.72	1.00	1.52	5.20	9.54	19.18	30.06	39.44
$(10^6$ km)	57.9	108.2	149.6	227.9	778.3	1427.0	2869.6	4496.6	5900
Minimum distance from sun (AU)	0.31	0.72	0.98	1.38	4.95	9.00	18.28	29.79	29.59
$(10^6$ km)	45.9	107.4	147.1	206.7	740.9	1347	2735	4456	4425
Maximum distance from sun (AU)	0.47	0.73	1.02	1.67	5.45	10.07	20.08	30.33	49.30
$(10^6$ km)	69.7	109.0	152.1	249.1	815.7	1507	3004	4537	7375
Sidereal period (yr)	0.241	0.615	1.00	1.88	11.86	29.46	84.01	164.79	247.69
Synodic period (d)	115.9	583.9	—	779.9	398.9	378.1	369.7	367.5	366.7
Inclination of orbit to ecliptic plane[a] (degrees)	7.00	3.39	0.0	1.85	1.30	2.49	0.77	1.77	17.2
Orbital eccentricity[a]	0.206	0.007	0.017	0.093	0.049	0.056	0.047	0.009	0.250
Average orbital velocity (km/s)	47.89	35.03	29.79	24.13	13.06	9.64	6.81	5.43	4.74

[a]These are two of the six parameters that uniquely define the planet's elliptic orbit. A third gives the angle measured eastward along the ecliptic plane from the vernal equinox to the ascending node of the planet's orbit; a fourth, the angle between the ascending node and the perihelion point measured in the direction of the planet's motion; a fifth, the date of closest approach to the sun (time of perihelion passage); and the sixth, the semimajor axis (mean distance from the sun), or the sidereal period, which is related to the mean distance by Kepler's third law.

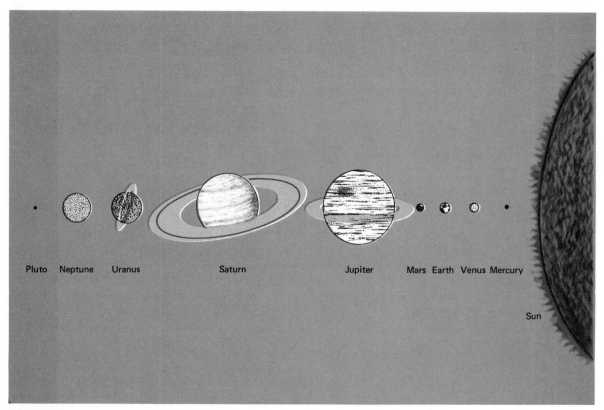

FIGURE 6.3
Relative sizes of the planets and the sun. This is a graphic illustration of the immensity of the sun compared with the planets and in turn the great size of the Jovian planets over that of the terrestrial planets.

TABLE 6.3
Terrestrial and Jovian Planets Compared

Average Characteristic	Terrestrial	Jovian
Distance from sun (AU)	0.9	16
Sidereal period	Hundreds of days	Tens of years
Spacing between members (AU)	0.4	8
Mass (earth = 1)	0.5	110
Radius (earth = 1)	0.7	7
Density (g/cm³)	5.0	1.2
Rotation period	Slow, in days	Rapid, in hours
Shape	Spherical	Oblate
Temperature	430 K	85 K
Atmosphere	Thin; CO_2, N_2, O_2	Thick; H_2, He, CH_4, NH_3
Surface	Cratered,[a] volcanic	None
Interior composition	Iron, silicates, oxides	Hydrogen, helium
Satellites	0.5	11

[a]Except the earth, whose extensive tectonic activity has obliterated its early cratered history.

TABLE 6.4
Planetary Physical Data

	Mercury	Venus	Earth	Mars	Jupiter	Saturn	Uranus	Neptune	Pluto
Equatorial radius (earth = 1), (km)	0.38 2439	0.95 6050	1.00 6378	0.53 3348	11.27 71,900	9.44 60,200	4.10 26,145	3.88 24,750	0.25? 1,500?
Mass (earth = 1)[a]	0.056	0.815	1.00	0.107	317.9	95.1	14.5	17.2	0.002?
Mean density (g/cm³)	5.42	5.25	5.52	3.94	1.31	0.69	1.19	1.66	1.0?
Rotation period[b]	58.65 d	−243.01 d	23.93 hr	24.62 hr	9.84 hr	10.23 hr	−15.5 hr	15.8 hr	6.4 d
Surface gravity[c] (earth = 1)	0.39	0.88	1.00	0.38	2.34	0.93	0.79	1.12	0.4?
Inclination of axis to orbital plane (degrees)	0.0	2.0	23.5	24.0	3.1	29.0	97.9	28.8	>50
Oblateness	0?	0?	0.003	0.008	0.065	0.091	0.025	0.020	?
Velocity of escape (km/s)	4.26	10.30	11.18	5.02	59.6	35.6	21.2	23.6	5.3?
Albedo[d]	0.06	0.72	0.39	0.16	0.70	0.75	0.90	0.82	0.3?
Solar energy received[e] (earth = 1)	6.7	1.91	1.00	0.043	0.037	0.011	0.003	0.001	0.0006
Magnetic-field strength	Weak	Not detectable	Moderate	Weak	Very strong	Moderate	?	?	?
Typical temperature (K)	440	730	288	218	120	90	60	60	50
Atmospheric gases observed (principal constituents)	He, H	CO_2, N_2	N_2, O_2, H_2O	CO_2, N_2, Ar	H_2, He	H_2, He	H_2, CH_4	H_2, CH_4	CH_4?
Number of known satellites	0	0	1	2	17	21	5	2	1

[a] If the planet has no satellite, the mass is found by mathematical analysis of gravitational disturbances. These disturbances would be the forces that the planet exerts on comets, asteroids, and space vehicles during near approaches. If the planet has satellites, the planetary mass may be found by applying Newton's modification of Kepler's third law (page 53) or from a spacecraft flyby. Mass of the earth = 5.98×10^{27} g.

[b] Minus sign indicates reverse or retrograde rotation.

[c] Surface gravity for earth = 978 cm/s².

[d] Fraction of incident sunlight reflected from the planet's surface or atmosphere.

[e] Solar energy received by earth = 1.36×10^6 erg/cm²·s.

THE SATELLITES

Of the 49 or so satellites all but four belong to the Jovian planets. It seems likely that more will be discovered in the future; and in fact *many* more may eventually be found since Jupiter and Saturn could gravitationally bind a lot of small bodies. Very small and faint satellites of the Jovian planets could easily escape detection by our present technology. Table 6.5 lists information about some of the larger natural satellites. Two of Jupiter's satellites, Ganymede and Callisto, and one of Saturn's satellites, Titan, are as large as, or larger than, Mercury.

Those satellites that are reasonably near their parent planet move in nearly circular orbits in the plane of their planet's equator and in the same direction as their planet rotates. The outer satellites usually have more eccentric orbits, which are more highly inclined to the equatorial plane of their planet. The four outer satellites of Jupiter, the most distant satellite of Saturn, and the inner satellite of Neptune have orbits that are reversed from the direction of their planet's rotation. It is possible that the reason for these differences is that the outer satellites were captured by the primaries after the planets and their inner satellite systems were formed.

RING SYSTEMS

One of the most exciting developments in planetary research in recent years has been the discovery of *ring systems* for Uranus and Jupiter. (Saturn's rings had been discovered with the introduction of the telescope into astronomy.) Rings are actually individual, small solid bodies in orbit about a planet in its equatorial plane. They are thus very small satellites of the planet. Uranus's and Jupiter's rings do not contain so many tiny satellites as do Saturn's; so they are much fainter than the ring system about Saturn and have managed to escape detection until recently. Although no ring system has yet been found for Neptune, it is possible that it also has a faint set of rings similar to those around the other three Jovian planets.

MINOR MEMBERS

The *asteroids*, or minor planets, are rocky bodies whose diameters vary from a few between 150 and 1000 kilometers down to thousands less than a kilometer across. Most asteroids are found between Mars and Jupiter, traveling around the sun in the same di-

> This world was once a fluid haze of light
> Till toward the center set the starry tides,
> And eddied into suns, that wheeling cast
> The planets.
>
> Alfred Tennyson

rection as the planets. However, many of them orbit the sun in the vicinity of the earth's orbit, with some in fairly elliptic orbits. In recent years the term "asteroid" has been expanded to include small objects, which are presumably not comets, located in the outer portion of the solar system. It is unlikely that their physical makeup is like that of those in the inner part of the solar system.

The *meteoroids* range in size from irregular solid bodies, called *meteorites* when they strike the ground, to tiny particles, called *meteors* if they merely flash through the atmosphere. As we go down the scale in size, the number of meteoroids increases very rapidly. They are composed of rocky material and are apparently related to asteroids and comets. All the meteoroids are satellites of the sun and are apparently moving in a wide variety of orbits, as best as we can determine.

Unlike the planetary bodies, most *comets* move around the sun in highly eccentric orbits with very long periods of revolution and at all angles of inclination. Some comets with short periods are regular visitors to the vicinity of the earth. The small masses of comets mean that it is possible for the larger planets, Jupiter in particular, to alter their orbits. Astronomers believe a comet to be a "dirty iceball" that is a conglomerate of icy materials mixed with rocky matter, while most of the asteroids and meteoroids are composed of a rocky material. The cometary composition is apparently characteristic of many bodies in the outer solar system. We shall have more to say in later chapters about the relationships of asteroids, meteoroids, comets, satellites, and planets to each other.

The *interplanetary medium* is primarily gas particles—mostly protons and electrons—that are ejected from the sun's atmosphere at several hundred kilometers per second. These subatomic particles form the *solar wind*. Some dust is there too, most of it being cometary debris. Despite huge numbers of gas and dust particles, interplanetary space has fewer

TABLE 6.5
The Larger Natural Satellites

Planet and Satellite	Average Distance from Planet's Center (10^3 km)	Period Revolution (d)	Radius (km)	Mass[a] (relative to moon)
Earth				
Moon	384	27.32	1,738	1.0
Jupiter				
Io	422	1.77	1,816	1.214
Europa	671	3.55	1,563	0.663
Ganymede	1,070	7.16	2,638	2.027
Callisto	1,880	16.69	2,410	1.463
Saturn				
Tethys	295	1.89	530	0.010
Dione	377	2.74	560	0.014
Rhea	527	4.52	765	0.034
Titan	1,222	15.95	2,575	1.849
Iapetus	3,561	79.33	730	0.026
Uranus				
Titania	439	8.71	840	0.016
Oberon	587	13.46	850	0.011
Neptune				
Triton	354	5.88	1,900	0.776

[a]Mass of moon = 7.35×10^{25} g.

bits of matter and is a better vacuum than can be made in a terrestrial laboratory.

6.2
Earth-Based Study of the Planets

These days astronomers do very little naked-eye study of the planets with a telescope. The primary function of a telescope is for direct photography, spectroscopic studies, or photometric work because these techniques leave a record that astronomers can study as needed. Also using a radiation detector like a photographic emulsion or a photoelectric device enables astronomers to make quantitative measurements, something generally not possible with eye observations. Planetary photographs have been collected from various observatories worldwide in an international center in France that now contains over 33,000 pictures.

PLANETARY PHOTOMETRY AND SPECTROSCOPY

As a reminder from Chapter 5, in photometry astronomers measure the amount of radiant energy, while in spectroscopy they study the composition by wavelength of white light. Photometric measurements provide information about the nature of reflecting materials, such as clouds in a planet's atmosphere or the surface of a planet or its satellite. Photography through color filters, which restricts the light to a narrow spectral region, and conventional spectroscopy give us clues about the chemical composition of a planet's surface, clouds, and atmosphere. Atmospheric constituents for the planet may be revealed by absorption lines or bands that are superimposed on the spectrum of the sunlight reflected from within the atmosphere, as shown in Figure 6.4. These planetary absorption features are sometimes difficult to separate in wavelength from absorption lines originating in the sun's atmosphere.

With Polaroid filters and other devices for measuring the polarization of light, astronomers can analyze a planet's surface and atmosphere by the manner in which reflected sunlight is polarized. To under-

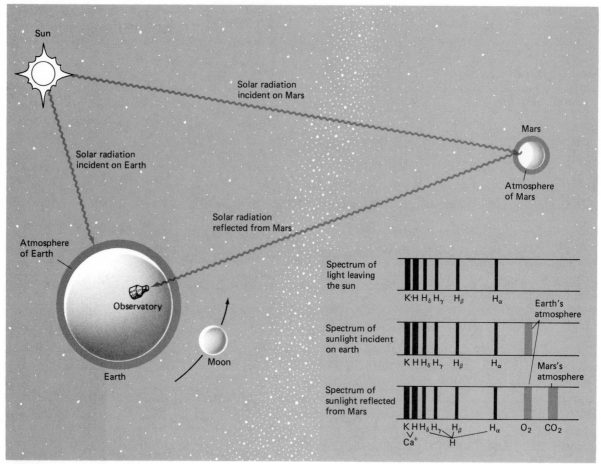

FIGURE 6.4

Reflection of sunlight from Mars. Incident solar radiation passes through Mars's atmosphere and is reflected toward the earth by its surface; the reflected solar radiation traverses the Martian atmosphere a second time on its way to the earth. At the lower right are idealized spectra for the solar radiation as it leaves the sun, after it has traversed the earth's atmosphere, and after it is reflected from Mars, having traversed the Martian atmosphere twice. The earth's atmosphere produces a molecular-oxygen (O_2) absorption band, while the Martian atmosphere produces a carbon dioxide (CO_2) absorption band.

stand the significance of *polarization* of light in this context, you should remember what happens when sunlight reflected from, say, a car windshield is observed through Polaroid sunglasses. The Polaroid lens does not transmit all the reflected light that is polarized, and thus the glare is less. From this example we can go on to study reflection from a variety of surfaces to show that the degree of polarization of reflected light is indicative of the nature of the reflecting material, such as a planet's surface.

THERMAL AND NONTHERMAL RADIATION

Thermal radiation from the planets, which is due to the fact that the planets are hot, can be studied far into the infrared with today's heat-sensing instruments. This radiation is blackbody radiation, caused by the random thermal motion of the particles that compose the outer parts of the planets. Thus astronomers may apply Planck's law, the Stefan-Boltzmann law, and Wien's law to such radiation. The data obtained from infrared radiation provide important information on surface and atmospheric temperatures and, indirectly, chemical composition.

Nonthermal radiation is radiation due to physical processes other than that involved in producing blackbody radiation (page 80). That is, it owes its ex-

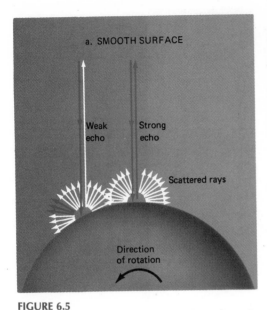

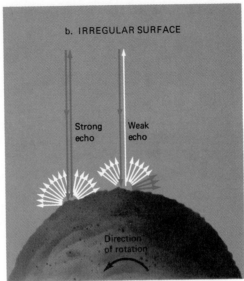

FIGURE 6.5
Mapping by radar. (a) The signals are shown bouncing off a smooth surface. Only the center of the curved surface reflects a strong echo (colored arrow) back to earth. Where the surface curves away, the rays bounce off at an angle so that the planet's edges return only weak signals to the earth. (b) If, however, the radar rays bounce off irregularities, such as mountains, the patterns are changed. In what is otherwise an area of weak echoes strong echoes bounce back from a surface projecting outward, as shown on the left of the diagram. Radar rays striking a surface projection at an oblique angle, as shown to the right, bounce their strongest signals off to one side, and only weak scattered signals return to the earth. Also, if the planet is rotating about an axis perpendicular to the diagram, there will be a Doppler shift to longer wavelengths of the signal reflected from the surface on the left side, no shift for the center, and a Doppler shift to shorter wavelengths on the right side.

planation to some other fact than that the planet is hot. For example the light produced in lightning is nonthermal radiation. The planets' normal thermal radiation and any nonthermal radiation present are often observable with radio telescopes in the regions of millimeter, centimeter, and meter wavelengths.

RADAR MAPPING OF PLANETARY SURFACES
Pulsed signals sent and received by radar have given us relief maps of Mercury, Venus, and Mars. This is important and somewhat remarkable for Venus since a thick cloudy atmosphere prevents visual obser-

vations of its surface. The signals are reflected from the planet's surface and return to the earth (or an orbiting satellite) as an echo modified by the surface terrain (Figure 6.5). The returning wave train from different portions of the surface is analyzed by a computer and then can be used to construct a topographic map of the planet's surface (see Figure 8.4).

Analyzing the Doppler shift in the reflected radar signal caused by the rotation of the planet has given us the period of rotation for Mercury and Venus; it could not have been found for Venus by optical methods because of cloud cover. The radio wave reflected from the part of the surface that is approaching the earth has its wavelength slightly decreased because of the Doppler effect. The echo from the side of the planet that is receding from us is slightly increased in wavelength, as explained in Figure 6.5.

6.3
Overview of the Planets

As long as humanity has been aware of the sky and the objects in it, various cultures have recognized the five bright planets—Mercury, Venus, Mars, Jupiter, and Saturn—as something other than stars. The number of known planets had not changed throughout history until the invention of the telescope, after which Uranus was discovered in 1781, Neptune in 1846, and Pluto in 1930.

In this section the discussion of the planets focuses primarily on general features that were discovered during earth-based study. In Chapter 7 we shall cover in detail the earth and moon as an introduction to the terrestrial planets. Chapter 8 on the terrestrial planets and Chapter 9 on the Jovian planets are devoted to a comparative discussion that has been made possible by the space program. In a preliminary way we raise some points here about the planets whose details will be covered in the appropriate sections of Chapter 8 and 9.

MERCURY

Although one of the brighter objects in the heavens, most people have never seen Mercury. Mercury is difficult to study from the earth because it is so close to the sun. Its maximum angular separation (greatest elongation) is only 28° on either side of the sun. At this favorable position for viewing its phase corresponds to a quarter moon; the full phase occurs at superior conjunction, when Mercury is almost in line with the sun (see Figure 6.6). Swift orbital motion keeps the planet visible low above the horizon for only a few days each year, immediately after sundown or before sunup. It is best seen when it is a morning star during the early part of the year or an evening star during the last half of the year.

Of the terrestrial planets (excluding our moon) Mercury is the least massive, being about 6 percent of the mass of the earth, and smallest in size, being less than twice the size of the moon. However, next to the earth, it has the highest mean density, a point we shall come back to later.

Mercury's rotation period is two-thirds of its orbital period; thus the planet completes three rotations during two orbital revolutions. This synchronization of its rotation and revolution, like that between the earth and the moon, is not accidental; it was apparently set up by the strong tidal pull exerted by the sun. Different parts of the body of Mercury are at slightly different distances from the sun. Thus they experience slightly different accelerations from the sun's gravitational attraction. Such a differential gravitational effect is known as a *tidal force*, and it is responsible for having slowed the planet's rotation, trapping it so that the ratio of its rotation to revolution period is three to two. As a result, the sun takes 88 days after rising on

the eastern horizon to cross Mercury's sky and set on the western horizon; meanwhile the planet completes one orbit of the sun.

VENUS

Venus, the second closest planet to the sun, is a yellow color; it is the third after the sun and moon in brightness in our night sky. Like Mercury, Venus goes through all the lunar phases, as Galileo first observed in 1609 (and shown in Figure 6.6).

Since it has a larger orbit than Mercury, Venus swings farther out from the sun as viewed from the earth, about 47°, or twice as far as Mercury. Venus remains visible as an evening star in the western sky or as a morning star in the eastern sky for weeks at a time. Although Venus comes slightly closer to the earth than Mars does, we cannot see features then because its dark hemisphere is turned toward us.

Venus's diameter, mass, and density are slightly less than those of the earth. Its mass is 80 or so percent of that of the earth. Venus possesses a mean density over 5 grams per cubic centimeter, suggesting that its internal structure is similar to that of the earth and Mercury.

The most striking feature about Venus is a cloud cover that totally hides the surface in visible and infrared radiation. The clouds themselves are almost totally featureless in the visible wavelengths, where the planet appears bland and featureless with a pale yellow color. In the ultraviolet portion of the electromagnetic spectrum the clouds possess features that we discuss in Chapter 8.

Venus's rotation was a mystery that eluded solution by optical or spectroscopic observations because of this cloud cover and the planet's slow rate of rotation. But Doppler shifts noted in radar observations solved the problem. The planet rotates in a retrograde direction, with its axis of rotation inclined only 2° from the perpendicular to its orbital plane. (*Retrograde* here means a direction of rotation reversed from that of revolution about the sun.) The period of rotation as determined from radar measurements is 243 days, 18 days longer than its orbital period. Because its revolution period is about 225 days, just slightly shorter than the rotation period, the Venusian day is 117 days long, with 58.5 days of sunlight and 58.5 days of darkness. Thus the sun rises on the western horizon and sets approximately twice during the Venusian orbit with respect to the earth.

MARS

Mars has such an eccentric orbit that the closest approach at opposition between earth and Mars comes every 15 to 17 years, when Mars is near perihelion in its orbit. At the last favorable opposition in August, 1971, Mars came within 60 million kilometers of the earth. At a time like that even the most casual observers of the heavens are struck by the planet's brilliant ruddy color, far outshining the brightest stars. The next favorable opposition will be in August, 1988.

Mars is a little more than half the earth's size, has about 11 percent of the earth's mass, and therefore has a mean density 75 percent of that of the earth. The Martian day lasts 24 hours and 37.4 minutes. Its axis of rotation tilts by 24° from the perpendicular to its orbital plane, giving the red planet seasons like those of the earth; but they last twice as long because the Martian orbital period is nearly 2 years.

Visual observers have made countless detailed maps of the numerous and varied surface features of Mars. Two observers are especially notable: the Italian astronomer Giovanni Schiaparelli (1835–1910) and the American astronomer Percival Lowell who mapped and named many Martian features. Through a telescope the red planet appears to have earthlike characteristics, such as white polar caps and large dark areas, which vary with the Martian seasons.

The recognition that Mars has polar caps dates back to the early 1800s. When Mars is closest to the sun, the south pole is inclined toward it. As a result the large southern polar cap recedes during the Martian summer, leaving behind a small residual cap or on some occasions none at all that can be seen from the earth. On the other side of the orbit the north pole is inclined toward the sun when the planet is farthest from the sun. The residual polar cap never quite disappears during the Martian summer in the northern hemisphere.

Mars has also for some time been known to have an atmosphere where vast yellow dust storms occur. Once or twice a Martian year (about 2 earth years) a storm grows to global proportions, enveloping almost the entire planet in a dense shroud of dust. White clouds in the Martian atmosphere have also been observed from the earth.

Although these various characteristics observed from here suggested great similarities to our planet, spacecraft revealed a surface topography that is as much like the moon as it is the earth. This quickly put

FIGURE 6.6
Phases of Mercury and Venus. (left) As defined in Figure 2.8, Mercury and Venus proceed during their orbiting of the sun through a succession of elongations that have special names. Accompanying this is a succession of phases. (right) Venus photographed with a constant magnification at various phases. At crescent phase Venus is fully six times larger than at full phase and shows its greatest brilliance.

FIVE PHASES OF VENUS

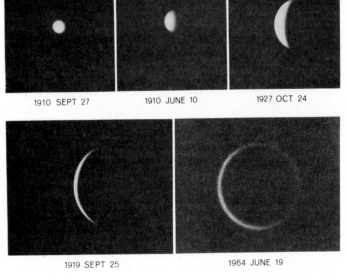

1910 SEPT 27   1910 JUNE 10   1927 OCT 24

1919 SEPT 25   1964 JUNE 19

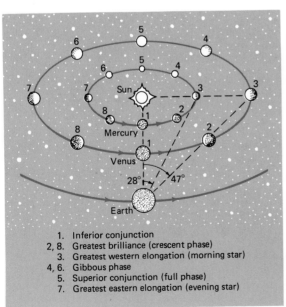

1. Inferior conjunction
2, 8. Greatest brilliance (crescent phase)
3. Greatest western elongation (morning star)
4, 6. Gibbous phase
5. Superior conjunction (full phase)
7. Greatest eastern elongation (evening star)

to rest the many long-running romantic notions that seasonal changes were due to vegetation and the inference that Mars might thus be the home of intelligent life.

FIGURE 6.7
Jupiter. (above) A direct photograph of Jupiter in red light taken with the 5.1-meter Hale reflector on Palomar Mountain. (below) An illustration of the effect of differential rotation on Jupiter. In the left figure an imaginary line that parallels a meridian of longitude is moving at a faster rate at the equator than near the poles, as shown by the length of the arrows. The center and right figures show what happens to the line after one rotation and two rotations, respectively.

JUPITER

Fifth planet from the sun, Jupiter is the largest and most massive of the planets in the solar system. In our night sky it glows with a bright, steady yellow light, outshining the stars. The mean diameter of Jupiter is about 11 times greater than the earth's, and Jupiter is more than 1000 times larger in volume than the earth, as one can see in Figure 6.7. However, Jupiter's mass is barely more than 300 times that of the earth even though it exceeds the combined masses of all the other bodies orbiting the sun. Thus its mean density is about one-fourth that of the earth. Because its axis is tilted only 3° from the perpendicular to its orbital plane, the planet has little seasonal change.

Not all portions of the visible layers of Jupiter, which appear as alternating dark and light bands parallel to the equator, rotate in unison. The equatorial region completes its rotation several minutes sooner than adjacent higher latitudes. This phenomenon is known as *differential rotation* and is possible in fluid media, such as gases. It is not something one expects a solid body, like the surface of a terrestrial planet, to do. Figure 6.7b shows the consequences of differential rotation to the outer layers. Jupiter's rapid 10-hour rotation and low density combine to flatten the planet about 6 percent in its polar diameter. This is again more characteristic of a fluid body that will readily deform than of a solid body that does not easily flow. The dark-band structure is composed of reddish and brown shades with irregular patches of gray, blue, and white clouds. The light zones are primarily yellow

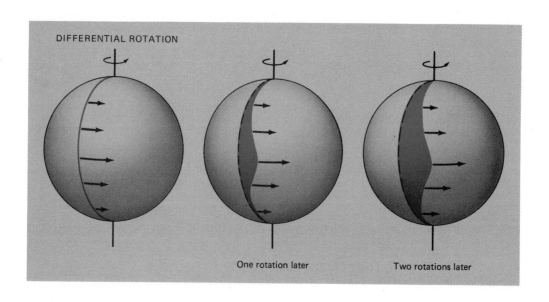

DIFFERENTIAL ROTATION

One rotation later   Two rotations later

FIGURE 6.8
Changing views of Saturn's rings. The rings are inclined 29° to the plane of Saturn's orbit.

in color. The entire band structure is constantly undergoing changes in color and intensity. Clearly what we are viewing are clouds in Jupiter's atmosphere and not a solid surface such as the terrestrial planets have. Most striking of all the atmospheric features is the Great Red Spot, which has been observed for at least 300 years. It is immense, being about four times the size of the earth.

In the early days of radio astronomy Jupiter was found to be an intense source of radio radiation. If this radiation were just part of the planet's thermal radiation, then Jupiter would have to be extraordinarily hot. Since it is not, the radiation must be due to nonthermal processes, such as free electrons spiraling about magnetic lines of force.

SATURN

Its rings make Saturn, sixth planet from the sun, one of the most remarkable objects in the heavens. Brighter than all the stars except Sirius and Canopus, it shines with a steady ashen color. Saturn is second among the planets in mass and size. The mean diameter of Saturn minus its ring system, is almost 10

times that of the earth, and its mass is about 100 times greater. Its density is the lowest of any planet, 0.7 times that of water. The small mean density leads to the often-quoted observation that, if you could find a lake large enough, Saturn would float in it, being lighter than water. Rapid rotation (a rotation period of a little over 10 hours) and an unusually low density give it more polar flattening than any other planet, about 11 percent.

Saturn is twice as far from us as Jupiter, but the markings that we can see on the noticeably flattened disk of Saturn faintly resemble the banded cloud structure of Jupiter's atmosphere (see Figure 6.8). The coloration is more restrained, and the details are less distinct. On rare occasions a bright spot may appear. Thus, as in the case of Jupiter, Saturn is a fluidlike body rather than a solid like a terrestrial planet. As for Jupiter, astronomers have detected weak radio emission in low-frequency bursts that are synchronized with Saturn's 10.2-hour rotation period.

Saturn's axis of rotation is inclined by 29° to its orbital plane. Since the plane of its rings is perpendicular to its rotation axis, the rings do not lie in the orbital plane and therefore present a varying aspect to

> It has generally been supposed that it was a lucky accident that brought this new star [Uranus] to my view; this is an evident mistake. In the regular manner I examined every star of the heavens, not only of that magnitude but many far inferior, it was that night its turn to be discovered. . . .
>
> William Herschel

the earth as the planet goes through its roughly 30-year orbital period. Figure 6.8 also shows geometrically how this occurs, along with photographs of the planet with its rings seen at four different angles to the line of sight. When seen almost edgewise, every 15 or so years, the rings almost disappear from sight, indicating that they are very thin compared to their radius. Most of Saturn's satellites orbit in the same plane as the rings, the planet's equatorial plane, and orbit outside the rings. As was shown in Table 6.5, Titan is one of the most massive satellites in the solar system, and it is one that has been known for some time to possess an atmosphere.

URANUS

"In examining the small stars in the neighborhood of H Geminorum I perceived one that appeared larger than the rest; being struck with its uncommon appearance . . . I suspected it to be a comet." So wrote William Herschel on the night of March 13, 1781, in his observing journal. Herschel and other astronomers first believed the newfound object to be a comet and vainly tried to derive a cometary orbit for it. It was almost a year before they realized that this was a new planet.

Uranus has a radius 4 times larger than that of the earth and a mass almost 15 times greater. Although Uranus is somewhat larger than the more distant Neptune, it is less massive by about 15 percent. Consequently it has a smaller mean density than Neptune but larger than that of Saturn. Its average density is slightly higher than that of water. In a large telescope the slightly flattened disk is a light apple green in color.

Nearly 3 billion kilometers from the earth, Uranus presents an almost featureless appearance. Although a few atmospheric features have been reported, none have been confirmed. As with Jupiter and Saturn, we are probably seeing clouds in its atmosphere rather than a solid surface.

Uranus's rotation has a peculiarity. Its axis is tilted 98° to the perpendicular of its orbital plane—that is, it lies on its "side," so that we see it rotate in the reverse direction barely. For Uranus the retrograde rotation is due to the peculiar inclination of the axis, while for Venus it is a true reverse rotation. When its axis is in our line of sight every 42 years (half the sidereal period), we observe either its sunlit northern or southern hemisphere, while the opposite hemisphere is dark. One-quarter or three-quarters of its period later (21 years or 63 years), its axis is at right angles to our line of sight, and we observe both the northern and southern hemispheres, as shown in Figure 6.9.

The two brightest satellites of Uranus, Titania and Oberon, were discovered by William Herschel in 1787, only 6 years after he discovered the planet itself. In all, the planet has five known satellites, which are visible in the infrared photograph in Figure 6.10. All the satellites move in nearly circular orbits that lie close to the equatorial plane of Uranus, the same plane as the ring system, and well outside the rings. In these respects Uranus is similar to Saturn. The ring system was accidentally discovered in 1977 from observations from an airborne telescope that was being used to remeasure Uranus's diameter and study its atmosphere as the planet passed over a background star. The expectations of future findings produced by such accidental events as the discovery of Uranus's

FIGURE 6.9
Hemisphere views of Uranus. Every 42 years or so Uranus presents first one polar region and then the other to the earth.

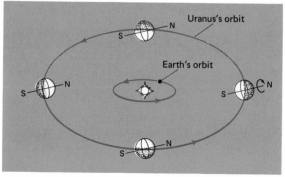

132

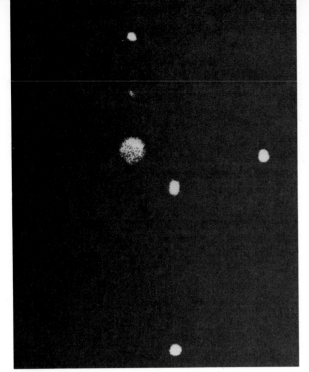

FIGURE 6.10
Uranus and its five known satellites. This infrared photograph from Mauna Kea Observatory in Hawaii shows the planet's large disk and its satellites, from top to bottom, Umbriel, Miranda, Oberon (right), Ariel (left, nearest Uranus), and Titania.

rings contribute to the excitement and allure of science among not only the general public but scientists as well.

NEPTUNE

After Uranus had accidentally been discovered, astronomers were long perplexed that, even allowing for the perturbations of Jupiter and Saturn, Uranus's orbital motion was less predictable than that of the other planets.

The discrepancy was finally resolved in 1845 and 1846 by two astronomers, John Adams (1819–1892) in England and Urbain Leverrier (1811–1877) in France. By a brilliant application of the law of gravitation they arrived independently at the conclusion that there must be a disturbing body beyond the orbit of Uranus.

Leverrier's results were communicated to Johann Galle (1812–1910) of the Berlin Observatory, who received the information on September 23, 1846. Within half an hour of searching Galle located the new planet among a group of eight stars whose positions had been charted on a recently prepared map. Recent historical research suggests that Galileo probably saw Neptune in December 1612 and January 1613, fully 233 years before Galle found it, but he did not recognize

that it was a planet. We know that Neptune passed extremely close to Jupiter during that time.

Looking at Neptune through a telescope, we see a slightly flattened, bluish-green, almost featureless disk. Observers at times have reported irregular, indistinct markings and a bright equatorial zone, although observations of the planet are very difficult to make and subject to some degree of doubt.

Neptune's diameter is about 3.5 times that of the earth. Its mass is 17 times greater and its mean density is one-third that of the earth. Neptune is about 30 AU from the sun or 30 times the earth's distance. Thus the angular diameter of the sun is one-thirtieth of what it is from the earth. For us the sun has an angular diameter of 0.5°, or 30 minutes of arc, so that from Neptune its angular diameter is 1 minute of arc. We have tried to illustrate this very pronounced difference in Figure 6.11. At a distance of 30 AU Neptune's orbital, or sidereal, period is almost 165 years. Thus it has yet to complete one orbit of the sun since its discovery in 1846.

The larger of Neptune's two satellites, Triton, was discovered less than a month after the planet itself. It orbits Neptune in about 6 days in a direction opposite to the planet's eastward rotation. The orbital plane in which Triton moves is inclined to the equatorial plane of Neptune. Triton appears to be somewhat larger than the moon, but its mass is only about 80 percent that of the moon, producing a lower mean density.

The smaller satellite, Nereid, takes nearly a year to swing around Neptune in a highly elongated ellipse, varying between about 1.5 million and almost 10 million kilometers from the planet. Nereid's orbital plane is also inclined to Neptune's equatorial plane. Neptune's satellites are distinctly different from those of Uranus. Their inclined orbital planes and Nereid's elongated orbit continue to prompt speculations on their origin.

PLUTO

Spurred by the success of the discovery of Neptune, astronomers searched for evidence for even more distant planets. Percival Lowell (1855–1916), founder of the observatory bearing his name in Flagstaff, Arizona, was convinced by his calculations begun in 1905 that minute discrepancies still complicated the orbit of Uranus. (Neptune had not been observed long enough to provide useful data.) He concluded that the irregularities might be caused by a planet beyond Neptune.

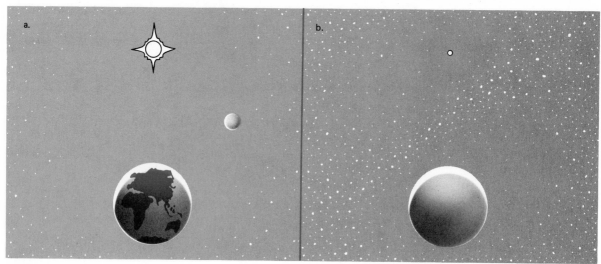

FIGURE 6.11
The sun seen from Neptune. In (a) is roughly the way the sun looks from slightly above and behind the earth, with the moon to the right of the earth-sun line. In (b) from slightly above and behind Neptune we see the sun more like a bright star than the object we see from the earth. Since the sun has an angular diameter of 30 minutes of arc as seen from the earth at 1 AU, the sun will have an angular diameter of 1 minute of arc as seen from Neptune, which is 30 AU from the sun. We have positioned ourselves approximately four times farther behind Neptune than we were for the earth so that they will have the same angular diameter (Neptune's diameter is almost four times that of the earth). Since we are outside the atmospheres of both planets, it would be possible to see stars in the vicinity of the sun.

Several years of intermittent and unproductive search passed. Then in January 1929, the Lowell Observatory acquired a 13-inch photographic refractor and put a young assistant, Clyde Tombaugh, to work on a new search. After a year of photographing star fields along the ecliptic and later all over the sky, Tombaugh made the historic find on photographs taken in January 1930.

Pluto's great distance from the sun (a semimajor axis of about 40 AU) and its very long sidereal period of almost 248 years have made it a difficult planet to study. At the time of its discovery it was estimated that Pluto was approximately earthlike in size and mean density. Since then, with a longer period for studying the planet, estimates of its size and mass have decreased.

It is by means of Pluto's gravitational attraction for Neptune and—to a lesser extent—Uranus that astronomers have tried to estimate its mass. This means that they must know its orbit, which turns out to be the oddball among the nine planets. Pluto's large eccentricity carries it as close to the sun as about 30 AU and as far away as almost 50 AU, a variation of almost 20

AU, or about 3 billion kilometers. Thus during a portion of its orbit it is closer to the sun than Neptune is. In fact Pluto has been closer to the sun than Neptune since the winter of 1978 and 1979 and will remain so until the spring of 1999. If you refer to Figure 6.1, you can imagine Pluto now moving along that portion of its orbit that is north of the plane of the ecliptic (on the right-hand side). Pluto reaches aphelion while it is south of the ecliptic plane (on the left-hand side). At that point it will be about 14 AU below the plane of the ecliptic, which is a greater distance than Saturn is from the sun. Pluto's crossing of Neptune's orbital plane is done well above or below it so that there is no chance for the two planets to collide.

Pluto's brightness varies slightly, presumably because sunlight is reflected unevenly from its surface—perhaps because of light and dark areas. Photoelectric observations of these variations reveal that the period of Pluto's rotation is 6.4 days. The nature of the brightness variation also suggests that Pluto's axis of rotation is inclined relative to its orbital plane by an angle in excess of 50°. There is a good chance that Pluto's satellite orbits in the equatorial plane of the planet. If so,

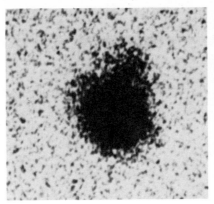

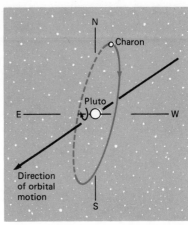

FIGURE 6.12
Presence of satellite orbiting close to Pluto. (right) The bulge at the top of the image of Pluto is probably a satellite, Charon. The speckled appearance of the photograph is due to the graininess of the photographic emulsion. (far right) Although Charon has not been clearly resolved, the diagram shows the most likely aspect of Charon's orbit.

then, as shown in Figure 6.12, Pluto's axis of rotation is almost in its orbital plane like that of Uranus.

The planet's small disk, measured with difficulty even by a large telescope, is less than one-fourth the earth's diameter. Pluto's mass is at most a few percent that of the earth's, which leads to a best guess for the mean density of about 1 gram per cubic centimeter. This prompts the belief by astronomers that Pluto is composed primarily of water in a solid form.

Recent infrared observations suggested a surface composition dominated by ices, primarily solid methane. Pluto's low surface gravity and extremely low temperature mean that its atmosphere is a tenuous one of methane and possibly neon; indeed, recent infrared observations have spectroscopically detected methane in Pluto's atmosphere. However, its observed reflectivity is not consistent with a prediction of methane ice, or frost, alone. That is, the dark patches are possibly a rocky, silicate material.

Pluto appears to have a satellite, Charon, as shown in Figure 6.12, which was taken in June, 1978. (The discovery was made during the examination of photographs taken as part of the routine task of refining data on the planet's orbital motion.) In Figure 6.12 the image of Pluto is elongated, with a bulge that is hardly a mountain on the surface. Although the photographs do not clearly resolve the satellite, its estimated diameter is around 900 kilometers.

If Pluto and its satellite have equal mean densities, the satellite is about 5 to 10 percent of the mass of Pluto, which would make it by far the largest satellite in the solar system in comparison with its parent planet (the moon is about 1.2 percent of the earth's mass). The satellite orbits Pluto in an approximately circular orbit at an estimated distance of 20,000 kilometers from the center of the planet. The orbital period is 6.4 days and appears to be inclined to Pluto's orbital plane, as Figure 6.12b shows.

Is it possible that even more remote planets lie beyond Pluto? As we try to show in Chapter 10, there is no physical reason that other objects like Pluto cannot exist. In fact for 13 years after the discovery of Pluto Tombaugh continued to search for more trans-Neptunian planets. His and subsequent efforts have shown that, if one or several are out there, they are extremely faint, and it is unlikely that they could be anything like the size of Jupiter.

6.4 Exploration of the Solar System

WHY SPACE EXPLORATION?
It is difficult in a few short paragraphs to do justice to the drama, excitement, and intellectual achievements of the space program. We have all become somewhat blasé about its accomplishments. But we should remember that only about 50 years have elapsed between the discovery of Pluto and *Voyager 1* and *2*'s close-up views of Jupiter and Saturn. It has been a little more than 20 years since the first Venus probes

> Just as Galileo could not have foreseen the advancement in our knowledge initiated by his discoveries of the four Jovian moons in 1610, neither can we fully comprehend the scientific heritage that our exploration of space is providing future generations.
>
> Edward C. Stone

by the Soviet Union in 1961 and the United States in 1962 propelled astronomy into the realm of physical exploration of the planets. Spacecraft have now visited all 5 naked-eye planets and one is on its way to the first of the "discovered" planets (Uranus). Also we have landed spacecraft on Mars and Venus and carried out experiments on their surfaces. We have brought back samples from the moon and could do so from Mars, asteroids, and comets. We have photographed and mapped the surface of Mars and with an orbiting radar system have mapped the surface of Venus through its veiling clouds. Although there are no immediate plans to do so, we could build and orbit about the earth a permanently manned space station from which to explore the solar system.

But what were the reasons that propelled the United States into space exploration? Of course the pursuit of knowledge was one. Unfortunately, to many people knowledge for its own sake is not a sufficient rationale for the expenditures that have been made in the space program. Basically, there are four dominant reasons[2]: knowledge, vision, applications, and national prestige.

The growth of *knowledge* concerning the solar system has been overwhelming, leading many observers of science to refer to the last 20 years of solar-system astronomy as its golden age. (The next three chapters will cover many of the new—in some cases startlingly new—findings.)

BEGINNINGS OF SPACE EXPLORATION

Concrete planning in space exploration that was to include planetary exploration can be said to have begun in the period 1957 to 1959. The first coherent plan for planetary exploration was produced by NASA's Jet Propulsion Laboratory in 1959. This plan contained the lunar-exploration programs *Ranger,* a hard-landing spacecraft, and *Surveyor,* a soft-landing craft. Then in 1961 came plans for the *Apollo* manned-landing program and the lunar-orbiter photomapping program. Plans for more ambitious ventures followed that would ultimately lead us to explore all the solar system. Some of these plans have been executed, but others have been dropped for various reasons. The results of many of these missions will be discussed in later chapters. But before we get to that, has there

been anything like a grand strategy for exploring the planets? The answer is that in a very real sense there has been.

GRAND STRATEGY FOR EXPLORING THE PLANETS

What can be considered a grand strategy for planetary exploration began to emerge in the 1960s. It was not, of course, conceived fully developed but has evolved from NASA-sponsored groups of scientist and engineers most from universities and from the scientists composing the Committee on Planetary and Lunar Exploration of the National Academy of Sciences Space Science Board. The basic concept is that of a four-stage program, each successive stage building on its predecessor. The four stages are reconnaissance, exploration, intensive study, and exploitation (or utilization).

By *reconnaissance* we mean the effort to characterize a planet. The reconnaissance phase has been completed for the terrestrial planets and will be partially completed for the Jovian planets after *Voyager 2* makes its close passage by Uranus and Neptune. In the *exploration* phase we hope to determine much more about the physical state of a planet and to derive some understanding of the physical processes that have shaped and are shaping its evolution.

The *Voyager 1* and *2* flybys are very sophisticated spacecraft so that they not only have performed a reconnaissance mission but also have taken a major step toward exploration of Jupiter, Saturn, and their satellites. *Pioneer Venus,* to be described in Chapter 8, launched the global-characterization program for Venus. Figure 6.13 shows some of the spacecraft used by NASA in planetary exploration. The *Viking* mission represents the exploration phase of Mars.

The third stage in planetary study is referred to as

---

[2]According to Noel W. Hinners, of the National Aeronautics and Space Administration.

> Solar system exploration must be viewed with two minds, one asking why it exists, and the other how it has been accomplished. To confuse the two, as frequently happens, makes it difficult to understand the problems of the future.
>
> Noel W. Hinners

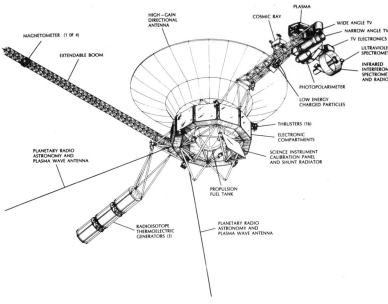

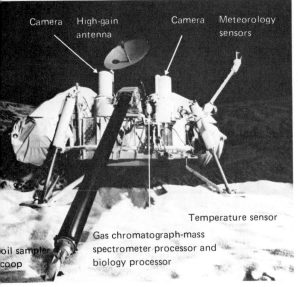

FIGURE 6.13
Space observatories for the exploration of the planets. (top left) An artist's conception of the *Pioneer Venus* orbiter and multiprobe as they approached Venus in 1978. (left) *Viking 1*'s lander showing its array of scientific instruments for the study of Mars' surface conditions. (above) An illustration of the *Voyager* spacecraft whose extensive studies of Jupiter and Saturn opened a new era in our understanding of the Jovian planets.

the *intensive-study* phase, which seeks "to define or refine the remaining scientific questions of the highest order that have been revealed by reconnaissance and exploration and that can be studied in depth."[3] These kinds of study are of course those done from onsite inspection (not necessarily manned). Only the moon and Mars have been subjected to intensive study—and this still in a limited fashion.

The final phase is probably not truly part of a scientific study. Known as *exploitation*, or *utilization*, or *comprehensive understanding*, this fourth phase yields scientific data as a by-product. It is a level that has been reached only for the earth although there are proposals and plans for establishing a permanent base or bases on the moon, which would include some for

[3]Noel W. Hinners.

mining and manufacturing. Besides the moon there are no plans for exploitation of any celestial body that go beyond the talking stage.

THE FUTURE FOR SPACE EXPLORATION
The past for planetary and solar-system exploration has been spectacular, both from a scientific perspective and for its inspirational value. It has not been accomplished without hard work, some grief, false starts and hopes, and quick modifications. This is in part because our methods for establishing goals, planning, and funding through the federal government are not really geared to a time scale for planetary exploration that can involve 10 to 15 years from conception to launch. In this regard the problems in planetary exploration are not much different from other areas of science, in that the length of time between conception of a scientific experiment and its execution grows steadily longer. It is difficult to carry out complicated scientific programs spanning many years, during which the level of funding shows dramatic fluctuations.

SUMMARY

The solar system. The scale of the solar system is defined in relative terms (astronomical units) and in absolute terms (using radar techniques). Computations of the size of the solar system are based on Kepler's third law. The nine planets and many of the minor bodies revolve about the sun in roughly circular orbits, moving in the same direction and nearly the same orbital plane.

The sun is the central and principal object of the solar system. The sun contains 99.86 percent of the mass of the solar system and, because of its great mass, the sun dominates the dynamics of the solar system. The sun also steadily radiates great amounts of electromagnetic energy and intermittently emits streams of high-energy subatomic particles.

The nine major planets had a common origin but different chemical, physical, and geological histories. The matter that formed the solar system can be categorized as either rocky, icy, or gaseous. On this basis, the inner group of solid planets, known as the terrestrial planets, are composed of rocky materials; the outer group, known as the Jovian planets, are primarily icy and gaseous bodies.

Satellites revolve about the planets. All but four of the 49 known satellites belong to the Jovian planets. Saturn, Jupiter, and Uranus also have ring systems composed of icy or ice-coated particles.

Minor members of the solar system are asteroids, meteoroids, comets, and the so-called interplanetary medium of dust and gas particles.

Earth-based study of planets. The physical and chemical properties of planets are studied by means of photometry (measuring the amount of radiant energy) and spectroscopy (determining the particular wavelengths composing white light). Radiated energy may be emitted as part of normal thermal or selective nonthermal processes. Studies of thermal radiation give information about surface and atmospheric temperatures, densities, and chemical composition. Studies of nonthermal radiation can provide information about other physical processes. The surfaces of Mercury, Venus, and Mars have also been mapped by radar techniques.

Overview of the planets. The planets divide by physical and chemical properties and location into two groups: the terrestrial planets and the Jovian planets with Pluto fitting neither group. The terrestrial planets—Mercury, Venus, earth, Mars, and to some extent the moon—are small rocky bodies with large mean densities, slow rotation, thin or no atmospheres, and few if any satellites. They are the ones closest to the sun. By contrast the Jovian planets—Jupiter, Saturn, Uranus, and Neptune—are large planets composed of light elements with small mean densities, rapid rotation, thick atmospheres, and many satellites. They are widely separated from each other and are located far from the sun.

Exploration of the solar system. The vast space exploration program of the United States and other countries has been motivated by the search for knowledge, a vision of exploration as a good in its own right, the expectation of practical benefits, and interest in maintaining national prestige. The strategy used so far for planetary exploration can be divided into the four phases of reconnaissance, exploration, intensive study, and exploitation or utilization.

KEY TERMS

asteroid
center of mass
comet
differential rotation
icy materials
interplanetary medium
Jovian planets
meteoroids

nonthermal radiation
polarization
radar mapping
ring systems
rocky materials
terrestrial planets
thermal radiation
tidal force

CLASS DISCUSSION

1. Why is the numerical value of the astronomical unit in kilometers important? How is it derived?

2. What does the story of the discovery of the trans-Saturnian planets tell us about the "scientific method" as applied to Newton's law of gravitation?

3. How do the terrestrial and Jovian planets differ in size, mass, rotation, distance from the sun, satellites, and general appearance?

4. How is the mass of the solar system divided among its various components: the sun, planets, satellites, minor members, and the like?

5. Briefly describe the approach to planetary exploration using spacecraft. How far have we gone in exploring each of the nine planets?

6. How do you feel about the tax dollars spent for planetary exploration? Is it too much or too little? What merit or lack of merit do you attribute to past and future space efforts?

READING REVIEW

1. How is the astronomical unit defined? What is its value in kilometers?

2. Which planet has the most extensive ring system?

3. Rank the planets by mass from the largest to the smallest.

4. What is the difference between thermal and nonthermal radiation from the planets? What does each tell us about a planet?

5. Which are the seven largest satellites, and how do they compare in size with Mercury? With Pluto?

6. Who is credited with discovering Uranus? Neptune? Pluto?

7. Rank the planets by the length of their rotation periods from shortest to longest.

8. What is a tidal force? Of what importance is it in understanding the motions of the planets?

9. Which bodies in the solar system move along the most elongated orbits?

10. What phase in planetary exploration do *Voyager 1* and *Voyager 2* fill? To which planets are they aimed?

SELECTED READINGS

Beatty, J. K., B. O'Leary, and A. Chaikin. *The New Solar System*. Sky Publishing, 1981.

Callatay, V. de, and A. Dollfus. *Planets: A History of Man's Inquiry through the Ages*. University of Toronto Press, 1976.

Hanle, P. A., and V. D. Chamberlain. *Space Science Comes of Age*. Smithsonian Institution Press, 1981.

Newell, H. E. *Beyond the Atmosphere*. NASA SP-4211, 1981.

Whipple, F. L. *Orbiting the Sun*. Harvard University Press, 1981.

7

The Earth-Moon: A Double Planet

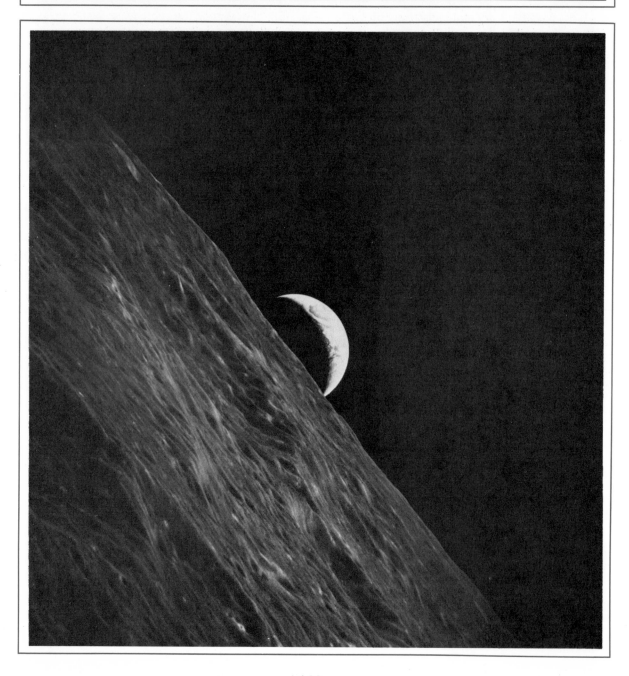

The title of this chapter suggests that the earth-moon system is more than a planet and its satellite. It is probably not unreasonable to classify the system as being more like a double planet for two reasons: First, as was shown in Chapter 6, the moon is not significantly smaller than the planet Mercury; second, the ratio of the moon's mass to that of the earth's is considerably larger than the typical planet-satellite combination in the solar system and is second only to that of Pluto and its satellite, Charon. Thus the earth-moon system is somewhat uncharacteristic in the solar system, and its origin and evolution may have had a somewhat different track from that of the other terrestrial planets.

One of the most significant concepts in our view of the world began only in the eighteenth century; in a period of about 100 years it revolutionized scientific thinking. What was that concept? It was the concept of change, or evolution. Although you may have encountered evolution in connection with biological evolution, the concept is much broader. Today our world view is one dominated by a knowledge that the universe, the stars, the earth, and life upon the earth's surface are gradually, sometimes rapidly, evolving in directions shaped by natural processes governed by the laws of physics and chemistry. Throughout this chapter and those that follow we shall try to sketch the processes of evolution as well as its results.

TABLE 7.1
Physical Data for the Earth

Quantity	Value
Equatorial radius	6,378 km
Polar radius	6,357 km
Mean radius	6,371 km
Mean circumference	40,030 km
Volume	1.08×10^{27} cm^3
Mass	5.98×10^{27} g
Mean density	5.5 g/cm^3
Surface acceleration by gravity	980 cm/s^2
Escape velocity[a]	11.2 km/s
Equatorial rotational velocity	0.46 km/s (1,040 mi/hr)
Lengthening of day	1.5×10^{-3} s/century
Surface area:	5.10×10^{18} cm^2
Land area	1.49×10^{18} cm^2 (29.2%)
Ocean area	3.61×10^{18} cm^2 (70.8%)
Geomagnetic-pole locations:	
North Pole	76 °N 101°W
South Pole	66 °S 140°E
Magnetic field	≈ 0.5 gauss

[a]Minimum velocity needed to overcome the gravitational attraction of a body of mass M and radius R, which is the condition that the escaping body's kinetic energy equals its potential energy; i.e., $\frac{1}{2} mv^2 = (GM/R)m$, or $v = \sqrt{2GM/R}$.

7.1
The Interior of the Earth

SIZE AND SHAPE OF THE EARTH

Measurements in different places on the earth reveal that the number of kilometers in 1° of latitude increases slightly from the equator toward the poles. These measurements indicate that the earth is shaped like an oblate spheroid, with the longer diameter in the equatorial plane and the shorter one in the polar direction (Table 7.1).

The rotation of the earth is primarily responsible for its shape in that the rotation causes the body of the earth to flow from high latitudes toward lower latitudes, forming an equatorial bulge. Because the earth is not a perfect sphere, its gravitational field is not the same in all directions. These variations affect the motions of artificial satellites. From unanticipated changes in satellite orbits, called *orbital perturbations,* we know that our planet is slightly pear-shaped. The stem portion at the North Pole is about 19 meters farther from, and the bottom portion at the South Pole about 26 meters closer to, the center of the earth.

ROTATION OF THE EARTH

Several phenomena were used historically to demonstrate the rotation of the earth besides the rising and setting of the sun and stars. One of the most vivid was the pendulum experiment devised in 1851 by the French physicist Jean Foucault (1819–1868). He hung an iron ball on a long wire from the dome of the

◀ Earthrise of a crescent-phase earth as seen from the back side of the moon. This photograph was taken by the crew of *Apollo 17* in December 1972.

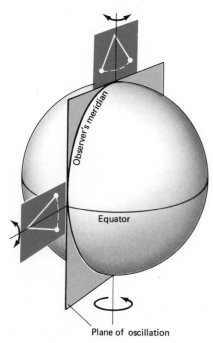
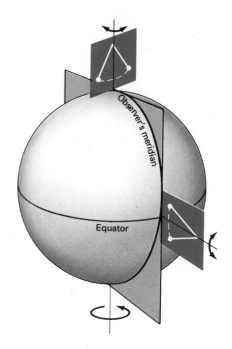

a. Orientation when pendulums started

b. Orientation after 90° rotation by the earth

FIGURE 7.1
Foucault pendulum experiment. (a) A scientist sets pendulums in motion, one at a pole and the other at the equator. The only force acting on the pendulums is gravity, which is directed toward the center of the earth. (b) After 6 hours the earth has turned through an angle of 90° relative to you as an observer outside the earth. For the scientist the plane of oscillation for the pendulum at the pole is now perpendicular (90°) to the observer's meridian of longitude. The rate at which the two planes change orientation relative to each other is 15° per hour clockwise in the northern hemisphere and counter-clockwise in the southern hemisphere. For the pendulum at the equator the plane of oscillation rotates with the meridian plane; that is, the rate of change between the two planes is 0° per hour. For latitudes between 0° and 90° the rate of rotation is between 0° per hour and 15° per hour. For example, the rate at San Francisco is 9.2° per hour and requires 39 hours for a full rotation.

Panthéon in Paris. Underneath it was a large circular table with a ridge of sand along its edge. As the pendulum swung, a pin attached to the bottom of the ball would make a mark in the sand.

After the pendulum had carefully been set into planar motion, it was apparent from the marks in the sand that it was deviating slowly in a clockwise direction. In actuality, because of the earth's rotation, the spectators and the building were turning underneath the plane of oscillation of the pendulum.

Once a plane of oscillation has been established for a pendulum, an external force is required to change the plane's orientation. Figure 7.1 illustrates what would have happened if the experiment had been performed at either a geographic pole or the equator.

The constant friction generated by the lunisolar tides (mainly near the shores and in the shallow seas) has slowed the earth's rotation. As a result the day has lengthened in several billion years from an estimated several hours to our present 24 hours. The slowing down of the length of the day is not uniform; a number of irregularities have been determined. The conversion of the earth's rotational energy into heat by tidal friction will continue indefinitely.

TIDES IN THE EARTH

The moon's gravitational pull on the oceans produces two tidal bulges on opposite sides of the earth in line with the moon. Why two high-water tides? Take an

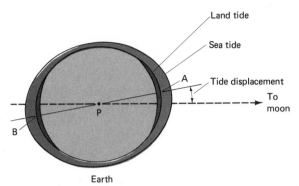

FIGURE 7.2
Polar view of the earth's tidal bulges produced by the moon's gravitational pull. The rotating earth carries the tidal bulge (marked by the line APB) slightly forward of the moon's position in its orbit.

idealized earth entirely surrounded by water, as in Figure 7.2. The moon's gravitation pulls harder on the water closest to it than it does on the water on the opposite side of the earth. Compared with the earth's center this *tidal force* causes an acceleration of the oceans on the side nearest to the moon larger than that on the opposite side of the earth. This causes water to move toward the moon on the near side and to recede from it on the far side, compared with the earth as a whole. Consequently water piles up in the form of an ellipsoid whose long axis is directed toward the moon. Midway between the high tides are the low tides.

Earth's rotation underneath the tidal bulges results in alternating high and low tides in the ocean twice each day. Because there is a slight lag before the oceans fully adjust to the moon's tidal force, the tidal bulges are dragged by the rotating earth somewhat ahead of the line joining the centers of moon and earth (Figure 7.2).

The sun also contributes to the tides, but only half as much as the moon does because of its much greater distance, despite its larger mass. When the sun and moon are in line, as at new or full moon, their combined gravitational pull is strongest, producing the largest tides.

If the sun and moon are pulling the earth, why doesn't the land move too? It does because the land is not absolutely rigid. However, land has a greater internal strength than water, and therefore, the land tides are very small; but approximately every 12 hours earth's solid surface rises and falls a few centimeters at any given place. Tidal motions are also evident in the earth's atmosphere, which is even less rigid than the oceans.

Tidal forces and their resulting effects occur in many different situations involving astronomical bodies. We shall encounter them at several points later in the book.

EARTH AND MOON: THEIR DYNAMIC RELATION

The mean distance between the moon's center and the earth's is about 400 thousand kilometers. Recently that distance has been measured to within several centimeters by timing the round-trip of a laser beam bounced off reflectors left on the moon by the *Apollo* astronauts. The moon travels around the earth in an ellipse of small eccentricity, with the earth at one focus. Some of the physical parameters of the lunar orbit are given in Table 7.2.

The point that orbits the sun annually according to Kepler's laws is not the geographic center of the earth. It is a point on the line joining the earth and the moon and is known as the *center of mass*. One can think of the center of mass as the center of balance of an imaginary rod supporting the earth at one end and the moon at the other, as shown in Figure 7.3. The center of mass for the earth-moon system lies inside the earth, since the mass of the moon is only about 1 percent that of the earth.

The moon turns once on its axis in the same time that it completes one orbit around the earth so that the same hemisphere is always toward us. This is rela-

That gentle Moon, the lesser light, the Lover's
    Lamp, the Swain's delight,
A ruined world, a globe burnt out, a corpse
    Upon the road of night.

Sir Richard Burton

TABLE 7.2
The Lunar Orbit

Orbital Parameter	Value
Maximum distance from earth (apogee)	406,700 km
Minimum distance from earth (perigee)	356,400 km
Mean distance from earth	384,401 km
Sidereal period (sidereal month)	27.32 d
Synodic period (synodic month)	29.53 d
Average orbital velocity	1 km/s
Inclination of orbit to ecliptic	5°8′
Eccentricity	0.055

tively easy to demonstrate to yourself. Walk around a stool, continually facing it; next walk around the stool, keeping your head and body pointed toward the same direction. In the first instance you rotated once while you revolved once, just as the moon does; in the second you did not rotate about your axis. If the moon did not rotate, we could see all its sides during the month. That the moon's rotation period is equal to the period of its orbital revolution (27.3 days) is not accidental. Tidal forces between the earth and the moon over the eons have equalized the rotation and revolution periods.

Originally both bodies were probably much closer, perhaps only 5 to 10 percent of their present distance,

and were rotating more rapidly. The earth's day was then a few hours long and its month, or the moon's orbital period, much shorter than now. Because of the earth's greater tidal force, the moon's rotation has slowed more rapidly than the earth's has.

Some of the earth's rotational energy is gradually transferred by the lunar tides to the orbiting moon so that the moon recedes from the earth several centimeters every year. Why? The moon's tidal force has a braking effect on the earth, which decreases its angular momentum, or quantity of rotational motion. To conserve the total angular momentum of the earth-moon system, the angular momentum in the moon's orbital motion must be increased. Hence it is accelerated ever so slightly in its orbit, spiraling outward from the earth.

As the moon recedes, the month must lengthen, according to Kepler's third law. Eventually the earth and the moon will face each other with equal periods of rotation and revolution (about 47 days) at a distance of about 560,000 kilometers. But the calculated time for this event to happen, several tens of billions of years, far exceeds our estimates of the earth-moon system's probable life span.

STRUCTURE OF THE INTERIOR OF THE EARTH

From what we have learned about the rotation of the earth, it is apparent that the earth is not absolutely rigid. The body of the earth will deform when subjected to various forces, such as mechanical waves. Geophysicists have a natural tool for probing the planet's internal structure: *seismic waves*, which are generated by earthquakes and spread out in all directions from the site of the quake. From the manner of their propagation, their periods and amplitudes of vibra-

FIGURE 7.3
Center of mass of the earth-moon system. The center of mass is located such that the product of the mass of each body times its distance from the center of mass is a constant; that is, $M_⊕ d_⊕ = M_☾ d_☾$.

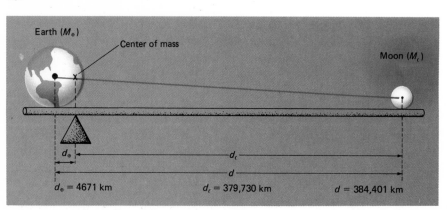

Earth ($M_⊕$)

Center of mass

Moon ($M_☾$)

$d_⊕$

$d_☾$

d

$d_⊕ = 4671$ km $d_☾ = 379,730$ km $d = 384,401$ km

tion, and their arrival time at various stations, scientists can deduce much about the earth's structure. There are now several artificial means of producing seismic waves.

Two kinds of seismic waves, pressure (*P*) and shear (*S*) waves, propagate inside the earth. The speed with which these waves travel through the earth (between 5 and 15 kilometers per second) depends on the material's density, compressibility, and rigidity. The particles of the earth that transmit the *P* waves vibrate back and forth in the direction in which the wave propagates, similar to the way sound waves are propagated through air. The *S* waves, which move at about half the speed of the *P* waves, cause the particles that transport the disturbance to vibrate perpendicular to the direction of the waves' propagation, as waves on a string do. Unlike the *P* waves, *S* waves cannot propagate through liquids, which damp their vibrations. As the *P* and *S* waves move downward through the earth, their speed increases with the increasing density of the material they are traversing. They are refracted or reflected on reaching a boundary between two distinctly different layers. By tracking the path of these waves, geophysicists can produce a picture of the earth's interior, showing a layered structure like that of an onion.

A model of the earth's structure is shown in Figure 7.4. At the center is a hot, highly compressed *inner core*, presumably solid and composed mainly of iron and nickel. Surrounding the inner core is an *outer core*, a molten shell primarily of liquid iron and nickel with lighter liquid material on the top. The outer envelope beyond the core is the *mantle*, of which the upper portion is mostly solid rock in the form of oliv-

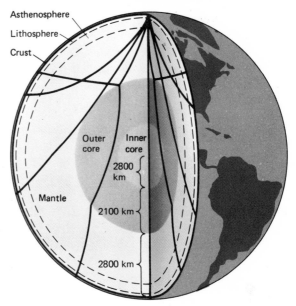

FIGURE 7.4
The earth's internal structure as revealed by seismic waves. The principal layers are labeled. Information about these layers is obtained by analyzing data from seismic recordings of earthquake vibrations passing through the earth. The velocities of the waves and their paths change abruptly at the interfaces between adjoining layers, as shown in the diagram.

ine, an iron-magnesium silicate, and the lower portion chiefly iron and magnesium oxides. A thin coat of metal silicates and oxides (granite), called the *crust*, forms the outermost skin. A summary of the physical characteristics of each of these zones is given in Table 7.3.

TABLE 7.3
Earth's Interior

Layer	Depth Below Surface (km)	Fraction of Mass Interior	Density Range (g/cm³)	Pressure (10⁶ atm[a])	Approximate Temperature (K)	Main Composition
Crust	0– 30	1.00–0.99	2.6– 3.3	0 –0.01	290– 700	Continents: granite or water
Mantle	30–2900	0.99–0.32	3.4– 9.7	0.01–1.3	700–4300	Basaltic rocks: silicates and oxides of iron and magnesium
Core:						
Outer	2900–4980	0.32–0.03	9.7–12.5	1.3 –3.2	4300–6000	Molten iron and nickel
Inner	4980–6370	0.03–0.0	12.5–13	3.2 –3.7	6000–6400	Solid iron and nickel

[a]Units are those of the typical pressure of the earth's atmosphere measured at sea level.

HOW OLD IS OUR PLANET?

In 1654 Ireland's Archbishop Ussher calculated that the earth was born on October 26 in the year 4004 B.C. He had worked out the date of creation from the Bible's chronology of the generations of patriarchs. Late in the nineteenth century scientific estimates of the earth's age rose to 50 million years. By the turn of the century even this figure was far too low to match growing geological evidence. The discovery of natural radioactivity at the start of the twentieth century freed the dating of the earth from theology and from rational but still inadequate scientific methods (see the accompanying box). Estimates of the age of the earth continued to go up from 1900 till now, when they have stabilized at about 4.6 billion years. Rocks collected during the lunar-landing missions in the *Apollo* program reassuringly yield the same age for the moon.

7.2
The Surface of Planet Earth

OUTER LAYERS OF THE EARTH

Earth's crust and the uppermost part of the mantle are known as the *lithosphere*. This is a fairly rigid zone that extends about 100 kilometers below the surface of the earth. The crust extends some 60 kilometers or so under the continental surface, but only about 10 kilometers below the ocean floor. The continental crust has a lower density than the oceanic crust does. It is primarily a light granitic rock rich in the silicates of aluminum, iron, and magnesium. In a simplified view the continental crust can be thought of as layered: On top of a layer of igneous rock (molten rock that has

DATING THE EARTH

The most accurate method we now have of dating the earth is based on knowledge of radioactivity. Nuclei of radioactive elements, such as uranium and thorium, spontaneously break down into lighter nuclei at measurable rates, emitting alpha particles (helium nuclei), electrons (beta particles), and gamma radiation. Their decay ends in a stable isotope (Table 7.4) of a lighter element, such as lead. The time required for half of the initial number of nuclei to decay into the final products is known as the element's *half-life*. Some examples are given in Table 7.4.

How old is a rock? Scientists start by determining how much of the final stable element has formed and how much of the parent element remains from which the daughter element formed. We combine that figure with the known rate of disintegration and some assumptions about the initial ratios.

For example, half of a sample of uranium 238 decays in 4.51 billion years into lead 206. If we found in some rocks a 50/50 ratio of uranium 238 to lead 206, the age of the rocks would be about 4.5 billion years (if no lead 206 had been present at the beginning). The oldest meteorite specimens suggest that some primitive lead 206 probably was present when the solar system was formed. The oldest rock formations we know, in Isna, Greenland, are 3.7 to 3.8 billion years old. Allowing for some primordial abundance of lead in the earth's rocks, geophysicists obtain 4.6 billion years for the age of the earth.

Scientists use radioactive dating of fossil-bearing rock to piece together the story of the earth's earliest inhabitants and their evolution. Thus the fossil record, which extends back about 3.5 billion years, can be used as a dating tool once it has been calibrated in this way. Another method is relative dating of rock layers, which depends on the fact that the most recent layers are on top of older sedimentary-rock beds. Successive forms of fossils in sedimentary layers of known age can also be used to date similar sedimentary layers elsewhere.

TABLE 7.4
Half-lives of Some Radioactive Nuclei Used in Geological Dating

Radioactive Nucleus	Half-life (10^9 yr)	Final Stable Nucleus in Decay Sequence
Uranium 238	4.51	Lead 206
Uranium 235	0.71	Lead 207
Thorium 232	13.9	Lead 208
Rubidium 87	49.8	Strontium 87
Potassium 40	1.31	Argon 40

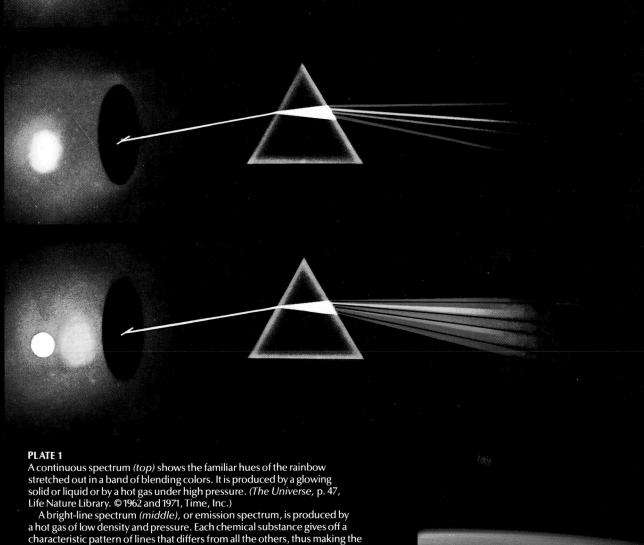

PLATE 1

A continuous spectrum *(top)* shows the familiar hues of the rainbow
stretched out in a band of blending colors. It is produced by a glowing
solid or liquid or by a hot gas under high pressure. *(The Universe*, p. 47,
Life Nature Library. © 1962 and 1971, Time, Inc.)

A bright-line spectrum *(middle)*, or emission spectrum, is produced by
a hot gas of low density and pressure. Each chemical substance gives off a
characteristic pattern of lines that differs from all the others, thus making the
identification possible. *(The Universe*, p. 47, Life Nature Library. © 1962 and
1971, Time, Inc.)

A dark-line spectrum *(bottom)*, or absorption spectrum, is caused by the
presence of a cooler gas in front of a source producing a continuous spectrum. The cooler gas absorbs light in those parts of the spectrum in which it
would emit bright lines if it were hot enough to radiate. The positions of the
dark lines thus provide a clue to the composition of the gas. *(The Universe*,
p. 47, Life Nature Library. © 1962 and 1971, Time, Inc.)

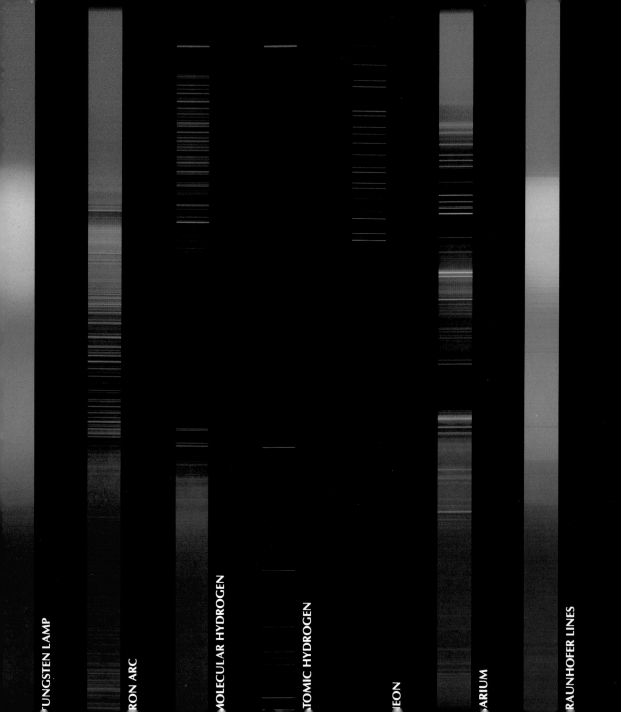

TUNGSTEN LAMP

RON ARC

MOLECULAR HYDROGEN

ATOMIC HYDROGEN

EON

ARIUM

RAUNHOFER LINES

PLATE 6
Surface of Mars as seen by *Viking 1* in June 1976 from 560,000 kilometers. Clearly visible are the Tharsis Mountains and a row of three huge volcanos rising about 20 kilometers above the surrounding plain. The volcano Olympus Mons is toward the top of this picture; the large, nearly circular feature near the bottom is the impact basin, Argyre. The white patch to the left of the southernmost Tharsis volcano is possibly surface frost or ground fog. (NASA)

PLATE 7
Martian surface. *Viking 1* obtained this picture on July 24, 1976. Part of the spacecraft's gray structure is in the foreground. Orange-red surface materials cover most of the surface, apparently forming a thin veneer over darker bedrock. The sky has a reddish cast, probably due to scattering and reflection from reddish sediment suspended in the lower atmosphere. (NASA)

PLATE 8
Ultraviolet photos, taken by the *Pioneer Venus Orbiter* during a period of 38 hours in May 1980, show the pronounced Y-shaped cloud formations produced by high-velocity winds on Venus. (NASA)

PLATE 9
Surface of Venus as photographed by the Soviet Union's *Venera 14*. (NASA)

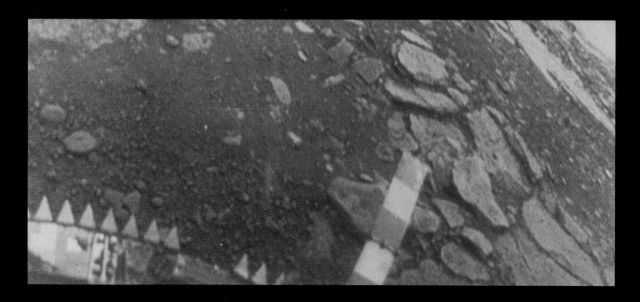

PLATE 10
Jupiter as seen by *Voyager 2* in June 1979. This picture, taken from about 24 million kilometers, shows the Galilean satellite Io to the right; the shadow on the left of Jupiter is Ganymede, the largest of the satellites. Clearly visible are the intricate and colorful patterns in the atmosphere of Jupiter. (NASA)

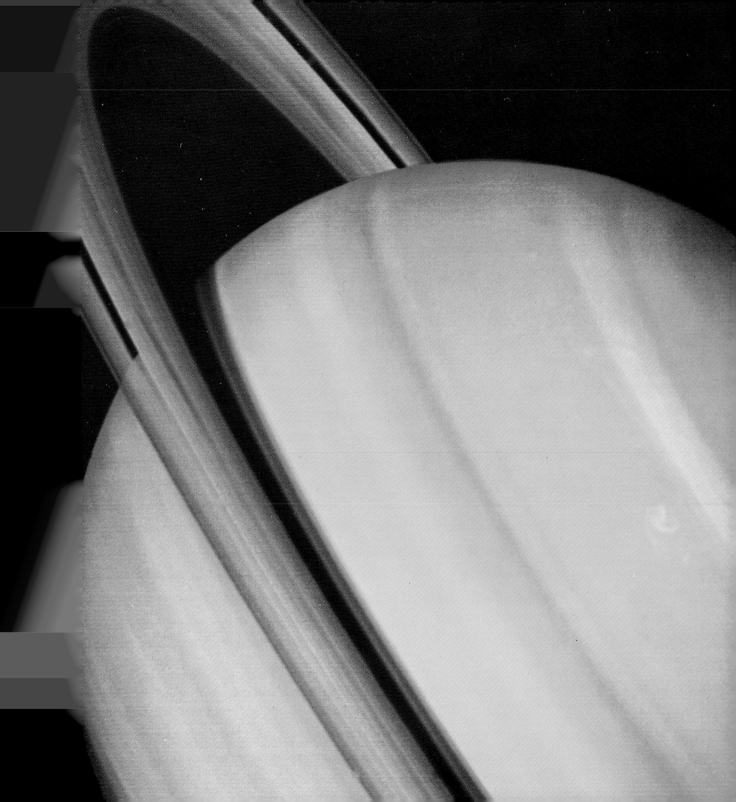

PLATE 12
Montage of the Saturnian system prepared from an assemblage of images
taken by *Voyager 1* during its Saturn encounter in November 1980. (NASA)

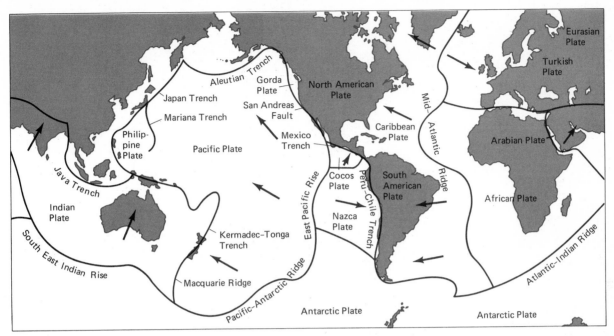

FIGURE 7.5
Mosaic of the earth's lithospheric plates. The surface is divided into approximately a dozen major plates and some minor ones. The plates are in constant motion with new plate material being added from midocean ridges at the trailing edge of the plate. The leading edges of some plates are being forced under other plates, forming the oceanic trenches.

hardened) lies a thin layer of sedimentary rocks (rocks formed by sediment and fragments that water deposited); there is also a soil layer deposited during past ages in the parts of continents that have had no recent volcanic activity or mountain building.

Sandwiched between the lithosphere and the lower mantle is the partially molten material of the asthenosphere, about 150 kilometers thick. It consists primarily of iron and magnesium silicates that readily deform and flow under pressure.

In efforts to date various regions of the continents, geochemists have shown with radioactive-dating techniques that the oldest rock formations on the continents are between 3.8 and 3.5 billion years old. For North America the oldest part, called a *continental shield*, is a crescent-shaped region bordering the west and south sides of Hudson's Bay. A younger crescent lying roughly to the west and south surrounds this oldest region, and the westernmost and southernmost parts of the continent are even younger. A somewhat similar pattern exists for the other continents.

The inference is that the continents are not original with the formation of the earth 4.6 billion years ago but are a secondary aspect that continues to grow. We know that the continental margins, particularly the western edge of North America, are new additions to the continents. The coastal regions are building due to the deposition of sediments washed down by rivers from the interior of the continent. In striking contrast the oldest known parts of the oceanic crust are about 200 million years old or almost twenty times younger than the oldest parts of the continents.

PLATE STRUCTURE OF THE LITHOSPHERE: CONTINENTAL DRIFT

The idea that the continents drift relative to each other was one that few hurried to accept when the German geologist Alfred Wegener (1880–1930) proposed it in 1912. Yet recent research has revealed a variety of evidence showing that the lithosphere is indeed segmented into about a dozen or so major plates of different sizes, as shown in Figure 7.5. Floating on the earth's mantle, they move slowly, carrying the continents with them at an average rate of several centimeters each year. This motion is known as *plate tectonics,* or more popularly as *continental drift.*

Exploration of the ocean floor has shown the existence of a number of midocean ridges that rise several

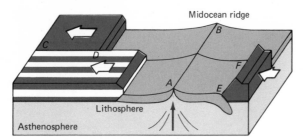

FIGURE 7.6
Boundaries of the lithospheric plates. Shown in an idealized fashion, three kinds of plate boundary are shown. The first is the midocean ridge *AB*, out of which new material is being forced from the asthenosphere to solidify on the trailing edge of the four plates shown. The second type of boundary is that marked by *CD*, which is two plates sliding past each other. In this case (but not in every case) they are going in the same direction with different speeds. The final boundary is that marked *EF*, along which one plate is forced down underneath the other. This type of boundary results in an oceanic trench; the plate being forced into the asthenosphere melts on the leading edge, thus recycling material.

FIGURE 7.7
Schematic view illustrating tectonic activity. The diagram shows one model of convection currents proposed to explain how the plates are moved. Warmer material moves upward from the asthenosphere toward the ridge and spreads out laterally as it cools, forming new crustal-plate material. The plates, carrying the continents with them, ride on this convection flow. The cooler material descends at the trench and is reheated within the asthenosphere, completing the cycle of convection.

kilometers above the ocean floor and are thousands of kilometers long. They mark one type of plate boundary. For example, the Mid-Atlantic Ridge separates the North and South American plates from the Eurasian and African plates, while the East Pacific Ridge separates the Pacific plate from the North American, Cocas, and Nazca plates (Figure 7.5). Midocean ridges and other types of plate boundary are illustrated schematically in Figure 7.6. It appears that lava, forced upward from the asthenosphere into a midocean ridge, pushes out laterally from the ridge to form new plate material, which gradually cools, thickens, and solidifies at the trailing edge of the older plate material (Figure 7.7) Rock samples from as far down as 8 kilometers below sea level verify that earth's youngest volcanic rocks are those found near these midocean ridges.

We have further confirmation that the plates move from the shape, geological structure, and fossil record of the continents. Still more evidence comes from rocks. Igneous rocks with similar magnetic fields, which were frozen at the time the rocks solidified, have been found at continental margins that are now widely separated.

Another line of evidence derives from the heat flow out of the earth's interior. Compared to the energy falling on the earth from the sun, the interior flow is scarcely a trickle: The heat conducted through an area the size of a football field is roughly equivalent to the energy given off by three 100-watt light bulbs. Yet over the 4.6-billion-year history of the earth this trickle of

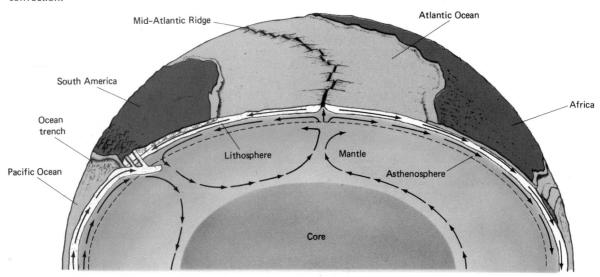

Man makes a great fuss
About this planet which is only a ball-bearing
In the hub of the universe.

Christopher Morley

energy has contributed to the work of making continents drift, opening and closing ocean basins, building mountains, and causing earthquakes. The geographic variation in the heat flow from the interior is not great, but the global variation shows that the major oceanic ridges are high-heat-flow zones while the older continental shields and sedimentary regions are low-heat-flow zones.

How are the plates transported across the mantle? It appears that they are driven by the horizontal flow of convective currents within the mantle, circulating in the upper, softer portion of the mantle, as illustrated in Figure 7.7. Often the leading edge of one plate is pushed downward and forced into the mantle, to create a deep trench. This process can form a coastal mountain belt, like the Andes, on the overriding plate. As the other plate descends over millions of years, it heats up and becomes part of the general circulation in the asthenosphere. Plates separate along midocean ridges. Most of the great geologic processes—volcanic activity, mountain building, formation of ocean trenches, earthquakes—are concentrated on or near plate boundaries.

A CHANGING FACE FOR THE EARTH

About 200 million years ago the last mass movement of the continents began (Figure 7.8). Earth then had only one consolidated land mass, today called Pangaea. It is believed that this supercontinent accumulated from migrations produced by previous drifting. Some 20 million years later sea-floor spreading had separated the supercontinent into two segments: Laurasia in the north and Gondwana in the south. About 45 million years later the North Atlantic and Indian Oceans had widened and South America had begun to separate from Africa, while India drifted northward. During the next 70 million years the South Atlantic Ocean widened into a major ocean, the Mediterranean Sea began to open up, and North America just began to separate from Eurasia.

A computer-generated projection for the next 50 million years suggests that the Atlantic and Indian oceans will enlarge, and the Pacific will contract. Australia will continue drifting northward toward a possible collision with Eurasia. Africa's northward movement will doom the Mediterranean. In 10 million years Los Angeles, which is part of the Pacific plate, will have come abreast of San Francisco, which is sitting on the North American plate, and from there will eventually slide into the Aleutian Trench.

Average plate motions are on the order of 5 to 6 centimeters per year so that the reshaping of the earth's face is quite dramatic when one considers the age of the earth. In about 2 billion years the gradual cooling of the earth from heat loss will mean that the asthenosphere will flow less readily and that the plate-motion phase of the earth's evolution will probably come to an end. Thus the earth will enter a new phase, in which the plate motions of the earth's lithosphere are not responsible for most of the large-scale terrain features. Large mountain ranges, like the Himalayas, will no longer be uplifted, and they will erode away over millions of years.

GEOGRAPHY OF THE SURFACE

As far as surface geography is concerned, there appear to have been two major terrain-shaping mechanisms at work on the earth (and for that matter presumably on the moon, Mars, Venus, and Mercury, the terrestrial planets). These are *impact cratering* by meteoric bombardment and *thermal-tectonic activity* due to an outflow of thermal energy from the deep interior of the earth. Erosion by wind, water, and life and tectonic activity (deformations and motions of the crust), with its accompanying crustal strains and slippages, are the dominant mechanisms now (only on the earth). They have all but erased the results of the impact-cratering phase in the earth's history, except that remnants of the last of that phase remain in nearly a hundred ancient impact structures, some of which

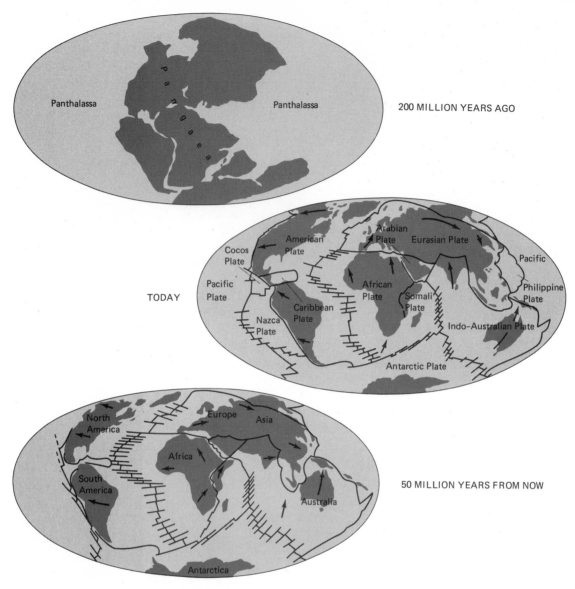

FIGURE 7.8
The drifting continents. Two hundred million years ago a single supercontinent—which geologists now call Pangaea, "all lands"—was washed by a universal ocean, Panthalassa. In the last 65 million years almost half the oceans were created. Greenland split from North America, which completed its cleavage from Eurasia as the Atlantic widened, and northward-drifting India collided with Eurasia, thrusting up the Himalayas. Australia was then joined to Antarctica. Thirty million years ago the eastward movement of Africa, after splitting from Gondwana, stopped. South America was still moving westward, widening the South Atlantic. In the future the Atlantic and Indian Oceans will continue to widen and the Mediterranean to shrink. The Pacific Ocean will narrow, and California west of the San Andreas Fault will be detached from the mainland and slide toward the Aleutian Trench.

are as large as the largest visible ones on the moon. (Figure 7.9). It is estimated that on the earth tectonic activity with its accompanying volcanic activity dominates better than 90 percent of the present terrain, with not more than 10 percent of the cratered terrain remaining. Present evidence suggests that surface evolution on the other terrestrial planets, as revealed by various space missions, has not been so heavily influenced by thermal-tectonic activity as that for the earth.

FIGURE 7.9
A 220 million-year-old impact crater in Brazil. The crater consists of a 3-kilometer-diameter inner ring and a 12-kilometer outer ring.

7.3
The Earth's Atmosphere

ROLE OF THE ATMOSPHERE: GREENHOUSE EFFECT

If the earth had no atmosphere, life would not exist here. The insulating blanket of air surrounding us maintains a temperature range favorable for life because the sun's radiant energy, which is primarily in the visible wavelengths, is absorbed by and warms the ground, which in turn reradiates energy in the infrared region of the electromagnetic spectrum. (The reradiated energy is in the infrared because the warming of the surface by sunlight maintains it at a temperature of not quite 300 K.) The reradiated infrared photons' passage outward into space, however, is restricted by carbon dioxide and water-vapor molecules in the atmosphere; they absorb the energy and reradiate much of it back to the surface of the earth. This is called the *greenhouse effect,* after the similarity of the action to that of the glass in a greenhouse, which prevents heat radiation produced inside the greenhouse from escaping.

Without its atmosphere the earth's average temperature would be about 20° to 30° lower than its present value of 15°C (288 K). Since water would be frozen at that temperature, it could not effect the development and maintenance of life as it does in a liquid form. Worldwide circulation of the atmosphere also transports thermal energy and helps to moderate the extremes in temperature that would otherwise exist.

Moreover, the upper atmosphere is important for our survival. It protects us from harmful ultraviolet and X-ray radiation from the sun, vaporizes meteoroids entering the atmosphere, and absorbs most of the incoming highly energetic subatomic particles that we call *cosmic rays.* Finally the atmosphere creates the soft blue appearance of the sky: atmospheric gases scatter the photons of sunlight in the blue region much more efficiently than they do photons of longer wavelengths.

In all, the earth's atmosphere plays a very vital role beyond the obvious one of providing the oxygen that we breath. One of mankind's most important challenges is in understanding all aspects of our atmosphere and preserving it; for our continued existence depends upon that knowledge.

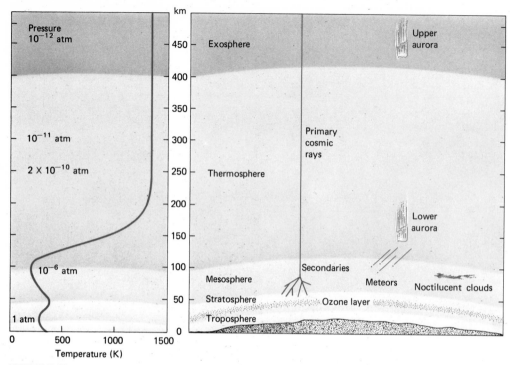

FIGURE 7.10
Structure of the earth's atmosphere. The various regions are differentiated by their distinct thermal, physical, chemical, and electrical properties.

PHYSICAL PROPERTIES OF THE ATMOSPHERE

The total mass of the atmosphere is about one-millionth of the total mass of the earth. It has several layers, which are shown in Figure 7.10, each with distinctive thermal, physical, chemical, and electrical properties. Approximately half the atmosphere is contained in the first 5.6 kilometers, and 99 percent of it lies below 35 kilometers.

Our weather takes place in the bottom layer, called the *troposphere*. Eleven kilometers up, the temperature drops to −55°C. Above this region lies a 40-kilometer-thick layer, the *stratosphere*, where the temperature slowly rises, reaching a maximum of about 0°C at 50 kilometers, and somewhat below this altitude an absorbing layer of ozone screens out most of the incoming ultraviolet radiation. Within the next layer up, the *mesosphere*, the temperature rapidly drops to a minimum of −85°C at its upper limit, 90 kilometers.

Above the mesosphere is the *thermosphere*. Here the still more dangerous X rays and gamma rays are effectively filtered out by molecular oxygen and nitrogen and by their dissociated atoms at even higher altitudes. The temperature climbs steadily throughout the thermosphere and into the *exosphere*, the atmospheric fringe several hundred kilometers above sea level.

CHEMICAL COMPOSITION OF THE ATMOSPHERE

Up to about 90 kilometers gravitational settling causes no significant separation of the atmospheric gases by atomic weight. No separation occurs because the atomic and molecular constituents are mixed by air currents and random thermal motion. The chemical composition of the atmosphere therefore remains nearly uniform, with 77 percent nitrogen, 21 percent oxygen, nearly 1 percent argon, 0.03 percent carbon dioxide, and almost 1 percent water vapor (which varies up to several percent in the troposphere). The atmosphere has minute traces of other gases, including neon, krypton, xenon, methane, ammonia, nitrous oxide, carbon monoxide, and ozone.

Above about 90 kilometers or so the constituents are not well mixed; the heavier molecules and atoms settle toward the bottom, the lighter ones diffusing to the top. At extreme heights a rarefied layer of helium extends from about 600 to 1000 kilometers, topped by a very tenuous hydrogen layer that merges into interplanetary space.

The chemical composition of the atmosphere is not static. The present composition results from a balance between those processes that introduce a particular molecule into the atmosphere and those that remove the molecule from the atmosphere, as shown in Figure 7.11. Probably the most meaningful example is that of oxygen since it is essential for our existence.

Atmospheric oxygen is almost entirely produced in photosynthesis, primarily by green plants in shallow seas and to a lesser extend by plant life on land. A little oxygen comes from the direct dissociation of atmospheric water molecules by ultraviolet photons from the sun. Oxygen is chemically quite an active molecule, combining readily with a number of different atoms. Thus the formation of oxides in rocks removes oxygen from the atmosphere. Breathing by animal life also depletes atmospheric oxygen. If the supply of oxygen were shut off, it would take only a few tens of

FIGURE 7.11
Cycle of atmospheric gases. The earth's atmosphere balances mechanisms introducing chemical components and mechanisms taking them out of the atmosphere. Outgassing from volcanoes, for example, releases water vapor, carbon dioxide, nitrogen, and other molecules into the atmosphere, while the formation of carbonate rocks removes carbon dioxide. Oxygen is present in the atmosphere because of photosynthesis by plant life, and it is removed by a variety of processes, including respiration by animal life. No attempt has been made to include all production and removal mechanisms.

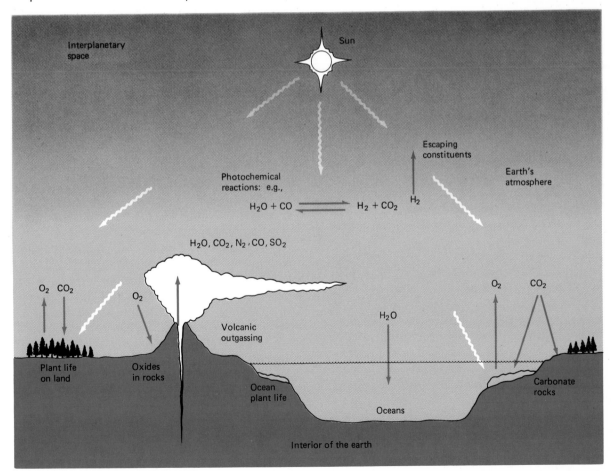

thousands of years to remove the major portion of oxygen now existing in the atmosphere.

The abundance of the other molecules in the atmosphere is also controlled by various "production and destruction" processes. And as in the evolution of the earth's surface, the atmosphere has also changed with time. Clearly, if atmospheric oxygen is due to the existence of life, then oxygen would not have been present prior to the emergence of life. The origin of the primitive earth's atmosphere is probably the result of outgassing by volcanoes and the escape of gases from the crust. The gaseous emission from present-day volcanoes includes water vapor, carbon dioxide, nitrogen, inert gases, and small amounts of methane, ammonia, and sulfur compounds. It is estimated that on a lifeless earth with no significant amounts of liquid water the dominant atmospheric constituent would be carbon dioxide in a very dense atmosphere. This estimate is based on the fact that a large amount of carbon dioxide is trapped in carbonate rocks on the earth's surface. Carbon dioxide is the main component of the atmospheres of Venus and Mars, with the Venusian atmosphere's being some hundred times denser that ours is. About 2 billion years ago the transition began to an oxygen-nitrogen atmosphere. The amount of oxygen grew from a trace to the present 21 percent as a result of the development of oxygen-producing photosynthesis by green plants.

IONOSPHERE

Within the earth's atmosphere are layers in which the concentration of free electrons is above the average atmospheric value. These layers constitute the *ionosphere*. The electrons are due to the ionization of atmospheric molecules and atoms by solar ultraviolet and X-ray photons.

Radio waves of certain wavelengths, for example the AM band of conventional broadcasting, transmitted by ground stations are reflected between the ionosphere and the earth's surface. This makes possible long-distance communication between stations that are not along a direct line of sight because of the earth's curvature. Radio wavelengths greater than about 10 meters, or frequencies less than about 30 megahertz, are turned back by the ionospheric layers. Shorter wavelengths, or higher frequencies, such as radar signals, pass through the ionosphere into space with little or no bending.

7.4
Magnetosphere of the Earth

A MAGNET INSIDE THE EARTH?

When it became known that the interior of the earth is hot, it was obvious that the earth's magnetic field could not be a permanent magnet. This is because heating disorients various parts of a magnet destroying the ability to produce a coherent magnetic field. Thus arose the puzzle of where the earth's magnetic field comes from. Scientists now believe it to be caused by circulation of liquid metal in the outer core: If friction can ionize the metal atoms, then the flow of ionized material becomes an electric current, which produces the magnetic field. Such a mechanism is known as a *dynamo*, a device that converts mechanical energy of motion into electrical energy. Thus the earth is more of an electromagnet than a permanent magnet.

In appearance the magnetic field of the earth resembles that of a bar magnet inclined slightly to the earth's axis of rotation. The magnetic lines of force run between the north and south polar regions of the earth, much as the pattern formed by iron filings sprinkled around a bar magnet does. The intensity of the magnetic field decreases away from the earth's surface, but the magnetic field can still be measured many tens of thousands of kilometers out in space.

However it began, the earth's magnetic field has changed polarity (the north magnetic pole becomes the south magnetic pole and vice versa) many times over geologic time. Scientists trace the history of these changes by studying the magnetism frozen into rocks of different ages: Iron particles in molten lava beds align themselves along the lines of the existing magnetic force, and after the rocks solidify, they retain the orientation of the magnetic field indefinitely. Such rocks show that magnetic reversals have come at intervals as short as 35,000 years. Why the reversals? We do not know. One suggestion is that they may be related in some way to changes in the earth's rotation or in the fluid state of its outer core.

FAR MAGNETIC FIELDS

The *magnetosphere* is that part of the magnetic field surrounding the earth that exerts a force strong enough to control the motions of charged subatomic particles entering the field. Exerting a strong force even 50,000 kilometers away from the earth's surface, this magnetic field protects us from bombardment by many of the charged subatomic particles traveling through space at speeds that are a significant fraction of the speed of light.

From satellites monitoring the magnetic field we have learned much about the magnetosphere's strength, direction, and composition. A cutaway section of its structure is shown in Figure 7.12. It has several concentric zones; the principal ones are zones of high subatomic-particle densities known as the *Van*

Allen radiation belts (named after American physicist James Van Allen, (1914–   ) who discovered their existence in 1958 from *Explorer* satellite data). The Van Allen belts encircle the planet in two doughnut-shaped regions about 3,000 and 17,000 kilometers from the earth's surface.

Charged particles, mainly protons and electrons, populate the magnetosphere's radiation belts. Most of these subatomic particles are ejected from the sun as a reasonably steady flow of matter in the plane of the ecliptic known as the *solar wind*. When the solar-wind particles encounter the earth, they are either diverted away from it or trapped by its magnetic field.

FIGURE 7.12
Cross section of the magnetosphere in the plane of the earth-sun line and earth's magnetic axis. Earth's magnetic field presents an obstacle to the plasma flow of the solar wind and is shaped by the encounter with the solar wind, which gives rise to the frontal, bow shock wave. The magnetosheath is the region of compressed subsonic plasma flow behind the shock. The magnetopause marks the thin boundary that separates the magnetosheath from the magnetosphere. Trapped charged particles are concentrated in the inner and outer Van Allen radiation belts (dark-color regions). Within the lighter shaded areas the particles are less stable and tend to be precipitated into the upper atmosphere, where they produce auroras.

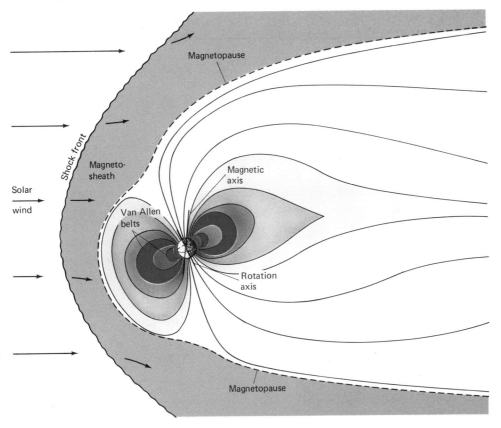

The collision of solar-wind particles with the earth's magnetosphere creates a shock wave that distorts and compresses the magnetic field on the sunlit side and stretches it into a long tail on the night side. (A *shock wave* is a large-amplitude compression, such as the sonic boom made by a jet plane.)

INTERACTION OF MAGNETOSPHERE AND ATMOSPHERE

Charged particles constantly spill out of the outer Van Allen radiation belt and fall into the auroral latitudes of the earth's atmosphere. There they collide with atoms of oxygen and nitrogen and stimulate these gases to radiate pale greens and occasional bright reds in patches or across the whole sky. These are the *auroras,* called the northern lights in our hemisphere (see Figure 7.13). They are most often seen in zones between 65° and 70° north and south magnetic latitudes. Because the subatomic particles enter the atmosphere easily when the solar wind is more intense, more auroras color our night skies during the height of the 11-year sunspot cycle (see Section 11.3), when the sun is emitting more subatomic particles.

COSMIC RAYS

In the early part of this century scientists found evidence for some kind of radiation entering the earth's atmosphere from outer space. Believing the radiation to be electromagnetic in nature, they called it "cosmic rays." Shortly thereafter cosmic rays were shown to consist of subatomic particles rather than electromagnetic waves; but the name stuck, and scientists still refer to them as rays. They should not be confused with very small meteoric particles that also enter the atmosphere.

Cosmic rays consist primarily of protons and helium nuclei (or alpha particles) with some heavier nuclei and a few electrons. They appear to be coming from all directions in space in about equal numbers. The kinetic energies of the cosmic-ray particles cover a very wide range, with the high-energy ones among the most energetic known in nature. There is a low-energy—that is, a low-velocity—component of cosmic rays that we know to be coming from the sun. Thus scientists divided the cosmic rays into a Galactic cosmic-ray component coming from outside the solar system and a solar cosmic-ray component.

When cosmic-ray particles encounter the magnetosphere, the lowest-energy ones are trapped in the

FIGURE 7.13
Northern lights over Canada. Charged particles from the sun crash through the upper atmosphere near the magnetic poles and cause it to glow like a neon sign.

earth's magnetic field, while those with a little larger energies are channeled by the magnetic field to enter the earth's atmosphere at high magnetic latitudes (which is approximately the same as high geographic latitudes). The higher-energy cosmic rays from Galactic space penetrate the magnetosphere almost as if it were not there, striking the nuclei of atmospheric molecules, from which enormous showers of subatomic particles rain down on the surface of the earth. As we go about our daily lives, we are almost continually pierced by these cosmic-ray-shower particles.

The dynamics of these cosmic-ray showers and primary cosmic-ray particles have told us a lot about the earth's magnetic field and the magnetosphere. Charged particles, such as the cosmic-ray particles, are controlled in their motion by magnetic fields and thus are natural probes to reveal the intensity and direction of the magnetic field. We shall return to the cosmic rays in Chapter 16, as we discuss our Galaxy.

7.5 Our Nearest Cosmic Neighbor: The Moon

As stated at the beginning of this chapter, we could think of the earth-moon system as a double planet: The moon is more nearly comparable in size to the earth than the other satellites are to their planets, with the exception of Pluto and Charon. Yet the moon would not quite span the width of the United States, and its mass is roughly 1 percent of that of the earth. Table 7.5 summarizes the moon's physical properties.

LUNAR EXPLORATION

The lunar-exploration program that began in 1964 with unmanned craft and culminated in six manned *Apollo* landings between 1969 and 1972 has provided us with a priceless legacy of lunar materials and data. Lunar rocks (see Figure 7.14) have been collected from nine different locations, six by the United States and three by the Soviet Union (the most recent being August, 1976). The samples returned amount to more than 2000 individual samples, weighing about 382 kilograms (843 pounds).

Five instrument packages were left on the lunar surface, the last surviving one operating until October 1977. The seismometers in these packages detected meteoric impacts and many lunar quakes during their operating life span of about 8 years.

The *Apollo* program also carried out an extensive effort to photograph and analyze the lunar surface. The result is maps of some parts of the moon better than those of some areas on earth. X-ray and radio-activity studies from orbit have yielded estimates of the chemical composition of about one-quarter of the lunar surface, an area about the size of the United States and Mexico together.

THE LUNAR SURFACE

Because of its small mass, the moon's history has been vastly different from the earth's. With a small mass comes a weak gravitational attraction; as a result the moon retains almost no atmosphere. It has no surface water, either free or chemically combined in the rocks

TABLE 7.5
Physical Data for the Moon

Quantity	Value or Description
Mean radius	1738 km (27% of earth's equatorial)
Rotation period	27.32 d
Inclination of axis	6.5°
Mass	7.35×10^{25} g ($\frac{1}{81.3}$ that of earth)
Average density	3.3 g/cm³
Surface acceleration by gravity	162 cm/s² ($\frac{1}{6}$ that of earth)
Escape velocity	2.4 km/s
Atmosphere	Nearly absent; traces of helium, argon, and possibly neon
Water	None present in lunar samples; interior may hold some
Average surface reflectivity[a]	7%; lowest in maria (approximately 5%); highest in cratered regions (approximately 10 to 15%)
Maximum day temperature[b]	130° C
Minimum night temperature[b]	−170° C
Magnetic field	No general field; weak local fields over the lunar surface but less than 1% of earth's general field

[a]Derived from photographic and photometric color measurements of the sunlight reflected from the moon's surface.
[b]Found from terrestrial observations and from temperature gauges placed on the moon's surface.

FIGURE 7.14
Astronaut David Scott on the moon, preparing to photograph a lunar rock sample.

or *rilles*. The kinds of surface features on the moon are outlined in Table 7.6.

Over eons of time small meteoroids have pulverized the lunar surface, leaving a dusty layer, some 1 to 20 meters deep, covering the lunar terrain. Known as the *regolith*, it is the lunar "soil" on which the astronauts left their footprints. Since this soil contains no water or organic matter, it is totally different from solids formed on the earth by water, wind, and life. It could only have been formed over billions of years on the surface of an airless body.

More than just bits of ground-up lunar rocks, the regolith has also been exposed to cosmic-ray particles, subatomic matter flowing from the sun, and a fine dust from interplanetary space. Without an atmosphere to shield it the layers of the regolith contain both the record of lunar events and that of events in the larger solar system.

CRATERS ON THE LUNAR SURFACE

A tremendous number of impact craters pit the moon, evidence of cataclysms that altered the crust during its past. More than 30,000 are visible by telescope. The total, down to bushel-basket size, may well exceed a million. The great walled plains, or supercraters with low profiles, like Clavius or Grimaldi (Figure 7.16) have structures similar to those of the maria but on a smaller scale. Their diameters are between 200 and 300 kilometers.

Next in size on the moon's front side are some three dozen impact craters from 80 to 200 kilometers in diameter. A third of them have conspicuous light-colored streaks, called *rays*, radiating outward in all directions up to several hundred kilometers long, such as the well-known ray craters—Tycho, Copernicus, Kepler, and Aristarchus. Many of the small secondary craters, as well as the ray systems, were apparently formed by a rain of debris ejected from the primary crater after a large body struck the surface.

Impact craters are reasonable circular, with the interior rim steeper than the outer rim. The larger craters have terraces on their inner walls and frequently have a fairly smooth floor from which a few low peaks rise (Figure 7.16). Beyond the craters the terrain is hummocky and overlain with the ejected material from the cratering activity. Even moderate-sized craters, like the larger ones, have high walls. In these craters having a central peak, the peak is believed to have been created by the elastic rebound of rock from

(as in earth rocks), although some water may be trapped under its surface. It also has no general magnetic field, but its rocks suggest that a strong one existed in the very distant past. However, the moon is far from a simple, featureless satellite.

Galileo's subdivision of the lunar surface into *maria*, the low-lying, almost circular dark regions in Figure 7.15, and *terrae*, the rough, cratered highlands, is still significant in terms of lunar history and terrain-shaping processes. The maria are covered with layers of basaltic lava similar to the lavas that erupt from terrestrial volcanoes in Iceland, Hawaii, and elsewhere. The highlands are a lighter-colored rock that are older than the rocks of the maria.

Even with earth-based telescopes observers over the years have recognized a variety of other surface features, such as a range of sizes of impact craters, rugged mountain ranges, and deep winding canyons,

FIGURE 7.15

Photographic map of the moon, showing dark maria and lighter highlands. This photograph is a combination of two photographs taken at first and third quarter. Note that on the left half the shadows in craters are on the left side, while on the right half shadows are on the right side of the craters.

Mare Crisium—Sea of Crises
Mare Foecunditatus—Sea of Fertility
Mare Frigoris—Sea of Cold
Mare Humorum—Sea of Moisture
Mare Imbrium—Sea of Showers
Mare Nectaris—Sea of Nectar

Mare Nubium—Sea of Clouds
Mare Serenitatis—Sea of Serenity
Mare Tranquillitatus—Sea of Tranquility
Mare Vaporum—Sea of Vapors
Oceanus Procellarum—Ocean of Storms

TABLE 7.6
The Moon's Surface

Surface Feature	Description
Maria (seas)	Dark-colored, relatively smooth basins, roughly circular, diameter about 300 to 1000 km; wrinkled by serpentine ridges, indented with a few shallow craters; 16 on side facing earth, including one "ocean"; 4 on back side; cover about 17% of surface
Craters	Millions, from microscopic size up to about 250 km in diameter; more on back side
Ray systems	Several radiate from prominent craters; mostly on front side
Mountains	Several dozen on side facing earth, including mountain ranges as well as isolated peaks
Rilles (crevices)	More than 1000 on front and back, both straight and winding
Ridges	Narrow, wrinkled projections rising to hundreds of meters and many kilometers long; extend across basins
Scarps	Occasional abrupt discontinuities as steep slopes
Domes	Volcanic blisters up to about 15 km in diameter dotting the maria

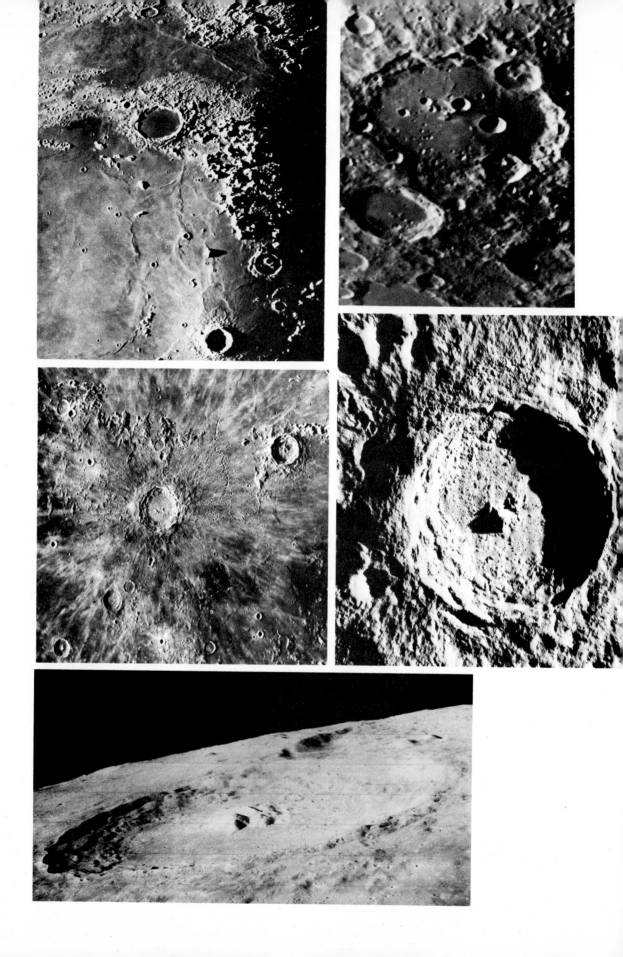

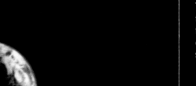

PLATE 4
The earth and moon seen from *Voyager 1* in September 1977 when it was 11.66 million kilometers from earth. The moon at the top has been artificially brightened compared to the crescent-shaped earth. (NASA)

PLATE 5
Surface of the moon. Scientist-Astronaut Harrison H. Schmitt is shown examining a huge, split boulder at the Taurus-Littrow landing site of *Apollo 17.* (NASA)

◄ **PLATE 3**
Visible-region spectra, continuous spectrum for a tungsten lamp, and the absorption spectrum for the sun. (Courtesy of Bausch & Lomb)

FIGURE 7.16

◀ Some famous lunar impact craters. The many craters that pockmark the moon's surface attest to the violence of the moon's past, when it underwent a series of asteroidal bombardments that petered out about 4 billion years ago. Plato is the large smooth-bottomed crater in the top left photograph. It is immediately above the center of the photograph here, which shows the eastern section of Mare Imbrium. Close to the terminator, near the right edge center, is the Alpine Valley, appearing as a long dark streak across the Alps Mountains. Clavius (top right) is the largest crater on the near side of the moon, about 240 kilometers in diameter. The crater Copernicus (center left) is about 80 kilometers across and clearly shows the diverging ray system associated with it. Notice the terraced walls on the left side of the crater Tycho (center right). (bottom) In the foreground is crater Theophilus, which is about 100 kilometers across. Cyrillus is the crater in the background. Theophilus is younger and has a sharper, more distinct appearance than Cyrillus.

Top row: Crater Plato; crater Clavius

Center row: Crater Copernicus; crater Tycho

Bottom: Craters Theophilus and Cyrillus

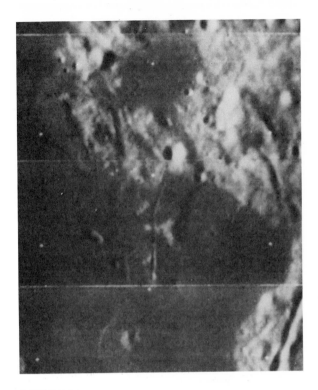

FIGURE 7.17

Volcanic craters and volcanic domes on the moon. (above) *Orbiter 4* photographed a small shieldlike lunar volcano, which is about 12 kilometers in diameter. (below) A lunar volcanic dome that is approximately 19 kilometers in diameter. Lunar domes resemble those in northern California and Oregon. They may represent the eruption of lavas that are more viscous than typical basaltic lavas.

below the surface after the initial impact. Others have bare floors, presumably because they were flooded with lava; the crater Plato is a good example.

The impact craters are not all of the same age, as one can see in Figure 7.16 with the crater Clavius. The rim of the major crater is eroded and worn, whereas the half dozen or so small ones in the center have sharper and higher rims. Clearly the impacting bodies that produced the small craters superimposed on the rim of the large crater must have fallen more recently than that one which formed the large crater. An inspection of crater photographs shows that they have not only a spectrum of size but also a range in erosion and wear; that is, they are not all the same age. As an example, the craters Copernicus and Tycho are about 600 and 200 million years old, respectively.

Volcanically produced craters, formed mostly during the moon's early history, are present but in smaller numbers than those of impact origin. Volcanic craters are not always circular. Their rims slope at about the same angle, both inside and outside, and their rims and floors are shallower than those of impact craters. Several volcanic domes and other evidence for volcanic activity can be seen in Figure 7.17. Thus thermal activity as a terrain-forming mechanism has not been

absent from lunar history, although it is markedly less important than on the earth.

OTHER LUNAR SURFACE FEATURES

Highlands constitute about 80 percent of the surface of the moon. Although fractures are observed in the lunar crust, there is no evidence of folded mountain belts (as on the earth) nor other indication of thermal-tectonic activity. Lunar mountain ranges were apparently produced in conjunction with the formation of the impact basins that are now seen as maria. Most mountain chains are on or near the periphery of the roughly circular maria. The mountains bordering the maria rise more steeply on the side facing them than on the other side. Many have lofty peaks, occasionally rising over 7000 meters above the surrounding plains.

Beyond the eastern edge of Mare Imbrium a narrow valley cuts across the lunar Alps Mountains. This feature has long been known from photographs taken from the earth. From photgraphs taken by an orbiting spacecraft we now know that the Alpine Valley is a deep trough some 3 to 10 kilometers wide and over 100 kilometers long. Narrow channels (rilles), which resemble chasms or gorges, cut many kilometers across the lunar terrain, frequently without interruption. Running lengthwise down the middle of the Alpine Valley is a very conspicuous rille, as can be seen in Figure 7.18. Rilles may be lava channels, part or all of which were roofed when filled with flowing lava. Now these tubes have collapsed and are partly choked with rubble from the days of active lava flows.

FAR SIDE OF THE MOON

Topography on the far side of the moon is strikingly different from that on the near side (see Figure 7.19). Craters are everywhere, but few have steep slopes. The face averted from the earth has no extensive mountain ranges and no large lava-flooded basins comparable to Mare Imbrium on the near side al-

FIGURE 7.18
The Alpine Valley. Note the rille running centrally along its floor. The valley is about 130 kilometers long and up to 10 kilometers wide.

though some small basins exist on the far side. The far side thus lacks the near side's extensive lava flooding.

The moon's center of mass is displaced from its geometric center about 2 kilometers earthward. One consequence is that the lunar crust facing the earth is about half as thick as that of the far side. Perhaps this variation explains why the near side has more volcanic activity, which accounts for its large deposits of dark mantling material. It may also help explain why the basins on the far side are only partially filled.

> Art thou [the moon] pale for weariness
> Of climbing heaven, and gazing on the earth,
> Wandering companionless
> Among the stars that have a different birth?
>
> Percy Bysshe Shelley

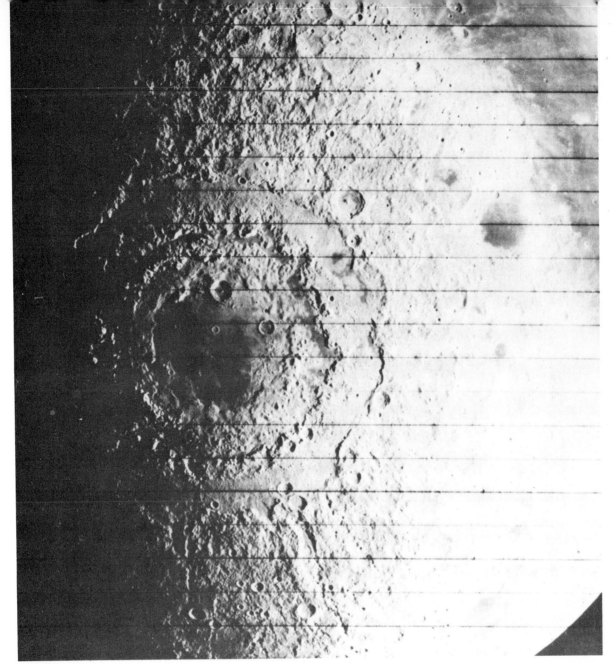

FIGURE 7.19
Mare Orientale basin photographed on the back side of the moon. The basin's outer scarp is about 930 kilometers in diameter. The bull's-eye basin is in an excellent state of preservation, with very little encrustation, erosion, or disfigurement. The large dark portion in the upper right is Oceanus Procellarum, which is also visible on the moon's near side.

7.6
Evolution of the Lunar Surface

CHEMICAL COMPOSITION OF THE SURFACE

The moon appears to have formed from the same chemical elements, although in somewhat different proportions, as those that formed the earth. It has less iron, more of the substances that are hard to melt (refactory materials)—such as calcium, aluminum, and titanium—and less of the easily vaporized atoms and molecules (volatile compounds)—such as sodium and potassium—than the earth does.

The most common surface rocks are anorthosites (silicates mainly of aluminum and calcium), iron-rich basalts from the maria, and thorium-rich and uranium-rich rocks. No traces of water and no organic compounds, the indicators of living processes, were discovered in any of the lunar samples. In fact the *Apollo* lunar rocks contain only tiny amounts of carbon and the carbon-based compounds from which life originates. With no water or oxygen present the minerals in the lunar rocks could not react with water to form clays or rust, nor did the iron react with oxygen to form oxides. The lunar highlands, which cover about four-fifths of the lunar surface, are the oldest preserved terrain. They consist mostly of material rich in aluminum and silicon and lighter than the mare basalts.

THE FIRST 2 BILLION YEARS OF THE MOON

Radioactive dating of the lunar rocks points to an origin much like that of the earth. Most of the lunar material probably came from accreting planetesimals (small solid bodies) that were part of the contracting material forming the solar system. Nearly all the original crust was lost when the moon underwent a global melting to a depth of at least several hundred kilometers, followed by chemical differentiation or separation shortly after the moon's formation. A few of the rock samples returned to the earth are about 4.6 billion years old, the same age as the earth (even though earth rocks no longer exist that are older than 3.5 to 3.8 billion years).

The highland areas are apparently about 4.0 to 4.3 billion years old. After its formation, the global crust of the moon was continually modified by the impact of material from elsewhere in the solar system. The cratering record preserved in the early crustal units represents a distinct phase of intense cratering, which began to decline rapidly about 3.8 billion years ago (see Figure 7.20). Although volcanic processes may have operated during this early period, the surface history of the moon is primarily that due to cratering. As mentioned earlier, this phase in the earth's history has been almost completely erased. Impact cratering continues today, but at a drastically reduced rate from what it must have been billions of years ago.

The next stage in lunar history was dominated by the formation of dark mare plains, which cover about 17 percent of the lunar surface, favoring the earthward side. These structures are relatively thin ponds of basaltic lava totaling less than 1 percent of the volume of the crust. *Apollo* rocks from maria suggest that the major outpouring of lava occurred between 3.9 and 3.2 billion years ago. Although some mare deposits may be as young as 2 billion years, there appears to have been no extensive igneous activity on the lunar surface for the last 3 billion years. Thus the shaping of the present lunar terrain is almost the opposite of that of the earth's—the moon dominated by cratering and the earth by volcanic and tectonic activity.

7.7
Internal Structure of the Moon

LUNAR MAGNETIC FIELD

There is no general lunar magnetic field as large as approximately one-ten-thousandth that of the earth, which seems to indicate that the moon does not now have the molten iron-nickel core comparable to that of the earth believed necessary to produce a magnetic field. But evidence suggests that the moon may have had a stronger magnetic field early in its history. Random magnetic fields up to about 0.6 percent of the earth's field intensity were detected at different sites by the *Apollo* astronauts, but we do not know the reason for such magnetic anomalies.

SEISMIC ACTIVITY

From seismographs left on its surface we know that seismic events on the moon follow patterns different from those here on the earth. Moonquakes, rare meteorite impacts, and artificially produced vibrations (grenade explosions and crash landings of discarded spacecraft) are transmitted very slowly through the lunar material. They build gradually and then take up to an hour to subside.

Some seismic disturbances have been traced to geologic movements in the rilles; others, to occasional impacts of meteoroid swarms. Moonquakes

Every step we take upon mother earth should be done in a sacred manner; each step should be as a prayer.

"The Sacred Pipe," Black Elk's Account of the Oglala Sioux

FIGURE 7.20
Evolution of the moon's surface. The moon 3.9 billion years ago (left), when it was undergoing collisions with fragments of material left over after the formation of the solar system. The largest of these impact features were like Imbrium Basin in the upper left. Most of the surface was saturated with craters.

The moon 3.1 billion years ago (above) after the cratering barrage ended. Shattering of the crust allowed magma from within the moon to reach the surface and to fill large basin floors. This continued over the next half a billion years or so, forming the maria.

The moon today (left), with newer craters, such as Copernicus and Tycho, whose bright rays overlie the ancient surface. Thrown out into the maria, material ejected from recent cratering tends to lighten the maria's contrast with the highlands.

frequently coincide with tidal stresses triggered by the varying distance between moon and earth. They occur at depths of 600 to 900 kilometers, much deeper than earthquakes. About 80 sources for these deep moonquakes have been discovered so far. But compared with the earth's seismic activity, the moon's is fairly subdued; the whole moon releases less than one-ten-billionth of the earth's earthquake energy.

About 35 shallow quakes, presumably tectonic events, have been detected. Thus if the moon is expanding or contracting, it is doing it extremely slowly. In the last 3 billion years thermal and geological activity has been relatively rare. As we have mentioned, most volcanic activity appears to have ceased about 3 billion years ago, but some minor activity may still be going on.

MODEL FOR THE INTERIOR

Seismic data tells us that the crust is about 60 kilometers thick, twice the thickness in kilometers of the earth's crust. The basaltic material of the maria was formed by the impact of large bodies, which provided the energy to melt partially the iron-rich regions below the crustal layer. Heat flow from the deep interior through the lunar crust is no more than about a third of that for the earth. Thus thermally driven processes cannot be so important as for the earth.

The moon's mantle, nearly 800 to 1000 kilometers deep under the lunar crust, is uniformly structured. Most of it may be pyroxene and olivine, minerals containing silicon, oxygen, calcium, magnesium, and iron. The seismic data reinforce the view that the moon's core is unlike earth's metallic core. Probably the lunar core consists of partly molten silicates, with a small metallic center, as shown in Figure 7.21. At present scientists can neither rule out the existence of a small iron core nor prove that one does not exist.

Only a few decades ago the moon was a light in the sky that, even though near, was still part of the remote cosmos. Now it is a place that has been visited by human beings and studied to such a close extent that it seems no longer to be a cosmic body. What is its future role in the affairs of man? There are proposals to establish permanent bases on the moon for astronomical observatories, mining operations, ore refining, or manufacturing; and it is not impossible that the

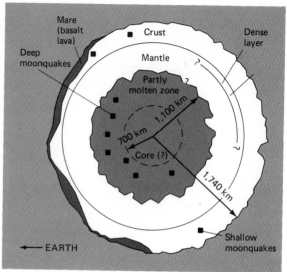

FIGURE 7.21
Schematic diagram of one proposed model of the moon's interior. The moon is not uniform but is apparently divided into layers as is the earth. There is probably a crust of about 60 kilometers and a mantle of some 800 to 1000 kilometers. Although still unknown, there may be a small iron core surrounded by a partially molten zone.

moon may become the departure station for manned exploration of the solar system. But from any point of view it is no longer a strange and distant world.

SUMMARY

The interior of the earth. The shape of earth is an oblate spheroid with the larger diameter in the equatorial plane. This equatorial bulge is a result of the rotation of the earth. In addition to bulges of solid portions of the earth's crust, the moon's gravitational attraction produces two tidal bulges in the oceans and atmosphere. The moon's braking effect on the earth is also slowing earth's rotation. The study of seismic waves reveals that the interior of the earth is not rigid. The hot, highly compressed inner core of solid iron and nickel is surrounded by a molten outer core of liquid iron and nickel. The outer envelope sur-

rounding the core is the mantle. A thin layer—the crust—overlies the mantle and contains the continents and ocean basins.

Surface of planet earth. Earth's crust and outermost part of the mantle constitute the lithosphere. Underlying it is the asthenosphere, a partially molten layer that flows under pressure. The lithosphere resting on the asthenosphere is thus not rigid but is actually broken into a dozen major lithospheric plates that have been rafted across the earth's mantle over the 4.6 billion years of earth's history. The plate boundaries are regions of intense earthquake and volcanic activity.

The present conformation of land masses is the result of the breakup of a single supercontinent about 200 million years ago. Because plate movements are driven by the outflow of heat from the earth's interior, over the next 2 billion years or so, when the earth has cooled further, the plate movements will slow and eventually cease. Other forces that shaped the surface features of earth are impact cratering by meteorites early in the earth's history. That phase has been virtually erased by more recent thermal-tectonic activity.

Earth's atmosphere. Earth's atmosphere insulates the earth by the greenhouse effect and blocks harmful electromagnetic and particle radiation from reaching the surface. The atmosphere has several layers, each with distinctive thermal, chemical, physical, and electrical properties. The chemical composition of the lowest 90 kilometers of the atmosphere is primarily nitrogen (77 percent) and oxygen (21 percent). Above 90 kilometers, there is a chemical separation with the heavier elements settling and the lighter molecules rising. The relatively large concentration of oxygen now in the atmosphere is the result of oxygen-producing plants, beginning about 2 billion years ago. The amount of each constituent is controlled by various production and destruction processes.

Magnetosphere of the earth. The earth acts as an electromagnet. Motion of electrical currents through the earth's outer core generates the magnetic field. Magnetic lines of force run between the poles and extend outward from the earth, forming the magneto-sphere. This magnetic zone controls the motion of high-energy particles of the solar wind and cosmic rays, trapping many in the Van Allen radiation belts.

The moon. The lunar exploration program has given scientists samples of lunar material to study and strengthened the belief that the earth and moon are the same age, about 4.6 billion years. The lunar surface presently differs from earth's in having no atmosphere, no water, no general magnetic field, no active volcanoes. The moon's surface is marked predominantly by two types of physical features: lava-flooded basins and mountainous highlands. The lunar terrain is pock-marked by hundreds of thousands of impact craters of all sizes. The face of the far side of the moon differs from that of the near side.

Evolution of the lunar surface. Although the same elements appear on earth and moon, the proportions are different. The original crust was lost when the moon underwent global melting. Intense cratering, occurring later, declined significantly about 3.8 million years ago. Maria formed in the next phase when lava outpourings occurred 3.9–3.2 million years ago.

Internal structure of the moon. Seismic activity is occasional, but of a low order of magnitude. Most volcanic activity ceased about 3 million years ago. The conclusion is that the lunar interior is rigid and lacks sufficient thermal circulation required for the dynamic events of earth's interior. The moon may or may not possess a small iron-rich core.

KEY TERMS

aurora
continental shield
cosmic rays
crust
greenhouse effect
igneous rock
impact cratering
inner core
ionosphere
lithosphere
magnetosphere

mantle
maria
outer core
plate tectonics
sedimentary rock
seismic waves
shock wave
solar wind
thermal-tectonic activity
Van Allen radiation belts

CLASS DISCUSSION

1. How have scientists been able to determine the structure of the earth's interior? What comprehensive picture do we have of the interior?

2. What is meant by thermal-tectonic activity? Why does it occur? What is continental drift? What role do midocean ridges play?

3. Is the surface of the earth old or young compared with the estimated age of the earth? How do we know? Will the surface change at any time in the future?

4. How is the atmosphere of the earth structured? What influences its structure? Has its chemical composition and structure always been the same as now?

5. Describe the lunar surface compared with the earth's. What have been the mechanisms for change on the lunar surface? Is the lunar surface old or young?

6. Are both the near and the far sides of the moon the same? If different, how?

7. How is it that some lunar rocks brought back from the moon are older than the oldest terrestrial rocks if the earth and moon were formed at the same time?

READING REVIEW

1. Do all seismic waves move at the same speed? Are they all the same?

2. How thick is the earth's core compared to the mantle?

3. Are the continental and oceanic crusts of the same composition and same age?

4. Where do volcanoes occur on a lithospheric plate?

5. Is there any cratered terrain still to be found on the earth's surface?

6. Is the earth's atmosphere chemically homogeneous from bottom to top? If not, is any of it homogeneous?

7. Why does the clear sky of the earth sometimes have a whitish appearance instead of its usual blue color?

8. What are the ray systems observed around some lunar craters?

9. What kinds of activity are still present on the moon?

10. Is the composition of the lunar surface very different from that of the earth?

11. Does the moon have seismic events as the earth does?

12. How large is the lunar magnetic field compared with that of the earth?

SELECTED READINGS

Carrigan, C. R., and D. Gubbins. "The Source of the Earth's Magnetic Field," *Scientific American,* February 1979.

Cloud, P. "Beyond Plate Tectonics," *American Scientist,* July–August, 1980.

French, B. M. *The Moon Book.* Penguin Books, 1977.

NASA. *Mission to Earth: Landsat Views the World.* U.S. Government Printing Office, 1975.

Siever, R. "The Steady State of the Earth's Crust, Atmosphere and Oceans," *Scientific American,* June 1974.

Wood, J. A. "The Moon," *Scientific American,* September 1975.

8

The Inner Solar System: The Terrestrial Planets

Since the space age began in the late 1950s, more has been learned about the planets and some of their satellites than in all of preceding history. It seems probable that we shall have explored, as part of the space programs of several countries, much of our solar system by the end of this century.

By studying different planets, each with its own characteristics, we hope to learn how they have evolved to their present state through the same or different sets of dynamic processes; for the planets are the laboratories needed for observing processes beyond the range, in both time and extent, of our terrestrial environment. To understand our own planet better, we need a perspective that can only be acquired by comparative study of the other terrestrial planets. The full impact on earth's environmental problems of our increased understanding of the solar-terrestrial connection, the interplanetary environment, and the comparative study of the planets is yet to be realized. It is not an accident that we have evolved on the earth rather than on Venus or Mars, and we should be aware of that fact when we tamper with our environment.

In this chapter we consider what scientists now know about the surfaces, interiors, and atmospheres of the other terrestrial planets, having just studied these regions for the earth and moon.

8.1
Surfaces of the Terrestrial Planets

TERRAIN-SHAPING PROCESSES

In the last two decades two revolutions have occurred in our thinking about planets, particularly the surfaces of the terrestrial planets: the recognition that the lithosphere of the earth is divided into large plates moving relative to each other (plate tectonics) and the results of planetary exploration from the United States and Soviet space programs.

Almost everywhere across the surfaces of the moon, Mercury, and Mars, we see evidence for impact cratering and volcanic plains, with the relative

◄ Sunrise on the planet Venus, a photograph taken from the *Pioneer Venus* spacecraft in orbit about the planet.

percentage of the two terrain-shaping mechanisms varying from one body to the next. And from radar studies of Venus it appears that evidence is there for both processes. For the earth, however, scientists have a very different picture: It is a planet dominated by tectonic and thermal activity.

Occurring some 3.2 to 4 billion years ago, a period of heavy bombardment produced much of the impact cratering seen on the terrestrial planets. A by-product of that period of heavy impact cratering was lava flooding of immense impact basins to form maria, impact basins filled by lava flooding. This is essentially why the surfaces of the moon and Mercury look as they do. For the larger of the terrestrial planets—earth, Venus, and Mars—the second major terrain-shaping mechanism has remade almost all of the surface in the case of the earth but only a part of the surfaces of Venus and Mars. This mechanism is the tectonic-thermal activity of lithospheric convection. The surface is fractured, deformed, and in the case of the earth replaced by erosion from water, wind, and life, over billions of years. Heavily cratered terrain is generally believed to be the oldest type of surface. Its preservation implies a relatively stable history for a crust. Volcanic material can reach a planet's surface to develop plains or volcanoes only if the crust can be broken, allowing lava to flow onto the surface. Of the terrestrial planets the earth shows far and away the greatest volcanic activity, with presumably Venus next (we really don't know at this point), followed by Mars, Mercury, and the moon.

MERCURY: LIKE A DRIED AND CRATERED APPLE

Observations from the earth had hinted that Mercury might look like our moon. But it was the *Mariner 10* flyby on March 29, 1974, and the two subsequent flybys on September 21, 1974, and March 16, 1975, that showed that the planet is indeed heavily cratered (Figures 8.1 and 8.2) like the moon. Although from photographs the overall surface of Mercury is remarkably similar to that of the moon, there are significant differences—differences that suggest a somewhat different surface evolution from that of the other terrestrial planets.

The surface of Mercury is pockmarked with craters ranging from hundreds of meters up to hundreds of kilometers. Some of the bright craters have extensive ray systems like those on the moon. Compared to the moon, Mercury and Mars are deficient in craters in the

FIGURE 8.1
Comparison of cratered terrain on Mercury (top left), the moon (top right) and Mars (right). All three photographs are at about the same scale and are seen under about the same lighting conditions. Compared to the moon, Mercury and Mars have fewer small craters in the 10- to 50-kilometer range. These two planets have more extensive intercrater plains than the moon has, but the distribution and appearance of Mercury's craters are more like those of the moon than of Mars. Note that the Martian craters are more eroded or degraded than are those of Mercury or the moon.

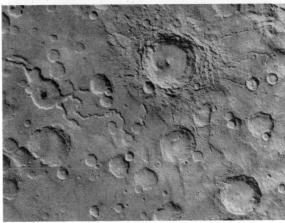

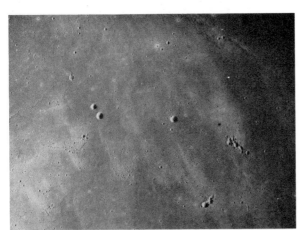

FIGURE 8.2
Maria on the moon (left), Mercury (bottom left), and Mars (bottom right). The field of view in each photograph is about 100 × 140 kilometers with comparable illumination and viewing angle; resolution, however, ranges from about 50 meters for the moon to 500 meters for Mercury.

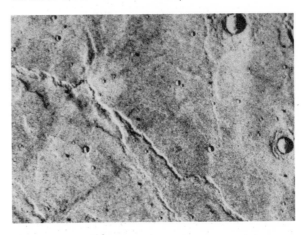

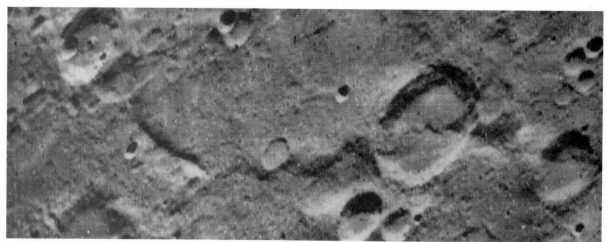

FIGURE 8.3
Scarp on the Mercurian surface. As much as 3 kilometers in height at points along its almost 500-kilometer length, this scarp is known as Discovery Scarp. Note that the scarp cuts across several craters; this transection indicates that the scarp is younger than the craters are. It probably formed when the crust of Mercury shrank during cooling and solidification.

range of a few tens of kilometers, and in their place we see the extensive intercrater plains of Figure 8.1.

Like the moon, both Mercury and Mars have dark maria, as shown in Figure 8.2; while those for Mars and the moon are almost indistinguishable, the maria on Mercury exhibit small but significant differences. Twenty or so of Mercury's maria are several hundred kilometers across, while Caloris, the largest basin is more than 1000 kilometers wide. Its interior surface resembles the famous Orientale Basin on the moon (see Figure 7.19). The basins on Mercury were created during the heavy bombardment period by a period of volcanism that filled them with lava. Mercury and the moon have remained virtually unaltered since this period of lava flooding 3 to 4 billion years ago. Mercury is also like the moon in a way not yet fully explained for either body: The craters cluster in one region, and the maria in another.

But there are some conspicuous differences between Mercury and the moon. Craters on the lunar highlands are densely packed, with rims of young craters overlying old craters, and the mare regions are sharply bounded. On Mercury, by contrast, craters are often interspersed with relatively smooth plains, giving the terrain a speckled appearance. Mercury has relatively few craters in the range of a few tens of kilometers, as mentioned above. Scarps or cliffs a few kilometers high and often hundreds of kilometers

long are distributed widely over the old and heavily cratered regions (Figure 8.3). They cut across plains and craters alike. Unlike the scarps on the moon, they may have been formed when the crust of Mercury cooled and shrank, like wrinkles on the skin of an old apple.

SPACE MISSIONS TO VENUS: A RUSSIAN SPECTACULAR

The Soviet Union has launched 14 *Venera* spacecraft toward Venus since 1967, with 12 of them descending through the atmosphere to the surface. *Venera 7* in 1970 and *Venera 8* in 1972 made the first successful, although shortlived, landings. *Venera 9* and *10* in October, 1975, landed on the daylight side and transmitted back to earth the first pictures of Venus's surface before succumbing after about 1 hour to the high temperature and pressure of that hostile environment. *Venera 11* and *12* landed in late December, 1978, surviving about 100 minutes with apparently no pictures transmitted. On March 1 and 5, 1982, *Venera 13* and *14* landed and took four pictures before breaking down after a couple of hours under the hellish conditions of the surface.

The United States has launched three Venus flybys: *Mariner 2* in 1964, *Mariner 5* in 1967, and *Mariner 10* in 1973. Our most ambitious effort was the two *Pioneer*

Venus craft launched on May 20, 1978, and August 8, 1978. The first member of the *Pioneer Venus* fleet was an orbiter, shown in Figure 6.13, which went into orbit around the planet. The rest of the fleet, which consisted of one large and three small probes and the mother ship that carried the probes, went immediately to the surface to widely separated landing points. Although none of the probes was designed to survive on the surface after impact, one small probe survived 1 hour and 8 minutes.

VENUS'S HIDDEN SURFACE:
A MASTER DETECTIVE STORY

Venus's surface is hidden by total cloud cover. Radar studies from the earth show that the planet's surface possesses geological features suggestive of impact cratering, volcanism, and tectonic activity. *Pioneer Venus* radar studies confirm the earth-based observations, and a contour map produced from them is

shown in Figure 8.4. While 65 percent of the earth's surface is ocean basin lying on the average about 5 kilometers below sea level, fully 60 percent of Venus's surface lies within 0.5 kilometer of the mean radius of the planet, and only 5 percent is more than 2 kilometers above the mean radius. Despite this limited spread in elevation, the maximum distance between the highest and lowest points is about 13 kilometers, which is comparable to that of the earth.

The highland area, Ishtar Terra, in Figure 8.4 is unlike anything seen on the moon, Mercury, or Mars. It is larger than the continental United States and stands several kilometers above the mean planet radius. In elevation it is similar to the Tibetan Plateau on earth but about twice its size. Maxwell Montes on the eastern end of Ishtar Terra contains the highest point on the surface of Venus—about 11 kilometers above the mean radius, or 2 kilometers higher than Mount Everest is above sea level. The Venusian mountains containing Maxwell Montes rise out of the Ishtar Terra plateau in the same way that the Himalayas stand on

FIGURE 8.4
Contour map of the Venusian surface produced from radar measurements by the *Pioneer-Venus* orbiter. The shading indicates surface elevations above and below a mean radius of 6051 kilometers. Of the terrain covered about 20 percent is lowland plains, 70 percent rolling upland, and 10 percent highlands. The greatest elevation is a mountain peak (between longitude 0° and 10°, latitude 60° and 70°) that rises about 11 kilometers above the mean radius, or 2 kilometers greater than the height of Mount Everest above sea level.

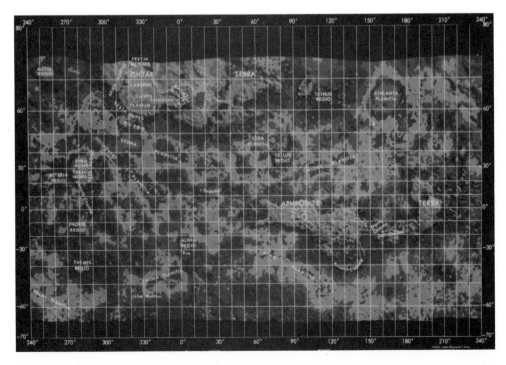

FIGURE 8.5
An artist's impression of the part of Ishtar Terra containing Maxwell Montes, portrayed with enhanced vertical elevations. Although the heights are exaggerated in proportion to distances across the plateau, one can get some sense of this geological feature. The North American continent without water in the ocean basins might look similar.

the Asian plate. An artist's rendering of this continent-like feature is shown in Figure 8.5. The other continentlike region, Aphrodite Terra, is also comparable in size to the continental United States.

Radar observations of the hidden surface disclose some almost circular structures between a few tens of and 1000 kilometers in diameter, which may be impact craters and basins. Other large features include a 1000 kilometer-long trough a few hundred kilometers wide and a few kilometers deep, which is similar in scale to the Martian Valles Marineris (partly shown in Figure 8.8). Maxwell Montes is a large, low, circular dome some hundreds of kilometers across, with a central depression 100 kilometers or so in diameter, similar in many respects to a volcanic peak. If truly a volcano, it is about 25 percent larger than Olympus Mons on Mars (see Figure 8.9). There are two other regions of suspected volcanic activity: Beta Regio and the eastern end of Aphrodite Terra (Figure 8.4).

Thus the possibility of impact craters and basins suggests a surface billions of years old, while the possibility of volcanoes, rift valleys, and plateaus point to youthful parts of the surface that may be only millions of years old. Venus apparently has not preserved as much of its early surface history as has the moon and Mercury, but it may be more like Mars than it is like the earth.

The two Soviet landers, *Venera 9* and *10,* sent back the first photographs of the Venusian surface in 1975, while *Venera 13* and *14* provided four more, with at least one in color. Sunlight filtering through the cloudy atmosphere supplies enough light to make the surface look like a dark, overcast day on the earth. The atmospheric color is decidedly orangish since the blue wavelengths of incoming sunlight are absorbed and scattered by the clouds. The *Venera 13* view in Figure 8.6 shows a rock-strewn plain with a dark fine-grained material interspersed between the rock outcroppings. Although having a number of noticeable differences, the view roughly resembles Martian terrain (see Figure 8.10). *Venera 14* landed on a terrain different from that for *Venera 13*. Its view was of a plain of broken rock layers that extend to the horizon. Samples were collected by both spacecraft and analyzed. Their composition suggests that the material is basaltic rock, an igneous rock extruded from the interior and generally silicon-poor and metal-rich. Such basaltic rock is common on the earth and moon.

MARS AND PROJECT *VIKING:*
SUCCESS BEYOND OUR WILDEST DREAMS

Once many astronomers thought that the seasonal changes of Mars's surface seen from earth were due to growing and dying vegetation. The photographs transmitted to the earth by the *Mariner* and *Viking* flybys, orbiters, and landers between 1965 and 1976 revealed a waterless, cratered planet that in many ways was similar to but in many other ways was different from both the moon and the earth.

The fine, delicate streaks called "canals" and sketched by observers on early Martian maps are illusory. The *Mariner* and *Viking* pictures revealed these canals to be nothing more than dark-floored craters or irregular dark patches aligned by chance and linked unconsciously by early observers into lines that looked like canals.

Viking 1 and *2* were two of the most sophisticated pieces of technical hardware in the space effort. Both arrived in the vicinity of Mars in the summer of 1976 after a 10-month journey. From orbit each detached a lander that sat down on the surface of Mars and began to collect information. Although the two orbiters and one of the landers have ceased to operate, analysis of their data will continue for many years to come. All outlived their expected useful life and have provided us with a wealth of information about the surface terrain, atmosphere, biological activity, and satellites of the red planet Mars.

THE MARTIAN SURFACE: A RED, DUSTY WORLD

Like the moon and Mercury, Mars has a different topographic pattern in each hemisphere, but it is more diverse and complex than either the moon or Mercury. Mars's northern hemisphere is generally lower than the mean radius of the planet by a couple of kilometers, possesses few craters, and has been altered by intense volcanic activity and subsequent lava flooding. The extensive lava flooding occurred at various times after the cessation of the heavy bombardment period in the early solar system. On the other hand the southern hemisphere has a densely cratered surface, averaging a couple of kilometers greater than the mean radius. Its crust has not changed appreciably throughout the planet's life.

There are about 16 smooth circular basins, or maria, on Mars. One that has long been observed from the earth is Hellas, an almost craterless basin about 1600 kilometers wide, or about one and a half times the size of the largest lunar sea, Mare Imbrium. The small number of impact craters on the maria suggests

FIGURE 8.6
From the top down, surface close-ups of the moon, Venus, earth, and Mars. The *Venera 13* photograph of Venus's surface shows rocky terrain (March, 1982). Perhaps volcanism formed these rocks; this would be in accord with radar views that show evidence of tectonic activity. In all four frames, chemical weathering has played a minor role with most of the fine-grained material the result of impacting or wind (except for the moon).

PERCIVAL LOWELL   (1855–1916)

Percival Lowell came from a distinguished New England family. His younger brother, Abbott, was president of Harvard University and his sister, Amy, was a well-known poet and critic. After his graduation from Harvard in 1876, with distinction in mathematics, he traveled for a number of years throughout the Far East before settling down to a career in astronomy. He was particularly interested in Mars and its "canals," of which drawings by the Italian astronomer Schiaparelli had received wide public attention.

In 1894 he founded the Lowell Observatory at Flagstaff, Arizona. Its altitude of some 7000 feet and its dry desert air made it an excellent observing site for the study of Mars, which was then close to the earth.

During 15 years of intensive study of Mars, whose surface markings he drew in intricate detail, Lowell pressed for acceptance of the network of several hundred fine, straight lines he drew on his maps and their intersection in a number of "oases." Lowell concluded that the bright Martian areas were deserts and the dark ones patches of vegetation and that water from the melting polar cap flowed down the canals toward the equatorial region to revive the vegetation. He supposed the canals to have been constructed by intelligent beings who once flourished on Mars. Lowell published his views in two books: *Mars and Its Canals* (1906) and *Mars As the Abode of Life* (1908). The canals, as it turns out, are mostly chance alignments of dark patches that the eye, at the limit of resolution, tends to string together into lines.

Lowell's greatest contribution to planetary studies came during his last 8 years, which he devoted to the search for a planet beyond Neptune. He first analyzed the discrepancies between the observed and the calculated positions of Uranus after making allowance for the perturbations of Neptune. But on examining photographs of the region of the sky in which he predicted the new planet might be, he found no such object. The search continued for a number of years after his death at Flagstaff in 1916; the new planet, named Pluto, was discovered by Clyde Tombaugh in 1930. It is perhaps fitting that the first two letters of Pluto are Percival Lowell's initials even though he was not its discoverer.

that they formed after the heavy bombardment period ceased in the inner solar system some 4 billion years ago. Estimates of age for them are about 3 billion years or less.

The abundance of craters in some of the regions of the southern hemisphere is comparable to that in the bright highlands of the moon. The similarities between the cratered Martian southern hemisphere and the lunar highlands (Figure 8.1) has prompted the speculation that the two are of about the same age. Thus almost half the surface of Mars is ancient terrain, with many of its landforms having remained essentially unchanged over the last 4 billion years.

From spacecraft photographs, one sees that Mars has a wide variety of channels in the regions of oldest terrain. They appear to have formed between 3 and 4 billion years ago, shortly after the end of the heavy bombardment period. Figure 8.7 is an oblique-angle photograph of a channel that has all the appearance (left center) of some drainage systems on earth. Thus it is speculated that water flowed on the Martian surface and is now held as subsurface ice. Since both polar caps contain water ice, it is not unreasonable to hypothesize that a time existed in the past when a thicker and warmer atmosphere permitted liquid water on the surface.

An important departure from the character of the major portion of the Martian terrain is the Tharsis ridge, shown in Figure 8.8. This area is different because of three large volcanoes, running diagonally along the crest of the ridge, and a spectacular isolated volcanic structure, Olympus Mons (Figure 8.9), which is similar to, but much larger than, Mauna Loa and Mauna Kea in Hawaii as seen from the bottom of the Pacific Ocean. In addition to these large shield volcanoes rising some 20 kilometers above the surrounding plains, there are flattish saucer-shaped volcanoes, some which lack impact craters on their slopes sug-

gesting that they are relatively young. Tharsis ridge is 10 kilometers or so above the average Martian radius; the volcanoes extend above it.

In the equatorial region lies the spectacular canyon Valles Marineris, cutting across the middle of a plateau. It is nearly 4000 kilometers long, up to 100 kilometers wide in some places, and at least 4 kilometers deep (Figure 8.8). At its western end lies a complex pattern of intersecting fault valleys. Valles Marineris runs radially away from Tharsis ridge and probably results from the faulting that accompanied the evolution of Tharsis since its formation. However, thermal-tectonic activity on Mars is much weaker than that on earth.

From orbit the dominant features of the region around the *Viking 1* lander are craters. From the ground there are only a few obvious craters to be seen in the immediate vicinity of the lander. If Mars were like the moon, then there should be visible several small craters, tens of meters in diameter. Their absence indicates that the Martian atmosphere is dense enough to burn up small meteoroids or material ejected from large impacts before they reach the surface; and that surface erosion is vigorous enough to obliterate small craters. Thus there is not the profusion of small craters seen on the moon.

The area photographed by the *Viking 1* lander in the Chryse region is a gently rolling landscape, yellow-

FIGURE 8.8
A mosaic of photographs of the Martian surface taken in 1980 by the *Viking 1* orbiter on an exceptionally clear day. At the far left are three immense shield volcanoes on Tharsis ridge to the southeast of Olympus Mons (Figure 8.9). Running about 4000 kilometers across the center is Valles Marineris, an enormous canyon that appears to be the result of faulting as in the formation of the earth's lithospheric plates.

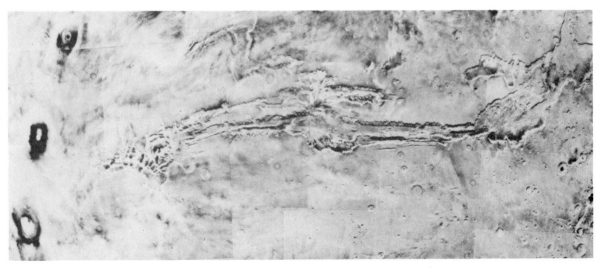

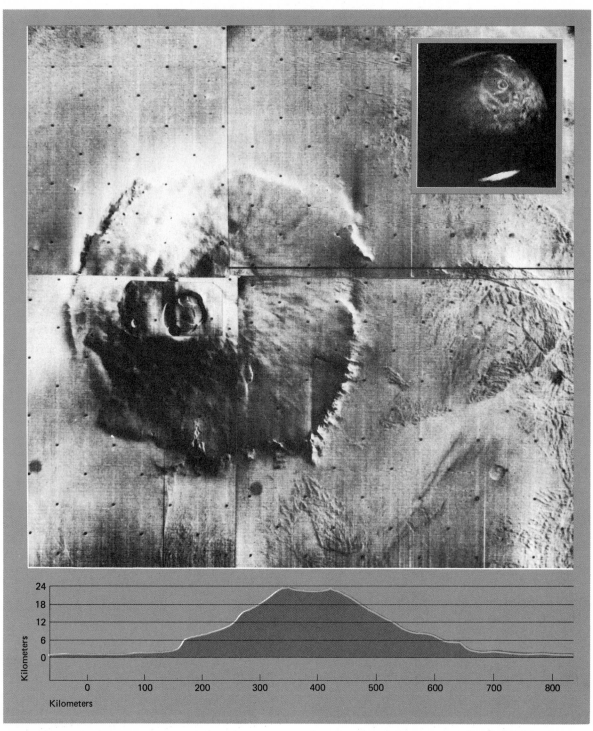

FIGURE 8.9
Mosaic of Olympus Mons on Mars. The supervolcano is nearly 600 kilometers wide at the base and 25 kilometers high and is capped by a crater 65 kilometers in diameter. It is about five times larger than the most massive volcanic cone on earth, Mauna Loa in Hawaii. At a great distance it appears as the small bright ring in the upper right of the inset photograph (taken by *Mariner 7*), in which the south polar cap is very conspicuous at the bottom. (The elevation diagram and mosaic of Olympus Mons are shown at different scales.)

FIGURE 8.10
View from the *Viking 1* lander, taken August 3, 1976. This remarkable picture shows a dune field with features similar to many seen on the earth's deserts. Cutting through the picture's center is the meteorology boom, which supports *Viking's* miniature weather station. The photograph covers 100° in azimuth. Just beyond the far right edge of the picture are several areas of exposed bedrock. The left side of the picture shows a large field of drifted material, probably deposited when the wind was blowing from north to south. The picture looks toward the east.

ish brown in color, strewn with rocks and dotted with drifts of fine-grained material. Within 30 meters or so of the lander lie several outcrops of bedrock, which in many ways resemble the semidesert regions of the American Southwest but without vegetation (Figure 8.10). Chemical analysis by the lander found the elements common in crustal rocks on the earth, but organic analysis failed to detect any organic molecules (those containing carbon, one of the essential ingredients for life). On earth, the soil of even the most dry, sterile-looking valleys contains many organic compounds.

The multilayered polar regions are still another type of Martian topography. A region near the south pole appears in Figure 8.11. The layered deposits probably hold appreciable quantities of frozen water mixed with dust beneath a carbon dioxide coating. Periodic changes in Martian climate may be responsible for the deposition of the successive layers of material.

EARTH'S SURFACE EARLY IN ITS HISTORY

As we discussed in Chapter 7, the evidence from the heavy-bombardment period in the inner solar system has been erased from the earth's surface. Even the oldest rocks, which are about 3.8 billion years old, are of no help in identifying this early impact cratering since they are mostly isolated outcroppings, many covered with ice. (However, later impact cratering still exists as discussed in Chapter 7.) The earth also has

basaltic plains, like the maria on the moon, Mercury, and Mars—not old ones, as on the other terrestrial planets, but very young. The major ones on earth are formed by the addition of new material to the lithospheric plates at the midoceanic ridges.

FIGURE 8.11
Layered terrain in the Martian polar regions. Taken by *Mariner 9*, the photograph shows a series of beds or layers about 10 to 50 meters thick overlying a pitted terrain (lower half of picture). The layers are apparently composed of ice particles and dust.

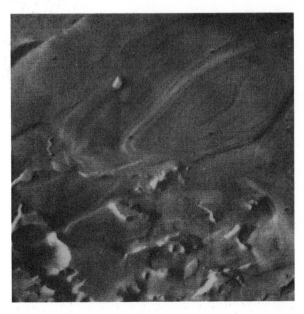

> The heavens themselves, the planets, and this center
> Observe degree, priority, and place,
> Insisture, course, proportion, season, form,
> Office, and custom, in all line of order.
>
> Shakespeare

It is unlikely that the earth ever went through an appearance like that of the moon, Mercury, and Mars, in which maria were formed from lava's flooding a huge impact basin. In all probability the thermal-tectonic activity of the earth's surface has always been too great to have allowed maria to form. Thus if we could watch a time-lapse movie of the evolution of the surfaces of all the terrestrial planets, it is unlikely that they would all start out the same and begin to depart from each other later. Rather, the surfaces of the moon and Mercury, which have thick lithospheres—thicker than those of Mars, Venus, and the earth—have not been fractured and then deformed by convection. This is because the moon and Mercury should have cooled very quickly, extinguishing any tectonic activity if it ever existed. And the earth has probably always shown vigorous tectonic activity, with Venus's less than that of the earth and Mars even less. Thus the moon and Mercury have the oldest surfaces; Mars's surface is old but with some youngish features; Venus one guesses to have a mixture of old and young; and the earth's surface is comparatively very young.

8.2
Internal Structure of the Terrestrial Planets

MODELS OF PLANETARY INTERIORS

Astronomers can construct a model of the interior of a planet from the planet's observed physical properties and from theoretical arguments about the physical laws governing its internal structure. What the astronomer needs to know is the planet's mass, density, shape, rotation rate, gravitational- and magnetic-field strengths, surface temperature, and chemical composition. The model would naturally be more ac-

curate if the astronomer had seismic data and rock samples, as we have for the earth and the moon.

Astronomers unfortunately do not have reliable numbers for each of the quantities listed above. Nevertheless, it is still possible to make some headway on determining the internal structure. The mean density of a planet—combined with its total mass, shape, and rotation rate—suggests its internal composition and distribution of mass. When constructing a *model of a planet's interior,* astronomers start by making the following assumptions:

1. Since the planet has a stable configuration (it is neither contracting nor expanding), the weight of matter due to gravity's pressing inward is balanced by the pressure from more central matter's pushing outward.

2. The pressure exerted by matter in the planet's interior depends on its density and temperature in a more complicated fashion than is true for a simple gas.

3. The flow of heat out of the deep interior determines what the decrease in temperature will be out through the body of the planet.

4. Finally the matter in the interior can change from solid to liquid, can deform and flow under pressure, and can form different types of mineral compounds.

Having constructed the model, astronomers can use it to predict how the temperature, pressure, and density vary from the planet's center to the surface. Planetary models essentially show us that, like the earth, the other terrestrial planets possess such layered zones as core, mantle, and crust. Examples of some interior models for the terrestrial planets are shown in Figure 8.12, which should be consulted when reading the following subsections.

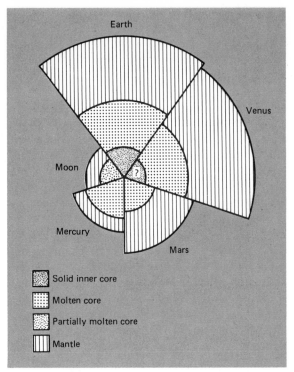

FIGURE 8.12
Models for the internal structure of the terrestrial planets, including the moon. The planets are drawn to the same scale. On this scale the crusts are not distinguishable; so they have not been included. The basic structure is a molten or partially molten core of iron-rich material enclosed by a silicate mantle. The earth and possibly Venus have an inner core of solid iron-rich matter.

WHAT SHOULD THE CHEMICAL COMPOSITION OF THE PLANETS BE?

Since scientists cannot directly determine the chemical composition of the deep interior of even the earth, how is it possible that astronomers can know the composition of distant planets? We don't know it by means of experimentation, but we believe we know something about it by means of a theoretical argument. The argument proceeds in the following fashion: Whatever composition the planets had at their birth has clearly not changed over the span of their lives since no significant influx of material from outside has been added to the planets' mass and no known processes are at work to change the original chemical composition into something else as there are in the case of stars.

What then was the original composition of the planets, and is it likely that it is not the same for each planet? As the solar system formed, it is probable that the sun was well along in the process and consequently reasonably hot by the time the planets began to form. Thus the temperature of the matter from which the planets would form was higher closer to the sun, probably 1000 K or so, and declined rapidly outward to about 100 K. Since that matter would have to be solid or solidlike, astronomers find that they can divide it into three groups on the basis of the ease with which they vaporize: The first group consists of the materials that are *gases* at temperatures above a few tens of degrees Kelvin, such as hydrogen and helium. Next are the *icy materials,* like methane, ammonia, carbon dioxide, and water (containing such light elements as carbon, nitrogen, and oxygen besides hydrogen), which do not vaporize until the temperature is over a couple of hundred degrees. Finally there are *rocky compounds,* such as iron and its oxides, silicates, and sulfides, which do not vaporize until the temperature is several thousand degrees. Hence close to the sun, where the terrestrial planets formed, the rocky materials would have been the only ones not in a gaseous form. Iron and the elements near it in the periodic table (see Appendix 4) should dominate the compositions of the terrestrial planets, as they seem to do, while lighter elements, like hydrogen, helium, carbon, nitrogen, and oxygen, should be the principal constituents of the Jovian planets, as evidence from their atmospheres confirms.

MERCURY: A LARGE IRON-RICH CORE

The values of its mass and radius imply that Mercury must contain a large fraction of iron, the only heavy element sufficiently abundant to account for the planet's high mean density (Figure 8.12). The iron and nickel content may be as much as 65 percent of the total mass of Mercury. By analogy with terrestrial, lunar, and meteoritic chemical abundances, we presume that silicates and oxides of iron are also prevalent.

An unexpected discovery by the *Mariner 10* mission was that Mercury has a shock front like the wave surrounding the bow of a ship plowing through water. Like the one for the earth in Figure 7.12, it is caused by the onrushing solar wind particles colliding with the planet's magnetic field. Thus it is apparent that Mercury has a magnetic field, which is about 1 percent as strong as that of the earth. The magnetic axis of its field almost coincides with Mercury's axis of rotation.

The magnetic field is intrinsic to the planet and is most likely the result of an internal mechanism that continuously generates the field in much the same way as the earth's does (pages 154–155). This is additional evidence for a large iron core. Chemical differentiation appears to have occurred very early in the planet's history, probably the first half billion years. Since then, the surface has been largely undisturbed by thermal and tectonic processes.

The model that has been derived for Mercury is one with a crust overlying a silicate mantle, which in turn surrounds a molten (or partially molten) iron-rich core. The core radius may be as much as 75 percent of the planetary radius, a percentage that is considerably greater than that for any of the other terrestrial planets, including the earth. Such a core should be adequate to generate the magnetic field observed by *Mariner 10*.

VENUS: SIMILAR TO THE EARTH?

Since Venus is the terrestrial planet closest to the earth in size and mass, it is reasonable to expect that it will generally be the planet most like the earth. Of critical importance in developing a model for its interior is Venus's chemical composition: As we have seen, from estimates of what chemical elements were likely to have been present at the time of formation of the terrestrial planets and from observations of the planets' influences on the motion of spacecraft, astronomers have made estimates for the iron and nickel content in the terrestrial planets; as percentages of total mass, the iron and nickel content is as high as 65 percent for Mercury, up to 38 percent for Venus, up to 33 percent for the earth and the moon, and only up to 26 percent for Mars.

Such an iron content for Venus suggests that it ought to have a molten core (like that of the earth), occupying about the same fraction of Venus's interior (Figure 8.12) as does the earth's (or significantly smaller than the core of Mercury). Overlying the mol-

Observe how system into system runs.
What other planets circle other suns.

Alexander Pope

ten iron-rich core is a silicate mantle and a crust on top of the mantle. It is possible that Venus (also like the earth) has an inner core of solid iron-rich material. Speculation for and against an inner core depends on estimates of the iron content of the planet.

MARS: MORE LIKE THE MOON

Besides the earth and moon, Mars is the only planet for which scientists have seismic data. Both *Viking* landers carried instruments to record quakes on Mars. Unfortunately, only the one on *Viking 2* worked, and on November 24, 1976, it appears to have detected its first Martian quake. If real, it suggests that the Martian crust has an average thickness of about 40 kilometers, with a maximum thickness of about 75 kilometers under the Tharsis ridge and a minimum thickness in the Hellas basin of about 10 kilometers. By comparison the earth's average crust thickness is 30 kilometers or so, and it covers a planet with nearly twice the radius of Mars. Hence the crust of the earth is about 0.5 percent of its radius, while that of Mars is about 1.2 percent. The moon's crust is about 4 percent of its radius.

The seismic activity on Mars, although somewhat more extensive than that of the moon, is much less than that of the earth. This, as noted, suggests a lithosphere (crust and outer portion of the mantle) a couple of hundred kilometers thick on a chemically differentiated body, with silicate mantle and iron-rich core of about 1500 kilometers radius. Thus of the four terrestrial planets Mars has the smallest percentage of iron and the smallest core for its size.

MAGNETOSPHERES OF THE TERRESTRIAL PLANETS

It has been known for many years that the earth has a magnetic field that extends far beyond the body of the earth to form a magnetic envelope called a *magnetosphere*. The earth's magnetic field is attributed to electrical currents flowing in the outer core; so the earth is somewhat like a giant electromagnet. If the other terrestrial planets also have electrical currents flowing in their deep interiors, then they too will have magnetospheres. Or we can turn the argument around: The spacecraft detection of a magnetic field tells us something about the interior of a planet; that is, it is sufficiently fluid for matter to flow in the deep interior to generate electrical currents.

As noted earlier, *Mariner 10* found evidence for Mercury's having a magnetic field, but it is much too weak to hold radiation belts such as the earth's Van Allen belts (see page 155).

Before the advent of planetary exploration with spacecraft, astronomers had suspected that the earth's sister planet, Venus, might have an internally generated magnetic field comparable to that of the earth. However, with the first few *Mariner* and *Venera* spacecraft it became evident that the magnetic field for Venus is much smaller than that of the earth. Venus has a well-developed bow shock formed by solar wind particles impinging on the planet's magnetic field; but that weak field, like that of Mercury, is too feeble to trap solar wind particles in a radiation belt. The weakness of Venus's magnetic field is probably due to the very slow rotation of the planet mentioned in Chapter 6.

In 1965 *Mariner 4* accomplished the first measurements of Mars's magnetic field. The intensity of the magnetic field is much less than that of the earth's field. There is a feeble bow shock formed between the onrushing solar wind particles and the field, but no radiation belts appear to exist.

In summary the earth is the only one of the terrestrial planets with a field intense enough to maintain radiation belts. This suggests that the earth is the only one possessing in its core a circulation pattern capable of generating a strong magnetic field.

8.3
Atmospheres of the Terrestrial Planets

WHY IS AN ATMOSPHERE THE WAY IT IS?

We are all aware of the characteristics of the earth's atmosphere and its consequent importance in maintaining life; Mercury and the moon, in contrast, have almost no atmosphere. How did the physical diversity of planetary atmospheres arise?

The nature of an atmosphere results from several factors:

1. The planet's distance from the sun along with its size and mass, which influence its ability to retain an atmosphere

2. Its chemical composition, which determines what processes go on in the atmosphere

3. Geological and chemical evolution of the planet's surface layers

4. Finally the atmosphere's interaction with biological life, if living organisms are in fact present

The planet's distance and mass are important in the following way: From basic physics we know that, the higher the temperature or the smaller the mass of a molecule (or both), the greater will be the average velocity of gas particles in the atmosphere. Since a particle moving outward might gain enough kinetic energy after colliding with another atmospheric particle to escape into space, can we determine the conditions necessary for the escape of a particular atom or molecule from a planet's gravitational field? Using the planet's temperature, mass, and radius and the masses and thermal velocities of its atmospheric constituents, astronomers have reached the following conclusion about the escape of different constituents: If a molecule's velocity is near a third of the velocity needed for escape, about half of that chemical species will escape from the atmosphere within weeks. For a planet to preserve various molecular components of its atmosphere indefinitely, the mean velocity of the gases must be less than a tenth of the velocity of escape.

The massive Jovian planets, with their large escape velocities (several tens of kilometers per second), have held their primeval atmospheres of hydrogen and helium, while the less massive terrestrial planets, with smaller escape velocities (several kilometers per second), have lost these light gases. Venus, the earth, and Mars have managed to retain atmospheric water molecules as well as some heavy gases. Mercury and the moon lack any appreciable atmosphere since these two bodies have small masses; moreover Mercury is close to the sun.

In spite of the noble-gas abundance in the sun, the surprising scarcity of neon, argon, krypton, and xenon on the terrestrial planets suggests that they may not have retained their original atmospheres. A secondary atmosphere for Venus, the earth, and Mars may have formed out of gases escaping from their interiors during volcanic eruptions.

The second factor affecting a planet's atmosphere is the atmosphere's chemistry, which for the terrestrial planets is very different from that of the Jovian planets. Because the terrestrial planets formed from rocky materials—such as iron and iron silicates and metal oxides—their early atmospheres should have been largely composed of such gases as carbon dioxide, nitrogen, and some water. Venus and Mars still have that kind of atmosphere. Planets like Jupiter and Saturn are more nearly like the sun in chemical composition than like the terrestrial planets. Depending upon the chemical composition and chemical activity of the planet's atmosphere, the greenhouse effect (discussed in Section 7.3) will be more or less effective in trapping incoming solar radiant energy. This causes a warming of the low atmosphere and the surface, which will influence much of the atmospheric chemistry. Venus is a good example of the long-term consequences of the greenhouse effect. Estimates are that the mean surface temperature of Mars, the earth, and Venus are about 5, 35, and 500 K warmer, respectively, than they would be without the greenhouse effect.

The third factor, the ways geological and chemical evolution affect a planet's atmosphere, is important in the case of the earth, Venus, and Mars. Outgassing from these planets' interiors in their early history consisted mainly of water vapor, carbon dioxide, and nitrogen, in approximately the same proportions that we observe in terrestrial volcanic gases today. On earth water vapor condensed to form the oceans, but nitrogen remained in a gaseous state. Most of the carbon dioxide combined with silicate rocks of the crust to form carbonate rocks, such as limestone, a reaction that occurs most efficiently in the presence of liquid water. If it could be released from crustal rocks along with the small amount dissolved in the oceans, carbon dioxide in the earth's atmosphere would equal about one-half of the amount in the dense atmosphere of Venus.

If at some earlier time water vapor condensed on Venus, the high surface temperature due to the greenhouse effect prevented water from remaining in a liquid form and kept carbon dioxide in a gaseous form. Most of the water vapor apparently dissociated into hydrogen and oxygen by absorbing ultraviolet sunlight. Hydrogen escaped, and the heavier oxygen may have combined with crustal rocks to form oxides. On Mars the outgassing of water vapor, carbon dioxide, and nitrogen was probably less complete than it was on the earth; yet carbon dioxide forms the largest part of the atmosphere of Mars. Mars appears at present to be in a cold phase, and a large amount of water is

apparently stored in the polar caps and under the surface of the planet as permafrost.

Finally we need to consider the effect of biological life, if any, on an atmosphere. Living organisms on a planet are bound to affect its atmosphere if the interaction between its biosphere (the zone in which life exists) and atmosphere is anything like that on earth: Large expanses of liquid water will moderate a planet's climate and can provide an environment conducive to the development of life if there is adequate protection from solar ultraviolet radiation; conversely most of the earth's free oxygen, so necessary to animal life, comes from photosynthesis (see Section 10.3). The oxygen is constantly replenished by green plants, plankton, and some bacteria. When living organisms became able to extract carbon dioxide from the atmosphere, they helped save the earth from the heat death that Venus has apparently experienced.

MERCURY: A LITTLE HELIUM AND HYDROGEN

The atmosphere on Mercury is very tenuous and is approximately a million billion times less dense than ours. Mercury's atmosphere seems to be supplied and constantly replenished by the solar wind. Helium and a couple of percent of atomic hydrogen have been identified as its principal constituents. The abundance of other atomic or molecular species, if they are present, is insignificant. No evidence has been found for atmospheric modification of any landform.

Just after Mercury formed some 4.6 billion years ago, gases like carbon dioxide escaping from the interior of the planet may have temporarily created an atmosphere of some extent. But soon after forming it would have escaped into space and vanished. This is because as mentioned above, the planet is not massive enough to hold much of an atmosphere at such a small distance from the sun, and probably was endowed with less gaseous material during its formation than the other terrestrial planets.

ATMOSPHERE AND CLOUDS OF VENUS

The atmospheric pressure at the surface of Venus is about 90 times greater than that of the air in your room. Analysis of the lower atmosphere suggests that it is about 96 percent carbon dioxide and about 3.5 percent nitrogen, with the remainder water vapor and some others (Table 8.1). Because of the high surface temperature (about 730 K), the carbon dioxide was

TABLE 8.1
Comparison of Principal Atmospheric Constituents

Constituent	Molecular Weight	Percentage by Volume		
		Venus	Earth	Mars
CO_2	44.0	96.4	0.03	95
N_2	28.0	3.41	78.1	2.6
O_2	32.0	<0.01	20.9	0.15
H_2O	18.0	0.14	≈1, varies	≈0.03, varies
Ar	39.9	<0.001	0.9	1.6
CO	28.0	0.002	?	0.6
Kr	83.8	?	0.0001	Trace
Xe	131.3	?	0.00001	Trace

apparently not depleted, as it was on the earth, by reacting with the primitive rocks to form carbonates and limestones and by absorption by water. Above 150 kilometers atomic oxygen is the most abundant species. And finally a huge cloud of hydrogen surrounds the planet far above the atmosphere.

More than almost any other aspect of Venus, the mysterious clouds that perpetually obscure the surface have been the subject of extensive speculation. In 1973 it was suggested that the clouds are composed of sulfuric acid droplets. The clouds begin around 46 kilometers above the planet's surface and seem to be confined to a fairly distinct layer, rising up to about 70 kilometers. Thin-haze regions lie above and below the cloud layer; the lower one has a surprisingly abrupt cutoff some 32 kilometers above the surface.

From the *Pioneer Venus* results we think that the clouds are indeed composed of sulfuric acid droplets and other particles, possibly free sulfur, so thick that during the probes' descent they appeared to be passing through a blizzard. Early analysis of the data also suggests that possibly several sulfur compounds also exist in the atmosphere. From the bottom of the haze down to the surface the atmosphere appears to be surprisingly clear.

If one assumes that Venus formed with about the same relative amount of water as the earth did, then the challenging question is what happened to it since it is not on the surface in pools nor in the atmosphere. Most likely it stayed in the vapor form because of the high temperature. Incoming ultraviolet photons from the sun could dissociate the molecule, with free hy-

drogen then able to escape over the planet's lifetime. Some water is also consumed in making the sulfuric acid droplets in the clouds, and a tiny amount of water vapor is still present in the atmosphere.

Why did this same escape of water not occur on the earth since its early atmosphere was probably similar in composition to that of Venus? It is probable that the advent of life and photosynthesis on the earth began to replace carbon dioxide with oxygen and prevent a substantial greenhouse effect; thus the water on the earth stayed primarily in pools on the surface. Moreover Venus receives about twice as much radiant energy from the sun as does the earth.

The atmospheric circulation is the same in both hemispheres. A vigorous, equatorial, east-west jet stream is quite evident in the upper atmosphere, moving around the planet in only 4 days, opposite to the direction of the planet's slow spin. The wind velocity decreases at lower altitudes until at the surface it slows to a gentle breeze. The lower atmosphere apparently circulates because of differences in solar heating between the equatorial and polar regions. Clouds rise near the equator, spiral toward the poles, and descend in what appears to be an almost continuous flow. But why such high winds reverse direction in the upper atmosphere we do not know.

MARTIAN ATMOSPHERE

As on Venus, carbon dioxide is the most abundant constituent of the thin Martian atmosphere, amounting to about 95 percent (Table 8.1). We know that the atmosphere also contains about 3 percent nitrogen, about 2 percent argon, lesser amounts of atomic and molecular oxygen, and traces of other constituents.

A small, daily, and seasonally variable amount of water vapor has been detected on Mars. The abundance of water in the atmosphere, however, is far too low for rain. Because the atmospheric pressure on Mars's surface is low (less than 1 percent of the earth's sea-level atmospheric pressure), water vapor cannot exist as a liquid on the open, flat ground, and rain could not fall even if water were more abundant.

Early-morning fog lying in craters and other low places is probably evidence of an exchange of water vapor between subsurface or surface ice and the atmosphere. The Martian atmosphere also possesses clouds, which are most probably condensations of water ice and carbon dioxide ice.

The warmest daytime temperature is around 30° C at the Martian equator, while the nighttime temperature drops to −130°C. Over the polar regions it is even colder. During summer the north polar ice cap gets up to only about −70°C—though very cold, not cold enough for the residual cap to be made of carbon dioxide ice. Thus it appears that it is water ice, which is consistent with finding more water vapor in the atmosphere at high latitudes near the poles.

One of the most exciting things to happen since the landing of the *Vikings* was the photographing of frost on the surface at the *Viking 2* site (Figure 8.13). The frost occurred during the northern winter, between May and November, 1977. The composition of the frost is not known; the air temperature was too warm for it to have been pure carbon dioxide ice, and the air was too dry for it to have been pure water ice. The best speculation is that it is some kind of mixture of carbon dioxide and water.

There is speculation that water ice may be a remnant of a denser atmosphere that Mars may have had in the first billion years or so of its existence. If that early atmosphere was a denser carbon dioxide (say 100 to 200 times more than at present) and water-vapor one, then it could have acted to trap infrared radiation in the greenhouse effect, and it would have been warm enough to contain substantial amounts of water vapor. This increased amount of carbon dioxide could easily have been provided by outgassing from the body of the planet early in its history. However, over time the formation of carbonate rocks removed carbon dioxide from the atmosphere and lowered both the temperature and pressure. In such conditions the

FIGURE 8.13
Frost at the *Viking 2* landing site. This photograph, taken September 13, 1977, when the temperature was about 174 K, shows patches of frost on the ground. It is thought that the frost is probably a mixture of water and carbon dioxide.

atmosphere could no longer retain much water, and water could not exist as a liquid on the surface. Thus the era of water erosion ended for Mars some time ago.

The skies at the locations of *Viking 1* and *2* are yellowish brown in color and seem to remain that way over the course of the Martian year. This color is probably due to dust particles suspended in the atmosphere below about 50 kilometers (Figure 8.14). Surface winds can stir the atmosphere sufficiently to hold dust particles. From the *Viking* data we know that the prevailing winds are westerly, as on earth, with velocities up to 70 kilometers per hour at the surface and over 360 kilometers per hour at altitudes above 10 kilometers. The winds are strong enough to create major dust storms that can engulf almost the whole planet and last for months. Undoubtedly the winds come from unequal solar heating among different portions of the Martian surface, driving air from high-pressure areas to low-pressure areas (as on the earth).

8.4
Satellites of
the Terrestrial Planets

MARTIAN SATELLITES

Only three satellites are known for the terrestrial planets. One is the moon, which we discussed in Chapter 7. The other two are the little satellites of Mars, which were discovered in 1877. Phobos, the inner one, and Deimos, the outer one, are potato-shaped bodies with cratered surfaces. Phobos orbits eastward, just as our moon does, and in the same direction that Mars rotates, in a period of 7.5 hours at a distance of about 6000 kilometers from the surface of Mars. This gives it an angular size, as seen from the surface of Mars, of about half that of our moon. Since it revolves about Mars much faster than the planet rotates, it rises on the western horizon and sets on the eastern horizon 5.5 hours later. This is counter to any other natural satellite in the solar system, as observed from its primary.

Phobos (Figure 8.15) is about 27 kilometers long, 21 kilometers high, and 19 kilometers wide. Both Phobos and Deimos have been shaped by high-velocity impacts, which appear to have sheared off large sections of each satellite. In addition, both have many craters but no ejecta or craters with central peaks, a reasonable feature since their gravitational attraction is very small. Material ejected during cratering impact simply escapes and does not fall back to the satellites' surface.

Phobos seems more heavily cratered than does Deimos, the largest crater, Stickney, being about 10 kilometers across. It also has mysterious, long, parallel grooves across a large part of its surface, as shown in Figure 8.15. They are a few hundred meters wide and a few tens of meters deep and may have been formed by the same impact that caused the crater Stickney.

Deimos is about half the size of Phobos — about 15 kilometers by 12 kilometers. It orbits Mars some 20,000 kilometers from the planet's surface in a period of 30.3 hours. Its angular size, as seen from Mars, is quite small, roughly equivalent to a quarter viewed at a distance of about 40 meters. Its orbital period is somewhat longer than the rotational period of Mars; so it rises on the eastern horizon and sets on the western horizon nearly 3 days later, while going through its phases twice.

The darkness of the surface of both satellites is probably due to carbon-rich and water-rich minerals, such as are found in black, crumbly meteorites known as *carbonaceous chondrites*. A number of the asteroids appear to have similar surfaces. This has led to the speculation that Mars's satellites may be captured asteroids, acquired early in the planet's life.

FIGURE 8.14
Atmospheric phenomena on Mars. *Viking 1* took this picture in August, 1978, which is midwinter in the southern hemisphere. On the horizon cloud streamers and a faint haze are visible.

8.5 Asteroids

DISCOVERY

On January 1, 1801, a Sicilian astronomer, Giuseppe Piazzi (1746–1826), accidentally discovered a faint object whose orbital motion was that of a body 2.8 AU from the sun. Although Piazzi thought it was a comet, others noted that it was located about where a major planet would be expected according to Bode's law (see Section 6.1). The object was later named Ceres, after the Roman goddess of agriculture. Shortly afterward, three more objects were discovered with orbits near 2.8 AU: Pallas in 1802, Juno in 1804, and Vesta in 1807. Since photographic techniques were introduced into astronomical research in the 1890s, nearly 3000 of these bodies have been discovered.

So instead of one planet in the slot at 2.8 AU, many small bodies orbit in the region between Mars and Jupiter. William Herschel called these objects *aster-*

FIGURE 8.15
Phobos as photographed by *Viking 1*. The surface of Phobos is covered with many impact craters, the largest being Stickney (not in this picture). The most striking feature is numerous long, parallel grooves that are 100 to 200 meters wide and 20 or so meters deep. The picture is about 11 kilometers across from top to bottom.

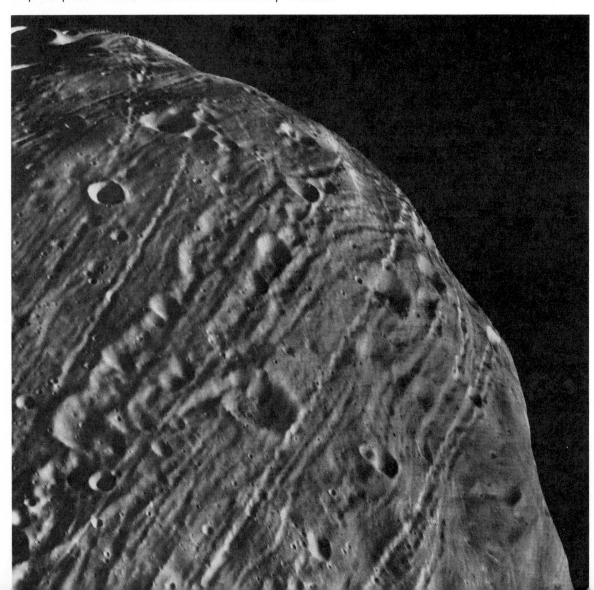

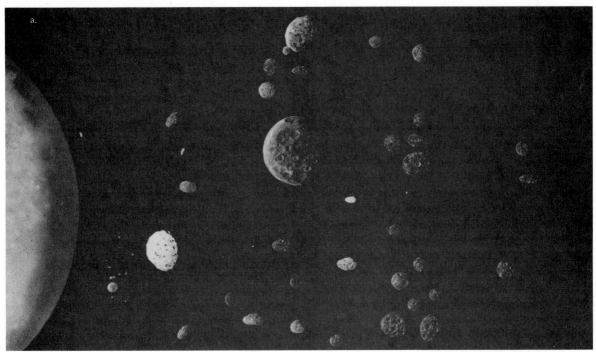

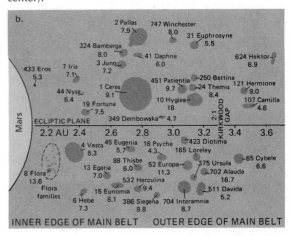

FIGURE 8.16

Asteroids. (a) This painting by Andrew Chaikin shows all 33 known asteroids larger than 200 kilometers. Correct in relative size and shape, they are contrasted with the limb of Mars on the left for size and are given their relative distance from the sun off to the left. In (b), the rotation period in hours is shown below the asteroids' names. Note particularly the suspected binary asteroid, Hektor (upper right), and the satellite to Pallas (upper center).

oids because in a telescope they looked like stars. Almost 95 percent of them have orbits between 1.6 and 3.3 AU, with periods from 2 to 6 years. Their orbits are more elliptic than are those of the planets and more inclined to the ecliptic. They also move in the same direction as the planets around the sun.[1]

For some time the asteroid Hidalgo was thought to have the largest orbit. Its orbital period is 14 years, with an aphelion just outside the orbit of Saturn. However, in October, 1977, a new asteroid, named Chiron, was discovered; it travels in a highly eccentric orbit (eccentricity = 0.38) at an angle of 6.9° to the plane of the ecliptic. It ranges between 8.5 and 18.9 AU, or roughly between the orbits of Saturn and Uranus, with a period of 50.7 years. Because of its great distance from the asteroid belt, there is a question whether it might be the first discovery in an outer zone of asteroids. Or possibly it is not even an asteroid like those between Mars and Jupiter but is something related to a comet.

Asteroids vary in size from Ceres (1025 kilometers) down to an estimated 100,000 that are no more than 1 kilometer in diameter and countless numbers of even smaller ones. All the asteroids together may add up to

no more than a few ten-thousandths of the mass of the earth. Ceres constitutes about 20 percent of the mass of all the asteroids (Figure 8.16). The six largest asteroids are listed in Table 8.2.

[1]An asteroid discovered in 1977 was found to be orbiting in the opposite direction.

CHARACTERISTICS

Photometric studies of the asteroids have for some time been interpreted as showing that they differ in size, shape, and rotation. All but the largest of the asteroids are too small to show a measurable disk. From the variation in their brightness, it has been assumed that most have somewhat irregular shapes with periods of rotation measured in hours. Recent evidence suggests, however, that the brightness variations for some are the result of two or more asteroids in mutual orbit about each other; that is, some asteroids are binary systems.

Their colors put nearly all asteroids into two categories: Some are bright reddish, a sign of silicates and iron-nickel or other metallic grains, and they populate mostly the inner part of the asteroid belt; but most have the darker neutral color of material containing various carbon-rich or water-rich compounds (carbonaceous) and occupy the outer part of the belt.

Collisions between two asteroids may produce effects ranging from craters (if a small one collides with a large one) to fragmentation of the two asteroids (if they are of comparable size). For example, if the body producing crater Stickney on Mars's satellite Phobos (Figure 8.15) had been a little larger, Phobos might have been broken into many small pieces. As it is, the grooves on Phobos may be large cracks produced by the impact.

8.6
Meteoroids, Meteors, and Meteorites

METEOROID DEBRIS

As much as 1000 tons of cosmic debris—billions of microscopic particles—pepper the earth daily. We are aware only of those weighing a significant fraction of a gram, which produce the so-called shooting stars that flash across the sky. All but a few are too small to leave luminous trails. These solid particles are called *meteoroids* before they encounter the earth. Those large enough to survive flight through our atmosphere and land are called *meteorites*. And the luminous trails of the smaller particles that are completely vaporized in the atmosphere are called *meteors*. In order of increasing size and brightness, meteors are classified as (1) telescopic and radio meteors, (2) visual meteors, and (3) fireballs, or bolides.

Our atmosphere slows incoming meteoroids and transforms their kinetic energy into radiant and thermal energy. A meteoroid passing through the atmosphere leaves a wide, dense column of electrons stripped from the atoms and molecules in its path. As the ionized atoms regain their electrons, they de-excite, emitting photons that make the momentary luminous trail we see from the ground as a meteor, or shooting star.

Anything that remains of the meteoroid slowly filters down through the air as dust and solidified droplets of melted meteoroid.

When meteoritic particles encounter the earth, they are moving anywhere from about 10 up to 72 kilometers per second, depending on their direction and the angle at which they strike the earth. The velocities convince us that meteoroids belong to the solar system, moving in independent orbits around the sun.

The normal observed rate for meteors is about 10 per hour over the entire sky. Why do we see fewer meteors before midnight than after midnight? During the evening hours we are on the back side of the earth, facing the direction opposite to earth's orbital motion, and we see only the swift meteoroids overtaking us from the rear. During the morning hours earth's rotation has turned us so that we are facing in the same direction as its orbital motion. Hence we see those meteoroids that we overtake and those that meet us head on.

METEOR SHOWERS

Several times a year we can see *meteor showers*, the swarms of shooting stars that dart from a small area in the sky. These showers can persist for hours or days. On such occasions the earth is passing through a large group of particles moving in ribbonlike fashion along an orbit around the sun. Perspective makes their tracks seem to diverge from a small spot in the sky called the *radiant*. The shower is named after the constellation in which the radiant appears. Some of the better-known showers are listed in Table 8.3.

Long ago astronomers found that some meteoroids travel in orbits much like those of some comets. They had found a link between meteor showers and the short-period comets (to be discussed in Chapter 9). The particle swarms may be debris left by evaporation and tidal disruption of comets. For example, on the night of November 13, 1833, watchers in the southern

TABLE 8.2
The Largest Asteroids[a]

Asteroid	Mean Distance from Sun (AU)	Diameter (km)	Mass ($10^{-4} M_\oplus$)	Rotation Period (hr)	Year Discovered	Type
Ceres	2.77	1025	1.96	9.1	1801	Carbonaceous
Pallas	2.77	583	0.36	7.9	1802	Peculiar carbon
Vesta	2.36	555	0.46	5.3	1807	Basaltic
Hygiea	3.14	443		18	1849	Carbonaceous
Interamnia	3.06	338		8.7	1910	Unknown
Davida	3.18	335		5.2	1903	Carbonaceous

[a]Data adapted from T. Gehrels, ed., *Asteroids* (Tucson: University of Arizona Press, 1979).

TABLE 8.3
Meteor Showers and Their Associated Comets

Name of Shower	Approximate Date of Maximum Display	Approximate Maximum Visual Hourly Count	Associated Comet	Period of Comet (yrs)
Lyrids	April 21	10	1861 I (Thatcher)	415
Eta Aquarids	May 4	10	Halley	76
Perseids	August 12	40	1862 III	105
Draconids	October 10	<500	Giacobini-Zinner	6.6
Orionids	October 21	20	Halley	76
Taurids	November 4	10	Encke	3.3
Leonids	November 16	10	1866 I (Tempel)	33
Andromedids	November 20	Low	Biela	6.5
Ursids	December 22	10	Tuttle	13.6

part of the Atlantic seaboard were awestruck as over 100,000 shooting stars per hour plummeted from the constellation Leo for 3 hours. The great display was produced when the earth encountered a swarm of meteors orbiting the sun in a period of 33 years and associated with comet Tempel (1866 I). The comet itself has long since vanished leaving the meteor shower as a remainder of its existence. The meteoric displays of 1866, 1899, and 1932 were progressively weaker; then on November 17, 1966, a fairly spectacular meteor shower was observed in the southwestern part of the United States.

With the passage of time the meteor stream—which is made up of conglomerates of fine dust, ices, and ice-covered particles—is strung out along the comet's orbit. This ribbon of particles typically aver-ages about 50,000 kilometers in cross section. Thus the earth must come fairly close to the meteor stream in order for us to see a meteor shower.

RECOVERED METEORITES

Most meteorites are discovered accidentally years after they fall. Of some three dozen meteorite falls weighing more than a ton only a few were seen descending. Not many falls are ever recovered: Most meteorites land in the oceans or in unoccupied places, where their fall is not likely to be observed. No known record tells of a community destroyed or an individual killed by a meteorite in spite of some close calls. Approximately 3000 meteorite specimens have been recovered and catalogued for study.

FIGURE 8.17
Aerial view of Barringer meteorite crater. The meteorite causing the crater weighed at least 30,000 tons and was considerably smaller than the 155 meter deep and 1300-meter wide crater.

Meteorites striking the earth have probably formed thousands of craters, but only 200 or so have been found. One great collision in 10,000 years is a conservative estimate, and at that rate at least 50,000 giant meteorites must have fallen on the earth in the past 500 million years. But the fossil craters left by many of these may lie buried and unnoticed in the earth's crust (see Figure 7.8). Probably most of them have been obliterated by weathering, erosion, and geological processes.

One that we know about, near Winslow, Arizona, is the Barringer meteorite crater (Figure 8.17), created by a meteorite weighing at least 30,000 tons. It struck the earth about 24,000 years ago and must have devastated all plant and animal life within a large area. The crater is over 1 kilometer across. Thirty tons of shattered iron fragments have been picked up within about 6 kilometers of the crater.

At 7 A.M. on June 30, 1908, a tremendous fireball flashed across the sky in Siberia. A great fall of flame brighter than the sun was seen leaping from a forested region near the Tunguska River. The sight of the fire was followed by the sound of an explosion powerful enough to level trees within 50 or so kilometers. Earth tremors were recorded on seismographs throughout Europe yet no large crater was formed, only many small ones. The most plausible explanation for the event is that a small comet (possibly part of comet Encke) or a large, fragile, stony meteorite struck the earth, dissipated its kinetic energy on the forest and the ground, and completely vaporized.

Three classes of meteorites have been established based on their chemical and metallurgical properties:

1. *Stony meteorites* are composed primarily of silicates of iron, magnesium, aluminum, and other met-

als. These generally have a relatively smooth, brown or grayish, fused crust indented with pits and cavities. Buried inside all but a small fraction of them are small pieces of glassy minerals, called *chondrules,* that apparently formed from molten droplets, presumably during the formation of the solar system.

One subgroup of the stones is the *carbonaceous chondrites,* which contain large amounts of carbon, water, and other volatiles that would have been driven off with the slightest heating above about 500 K. Therefore these are the most primeval samples of matter from the early solar system that we have. They are doubly interesting because they contain organic compounds, such as hydrocarbons, amino acids, and lipids. These biologically important compounds evidently formed in the primordial solar nebula without the assistance of living organisms.

2. *Stony-iron meteorites* are a mix of stone and iron. Their brownish crust sometimes contains pockets of the yellow mineral olivine. Inside the meteorite the iron may have a veinlike or globular structure.

3. *Iron meteorites* are almost exclusively composed of iron, with some nickel. They are easily identified by their characteristic pitted, brownish exterior and high density. Cut, etched, and polished, they usually have a peculiar crystalline pattern unlike any in terrestrial iron. They show evidence of melting and signs of other heating and cooling processes.

Stones are the most brittle kind of meteorite, and they are more fragile than the irons. Even though most falls are stones, more of the recovered meteorites are irons because they are relatively easy to identify and they resist weathering.

Those meteorites that have been dated by their natural radioactivity average tens of millions of years for the stones and 600 million years for the irons. These are their ages only since the breakup of the larger mass of which they were probably a part. The most ancient specimens are about 4.6 billion years old, the same age as the earth. The chemical and mineralogical sequences in the different classes of meteorites indicate that they share the same heritage as that of the rest of the solar system.

We are still not sure of the origin of meteorites. Are they the remains of comets? Perhaps, but the supporting evidence for this idea is not strong. Another line of speculation is that most meteorites may be descended from a few chemically differentiated asteroids, whittled down by repeated collisions early in the planetary

system's history. In such a case stony meteorites come from the original crusts, the stony irons from the intermediate parts, and the irons from the core. Regardless of our ability to understand their origins, it is evident that asteroids and meteorites are representatives of the unused building material from which the terrestrial planets formed at the birth of the solar system.

8.7 Interplanetary Medium

INTERPLANETARY DUST

The space between the planets is a vacuum by terrestrial standards, but it is not devoid of some *interplanetary gas* and small solid particles called *interplanetary dust.* The particulate matter consists of particles blown out from the sun's atmosphere by the solar wind, micrometeoric debris scattered by comets, and perhaps less plentiful granular powder strewn about by asteroid and meteoroid collisions.

We have learned about interplanetary dust from several sources. One is the *zodiacal light,* which is most easily observed in our Northern Hemisphere in spring after sundown in the west and in fall before dawn in the east. It appears as a faint pyramidal band of light tapering upward from the horizon along the line of the ecliptic. The spectrum of zodiacal light is a faint replica of the solar spectrum; it is produced by small particles lying in the plane of the planets' orbits that scatter solar photons in our direction.

Most direct evidence of interplanetary dust comes to us from spacecraft experiments. Electronic sensors on the skin of the spacecraft are arranged to count small dust particles as they strike the surface. From the numbers of impacts it is estimated that the average spacing between interplanetary dust particles is many meters. The total mass of dust particles is estimated to be about 10^{20} grams, or about a hundred-millionth of the mass of the earth.

INTERPLANETARY GAS

Most interplanetary matter is in the form of an ionized gas comprising the solar wind. It consists of an almost continuous stream of particles, mostly protons and electrons, flowing out from the sun's corona. As the solar wind moves forward, it forms an expanding spiral pattern due to the sun's rotation, and its velocity increases until, several solar radii from the sun, it equals the speed of sound in the plasma. Its velocity continues to increase as it flows outward, much as rocket gases are accelerated to supersonic velocities in a rocket nozzle. Near the earth the solar wind reaches a velocity of about 400 kilometers per second. Beyond the earth its speed remains very nearly constant.

At earth's distance the wind's density is down to about five protons and five electrons per cubic centimeter on the average, but it can rise on occasion to 100 particles per cubic centimeter. Compare that with the number of molecules in your room, about 3×10^{19} per cubic centimeter. The temperature of the wind particles is about 200,000 K in the vicinity of the earth. This is less than their approximately 1,000,000-K temperature when they were in the inner parts of the sun's corona. The density is so very low, however, that the wind transfers no appreciable quantity of heat to the earth.

The particles of the solar wind rush past the terrestrial planets and flow into the outer solar system. The outer portions are dominated by the four giant bodies—the Jovian planets—that we shall consider in Chapter 9.

SUMMARY

Surfaces of the terrestrial planets. For the terrestrial planets, the two major terrain-shaping processes have been impact cratering and thermal-tectonic activity. Cratered surfaces are presumably older terrain since evidence suggests that the intense period of cratering was some 4 billion years ago. The surface of Mercury is heavily cratered, resembling that of the moon but with some differences. This suggests that Mercury has had a history unlike that of the earth, which is dominated by thermal-tectonic activity. Mercury's surface features have remained relatively unchanged for 3 to 4 billion years.

Because of the obscuring cloud covers on Venus, direct observation of its surface is not possible. However, from a variety of measurements we can infer that both impact cratering and thermal-tectonic activity have shaped the surface of Venus. The younger portions of the surface may be on the order of millions of years old. Mars is more like the moon than the earth. Cratering and to some extent thermal-tectonic activity are responsible for the appearance of the Martian surface. Almost half the surface has remained relatively stable and unchanged for 3 to 4 billion years. What appear to be channels carved by flowing liquid are present on Mars, suggesting that a warmer, denser atmosphere long ago allowed liquid water to exist on the surface.

Earth, unlike the other terrestrial planets, possesses a relatively young surface that has been shaped by thermal-tectonic activity which has erased evidence of the earlier impact-cratering phase. The oldest rocks are at most 3.8 billion years old. Earth probably never resembled the moon, Mercury, or Mars since thermal-tectonic activity has probably always been vigorous with that for Venus being less active and that for Mars, even less significant.

Internal structure of the terrestrial planets. Astronomers can infer the internal structure of a planet from knowledge of such aspects as its mean density, mass, shape, rotation rate, and surface conditions. Extensive seismic data exist only for the earth and moon. Critical assumptions must be made in order to develop a model for a planet's interior. Such models for the terrestrial planets resemble a model for the earth with three layers—core, mantle, and crust. At the time of their formation the planets' chemical composition should have been determined by their distance from the sun. Closer to the sun, the lighter gaseous and icy materials vaporize at the higher temperatures, leaving as solids the iron-rich rocky materials. Mercury has an iron-rich core comprising 65 percent of the total mass of the planet while the percentage for Venus is 38 percent, earth 33 percent, and Mars 26 percent. Mars has the smallest core for its size.

Atmospheres of the terrestrial planets. Mercury and the moon have almost no atmosphere. Mars and Venus have atmospheres dominated by a carbon dioxide composition with a small percentage of nitrogen. Earth's atmosphere has primarily a nitrogen composition with about 21 percent oxygen, which comes from the biological development of the planet. The composition of a planetary atmosphere depends on the planet's distance from the sun, chemical composition of the interior through outgassing and the greenhouse effect, evolution of the body, and interaction with living organisms if they exist.

Satellites of the terrestrial planets. The two satellites of Mars and the single satellite of the earth are the only satellites of the terrestrial planets. The satellites of Mars are small irregular bodies of rocky material and may be captured asteroids.

Asteroids. About 3000 asteroids have been discovered that occupy the space between 1.6 and 3.3 AU from the sun in nearly the plane of the ecliptic. They are less than 1000 kilometers in size, most are irregular in shape, and they represent two broad classes of chemical composition: either a silicate-rich one or one of carbon-rich or water-rich compounds.

Meteoroids, meteors, and meteorites. More than a thousand tons of cosmic debris pepper the earth's atmosphere everyday. The smaller meteoroids, which are the majority, evaporate in the earth's atmosphere, leaving the long visible streak characteristic of the meteor; larger ones fall to earth as meteorites. About 300 meteorite specimens have been recovered for study.

Interplanetary medium. Gas and interplanetary dust occupy the space between the planets. The dust particles, small solid grains located in the orbital planes of the planets, scatter sunlight, giving rise to the faint glow of the zodiacal light. Interplanetary gas, which is primarily protons and electrons, constitutes the solar wind rushing out from the sun through the solar system. The density of the interplanetary medium is quite low.

KEY TERMS

asteroid
carbonaceous chondrites
evolution
greenhouse effect
icy materials
impact cratering
interplanetary dust
interplanetary gas
iron meteorite
magnetosphere

meteor
planetary atmosphere
planetary core
planetary interior model
planetary mantle
rocky compounds
stony-iron meteorites
stony meteorites
thermal-tectonic activity
zodiacal light

CLASS DISCUSSION

1. What are the physical processes that have shaped the surfaces of the terrestrial planets and the moon? Have all of these bodies followed identical paths in the evolution of their surfaces? What were the differences, if any?

2. What must an astronomer assume and know in order to construct a model of a planet's interior? How do the models for the terrestrial planets' interiors differ from the moon, if at all? Are the chemical compositions for the five bodies identical?

3. What factors affect a planet's atmosphere? What are the similarities and the differences in the terrestrial planets' present atmospheres?

4. How many satellites belong to the terrestrial planets? How do they compare to each other?

5. Are asteroids and meteoroids the same thing? If different, how do they differ? What, if any, is their relationship to comets?

6. Is there matter between the terrestrial planets? If so, what form does it take and how much is there?

READING REVIEW

1. On which of the terrestrial planets has thermal-tectonic activity been most significant?

2. Is the surface of Venus thought to be more like Mars or like the earth?

3. The space program of the Soviet Union has been most active with which of the terrestrial planets?

4. Do all of the terrestrial planets and the moon possess mare? Which ones do and which do not?

5. What evidence exists for water erosion on Mars? Did it occur recently or in the distant past?

6. Do all the terrestrial planets and the moon possess impact craters?

7. What is the dominant constituent of the terrestrial planets?

8. Are the present atmospheres of the terrestrial planets their original atmospheres?

9. What are the clouds on Venus made of? What is responsible for the clouds on Mars?

10. Do all the asteroids appear to be the same composition?

11. Do Phobos and Deimos have any similarities to the asteroids?

12. What is the typical rate (number per hour) at which meteors enter the earth's atmosphere?

SELECTED READINGS

Arvidson, R. E., L. M. Binder, and K. L. Jones. "The Surface of Mars," *Scientific American,* March 1978.

Leovy, C. B. "The Atmosphere of Mars," *Scientific American,* July 1977.

Murray, B. C. "Mercury," *Scientific American,* September 1975.

Pettengill, G. H., D. B. Campbell, and H. Masursky. "The Surface of Venus," *Scientific American,* August 1980.

Schubert, G., and Covey, C. "The Atmosphere of Venus," *Scientific American,* July 1981.

Wetherill, G. W. "Apollo Objects," *Scientific American,* March 1979.

9

The Outer Solar System: The Jovian Planets

Our knowledge of the inner part of the solar system (see Chapter 8) is considerably greater than that of the outer region: There are few or no data on many aspects of the outer parts of the system. And while the boundaries of the inner solar system are reasonably well defined (even if not well studied), the outermost limits are poorly defined: Is it possible that there are small, faint, distant planets beyond Pluto, awaiting discovery? What is the icy realm of the comets like?

The general information on the nine planets presented in Chapter 6 shows that there are distinctive differences between the terrestrial planets and the Jovian planets. For example, the four giant planets—Jupiter, Saturn, Uranus, and Neptune—contain 99.6 percent of the total mass of the sun's planets. And Jupiter and Saturn, with their large complements of satellites, are like miniature solar systems. Certainly the differences in the compositions of these bodies compared with the terrestrial planets indicate differences in the details of their formation.

9.1
Jovian Atmospheres and Clouds

SPACE MISSIONS TO THE OUTER SOLAR SYSTEM

The space age dawned in the outer solar system when the *Pioneer 10* and *11* spacecraft flew a first reconnaissance by Jupiter in December of 1973 and 1974. With gravitational assistance from Jupiter *Pioneer 11* went on to rendezvous with Saturn in September 1979, passing within about 21,400 kilometers of the planet. *Pioneer 10* and *11* will cross the orbit of Pluto (but far from the planet) sometime between 1987 and 1990 on their way out of the solar system.

In the late summer of 1977 *Voyager 1* and *2* (Figure 9.1) were launched toward Jupiter, Saturn, and their satellites. These spacecraft carried 11 different scientific instruments. For fear of radiation damage to measuring instruments, *Voyager 1* was targeted to pass within about 5 Jupiter radii in its March 1979 encounter with the planet; and *Voyager 2*, no closer than about 10 Jupiter radii in July 1979. Besides studying Jupiter, both spacecraft "looked at" its innermost satellite Amalthea and the Galilean satellites (Io, Europa,

◀ Composite NASA photograph of Jupiter and the Galilean satellites.

Ganymede, and Callisto). The Galilean satellites (shown in the photograph at the beginning of this chapter) were photographed with a resolution as good as that in *Mariner 9*'s photographs of Mars.

After departing from Jupiter, *Voyager 1* and *2* headed for an encounter with Saturn, using the gravitational assist of Jupiter for their acceleration toward Saturn. *Voyager 1* made its closest approach to Saturn in November 1980, *Voyager 2* following some 10 months later, in August 1981. *Voyager 2* also received a gravitational assist from Saturn for a rendezvous with Uranus in January 1986 and if all goes well, with Neptune in August 1989.

The next space mission to the outer solar system, *Project Galileo,* is currently scheduled for launch sometime in the next decade. This mission will send to Jupiter a 2-ton orbiter with a detachable probe that will enter the planet's atmosphere on the sunlit side. The orbiter will then go into orbit about the planet.

ATMOSPHERES OF JUPITER AND SATURN

There are many aspects of a planet's atmosphere that astronomers want to know about, such as chemical composition, temperature, density, cloud composition, winds, and how these change with height, position over the surface, and time. Such detail is not now available for the earth's atmosphere much less for the atmospheres of the other planets. But from *Voyager 1* and *2* we know far more about the atmospheres of Jupiter and Saturn than was available in the previous edition of this book (Figure 9.2). Probably the most fundamental aspect when trying to understand a planet's atmosphere is the vertical temperature structure, such as shown in Figure 7.10 for the earth and Figure 9.3 for Jupiter and Saturn. On the way up through Jupiter's and Saturn's troposphere's the measured temperature profile first declines and then increases into the stratosphere, where photons from the sun can be directly absorbed.

The first constituents of Jupiter's atmosphere to be identified were methane and ammonia in the 1930s. Some 30 years later the most abundant element, hydrogen, was identified and estimated to be 1000 times more prevalent than methane and ammonia. From these identifications estimates for the hydrogen, carbon, nitrogen, and oxygen abundances indicate that Jupiter's chemical composition (and similarly for Saturn) is more like that of the sun rather than like that of the terrestrial planets. In the 1970s and 80s, primarily

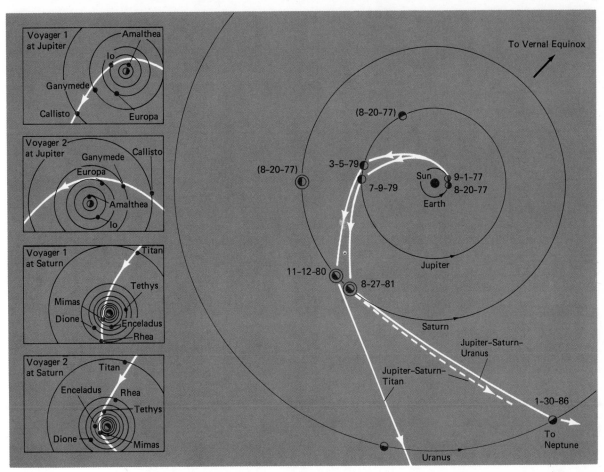

FIGURE 9.1
Space missions of *Voyager 1* and *2* to the Jovian planets. Relative distances are approximately to scale although the sizes of the planets and the sun are not to scale. The four insets show the details of the encounters with Jupiter and Saturn by *Voyager 1* and *2*.

through infrared observations, several additional molecules were found to be minute constituents of Jupiter's atmosphere. Many of these molecules are probably also present in Saturn's atmosphere, but Saturn is colder than Jupiter, so that some of these compounds probably freeze, forming solid crystals; thus they are not in a gaseous state capable of being observed spectroscopically. Table 9.1 summarizes the abundances of the major constituents of the atmospheres of Jupiter and Saturn (compare with Table 8.1 for the terrestrial planets).

Helium, the second most abundant element in the composition of the sun and presumably in Jupiter and Saturn, is not directly observable by spectroscopic means. From data from the *Pioneer* and *Voyager* mis-

sions indirect determinations of the helium abundance have been made. The values derived, 10 percent for Jupiter and 6 percent for Saturn, are consistent with the solar-composition hypothesis.

The most conspicuous aspect of Jupiter's and Saturn's atmospheres in visible light is their clouds. Knowing something about the vertical temperature profile and the chemical composition of the atmosphere has given astronomers clues to the basic constituents forming the clouds. For Jupiter and Saturn there appear to be three distinct cloud layers, as we have tried to show in Figure 9.3. The lowest is formed from water-ice crystals or possibly liquid drops, the next from ammonium hydrosulfide crystals (NH_4SH), and the highest from ammonia crystals (NH_3). The

FIGURE 9.2
Voyager photographs of Jupiter and Saturn. (top) Taken from about 20 million kilometers, the satellites Io, on the left above Jupiter's Great Red Spot, and Europa are seen passing in front of the planet. (above) Montage of photographs from *Voyager 1* and *2* showing Saturn, its rings, and seven of its satellites.

middle one can also be thought of as a compound of the more elementary molecules ammonia and hydrogen sulfide. All the molecules forming the basic cloud particles should lead to white particles; so other molecules are responsible for coloring the clouds, which are red, yellow, brown, blue, and white. The most likely coloring agent is the element sulfur, which forms a variety of colored particles depending upon its molecular structure. This has not been confirmed.

Infrared images of Jupiter and Saturn show that the cloud color correlates with altitude. Seen from the outside the blue clouds lie at the deepest levels in the atmosphere and are visible only through holes in the upper clouds (Figure 9.3). Brown clouds are the next highest, above which lie the white clouds; and finally the red clouds are the top layer. Compared to Jupiter the greater spread in altitude for the clouds in Saturn's atmosphere results from the smaller mass of Saturn, whose gravity is not so effective in compressing the atmosphere as is the more massive Jupiter.

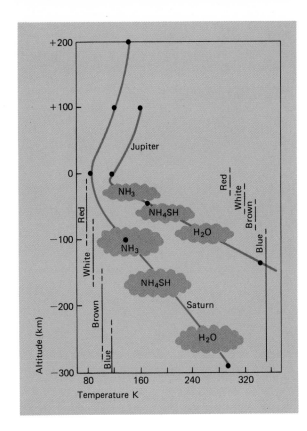

FIGURE 9.3
Atmospheric structure for Jupiter and Saturn.
Shown on the right is the temperature variation
for Jupiter, and on the left that for Saturn. Compared with Figure 7.9 for the earth, the minimum
in temperature (approximately zero altitude) is
where the troposphere ends and the stratosphere
begins. The dots along the temperature profiles indicate where the pressure is 10 times greater than
at the point above it. The pressure range is
roughly from 0.001 of, to 10 times, the earth's atmospheric pressure at sea level. Since Jupiter is
more massive than Saturn, it compresses its atmosphere more than Saturn does. Also shown are the
approximate locations of clouds formed of ammonia ice (NH_3), ammonia hydrosulfide crystals
(NH_4SH), and water ice (H_2O). All those clouds are
roughly white, so that the colors observed are produced by some coloring agent as yet unknown.
The approximate altitude range for each color is
shown by the vertical lines and are labeled red,
white, brown and blue. Thus red is highest and
coolest, while blue is lowest and hottest.

ATMOSPHERIC DYNAMICS FOR JUPITER AND SATURN

The alternating light and dark cloud bands that parallel
the equator of Jupiter and Saturn are constantly undergoing changes in color and intensity on a small
scale within the bands. Apparently this is because of
the formation or dissolution of clouds of differing
chemical compositions at different altitudes and latitudes. There are large-scale patterns, such as the
bands themselves and the Great Red Spot on Jupiter,
that last for years and sometimes centuries. This complex behavior involves the dynamics of the atmosphere for both planets.

The dominant observable motions in the atmospheres are alternating eastward (direction of rotation)
and westward winds that correlate with the colored
bands. As shown in Figure 9.4, Jupiter has five or six
eastward- and westward-moving wind streams in each
hemisphere, while Saturn has fewer but stronger wind
streams. Those winds are measured relative to each
planet's rotation. In the case of the earth there is only
one low-latitude westward wind stream, known as the
"trade winds," and one midlatitude eastward-moving

TABLE 9.1
Comparison of Principal Atmospheric Constituents for Jupiter and Saturn

Constituent	Molecular Weight	Percent by Number		
		Sun	Jupiter	Saturn
H_2	2	91[a]	90	·94
He	4	<9	10	6
H_2O	18	Trace	0.0001	?
CH_4	18	Trace	0.06	0.05
NH_3	17	Trace	0.02	0.02

[a]Hydrogen in the sun is predominantly atomic or ionized, with only a trace in molecular form.

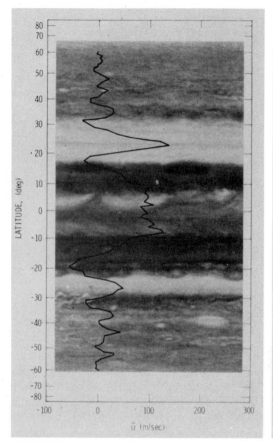

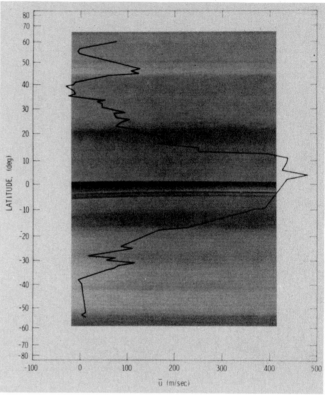

FIGURE 9.4
Correlation of horizontal wind speeds with the latitudes of the colored bands for Jupiter (left) and Saturn (right). The reference frame for measuring the wind speeds is the planets' rotation, and thus negative velocities are westward winds while positive velocities are eastward. For Jupiter there appear to be five or six eastward and westward wind streams in each hemisphere that definitely match the band structure. Although fewer wind streams are identifiable for Saturn, the velocities are greater for Saturn than they are for Jupiter. The black band across Saturn is the shadow of the rings along the equator.

jet stream. Jupiter and Saturn also have some vertical streaming.

Evidence suggests that these eastward- and westward-moving winds have been constant in latitude and velocity for the last 80 or 90 years. However, the cloud bands with which they correlate are changing, with small eddies between the wind streams sheared apart in 1 or 2 days. As one can see in Figure 9.5, eddy currents are deviations in what are otherwise alternating streams flowing east or west in the atmos-

phere. Where the steady winds have velocities up to 100 meters per second or so, the eddy velocities are a few tens of meters per second.

The cloud motions on a small scale are by no means orderly, as is evident in Figure 9.6. *Voyager* scientists were unprepared for the diversity and sometimes high state of turbulence in the cloud motions as photographed by the spacecraft. Surprisingly, photographs failed to reveal cloud features smaller than about 100 kilometers across. Narrow bands appear to coalesce and widen, while wide bands break apart. Material seems to be transferring between bands.

Conspicuous in the southern hemisphere of Jupiter is the oval Great Red Spot, which has varied both in size and intensity since its telescopic discovery three centuries ago. It measures about 14,000 by 40,000 kilometers. The sense of circulation of gas in the spot and other ovals is almost always counterclockwise in the southern hemisphere and clockwise in the northern. This indicates that they are high-pressure cells. White plumes of clouds race by the

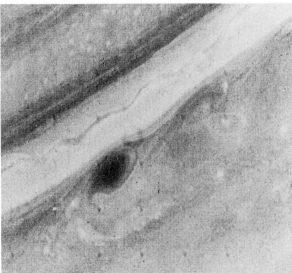

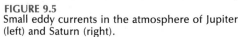

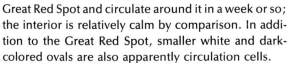

FIGURE 9.5
Small eddy currents in the atmosphere of Jupiter
(left) and Saturn (right).

Great Red Spot and circulate around it in a week or so; the interior is relatively calm by comparison. In addition to the Great Red Spot, smaller white and dark-colored ovals are also apparently circulation cells.

Saturn also has circulation cells in its atmosphere. Shown in Figure 9.7 is a brown oval photographed during the August 1981 flyby of Saturn by *Voyager 2*. It is not known whether the eddies and ovals on both Jupiter and Saturn extend as deep into their respective planets as do the wind streams. The long-term persistance of the winds and the short life for eddies and ovals are possibly related to the mass of material involved in the phenomena. Thus the winds probably extend deep into the planet, while the shorter-lived eddies are relatively shallow structures. However, this is still quite speculative.

The winds on the earth draw their energy from unequal heating by the sun between the equator and the poles, and in general the temperature decreases poleward at almost all levels of the atmosphere. On Jupiter the temperature difference between equator and pole is no more than 3 K, while for the much shorter distance of the earth it is more like 30 K. Even though the sun heats the equatorial region of Jupiter and Saturn more than it does the polar regions, just as it does for the earth, some mechanism must transport heat from the interior of the planet into the polar regions, reducing the temperature difference.

FIGURE 9.6
Close-up of Jupiter's clouds. This photograph was taken by *Voyager 1* on March 1, 1979 from a distance of about 5 million kilometers. The smallest features resolvable in the picture are about 95 kilometers across. The Great Red Spot is in the upper right corner along with a white oval. Note what appears to be a very turbulent region immediately to the west (left) of the Great Red Spot. There is also an intricate and involved structure that can be seen within the spot itself.

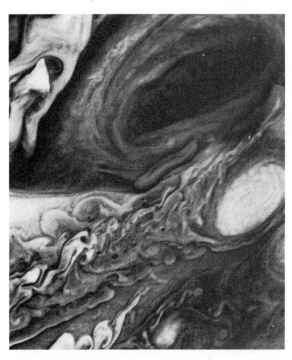

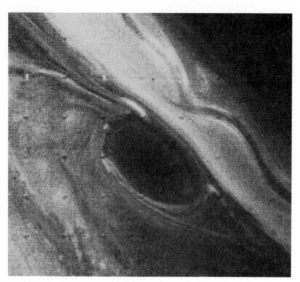

FIGURE 9.7
Brown oval in the atmosphere of Saturn.

URANUS AND NEPTUNE: ATMOSPHERIC STRUCTURE

The masses of Uranus and Neptune are 5 and 6 percent, respectively, that of Jupiter, while their mean densities are about equal to or larger than Jupiter's. This strongly suggests that, whereas Jupiter and Saturn are made primarily of hydrogen and helium, Uranus and Neptune contain greater percentages of oxygen, nitrogen, carbon, silicon, and iron. That is, these two planets are not likely to have solar compositions.

In the 1930s spectroscopic studies revealed methane in the atmospheres of Uranus and Neptune as in those of Jupiter and Saturn. Since then, hydrogen has been identified, helium has been inferred indirectly, and some other hydrogen-containing molecules are also known. Greatly distant from the sun, Uranus and Neptune are very cold, and thus a number of molecular combinations are probably frozen into a crystal or liquid-drop form.

Although alike in many respects, evidence shows that the atmospheres of Uranus and Neptune are not highly similar to each other or to those of Jupiter and Saturn. The atmosphere of Uranus appears to be cold and clear to great depths: In the upper atmosphere there rarely appear to be any haze and no observed clouds; but in the lower atmosphere, it probably becomes misty or hazy. Finally, the greenish cast to Uranus is thought to be due to the methane molecules in its atmosphere.

By contrast, the atmosphere of Neptune possesses a variable haze of unknown chemical composition. At times nearly half the planet's atmosphere is hazed over. This haze can dissipate and reform in a matter of weeks or even a few days. The haze is partly responsible for trapping solar radiation in the atmosphere so that Neptune's upper atmosphere is warmer than that of Uranus.

Important for Uranus is the fact that its axis of rotation lies almost in its orbital plane, causing regions near the poles to remain alternately in sunlight or darkness for periods approaching 42 years. What effect such a phenomenon has on the overall structure of the atmosphere and how much of a difference it produces between Uranus and Neptune is not known.

9.2
Internal Structure of the Jovian Planets

The general comments made about models of planetary interiors in Section 8.2 are applicable to the Jovian planets as well as to the terrestrial ones. Larger masses and the fact that the Jovian planets contain far more easily vaporized materials than the terrestrial planets do mean that the resulting internal structures for the Jovian planets do not resemble those of the terrestrial. Jupiter and Saturn are the only planets in the solar system composed primarily of hydrogen and helium (as is the sun): Only hydrogen and helium could give Jupiter and Saturn their mean densities of 1.31 and 0.69 grams per cubic centimeter, respectively, for the temperatures and pressures that characterize each planet. On the other hand, the mean densities of Uranus and Neptune are about twice that of Saturn even though their masses are much smaller. Thus their compositions are probably dominated by oxygen, carbon, and nitrogen, the third, fourth, and fifth most abundant elements in the solar system, even though their atmospheres contain significant amounts of hydrogen (and possibly helium).

JUPITER AND SATURN: THE SUNLIKE-COMPOSITION PLANETS

As was noted in Chapter 6, the rapid rotation of Jupiter and Saturn, coupled with their composition of low-density materials, argues that their internal structures are more fluid than solid. Another significant factor is that Jupiter and Saturn give off more heat than they receive from the sun. In the case of Jupiter it is about 1.5 to 2 times the amount from the sun, and for Saturn it is between 2 and 3 times the amount. Hence Jupiter and Saturn have internal sources of heat. It is extremely unlikely that the heat source is anything as exotic as that in the sun and the stars; Jupiter and Saturn are not small stars. But it is fair to say that they are more like the sun than like the earth, and they are clearly an intermediate type of body. The internal heat source is more likely the conversion of gravitational potential energy into thermal energy as the two planets contracted during their formation and after. In fact it is likely that they are still contracting but very slowly.

The best models for Jupiter's and Saturn's internal structure are shown in Figure 9.8, along with ones for Uranus and Neptune. Both Jupiter and Saturn have dense cores of rocky and icy materials—rather than compressed hydrogen and helium. The core is about 4 percent of the mass of Jupiter and 25 percent for Saturn, with temperatures in the range of 20,000 to 30,000 K and densities ranging from 2 to 20 grams per cubic centimeter. Surrounding the core is a layer existing under a pressure in excess of 3 million times the earth's atmospheric pressure. In it hydrogen and helium behave more like liquid metals than solids. The upper boundary of the metallic-liquid zone is rather abrupt, giving way to a molecular-liquid mantle of hydrogen and helium. Through both the metallic- and molecular-liquid zones, which are 96 and 75 percent, respectively, of the mass of Jupiter and Saturn, the temperature and density decrease. The molecular-liquid mantles gradually change to molecular gases, which are then the atmospheres of the two planets as shown for Jupiter in Figure 9.9.

URANUS AND NEPTUNE: SIMILAR YET DIFFERENT

Like Jupiter and Saturn, Uranus and Neptune have a three-layered structure, but unlike the solar-system giants, each layer is of quite different chemical composition. The core of each planet is probably a rocky (iron and silicates primarily) and icy (methane, ammonia, and water principally) material. For Uranus the pressure of overlying layers may not be sufficient to make the core solid, but it remains a thick viscous liquid with convective motions in it. On the other

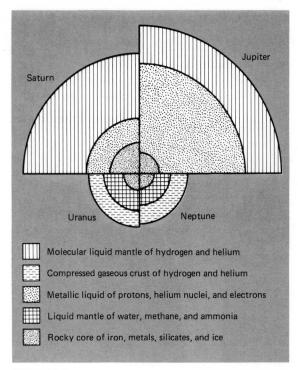

FIGURE 9.8
Models for the internal structure of the four Jovian planets.

Legend:
- Molecular liquid mantle of hydrogen and helium
- Compressed gaseous crust of hydrogen and helium
- Metallic liquid of protons, helium nuclei, and electrons
- Liquid mantle of water, methane, and ammonia
- Rocky core of iron, metals, silicates, and ice

hand, Neptune's greater mean density suggests that its core is solid.

Surrounding the core of each planet is a liquid mantle of water, methane, and ammonia, in which there may be some convective motions (for Neptune but not for Uranus). Finally, each planet has a thick crust of hydrogen and helium that is compressed by gravity into a very dense gas. The crusts gradually give way to low-density atmospheres. Thus like Jupiter and Saturn, these planets have no solid surface surrounded by a thin atmosphere as the terrestrial planets have.

Calculated models for the interiors of both planets suggest that their central temperatures are on the order of 7000 K. Since Jupiter and Saturn emit more radiant energy than they receive from the sun, is it not likely that the same is true for Uranus and Neptune? Yes, one might well expect that to be the situation for both; yet it is *not* true for Uranus and is true only for Neptune, which radiates about twice as much heat as it receives from the sun. Why this difference in what should be reasonably similar bodies? We really do not know.

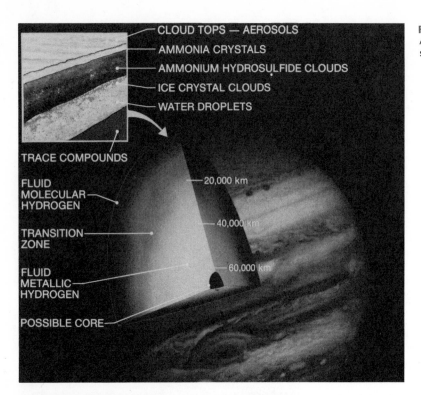

CLOUD TOPS — AEROSOLS
AMMONIA CRYSTALS
AMMONIUM HYDROSULFIDE CLOUDS
ICE CRYSTAL CLOUDS
WATER DROPLETS

TRACE COMPOUNDS

FLUID
MOLECULAR
HYDROGEN

TRANSITION
ZONE

FLUID
METALLIC
HYDROGEN

POSSIBLE CORE

20,000 km

40,000 km

60,000 km

FIGURE 9.9
Artist's conception of the internal structure of Jupiter.

9.3 Magnetospheres about the Jovian Planets

JUPITER'S MAGNETOSPHERE

Jupiter is the strongest radio emitter in the solar system after the sun. It emits both thermal and nonthermal radiation (see pages 126–127). At times its radio emission exceeds even the sun's in intensity. The nonthermal radiation is a type of *synchrotron radiation*, and it results from Jupiter's having a magnetic field and energetic, free electrons in radiation belts analogous to the earth's Van Allen radiation belts (see Section 7.4).

There are also occasional bursts having energies up to 10 million kilowatts. The bursts are more intense when the nearest Galilean satellite, Io, appears on one side of Jupiter as viewed from the earth. Why should the position of Io make a difference? We suspect that it is due to the motion of Io through Jupiter's magnetic field, disturbing the field and the electrons trapped in it.

Pioneer space probes ran into the bow shock wave formed by the solar wind's interacting with Jupiter's magnetic field as far out as 108 Jupiter radii. Data from the two *Pioneer* craft and the two *Voyagers* indicate that the boundary of the magnetosphere in the direction of the sun varies between about 50 and 100 Jupiter radii. The planet's inner radiation region is like earth's Van Allen belts but from 5000 to 10,000 times more intense.

Far out the magnetic field flattens into a disk extending several million kilometers from the planet, and its long tail, flowing out opposite to the direction of the sun, extends an unknown distance beyond the orbit of Saturn. The shape is influenced by the large centrifugal force that results from the planet's rapid rotation.

SATURN'S MAGNETOSPHERE

Saturn's magnetic field also defines a zone about it, or a magnetosphere in which it can control the motions of subatomic particles. The Saturnian magnetosphere is intermediate, in size and in the intensity of the magnetic field, between that of Jupiter and the earth. All three are based on a common framework of physical principles; yet each possesses its own distinctive character.

Prior to the late summer of 1979 astronomers could only speculate on the magnetic field and radiation belts around Saturn. During that summer *Pioneer 11* detected the boundary of the magnetosphere lying some 24 Saturnian radii from the planet (its rings extend about six radii from the center of the planet). Saturn's magnetosphere is apparently more disklike than that of the earth, which is more spherical but less so than Jupiter's larger magnetosphere.

Beyond Saturn we expect to find that Uranus and Neptune possess magnetic fields to create magnetospheres about themselves as do Jupiter and Saturn. However, until *Voyager 2* makes its pass by Uranus in 1986 and Neptune in 1989, we shall not have confirming evidence for their existence.

9.4
Ring Systems of the Jovian Planets

Of all the aspects of the Jovian planets their ring systems are among the most captivating. Galileo first observed what we know as Saturn's rings in July 1610, but it was not until 1655 that Christian Huygens proposed that they are a flattened disk of matter detached from the planet. Finally, in 1857 James Clerk Maxwell showed mathematically that they must consist of numerous tiny bodies in orbit about Saturn. This was experimentally demonstrated in 1895 from Doppler shifts showing that the ring particles move in Keplerian orbits, fastest close to the planet and slower farther away.

It can be shown with reasonable mathematical precision that particles swarming about a planet eventually form a thin ring system in the equatorial plane. This system is produced by the gravitational attraction of the planet and many gravitational interactions of the ring particles with each other. Satellites of the planet play an important role in sculpting the appearance of the ring system and in keeping it from spreading out in the equatorial plane. Also the ring system forms within several planetary radii of the surface of the planet and is not able to form at greater distances. Although the same basic principles underlie the three known ring systems, Saturn's rings are much more elaborate and complex than those of Uranus and Jupiter.

RINGS OF SATURN

The circular rings lie in a plane coinciding with Saturn's equator. During the 29.5-year period of the planet's revolution around the sun the rings are observed obliquely at different angles from the earth, as shown in Figure 6.8.

Three concentric rings have been known for some time and are labeled *A*, *B*, and *C* in order of decreasing distance from Saturn. The distance from the planet's surface is given in Table 9.2. The bright ring *B* is separated from ring *A* by a space of about 5000 kilometers, called *Cassini's division*. Next is the semitransparent ring *C*, the so-called crepe ring, which lies inside the inner edge of the *B* ring. An exceptionally faint *D* ring, which lies inside the inner edge of the *C* ring, has been found by *Voyager* investigations. Outside the *A* ring other faint rings, known as *E*, *F*, and *G*, have been identified. The vertical extent of all the rings is less than a couple of kilometers. Given their immense diameters, they are proportionally thousands of times thinner than a razor blade.

The composition of the ring particles is suggested by the way they reflect sunlight. Their infrared reflectivity indicates that they are water ice or at least covered with water ice. The particles are better reflectors of red light than they are of blue—which suggests that some other substance is mixed with the water ice. Ring particles vary in size from a few centimeters up to several meters. Each particle pursues its independent orbit around Saturn in accordance with Kepler's third law. The farther out from the planet, the lower are the particles' speeds where a solid ring would rotate fastest at the farthest point from the planet. The entire ring system lies within the critical distance called the *Roche limit*, equal to about 2.4 Saturnian radii. This limit is named after the nineteenth-century French mathematician Edouard Roche, who found that inside this limit the gravitational attraction exerted by a planet on two adjacent orbiting particles is larger than the attraction of the two particles for each other. Whether the rings were formed inside the Roche limit by the breakup of a satellite, comet, or other body or whether Saturn's gravitational force prevented primordial particles from coalescing to form a satellite is unknown.

Cassini's division and another known as *Encke's* division appear dark, suggesting an absence of particles. However, high-resolution photographs made by the *Voyager* spacecraft, as in Figure 9.10, surprised astronomers when they revealed that the three major

TABLE 9.2
Ring Systems of the Jovian Planets

Jovian Planet	Equatorial Radius of Planet (km)	Ring Designation	Average Distance of Ring (km)	Distance (km) Inner Edge	Distance (km) Outer Edge
Jupiter	71,900	Secondary 1	97,800	71,900?	123,700
		Primary	126,900	123,700	130,100
		Secondary 2	305,350	251,700	359,000?
Saturn	60,200	*D* ring	70,000?		72,600?
		C ring	82,800	73,800	91,800
		B ring	103,800	91,800	115,800
		A ring	128,400	120,600	136,200
		F ring	141,000		
		G ring	150,000		
		E ring	360,000?	240,000?	480,000?
Uranus	26,145	Ring 6	41,900		
		Ring 5	42,300		
		Ring 4	42,600		
		Ring α	44,800		
		Ring β	45,700		
		Ring η	47,200		
		Ring γ	47,700		
		Ring δ	48,300		
		Ring ϵ	51,200		

rings, *A, B,* and *C,* are made up of hundreds, if not thousands, of very narrow ringlets. Even Cassini's and Encke's divisions are crammed with ringlets, with something like 100 in Cassini's division alone (Figure 9.11). Apparently, particles in Cassini's division do not readily scatter photons in the backward direction so that they appear dark from the sunlit side.

The *Voyagers* provided evidence that some of the ringlets are not circular, while the *F* ring, shown in Figure 9.12, has knots, braids, and twists in it—which had not been predicted from gravitational theory. We are not sure what causes this strange behavior. Probably the most unexpected aspect found was wedge-shaped spokes orientated radially out from the planet in the *B* ring. Figure 9.13 shows the spokes from the sunlit side, where they appear dark, and looking back

FIGURE 9.10
Saturn's rings from 8 million kilometers. This computer-processed photograph taken by *Voyager 1* in November 1980 reveals approximately 95 individual ringlets in the *A, B,* and *C* rings. The threadline *F* ring can be seen at the far left with one of the planet's small satellites. The wide dark gap containing four ringlets is Cassini's division, each about 500 kilometers across.

FIGURE 9.11
Saturn's rings seen from the unlit side looking back toward the sun. The broad darkish band is the *B* ring with the bright *C* ring inside it toward the planet. Just beyond the dark *B* ring is the bright Cassini's division, which is dark from the sunlit side. Beyond Cassini's division is the moderately bright *A* ring with the very thin *F* ring just outside the *A* ring. This photograph was taken by *Voyager 1* moving away from Saturn at a distance of about 700 thousand kilometers.

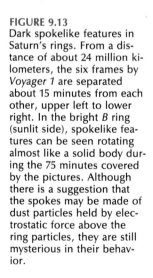

FIGURE 9.12
Transparency of Saturn's rings. In this *Voyager 1* photo taken after the spacecraft had passed the planet and was looking back toward the sun, one can see the shadows of the rings on the disk of the planet. The bright *C* ring is transparent enough so that it is lost across the face of the planet while the dark *B* ring is opaque and blocks a view of the planet. Cassini's division, also seen as a narrow, bright ring, is also transparent across the face of the planet.

FIGURE 9.13
Dark spokelike features in Saturn's rings. From a distance of about 24 million kilometers, the six frames by *Voyager 1* are separated about 15 minutes from each other, upper left to lower right. In the bright *B* ring (sunlit side), spokelike features can be seen rotating almost like a solid body during the 75 minutes covered by the pictures. Although there is a suggestion that the spokes may be made of dust particles held by electrostatic force above the ring particles, they are still mysterious in their behavior.

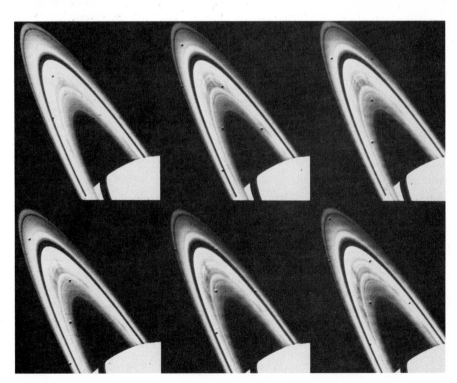

toward the sun, where they appear bright. They are perplexing in that, if produced somehow by the ring particles, Keplerian motion should dissolve the spokes in a short time; but they are seen to last close to 10 hours. The spokes are a mystery for which we may not have a satisfactory solution for many years.

DISCOVERY OF A RING SYSTEM FOR JUPITER

The notion that Jupiter possesses a ring system like that of Saturn was proposed some 20 years ago. *Pioneer 11* data were interpreted as consistent with the existence of a system of tiny satellites forming a ring about Jupiter. This was at best speculation, and it was only *Voyager 1*'s photograph of the Beehive star cluster that finally revealed the ring. The photograph showed a ring system extending some 0.7 to 0.8 Jupiter radii above the cloud tops of the planet. At most the ring is about 30 kilometers thick and 6,000 kilometers wide.

The particles composing the ring appear to be smaller on the average than Saturn's ring particles. Also unlike Saturn's ring particles, those of Jupiter's and Uranus's systems are quite dark. Thus they are not water ice or coated with water ice. The evidence suggests that they are probably silicate particles whose origin is not known.

There is also a diffuse disk, several times fainter than the bright ring, extending inside toward the planet. Surrounding the bright and diffuse rings is a faint halo some 20,000 kilometers thick. Saturn's rings are not embedded in such a halo.

URANUS'S RING SYSTEM: NARROW AND NEAT

Occasionally a planet will pass between the earth and a star. Such an event is called an *occultation* (from the Latin word meaning "hiding"). In recent years astronomers have carefully monitored these occultations since the time and place on the earth at which the occultation will be visible can be calculated. It requires a precise knowledge of the planet's orbit to make such a calculation, and the precision with which the prediction is confirmed by the observation in turn tells us how well we really know the orbit.

As the planet begins to occult the star, its atmosphere, which is partially transparent, covers the star first so that there is a gradual dimming of the star. If there were no atmosphere, the star's brightness would remain constant until the opaque body of the

Few missions of planetary exploration have provided such rewards of insight and surprise as the Voyager flybys of Jupiter. Those who were fortunate enough to be with the science teams during those weeks will long remember the experience; it was like being in the crow's nest of a ship during landfall and passage through an archipelago of strange islands. We had known that Jupiter would be remarkable, for man had been studying it for centuries, but we were far from prepared for the torrent of new information that the Voyagers poured back to Earth.

Thomas A. Mutch

planet cut off all light; the change would be sudden, not gradual.

In this manner astronomers aboard the Kuiper Airborne Observatory, an airplane fitted with an infrared telescope, flying high over the Indian Ocean discovered a ring system around Uranus on March 10, 1977. About a half hour before the occultation was to take place, the star's light dimmed unexpectedly for a few seconds, followed by four other dips in brightness minutes later. The sequence was repeated in reverse as the star passed beyond the disk of Uranus on the other side. Since the original discovery of five rings four less prominent rings have been identified, making a total of nine rings.

The rings appear to be very narrow, not more than 10 to 100 kilometers in width, and they lie close to the planet's equatorial plane. Hence the origin of the rings is closely related to that of Uranus since the planet's equatorial plane is almost perpendicular to its orbital plane. Six of the rings appear to be slightly elliptical, with the radii for all the rings lying between 1.6 and 1.95 planetary radii. All the rings are dark and have sharp edges. Since the ring particles are poor reflectors, it is hard to believe that they are coated with water (or ammonia or methane) ice. More likely they are a silicate- or carbon-bearing material.

SATELLITES OF JUPITER

At least 16 satellites are known to orbit Jupiter. The thirteenth was discovered in 1974 from the earth and the last three were found by *Voyagers 1* and *2*. Among the four largest, discovered by Galileo in 1610, Io and Europa are about the size of our moon while Ganymede and Callisto are larger than Mercury. The four Galilean satellites, Amalthea, and the three discovered by *Voyager* orbit within Jupiter's magnetosphere. As they orbit Jupiter, the Galilean satellites and Amalthea keep the same face toward the planet as the moon does with respect to the earth. The Galilean satellites, Amalthea, and the innermost small satellites form what can be called a "regular" satellite system in that they orbit in nearly circular orbits in Jupiter's equatorial plane and revolve in the same sense as Jupiter rotates. The outer eight are much smaller than the Galilean satellites and move in irregular orbits inclined at varying angles to Jupiter's equatorial plane. They compose a variable satellite system in most respects.

Before the *Pioneers* and *Voyagers* sailed past Jupiter and Saturn, astronomers were not generally aware of the many similarities between the Galilean satellites and the terrestrial planets. Even though there are compositional and internal structural differences, the Galilean satellites have and continue to evolve under processes similar to those operating in the terrestrial planets. Astronomers now give the Galilean satellites (shown in the chapter opening photo) a great deal of attention. Their mean densities in grams per cubic centimeter are as follows, in order of distance from Jupiter: Io, 3.41; Europa, 3.06; Ganymede, 1.90; and Callisto, 1.81. Hence Io and Europa, with size, density, and mass comparable to the moon, probably have a rocky, silicate-rich composition and structure similar to the moon while Ganymede and Callisto are lighter and made of a mixture of rocky and icy matter.

The photo essay in Figure 9.14 shows what astronomers suppose to be the internal structure of the Galilean satellites along with photographs of their surfaces obtained by the *Voyagers*. Clearly there are fascinating differences in their surface appearance. Europa, Ganymede, and Callisto have icy crusts with surfaces that are covered by ice or a mixture of ice and rocks many meters thick, whereas Io is quite different and a most exotic body.

IO: MOST VOLCANIC BODY OF THE SOLAR SYSTEM

Io (shown in Figure 9.14) surprised *Voyager* scientists by its surface appearance, which is a collage of mottled yellows, reds, and blackish browns. Passing very close to Io, *Voyager 1* was able to resolve features as small as a few kilometers. The satellite has a thin lumpy atmosphere, with sulfur dioxide as its primary constituent. The greatest excitement about Io is the positive identification of active volcanoes on its surface. Figure 9.14 shows an eruption occurring on the limb, with material being thrown up to altitudes of about 150 kilometers at velocities of about 1 kilometer per second. Such high speeds suggest that these are not earthlike volcanic eruptions, since the latter seldom exceed 0.1 kilometer per second. At least eight active volcanoes have tentatively been identified in the *Voyager 1* photographs, seven still erupting 4 months later, when *Voyager 2* arrived.

The orange color of the surface could well be due to sulfur compounds from the volcanic activity. Related to the discovery of volcanic activity is the fact that there are virtually no impact craters on the surface. There are also hot spots, up to 500 K, on the surface, whose typical temperature is 60 to 120 K. Finally, some 200 calderas, both with and without lava flows, dot the surface; the earth has only 15 or so. Thus the surface is active enough so that impact craters are either eroded away or filled in by volcanic debris in time periods as short as 1 million years. Io must possess the youngest surface of any solar-system body we have examined, and it is the only body be-

Then felt I like some watcher of the skies
When a new planet swims into his ken; . . .

John Keats

FIGURE 9.14
The Galilean satellites of Jupiter as photographed by *Voyager 1* and *Voyager 2*.

Top row:
Io is the innermost of the four Galilean satellites. This photograph was taken by *Voyager 1* on the morning of March 5, 1979, at a distance of 377,000 kilometers. Smallest visible features are about 10 kilometers across. The light colored areas are thought to be surface deposits consisting of mixtures of various salts, sulfur, and material from volcanic activity. The black areas are apparently volcanic cones and their associated lava flows. In the middle of the picture is a heart-shaped basin or plateau. The lack of craters is a good indication that Io is continually being resurfaced. Thus Io's surface is very young compared to that of the other Galilean satellites.

Voyager 2 took this picture of Io on the evening of July 9, 1979, from a distance of 1.2 million kilometers. On the limb of the satellite are two erupting volcanoes. The plumes extend about 100 to 200 kilometers above the surface. *Voyager 2* found seven of the eight volcanoes discovered by *Voyager 1* still erupting 4 months later. There are about 200 volcanic calderas on Io (compared to 15 for the earth). Material sprays out of the volcanoes in domelike formation because of the lack of any significant atmosphere on the satellite. Io certainly seems to be the most active object in the solar system as far as volcanic activity is concerned.

Center row:
A schematic is shown of the interiors of the Galilean satellites as presently understood and compared in size with the planet Mercury and the moon. Io and Europa are composed of rocky material with some ice on Europa. Jupiter flexes Io and heats it so that most water has escaped. Ganymede and Callisto are ice-rock mixtures as shown.

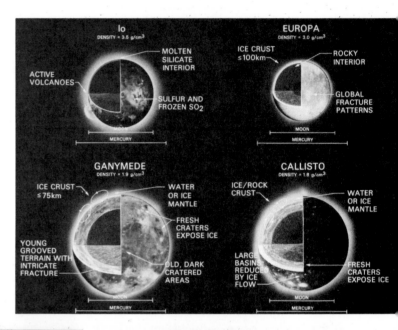

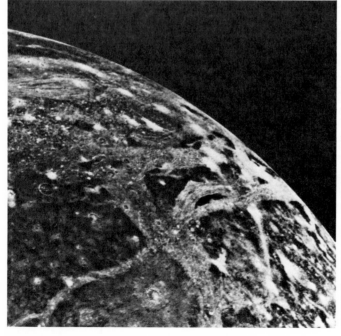

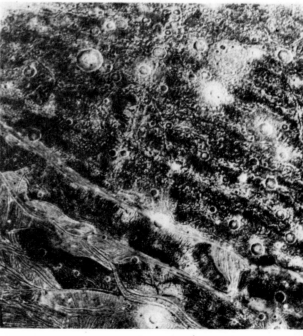

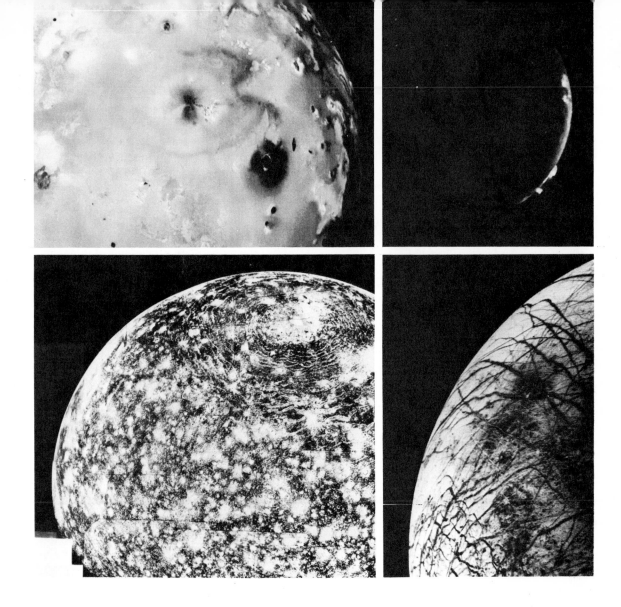

On the morning of March 6, 1979, from a distance of 202,000 kilometers *Voyager 1* obtained this photograph of Callisto, the fourth and most distant of the Galilean satellites. The dark surface has been very heavily cratered and thus is probably the oldest of those of the Galilean satellites. The bright spot near the limb is an impact basin with concentric rings around it. This probably shows the response of the icy surface to shock waves moving through the crust by the impact.

This picture of the second Galilean satellite, Europa, was taken by *Voyager 2* during a close encounter on the morning of July 9, 1979. The surface of Europa appears to be a crust of ice perhaps 100 kilometers thick. The complex array of streaks appear to be cracks in the crust that have been filled by material pushed up from below. Note that there seem to be no hills and valleys along the limb, indicating a relatively smooth surface. There is also a lack of impact craters in the crust. Thus Europa's surface is also fairly young but probably for a different reason from that for Io. No volcanic activity was seen on Europa by either *Voyager 1* or 2.

Bottom row:
Ganymede, Jupiter's largest satellite, was photographed in this picture by *Voyager 1* on March 5, 1979, from about 253,000 kilometers from the surface. The smallest features visible are about 2.5 kilometers across. Numerous impact craters can be seen on the surface, many with bright ray systems. The striking features are the light-colored bands traversing the surface, which contain alternate light and dark streaks.

Voyager 2, when about 150,000 kilometers from Ganymede, took this picture on July 9, 1979. In the foreground is the typical grooved terrain seen by *Voyager 1*. Note that it consists of intersecting bands of closely spaced, parallel ridges and grooves. This grooved terrain is somewhat less cratered than the dark terrain is. Therefore it is probably younger. Many of the craters have light-colored halos and rays, while some craters have dark halos. No good explanation exists for Ganymede's surface features.

sides the earth to show significant volcanic activity, which is actually greater than that of the earth.

The probable reason for the extensive volcanic activity is heating by tidal friction. Moving in an elliptical orbit, Io slightly shifts in and out in Jupiter's powerful gravitational field, with the result that it is flexed rapidly by tidal forces that release an enormous amount of frictional heat in its interior. This thermal energy, as it works its way to the surface, powers the volcanism. Io's volcanic activity may be akin more to the eruption of a hot-water geyser than to an explosive eruption.

EUROPA: A LOWELLIAN MAZE OF CANALS

In stark contrast to Io is the next Galilean satellite out from Jupiter, Europa, shown in the *Voyager 1* photograph in Figure 9.14. The satellite is moonlike in size, density, and mass but has a much higher reflectivity than the moon, which suggests an ice-rich surface. There is very little relief to the surface, which has been likened to a frozen ocean. The surface color is a lightly orange-hued off-white and has the least contrast of the four Galilean satellites. There appears to be a total lack of craters larger than about 50 to 100 kilometers across and very few smaller than that. The low number of craters is indicative of a young surface, something less than 3 billion years old.

What makes Europa the enigma that it is are the vast numbers of crisscrossing light and dark markings. The surface is a maze of these stripes such that at a distance Europa is reminiscent of Percival Lowell's imagined canals crisscrossing Mars and intersecting in oases. These features are tens of kilometers wide and some are thousands of kilometers long. In fact, some appear to extend halfway around the satellite and may be up to 3500 kilometers in length if they are truly continuous.

GANYMEDE: THE ICE GIANT

Voyager photographs of Ganymede (Figure 9.14) show a round disk with fairly distinct light and dark regions. Since the satellite is larger than Mercury but with a lower mean density, its composition is dominated by icy materials. The dark regions appear to be the oldest part of the surface in that they are heavily cratered while the light regions possess a smaller number of craters. On the terrestrial planets young craters are bright, often having a bright ray system of ejected material, while old craters are dark and lack ray sys-

tems. On Ganymede there are some dark-rayed craters that are relatively young; these craters suggest that ice-dominated soils can behave differently from the silicate-rich soils of the terrestrial planets.

Light regions on Ganymede display a number of features reminiscent of tectonic activity on the terrestrial planets and thus are presumed to be the youngest part of the surface. Dominating the light regions is a system of grooves that is very unlike the dark stripes on Europa. The grooves are parallel sets of ridges and troughs whose widths are up to tens of kilometers, and they may be a few hundred meters deep. Forming bands or sets of grooves, they wander for thousands of kilometers across Ganymede's surface to intersect in intricate patterns. The grooves have craters superimposed on them; so they are not very recent additions to the surface. The offset of grooves across what appear to be fault lines suggests that the breakup of dark crustal blocks is responsible for the grooves. Thus a global scale of tectonic activity may have existed at some time in the past on Ganymede.

CALLISTO: THE CRATERED SATELLITE

The outermost of the Galilean satellites is Callisto, which is shown in a *Voyager 1* photograph in Figure 9.14. It is a bit smaller and a little less dense than Ganymede, and it too probably had an ice-rock composition. Callisto has about 10 times as many craters as Ganymede. In fact craters dominate the entire surface and stand nearly shoulder to shoulder. Callisto is unique in having no plains or regions where later processes have obliterated the craters. It also appears to lack fractures in its crust, which Ganymede has.

Callisto has several large circular features surrounded by an almost concentric sequence of rings. The rings are raised, over 1500 kilometers in diameter for one and 500 kilometers for the other. For any features on Callisto there does not appear to be a significant difference in elevation, which is somewhat puzzling. This characteristic may be an indication of a relatively weak surface material that, because of icy composition, is unable to support much vertical relief.

SATELLITES OF SATURN

Like Jupiter's satellites Saturn's 17 satellites include a regular satellite system of 15 bodies moving in near-circular orbits in the planet's equatorial plane, their direction of motion the same as Saturn's direction of rotation. Iapetus, the second-outermost satellite, and Phoebe, the outermost one, form an irregular system with orbits inclined by 14.7° and 30°, respectively, to Saturn's equatorial plane. Phoebe's motion is retrograde, or opposite to Saturn's direction of rotation. Thus the 8 inner satellites of Jupiter, the 15 inner ones of Saturn, and all 5 of the known satellites of Uranus form regular satellite systems. Jupiter's 8 outermost satellites, the 2 outermost of Saturn and both of Neptune's form irregular satellite systems. It is probably more than coincidental that the ring systems are attached to the three planets having regular inner satellite systems. If the rings and regular satellites form as natural consequences of the formation of the planet, then what causes the irregular satellite systems? They *may* be captured bodies, such as pristine comets or icy-composition asteroids. Possibly in the next decade we shall have a better understanding of the differences after *Voyager 2* visits Uranus and Neptune.

Finally, whereas Jupiter has four planet-size satellites (the Galilean satellites), Saturn has only Titan that is that large. But it has four intermediate-size satellites, and Jupiter has none.

TITAN: THE ONE WITH AN ATMOSPHERE

Titan, shown in Figure 9.15, is another of the planet-size icy bodies of the outer solar system like the Galilean satellites of Jupiter. Estimates are that it must be 52 percent rock and 48 percent ice in order to have a mean density of almost 2 grams per cubic centimeter. It is unique—and thus highly intriguing—in that it is the only known satellite to possess a substantial atmosphere; Io's atmosphere is rarified and very spotty in its density. Above a unit of surface Titan's atmosphere contains more gas by a factor of 500 than does that of Mars and 10 times that of the earth.

From data provided by the close approach of the *Voyagers* (several thousand kilometers for *Voyager 1*) it seems that Titan and the earth are the only bodies in the solar system with atmospheres dominated by molecular nitrogen. Estimates for Titan lie between 82 and 94 percent for nitrogen, about 12 percent for argon, and a few percent for methane, which had been observed in 1944. In addition to these primary constituents there is a smattering of molecular hydrogen and hydrocarbon molecules.

In the very complicated atmospheric chemistry of Titan, methane clouds form some 3 kilometers above the surface and smog particles form from about 200

kilometers down to the surface. The smog is so thick that nowhere are there holes through which the surface could be seen by the *Voyager* cameras. Thus the satellite has a rather bland reddish-brown color, with the only markings a polar hood and a change in reflectivity at the equator.

In addition, the low and reasonably uniform surface temperature has prompted the speculation that there are oceans of liquid natural gas—that is, methane—on the surface. Methane ice clouds could also be in the low atmosphere, from which methane rain falls on occasion. Methane could apparently play the same role on Titan that water plays on the earth. Whether or not continents of icy soil and rocklike ice divide the methane oceans is pure speculation at this time. Nevertheless, Titan and the earth may be unique in the solar system in having substantial amounts of liquid covering their surfaces.

RHEA, IAPETUS, DIONE, AND TETHYS: INTERMEDIATE-SIZE SATELLITES

Rhea and Iapetus are the second and third largest of Saturn's satellites, with nearly identical radii of about 750 kilometers. The fourth and fifth largest are Dione and Tethys, with radii of about 550 kilometers. All four, shown in Figure 9.15, have about the same mean density of between 1 and 1.5 grams per cubic centimeter, and thus they are all presumably icy conglomerates. Iapetus is one of the irregular satellite system, while the others are part of the regular system. Iapetus is a strange body in that half of its surface is bright and the other half nearly black. The dark hemisphere is the one that appears to always face forward as the satellite orbits Saturn for reasons of which we are not sure.

Rhea shows a bright but bland leading hemisphere broken only by what appears to be a relatively young, large impact crater. The trailing hemisphere has bright swaths that cross a dark background with many craters. Dione, somewhat closer to Saturn, also has unequal leading and trailing hemispheres. Its leading hemisphere is uniformly bright, while the trailing one is dark with bright streaks somewhat like Rhea's. Tethys, the closest to Saturn of the intermediate-size satellites, is completely different: Its surface is heavily cratered and shows only small variations in brightness globally. Thus it does not have the hemispheric pattern of Rhea and Dione.

The *Voyager* spacecraft have given us a glimpse of new worlds in the satellites of Jupiter and Saturn that

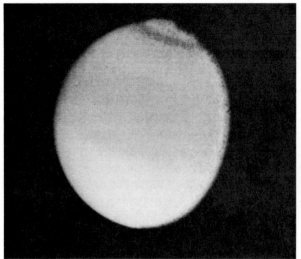

FIGURE 9.15

Saturn's larger satellites. The photo of Titan (top left), the largest satellite, was taken from a distance of 2.3 million kilometers and shows that its surface is completely hidden by its opaque atmosphere. Note that the southern hemisphere at the bottom is bright while there is a dark band near the equator and a dark ring at the north pole. The primary constituent of Titan's atmosphere is molecular nitrogen. The photo of Rhea (top right) shows a very old, heavily cratered region from a distance of 128 thousand kilometers. Iapetus (right), the outermost of Saturn's large satellites, shows bright and dark terrain from a distance of 1.1 million kilometers. Dione (bottom left) shows bright streaks and large impact craters in its icy surface. Tethys (bottom right) has a heavily cratered region at the upper right and a lightly cratered one at the lower right.

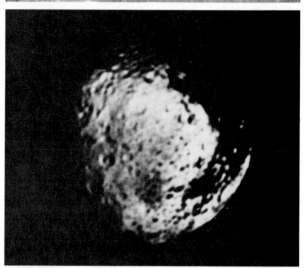

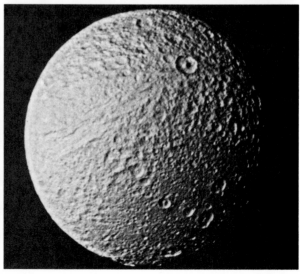

could not have been imagined from our earth-based studies. As is so often the case, the glimpse has raised more questions than it answers. Scientists who eagerly look to the future are hopeful that several new ventures to the outer solar system will be approved. However, the pace of scientific exploration can often be depressingly slow.

9.6
Comets:
Icy Messengers from the Past

DISCOVERIES AND APPEARANCE

Comets are some of the most spectacular bodies of the solar system and appear unexpectedly in all parts of the sky. They are often discovered accidentally in photographs taken by professional astronomers for other purposes or by amateur astronomers methodically searching for them. As an honor, comets are named after their discoverers. A record number appeared in 1977, 5 new comets and 15 old ones. Only once every other year or so does a comet become bright enough to be seen with the naked eye. A spectacularly bright comet appears about once or twice each decade.

Comets have four principal parts: nucleus, coma, hydrogen cloud, and tail (two types of tail may both be present). The usual telescopic comet appears as a small hazy object with a roundish nebulosity, called a *coma* (head), and occasionally a short *tail*. A brighter comet has more interesting features: an enlarged coma surrounding a small bright *nucleus* and a well-formed tail that points away from the sun, as shown in Figure 9.16. The size of the coma may vary from tens of thousands of kilometers to well over a million kilometers. Surrounding the coma is an immense cloud of hydrogen that spans millions of kilometers. The comet's mass is concentrated in the nucleus, which is the comet proper and which may be a fraction of a kilometer to several kilometers in diameter. Well-developed tails usually form when the comet is within the earth's orbit and can be millions to hundreds of millions of kilometers long. Very rarely will the tail be longer than the earth's distance from the sun. There can be two distinct types of tail, with both often present: One is a yellowish sweeping arc of dust, and the other is a long, straight, bluish tail of plasma.

We know that comets are flimsy structures of low density from the following evidence: They cannot be observed upon the solar disk when they pass in front of the sun; we can see stars through the tail and the outer portions of the coma; from night to night changes in brightness and size are observed in the coma and are even more pronounced in the tail; they are perturbed by the solar wind and by tidal, gravitational, or other disruptive forces.

ORBITS OF COMETS

The first positive evidence that comets are extraterrestrial objects was found by the sixteenth-century astronomer Tycho Brahe. He tried to find the parallactic displacement of the comet of 1577 relative to the background stars by comparing measurements from his observatory and other European centers. He decided that the object was more distant than the moon. A century later Isaac Newton demonstrated that comets are members of the solar system and move in elliptical orbits under the gravitational pull of the sun according to the law of gravitation.

Comets fall into two groups, long-period and short-period, depending on the period of orbital revo-

FIGURE 9.16
Changing appearance of the tail of Halley's comet during its next perihelion passage in February 1986. The tail points away from the sun because the material in the coma and tail is driven radially outward by the solar wind and the pressure of solar photons on dust particles of the comet. Normally the tail becomes prominent about the earth's orbital distance during approach to the sun. It lengthens and brightens approaching perihelion and fades as it recedes from the sun, when it can be quite different in appearance from approach. Halley's comet will be closest to the earth in November 1985 (0.62 AU) before perihelion passage, and 2 months after pehihelion in April 1986, when it is at 0.42 AU.

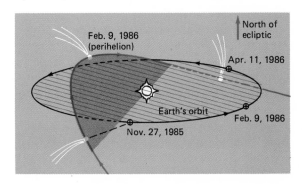

lution around the sun. Of the 600 or so for which information exists about 500 are long-period comets and approximately 100 are short-period comets with periods of less than 200 years. Those in the long-period group, which astonomers believe is the vast majority of all comets, travel in highly elongated ellipses inclined at all angles to the ecliptic plane. Their periods range from thousands to millions of years. The brighter members are usually among the most magnificent comets, with conspicuously long tails and well-defined nuclei. Some of these comets are approaching the sun for perhaps the first time. Several comets in the long-period category have grazed the outer parts of the sun. Frequently forces produced by the sun break them apart during this close encounter, and the fragments travel on as independent comets along orbits nearly identical with the parent's orbit.

A Dutch astronomer, Jan Oort, studying how cometary orbits are distributed, suggested in 1950 that a cloud of comets of not more than a few earth masses surrounds the sun at an average distance of 50,000 AU. (For comparison the nearest stars are over 300,000 AU from the sun.) Detached from this great reservoir (the number may be in the billions) by perturbations from nearby stars, a few begin to orbit the sun as long-period comets. Around 100,000 comets might have come close enough to the sun to be observable.

The second group of comets, now numbering over 100, orbits the sun at small or moderate angles of inclination to the ecliptic plane, nearly all in the same direction as the planets. Roughly, every third or fourth new comet discovered has a short period, ranging from 3.3 years to 200 years. Approximately half have their aphelion (the point in the orbit most distant from the sun) somewhere near Jupiter. When their orbital history is traced back mathematically, it is found that these objects have initially moved in long-period eccentric orbits, bringing them on one critical occasion into a chance encounter with Jupiter. The great planet's attraction has so modified their orbits that they now are part of Jupiter's short-period family of comets.

The nuclei of short-period comets have lifetimes of a few millennia since they lose about 1 percent of their mass on each perihelion passage. They eventually evaporate away their basic constituents of gas and dust leaving a rocky remnant that should be indistinguishable from an asteroid in appearance.

PHYSICAL AND CHEMICAL PROPERTIES

Today most astronomers agree that a comet is an icy conglomerate composed mostly of frozen water with some carbon dioxide and other ices and a little particulate matter and dust. This conglomerate forms the nucleus, which is surrounded by material vaporized by solar radiation to form the coma when the comet is close to the sun. As the comet approaches the sun, solar ultraviolet radiation breaks complex molecules down into simpler molecules of hydrogen, carbon, oxygen, nitrogen, and sulfur. These molecules are identified by the bright bands they produce in the emission spectrum of the comet. Emission lines of gaseous sodium are also present in the spectrum, along with some lines of iron, magnesium, and silicon when the comet comes very near the sun. The emission lines and bands are superimposed on the weak background spectrum of sunlight reflected from the cometary dusty material.

A typical bright comet's structure and appearance are illustrated in Figure 9.17. The comet's head plowing through the onrushing solar wind creates a bow shock wave. High-energy electrons in the solar wind ionize the molecular gases in the coma. Ultraviolet photons from the sun probably dissociate the hydroxyl radical (OH), releasing the hydrogen to form a

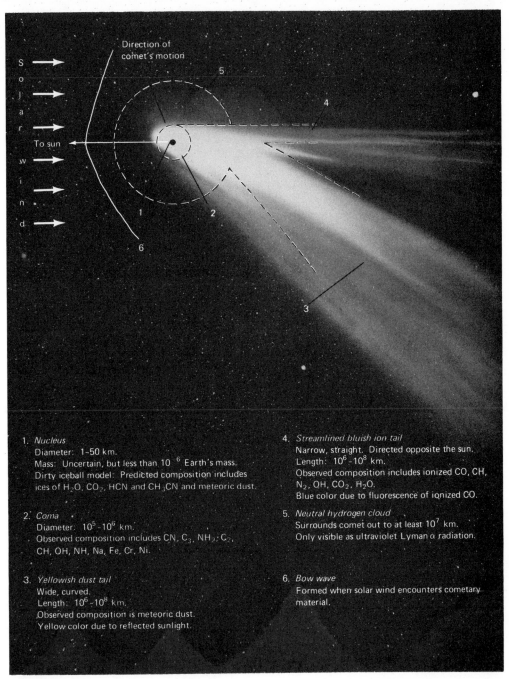

1. *Nucleus*
   Diameter: 1–50 km.
   Mass: Uncertain, but less than 10^{-6} Earth's mass.
   Dirty iceball model: Predicted composition includes ices of H_2O, CO_2, HCN and CH_3CN and meteoric dust.

2. *Coma*
   Diameter: 10^5–10^6 km.
   Observed composition includes CN, C_3, NH_2, C_2, CH, OH, NH, Na, Fe, Cr, Ni.

3. *Yellowish dust tail*
   Wide, curved.
   Length: 10^6–10^8 km.
   Observed composition is meteoric dust.
   Yellow color due to reflected sunlight.

4. *Streamlined bluish ion tail*
   Narrow, straight. Directed opposite the sun.
   Length: 10^6–10^8 km.
   Observed composition includes ionized CO, CH, N_2, OH, CO_2, H_2O.
   Blue color due to fluorescence of ionized CO.

5. *Neutral hydrogen cloud*
   Surrounds comet out to at least 10^7 km.
   Only visible as ultraviolet Lyman α radiation.

6. *Bow wave*
   Formed when solar wind encounters cometary material.

FIGURE 9.17
Schematic drawing of a comet superimposed on an actual photograph. This very bright comet was discovered by the Danish astronomer Richard West in November 1975 at the European Southern Observatory at La Silla, Chile. It was too far south to be observed in the northern latitudes until late February 1976, about the time it was rounding perihelion. Then it was bright enough to be seen with the naked eye near sunset — only the fourth comet to have been so visible this century. By March 1976 the nucleus had split into four different components, which separated from each other.

huge hydrogen cloud about the coma. Chaotic magnetic fields in the solar wind sweep the charged molecules away from the coma at high speeds, forming the narrow, bluish ion tail (item 4 in Figure 9.17). What causes the wide, yellowish curved tail (item 3 in Figure 9.17)? The sun's radiation (not the same as the solar

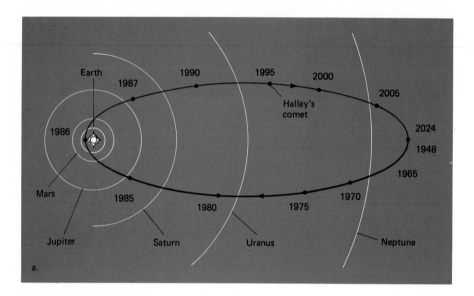

FIGURE 9.18
Orbit and position in the sky of Halley's comet. (a) The orbit, which has a period of approximately 76 years, a perihelion distance of 0.59 AU, and an aphelion distance of 35 AU. The comet is now moving between Uranus and Saturn and passes inside Saturn's orbit in 1984, reaching its next perihelion on February 9, 1986. (b) The position of the comet in the sky relative to the horizon for an observer located at 40°N latitude. In January 1986, Halley's comet will be approaching perihelion and conjunction with the sun (as Figure 9.16 shows). It will be an evening object in the western sky. The numbers following the dates are its anticipated apparent magnitude. Then in March 1986, after perihelion passage the comet is again visible as a morning object low in the southeastern sky. Finally, in mid- and late April it again becomes an evening object but this time visible in the south-southeastern sky. For most of the United States Halley's comet is never visible high above the horizon — less than 30° altitude — and the moon will be full on March 26, 1986, making it even more difficult to observe.

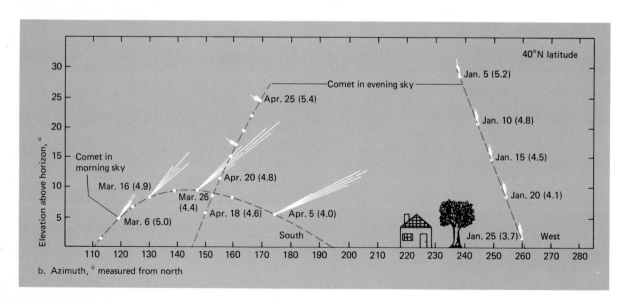

wind) pushes dusty material, which is flowing from the coma at different velocities, away from it at comparatively low speeds.

From their apparent structure and composition it seems probable that the comets are a link, or bridge, between the solar system and the interstellar medium. We shall defer a complete discussion of the interstellar medium until Chapter 14, but we may note here that many complex molecules are being discovered in dark clouds in the interstellar medium that, when frozen, could form icy structures, which are or may be similar to comet nuclei.

HALLEY'S COMET: PERIHELION PASSAGE IN 1985 AND 1986

When a bright comet appeared in the heavens late in the summer of 1682, Edmund Halley (1656–1742) calculated its orbit by using the method developed by his good friend Newton. He had done the same for the bright comets of 1531 and 1607 and was struck by the similarity in the three orbits. He concluded that the same comet must have made three revolutions in an elliptical orbit in a period of 75.5 years. He pre-

dicted its return in 1758, but he died 17 years before the comet returned as he had forecast: In March 1759 it passed perihelion, delayed on its approach to the sun by perturbations from Jupiter and Saturn. Because Halley recognized that this comet makes periodic returns, it was named posthumously in his honor.

Old records reveal that Halley's comet has been observed at every return since 239 B.C. Its appearance in A.D. 1066 is recorded in the historic Bayeux tapestry. Its orbit is shown in Figure 9.18. Halley's comet will next come near the earth in late 1985 and early 1986, as shown in Figures 9.16 and 9.18.

When Halley's comet returns, it will be the most studied comet in astronomical history. In addition to the major observatory programs, some of which have already begun, spacecraft from the European Space Agency, the Soviet Union (a joint venture with France), and Japan will greet it. Although NASA had planned a spacecraft rendezvous, it was never funded by Congress. A whole spectrum of scientific experiments are planned for both the spacecraft and the ground-based efforts.

Figure 9.19 shows Halley's comet at the time of its last perihelion passage in April 1910. It will be visible

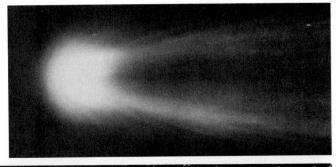

FIGURE 9.19
Halley's comet as photographed after its last perihelion passage on April 19, 1910. (below) Photograph taken May 13, 1910, 24 days after perihelion passage, at the Lick Observatory, showing its tail arcing 45° across the sky. The object to the lower left is the planet Venus. (right) Photograph taken at the Mount Wilson Observatory on May 8, 1910, 19 days after perihelion, showing only the head of the comet.

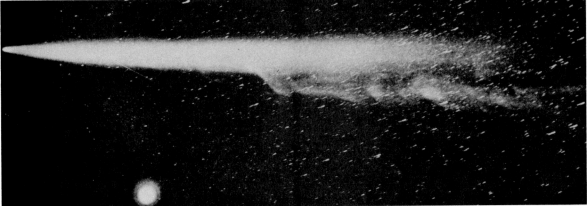

to the naked eye from the United States in January, March, and April 1986 as it passes perihelion on February 9, 1986. As shown in Figure 9.18b, it should be brightest in April but very low on the southern horizon. This should not deter millions of people from seeing what must be one of the most eagerly awaited events in this or any century.

SUMMARY

Jovian atmosphere and clouds. The most fundamental aspect in understanding a planet's atmosphere is the vertical temperature structure. Chemical constituents of Jupiter's and Saturn's atmospheres include primarily hydrogen, presumably helium, some methane, and some ammonia. In composition these atmospheres more closely resemble that of the sun than those of the terrestrial planets. Traces of other molecular compounds have also been identified in Jupiter. Some of these compounds may be solids in the colder atmosphere of Saturn, and that is why they have not been identified. Low-lying clouds of water-ice crystals, mid-level clouds of ammonium hydrosulfide, and high-altitude clouds of ammonia are the most conspicuous physical feature of the atmospheres of Jupiter and Saturn. Large-scale motions in the atmospheres of the two planets are responsible for light- and dark-colored bands of clouds, vigorous eastward and westward winds, turbulent motions, and the formation of circulation cells. Especially notable is the Great Red Spot in the atmosphere of Jupiter.

Little is known about the atmospheres of Uranus and Neptune, which contain comparatively more oxygen, nitrogen, carbon, silicon, and iron than do Jupiter and Saturn. Although similar to each other in some respects, there are differences with respect to each other and to Jupiter and Saturn. Neptune's atmosphere possesses a variable haze that is apparently not present in that of Uranus.

Internal structure of the Jovian planets. Because of the greater mass of the Jovian planets and a composition dominated by hydrogen, helium, and other light elements, the internal structure of these planets differs from that of the terrestrial planets. Around possibly a small, dense core of rocky and icy materials, Jupiter and Saturn have metallic-liquid and molecular-liquid layers for an internal structure overlain by extensive gaseous layers that are the atmosphere. The internal structures of Uranus and Neptune are different from those of Jupiter and Saturn because of their smaller masses and different compositions. Jupiter, Saturn, and Neptune radiate more energy than they receive from the sun. The source of energy is probably the slow contraction of these planets.

Magnetospheres about the Jovian planets. Jupiter is the strongest emitter of radio radiation in the solar system with the exception of the sun. This is the result of the magnetosphere, which is quite extensive. Saturn's magnetic field also defines a magnetospheric zone in which it can control the motion of subatomic particles. Whether or not magnetospheres exist about Uranus and Neptune will not be known with certainty until *Voyager 2* reaches them in 1986 and 1989.

Ring systems of the Jovian planets. Ring systems are formed of countless individual particles in orbit about the planet and shaped by interactions of the particles with each other and by the gravitational influence of satellites. Saturn's ring system is more extensive than that of Jupiter or Uranus. Rings of Saturn are composed primarily of water-ice or ice-coated rocky particles. The three major rings are composed of thousands of narrow ringlets. Saturn's ring system is far more complex than had been anticipated prior to *Voyager.* The ring systems of Jupiter and Neptune are fainter than Saturn's and are probably composed of silicates or carbon-rich material. Their structure appears to be quite different from that of Saturn.

Satellites of Jupiter and Saturn. Compared to the terrestrial planets, Jupiter and Saturn have many satellites; Jupiter has at least 16 and Saturn, 17. Jupiter's Galilean satellites resemble in many respects the terrestrial planets and have apparently been shaped by similar forces. Active volcanoes have been found on Io, although they do not much resemble those on earth. Io and Europa, with size, density, and mass comparable to that of the moon, are probably a rocky silicate-rich material similar in structure to that of the moon. While Ganymede and Callisto are as large as

Mercury, they have a low mean density and are probably composed of a mixture of rocky and icy materials. Saturn's satellite system contains the only satellite with an atmosphere: Titan is a rocky and icy mixture with an atmosphere dominated by nitrogen. Saturn also has four interesting, intermediate-sized satellites: Rhea, Iapetus, Dione, and Tethys.

Comets. Comets have four principal parts: nucleus, coma (head), hydrogen cloud, and tail. The two types of tail point away from the sun. Comets are an icy conglomerate of frozen water, methane, ammonia, carbon dioxide, and some small particulate matter and dust grains.

KEY TERMS

atmospheric composition
atmospheric temperature structure
circulation cell
cloud bands
comet nucleus
comet tail
cometary coma
convection
heat transport
ice-rich body

irregular satellite system
long-period comet
metallic liquid
nonthermal radiation
occultation
regular satellite system
ring system
Roche limit
short-period comet
synchrotron radiation

CLASS DISCUSSION

1. With regard to chemical composition, temperature structure, wind patterns, and clouds, are the atmospheres of the Jovian planets much alike? How do they compare with those of the terrestrial planets?

2. Assuming you could take a trip from far outside Jupiter into the giant planet's center, describe what you would see or be able to measure.

3. In terms of their overall structure, are the Jovian planets similar to the terrestrial planets? If different, how are they different?

4. What are the ring systems of the Jovian planets like? What are they composed of? Why

don't the terrestrial planets have ring systems? Why doesn't the ring matter coalesce to form a satellite?

5. Can you give a general description of the Jovian satellites that is reasonably valid? If some of the Jovian satellites are approximately the size of Mercury, should they be considered major bodies of the solar system?

6. Since comets are in very elongated orbits about the sun, one might suspect that they are ejected from the sun. Give as many reasons as you can why such a hypothesis is not true.

READING REVIEW

1. How many space missions have been sent to the outer solar system?

2. What causes the banded appearance of Jupiter and Saturn?

3. What is the primary chemical constituent of Jupiter and Saturn?

4. What is the Great Red Spot? On which planet is it located?

5. Are all of the Jovian planets self-luminous (that is, they emit more energy than they receive from the sun)?

6. How do we know that the rings of the Jovian planets are not solid disks?

7. Which of the giant planets has the most extensive ring systems?

8. Are Cassini's division and Encke's division gaps in Saturn's rings where no particles exist?

9. Which are the Galilean satellites and how do they compare with the moon in size and density?

10. Approximately how old is the surface of Io?

11. How large is a typical comet (the comet proper, not the tail)?

12. What is a comet made of?

SELECTED READINGS

Ingersoll, A. P. "Jupiter and Saturn," *Scientific American,* December 1981.

Owen, T. "Titan," *Scientific American,* February 1982.

Pollack, J. B., and J. N. Cuzzi. "Rings in the Solar System," *Scientific American,* November 1981.

Soderblom, L. A. "The Galilean Moons of Jupiter," *Scientific American,* January 1980.

Soderblom, L. A., and T. V. Johnson. "The Moons of Saturn," *Scientific American,* January 1982.

Yeomans, D. K. *Comet Halley Handbook.* Jet Propulsion Laboratory, 1981.

10

Evolution of the Solar System and Life

Now that we are acquainted with the nature of the solar system—with the exception of the sun—it is natural to ask about its origin, its variety of bodies, and finally why the earth is the home of life. In the preceding chapters we have touched at several points on the origin and evolution of the solar system, and it is now time to bring these scattered notions together into what astronomers currently believe happened some 4.6 billion years ago.

In discussing the origin and evolution of life on the earth and its prospects elsewhere in the solar system, we begin with ourselves since we are the only system of life for which we have any real information. From the earth we move to consider the solar system; and in the last chapter, Chapter 20, we shall consider the question of life beyond the solar system.

10.1
Origin of the Solar System

ARCHITECTURE OF THE SYSTEM

One of the most challenging questions the astronomer can ask is "How did the solar system begin?" Two sets of theories have been proposed in the past: those invoking an accidental catastrophic event, such as a near collision between a star and the sun, and those involving a natural noncatastrophic event, such as might occur in conjunction with the birth of a star. Chapter 14 covers the story of the birth of stars; so here we need only say that we know that stars are born, live out their lives, and die—just as the things of earth are not eternal. A historical approach to the explanations of planetary genesis is a good beginning, but first we should summarize the major characteristics of the solar system for which any theory of origin should account.

A sequence of natural forces evidently created and shaped the system somewhat along the lines revealed by the following clues, which suggest that the design most likely possessed a continuity in its processes and did not materialize through a sequence of unrelated, random events:

◄ Artist's illustration of rocky planetesimals forming in the inner part of the solar nebula near the developing sun.

Recall from Time's abysmal chasm
That piece of primal protoplasm
The First Amoeba, strangely splendid,
From whom we're all of us descended.

Arthur Guiterman

1. The planets are isolated from each other without bunching, and they are placed at orderly intervals (as Table 6.2 shows).

2. The planets' orbits are nearly circular, except for those of Mercury and Pluto.

3. Their orbits are nearly in the same plane; Mercury and Pluto are again exceptions.

4. All the planets and asteroids revolve around the sun in the same direction that the sun rotates (from west to east).

5. Except for Venus and Uranus, the planets also rotate around their axes from west to east.

6. The satellites of a planet can be divided into either a regular system with direct orbits approximately in the planet's equatorial plane or an irregular system with retrograde orbits inclined at various angles.

7. The terrestrial planets have high mean densities and relatively thin or no atmospheres, rotate slowly, and possess few or no satellites—points that are undoubtedly related to their smallness and closeness to the sun.

8. The giant planets have low mean densities, relatively thick atmospheres, and many satellites, and they rotate rapidly—all related to their great mass and distance from the sun.

9. Studies of chemical composition suggest that the terrestrial planets are small, dense, rocky bodies that are poor in hydrogen; the giant planets are large, rarefied, gaseous bodies that are rich in hydrogen; and most of the outer planets' satellites, comets, and Pluto are icy bodies.

NEBULAR HYPOTHESIS

The German philosopher Immanuel Kant speculated in the middle of the eighteenth century that the solar system had been formed out of a huge, rotating, gaseous nebula slowly contracting and condensing. Pierre Laplace (1749–1827), a celebrated French mathematical astronomer, expanded the idea in 1796, and it became the *nebular hypothesis.*

Laplace theorized that as the large, slowly rotating solar nebula of hot gaseous matter contracted, it rotated faster and faster, flattening into an equatorial ring. The physical principles involved here are the action of gravity and the *conservation of angular momentum,* which requires a spinning body to rotate faster as it shrinks. The angular momentum of a rotating body, a measure of its quantity of rotation, remains constant unless energy is taken out of rotation and put into some other form: If the radius of the body decreases, the rotational velocity must increase to compensate for the reduced radius; this is what we observe when we see a spinning ice skater rotate faster as she brings her outstretched arms closer to her body—her angular momentum is thus conserved.

Laplace supposed that when the centrifugal force acting on the outer rotating edge of the solar nebula exceeded the inward gravitational force of the nebular mass, a ring of gaseous matter was split off, eventually coalescing into a planet as is shown in Figure 10.1. The splitting process repeated itself, making concentric rings that formed into planets, while the central portion condensed to become the sun.

The theory has two major defects. First, whereas over 99 percent of the solar-system mass resides in the sun, 99 percent of the angular momentum of the system resides in the planets' orbital rotational motions. We might guess intuitively that the distribution of angular momentum in the forming solar system ought to roughly match the distribution of mass; the central mass could not have transferred this much momentum to the planets. Second, a hot gaseous ring of the type postulated would disperse into space and not pull itself together gravitationally to form a planet.

ENCOUNTER THEORIES

At the beginning of this century attempts to reconcile the nebular hypothesis with physical principles were temporarily abandoned. A different approach, the so-called *encounter theory*—which had been first conceived in 1745 by the French naturalist Georges Buffon

FIGURE 10.1
Laplace's nebular hypothesis for the origin of the solar system. (a) A contracting and spinning solar nebula of gaseous matter has flattened to a pancake shape. (b) As the disk contracts, it rotates faster according to the conservation of angular momentum, causing an outer ring to become unstable and separate from the rest of the disk. (c) The material in the ring ultimately coalesces into a planet, while the next ring prepares to detach and form the next planet, and so on.

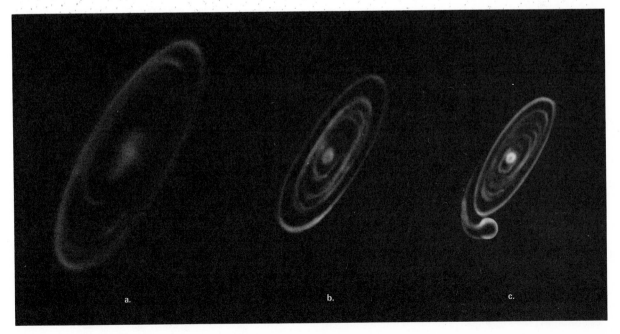

a.   b.   c.

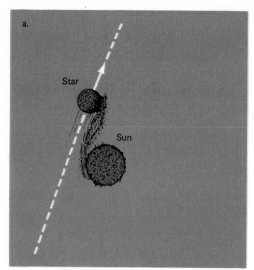
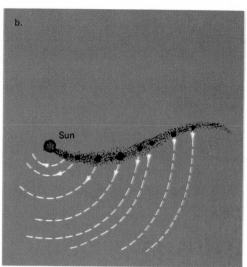

FIGURE 10.2
One version of the encounter theory. (a) The passing star pulls from the sun a gaseous filament that condenses into the planets. (b) The central section forms the giant, gaseous planets while the ends form small, terrestrial planets.

(1707–1788) when he proposed that material ripped off from the sun by collision with a comet had condensed into the planets—was taken by the American geologist Thomas Chamberlin (1843–1928) and the American astronomer Forest Ray Moulton (1872–1952). They suggested that giant eruptions were pulled off the sun by the gravitational attraction of a passing star.

Somewhat later another geologist-astronomer pair in England, Harold Jeffreys (1891–   ) and James Jeans (1877–1946), theorized that a cigar-shaped gaseous filament was pulled from the sun by the sideswiping action of a passing star, as in Figure 10.2. The middle section condensed into the Jovian planets, and the ends into the smaller planets.

The encounter theory accounts for the common direction of the planets' orbital motion and the sun's rotation as well as for the planets' nearly circular and coplanar orbits. In either version, however, this theory has serious failings in that solar matter, whether pulled or ejected, could not have acquired sufficient angular momentum nor hot gas have condensed into planets. Besides, the probability of a near encounter in our region of the Galaxy is vanishingly small, less than one in many millions.

PROTOPLANET THEORY: THE SOLAR NEBULA

By midcentury astronomers once more turned their attention to possible improvements in the nebular hy-

pothesis. A new factor was introduced in the form of the existence in the cool gaseous nebula of a small amount of dust, providing nuclei for the condensation of gas particles into larger aggregates that could accrete and solidify into the embryo planets. (The existence of dust particles in the interstellar gas clouds out of which stars are formed was accepted in the 1930s.) This modern version of the nebular hypothesis is called the *protoplanet hypothesis,* and it owes much of its recent revival to the power and scope of computer analyses. It was first formulated independently by Carl von Weizsacker (1912–   ) and by Gerard Kuiper (1905–1973) in 1945 and then extended and modified over the years by others.

The hypothesis begins with a fragment's separating from an interstellar cloud composed mainly of hydrogen and helium, with trace amounts of the other elements. With other fragments of the interstellar cloud presumably following a similar evolution, its central region, being somewhat more dense, collapsed more rapidly than its outlying parts. This formed the central portion of the *solar nebula,* whose outer portion contained a thin disk of solids within a thicker disk of gases. The original interstellar cloud must have been rotating, and as it fragmented, rotation was imparted to each fragment. Thus as the solar nebula contracted, it rotated more rapidly, conserving angular momentum.

The solar nebula grew by accretion as material continued to fall inward from its surroundings (Figure 10.3). Large-scale turbulence from gravitational instabilities ruptured the thin disk into eddies, each containing many small particles. These particles gradually built up into larger bodies by some combination of adhesive forces. Repeated encounters among them resulted in accretion of literally billions of still larger asteroid-sized aggregates called *planetesimals,* which orbited the center of the solar nebula. Mutual gravitational attraction led to further encounters and gradual coalescence into many roughly moon-size bodies, which in turn coalesced to form the planets.

Planetesimals must have differed in chemical composition, depending primarily on their initial distance from the sun as it formed. That is, as the central portion of the solar nebula contracted, the temperature rose to around 2000 K, hot enough to vaporize all compounds in the dust except the "high-temperature" metallic and silicate minerals in the inner portion of the disk, while the outer disk remained relatively cool. Planets that formed closer to the young sun, such as the terrestrial planets, would be expected to contain less of the volatile icy and gaseous materials and thus be richer in the rocky materials, as sketched in Figure 10.4.

TERRESTRIAL PLANETS' FORMATION

During and following the formation of the terrestrial planets, there was a catastrophic bombardment by the

FIGURE 10.3
Formation of planets from the solar nebula.

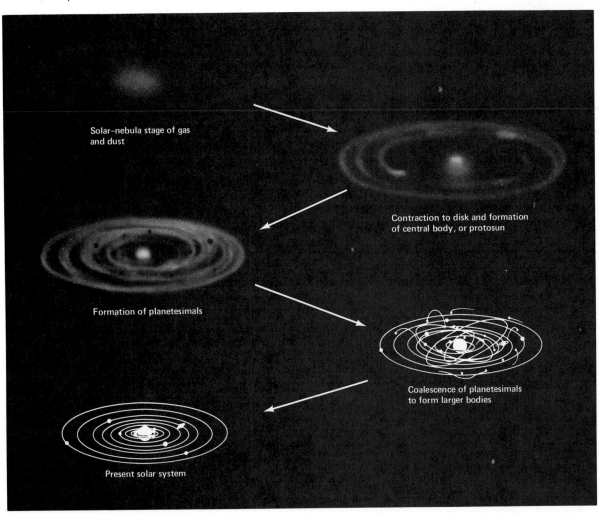

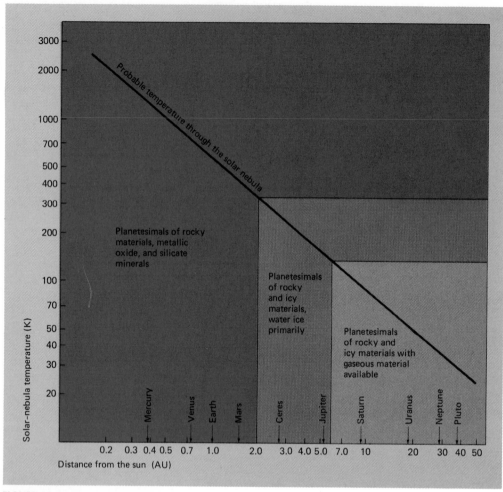

FIGURE 10.4

Chemical composition of the planetesimals. Temperature declined through the solar nebula as shown. In the dark-shaded region temperatures are so high that only the rocky materials can coalesce to form planetesimals. Farther out in the intermediate-shaded region, water could condense to form icy materials, which were far more abundant than the rocky materials also present in this zone. Planetesimals here were primarily icy bodies. The light-shaded region is where other ices would form, such as carbon dioxide and ammonia, and gaseous materials could also be accreted to the coalascing planetesimals if sufficient mass was present to hold the gas (as in the case of Jupiter and Saturn).

remaining rocky planetesimals that cratered these planets. The impacting material, coupled with intense radioactivity and subsequent gravitational concentration, produced sufficient heat to melt and chemically differentiate the planets into their presently layered structure (core, mantle, and crust). The atmospheres of the terrestrial planets were formed during this process and afterward by outgassing from impacting material and from the hot interiors of the planets.

In the asteroid belt between Mars and Jupiter the temperature of the solar nebula was lower so that carbon-rich and water-rich minerals could coalesce in the forming planetesimals. From about Jupiter outward, temperatures were even lower so that huge amounts of frozen water could accumulate with the rocky material in the planetesimals. At still colder temperatures other ices would have formed, such as ammonia and methane, giving those distant planetesimals a mixed composition of water, ammonia, and methane ice impregnated with a small amount of rocky matter.

JOVIAN PLANETS' FORMATION

Within the outer, cooler regions of the solar nebula the icy planetesimals collided, building larger bodies of ice and rock. If these bodies grew to a mass a few times that of the earth, then they drew in more hydrogen and helium from the surrounding interplanetary gas. Naturally capture and retention of gas was easier far from the sun, where the temperature was lower. Because of their great mass, they have kept very nearly the same relative proportion of hydrogen and helium to the heavier elements as the sun and the interstellar medium have. This is the most likely mode of formation for Jupiter and Saturn and why they are hydrogen-rich bodies. Uranus and Neptune were simply never massive enough to accrete hydrogen and helium to the extent that Jupiter and Saturn did. Thus oxygen, nitrogen, carbon, and silicon dominate their compositions. The comets are probably a fossil relic of the primordial icy planetesimals that existed in the outermost regions of the solar nebula.

In the regular satellites of Jupiter and Saturn we probably have a miniature version of planet formation. Contraction by Jupiter and Saturn at the time of their formation released a great deal of gravitational potential energy, heating them significantly. Jupiter was some 10 times brighter than at present so that the contracting planet raised the temperature of matter close to it, accounting for why the two inner Galilean satellites are rocky bodies, while the two distant ones are primarily icy bodies. This mimics the decline in density with distance from the sun found in the planets themselves.

FORMATION OF THE SUN

When the protoplanets were all formed, the nebula's central bulge rapidly collapsed into the protosun. Continued contraction raised its internal temperature from a few tens of thousands of degrees to several million degrees when the first stages of nuclear burn-

ing were initiated. (The nuclear-burning processes will be discussed in Chapter 13.) In the last stages of formation the sun may have had a much more intense solar wind, which presumably blew away much of the primordial gas and dust left over from the original interstellar cloud. But this point is still pretty much of a mystery.

A weakness in the protoplanet hypothesis is that it does not provide a completely satisfactory explanation for the observed distribution of angular momentum in the solar system. If the angular momentum of the planets could somehow be returned to the sun, its present slow rotation (like that of stars similar to it) of 2 kilometers per second would be increased to about 100 kilometers per second. The primitive sun apparently transferred most of its angular momentum to the planets as they were forming.

To explain this transfer of angular momentum, astronomers have proposed a braking action caused by magnetohydrodynamic forces on the sun as its magnetic field interacted with the ionized nebular gas in the disk. The magnetic lines of force spiraling outward from the rotating sun into the surrounding nebula would act as a magnetic drag on the spinning sun and serve as conduits, transferring angular momentum to the planetary disk.

There is a recent discovery of a relatively high abundance of some rare — by earth standards — isotopes in primitive meteorites. It has been proposed that the isotopic anomalies are due to the injection of matter from a supernova explosion into the solar system a few million years before the meteorites solidified. (A supernova is the explosion of a star in the last stages of its life.) Possibly the concussion from the explosion triggered the collapse of the interstellar cloud to form the solar nebula.

Regardless of the means of starting the formation process, planetary systems are believed to grow naturally from physical events that develop after an interstellar cloud has begun to contract. Finally, in just the

. . . then wilt thou not be loath
To leave this Paradise, but shalt possess
A Paradise within thee, happier far. . .
They hand in hand with wandring steps and slow,
Through *Eden* took thir solitarie way.

Milton

last few years astronomers have discovered large, cool, dust envelopes around infrared stars in the interstellar clouds of the Milky Way. Some of these objects may be in the early stages of nebular condensation visualized in the protoplanet theory. Hence new solar systems may be forming out in those distant interstellar clouds.

10.2
Origin of
the Earth-Moon System

FORMATION OF THE EARTH

Within a relatively short time after contraction of the solar nebula began the young earth had collected most of the matter that composes it today. Matter attracted by the growing earth collided with it, giving up its kinetic energy as heat. This energy, along with the energy resulting from the earth's gravitational contraction and emission by radioactive nuclei, heated the earth's interior.

In a few tens of millions of years the earth became molten; chemical differentiation followed. The heaviest elements, iron in particular, separated from the lighter elements oxygen and silicon (primarily in the form of silicates and oxides of iron and magnesium) and sank toward the center. The silicates (molecules containing both silicon and oxygen) and the oxides (containing oxygen) rose to form the mantle surrounding an iron-rich core. The lightest materials rose to the top and solidified as the crust.

In the Precambrian Era, about 4.0 to 4.5 billion years ago, the earth as a whole was cooling even though volcanic activity on the surface was intense. We believe that during this period an atmosphere of whose composition we are not certain was formed, probably from the gases carbon dioxide, carbon monoxide, water vapor, nitrogen, and possibly some hydrogen sulfide and hydrogen. As the earth cooled, these gases escaped from the interior during volcanic activity and water condensed, forming the oceans. But what of the moon during this period?

FORMATION OF THE MOON

Even with all our new information the moon's origin, like that of the solar system itself, is still shrouded in mystery. No proposal of either a catastrophic or a

noncatastrophic beginning for the earth-moon system is universally accepted.

The earliest concept, the *fission theory* (Figure 10.5), was that the earth was spinning rapidly and flattened to a dumbbell shape perhaps because of movement in the earth's molten core. The smaller end of the dumbbell broke away to become the moon, separating ever more from the earth because of tidal forces. The major objection to this theory is that the

FIGURE 10.5
Various theories for the origin of the earth-moon system: (a) The fission theory presumes that, because of rapid rotation, the earth broke apart, forming the moon, which spiraled outward. (b) In the condensation theory the earth and the moon formed in orbit about each other. (c) In the capture theory the moon formed in another part of the solar system and was later captured by the earth.

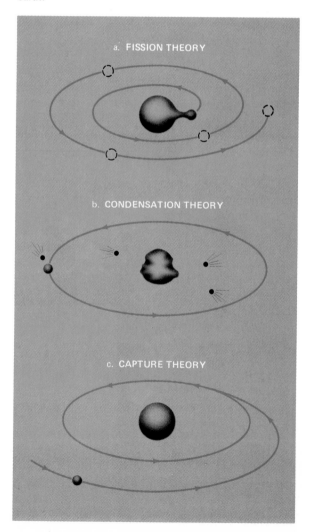

primitive earth could not have spun rapidly enough to promote fission through rotational instability.

Earth and moon may have formed by accretion from chemically related primordial planetesimals condensing out of the gas and dust of the solar nebula; this is the *condensation theory* (Figure 10.5). Colliding debris trapped in orbit around the growing earth accreted to form the moon. The two bodies are of comparable ages (4.6 billion years), a fact strongly backing this theory; the most recent chemical analysis of the lunar rock samples shows some chemical disparities between the earth and the moon so that the earth and the moon may or may not have evolved from the same parent material.

Finally, a third possibility is that a protobody originally was moving in a highly eccentric orbit around the sun. As it approached the earth almost on a collision course, it was disrupted by strong tidal forces, and part of the fragmented body became the earth's satellite—the *capture theory*. The problem with this idea is that capture is not easy to accomplish. Although it apparently can happen, some means must occur to remove some of the kinetic energy of the captured body. Just passing by a more massive body does not automatically lead to capture.

It seems to us appropriate to close this section with a quotation from Mark Twain's *Life on the Mississippi:* "There is something fascinating about science. One gets such wholesale returns of conjecture out of such a trifling investment of fact."

10.3
Chemical Basis and the Origin of Life on Earth

LIFE'S ESSENTIALS

The thin zone of land, water, and air in which life flourishes on the earth is known as the *biosphere.* Had the earth formed somewhat closer to the sun than it did, the greenhouse effect might have become dominant as it is on Venus, resulting in a hot, sterile, surface. Carbon dioxide in the earth's early atmosphere undoubtedly produced some greenhouse effect, but the effect declined with the reduction in the carbon dioxide content.

On the other hand had the earth formed farther away from the sun than it did, water would have remained frozen as subsurface ice or polar caps, and the surface of the earth would somewhat resemble the cold Martian landscape. Our earth, therefore, is at the right distance from our star, the sun, for development of an active biosphere. The chemistry of the only kind of life we know is built on the chemistry of the element carbon and the solvent water, supplemented by the biologically important atoms nitrogen, oxygen, phosphorus, and sulfur.

Carbon atoms bond easily with other carbon atoms, producing long chains to which other biologically significant atoms can bond: hydrogen, oxygen, and nitrogen. The resulting molecules, whether or not they are a part or product of living matter, are called *organic molecules.*

Water, because it can flow readily and remain in liquid form through a range of conditions, is the ideal solvent for many organic compounds. From water come hydrogen bonds, which give structural stability to strings of proteins, nucleic acids, and other long-chain carbon compounds. A liquid-water environment and moderate temperature make it possible for such long-chain carbon molecules to form, and they have become the biochemical basis of life as we know it.

Even with this biochemical basis life would not have been able to develop and sustain itself without proper temperature, a supply of nutrients, self-regulating mechanisms, and the sun's energy. The sun is the prime source of energy for driving the chemical reactions in the life cycle (Figure 10.6).

HOW DID LIFE BEGIN?

But how did life begin in what we now know as the biosphere? How did it derive from nonlife? (To discuss life elsewhere as we shall do later, we must start from a knowledge of the chemical and biological evolution that apparently led to life on our planet.) Biologists are not unanimous on all the factors in the definition of life, but most agree that life is different from nonlife: It evolves or changes (by chance mutations or otherwise) as time goes on while interacting with its environment in a unique way. But, most importantly, life is not a "thing"; it is a process made up of an unimaginably large number of complex chemical reactions that collectively produce all the characteristics we associate with life. For terrestrial life, evolution has led to intelligence, the aspect of extraterrestrial life in which we are most interested.

Until the mid-nineteenth century, most people believed in spontaneous generation: that life rose from nonliving matter—worms from mud, mice from refuse, maggots from decaying meat, and so forth. By careful experimentation Louis Pasteur (1822–1895) proved otherwise. Then Charles Darwin (1809–1882) set forth his theory of biological evolution. Darwin held that the more complex forms of life evolved from simple organisms over very long stretches of time, thus suggesting a unity for all earth's life. Considering the problem of how life came from nonlife, he speculated that in some "warm little pond," proteins and more complex organic molecules could have been synthesized from ammonia and phosphoric salts energized by sunlight, heat, or electricity.

Our contemporary ideas on the *chemical evolution of life* actually began in 1924, when the Russian biochemist Alexander Oparin (1894–   ) reintroduced chemical evolution as a necessary forerunner to biological evolution. In 1928 the English biologist J. B. S. Haldane (1892–1964) independently suggested an outline for chemical evolution, differing somewhat in detail from that of Oparin. Haldane's theory is still the basis of our understanding of chemical evolution, which we shall briefly outline in the remaining paragraphs of this subsection.

Earth's cooling and solidifying crust was racked by volcanic activity that presumably vented water vapor, carbon dioxide, nitrogen, hydrogen, and smaller amounts of other molecules that are easily vaporized.

FIGURE 10.6

Energy sources on the primitive earth that powered the chemical reactions of the life cycle. The primary energy source was ultraviolet photons from the sun. Other sources, such as natural radioactivity in rocks, lightning, meteoric impact, and geothermal activity, also must have played a part. Even mechanical-energy sources, such as ocean tides, geological activity, and landslides, contributed some energy in starting life processes.

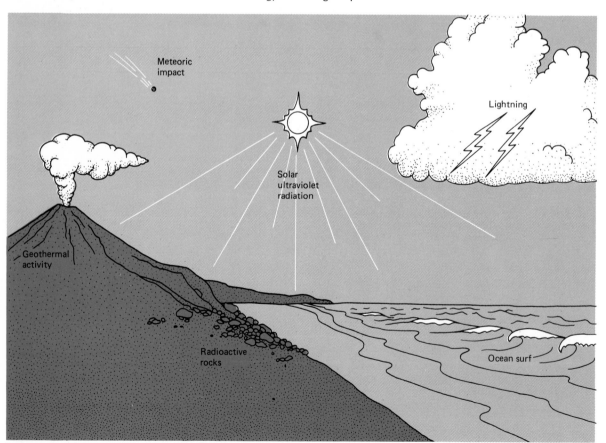

FIGURE 10.7
Outgassing can be seen at current volcanic sites.

These probably formed earth's early atmosphere. Today active volcanoes discharge large quantities of water vapor, carbon dioxide, and nitrogen, some sulfur, and traces of other gases (Figure 10.7).

Earth's magnetic field, which probably developed after the interior differentiated into its present layered structure, helped to keep this primitive atmosphere from being swept away by the solar wind. Subsequent cooling of the earth condensed water vapor, forming the warm seas and the shallow lagoons and pools that were destined to provide a haven for the development of organic compounds. From this primordial soup,

over a long time, the more complex organic molecules or biological macromolecules, such as proteins and nucleic acids — the basic ingredients of life — were fashioned. These were formed by energy from solar ultraviolet and visible light, electric discharges, and heat from radioactivity, volcanoes, and meteoric impacts. The probable course that biochemical evolution followed from the primordial raw materials to the first living organisms is outlined in Figure 10.8. The cell is the basic unit of life from which complex organisms, such as human beings are built. We investigate the cell and its components next.

FIGURE 10.8
Chemical evolution leading to life. The chemical content of living matter is approximately
78 percent water, 15 percent protein, 5 percent fat, and 2 percent carbohydrate.

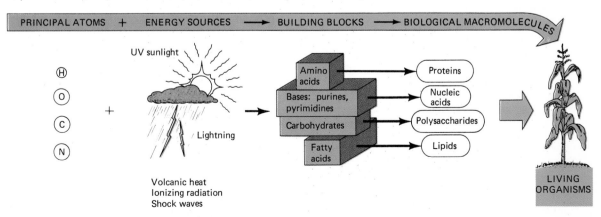

THE CHEMISTRY OF LIFE

An individual generalized *cell* is sketched in Figure 10.9. Modern biology has shown that the *cell* is in general the minimum organized unit of matter that displays those properties we collectively refer to as *life*. Collections of cells make up the organism. Cells vary in size from bird eggs to nerve cells several feet long to bacterial cells 0.1 micron in diameter. The typical human cell is about 10 microns across.

The cell has in it a water-based substance, called *cytoplasm,* surrounding the nucleus, which holds coiled, threadlike strands known as chromosomes. The *chromosomes,* which transfer hereditary characteristics to each generation of new cells, occur in pairs, with a fixed number in every cell of every species. (Human cells have 23 pairs.) The *nucleus* with its many chromosomes controls the cell's activities, structure, maintenance, and repair. It contains instructions for manufacturing the specialized cells (muscle, bone, liver, and so forth) and maintaining their functions to keep the organism alive. The cytoplasm outside the nucleus contains the complex organic molecules that are the raw material used in synthesizing amino acids and proteins, as directed by the *genetic code* in the DNA (*deoxyribonucleic acid*) molecule located in the chromosomes of the nucleus. The DNA molecule specifically determines the type of organism (human being or elephant, for example).

Amino acids are small, chemically reactive organic molecules, composed mostly of hydrogen, carbon, nitrogen, and oxygen, of which 20 are found in living organisms. They are arranged like beads and strung into long molecular sequences called *proteins*. A protein molecule may contain a hundred to a thousand amino acids. If each amino acid corresponds to a letter of the alphabet, then each protein is a sentence in a book of instructions. Because the 20 letters (amino acids) can be arranged in almost limitless ways to form a row (the protein), the possible number of proteins is enormous. Of these only a small percentage appear in living matter. The sequence of amino acids in the protein molecule determines its specific function. Proteins serve both as structural material and as controls, that govern chemical reactions in the cell. These chemical reactions give the cell properties that collectively can be called life. The proteins give any organism its distinguishing characteristics; synthesis of protein molecules is controlled by the genetic code in the DNA molecule in the chromosomes.

THE GENETIC CODE: DNA MOLECULES

The nucleic acid molecule, DNA, is very large and complex and is in the chromosomes of every living cell. One human cell has about 800,000 DNA molecules. Since there are many varieties of living organisms with many chromosomes, there are many different forms of DNA. The DNA has two functions: It carries the hereditary instructions for manufacturing proteins in the cell, and it passes genetic information on to daughter cells during cell division by making copies of itself (replication). The molecule takes the form of a double helix. A short untwisted segment of the DNA molecule is drawn in Figure 10.10.

Essentially DNA is a string of four kinds of subunit called *nucleotides,* linked like a twisted ladder, or a double helix. The nucleotides in turn are made up of bases (carbon-nitrogen compounds), sugars (carbon-hydrogen compounds), and phosphates (phosphorus-oxygen compounds). The sugars and the phosphates form the two handrails of the spiral ladder, with the bases, of which there are four, appearing together in definite pairs, forming the treads. The spiral is a right-handed spiral in which each tread is of the same size and at the same distance from the next and turns at a rate of about 30° between successive treads.

The genetic code, which specifies the hereditary message of life, is carried on the treads of the ladder

FIGURE 10.9
Structure of a cell. A million human cells could be placed on the head of a pin.

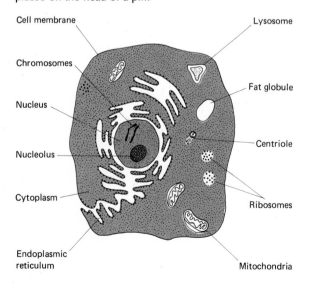

Cell membrane

Chromosomes

Nucleus

Nucleolus

Cytoplasm

Endoplasmic reticulum

Lysosome

Fat globule

Centriole

Ribosomes

Mitochondria

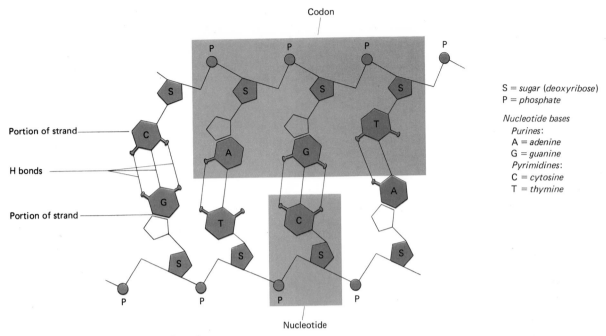

S = sugar (deoxyribose)
P = phosphate

Nucleotide bases
  Purines:
  A = adenine
  G = guanine
  Pyrimidines:
  C = cytosine
  T = thymine

FIGURE 10.10
Portion of a DNA molecule. Nucleotide bases form the treads of the ladder held together by hydrogen bonds. The rails of the ladder contain alternating sugar and phosphate stiffeners.

and is contained in the sequence of the four different nucleotides. Thus the language of heredity is written in an alphabet of only four letters. For each protein potentially capable of being formed a specific segment of the DNA molecule carries the information by which the 20 kinds of amino acid subunits in that protein are properly ordered during its synthesis. The number, type, and arrangement of the nucleotides in DNA determine the kind of organism that is created. A human cell has about 5 billion nucleotide pairs in the DNA of the 46 chromosomes. Different organisms have different sequences of nucleotides as well as different amino acid sequences in the proteins.

During cell division, replication of the DNA molecule begins with separation of the two spirals (Figure 10.11) by breakage of the hydrogen bonds between the bases, as shown in Figure 10.10. Each strand of the double helix directs the formation of a complementary strand to pair with it. This replication continues until the entire double-twisted configuration is complete. Each of the two newly formed helices contains one old strand and one new one. They migrate to opposite ends of the cell before the cell divides. Thus in the daughter cells are two daughter DNA mole-

cules, with nucleotide sequences identical to those of the parent DNA molecule in the parent cell.

This, then, is the biological process that defines life on earth. The genetic instructions for all organisms are written in the same chemical language. Indeed, such a shared hereditary language is one reason to believe that all organisms on earth are descended from a single ancestor, one single instant some 4 billion years ago, when all of earth's present life began.

BIOCHEMICAL EXPERIMENTS ON THE CHEMISTRY OF LIFE

Laboratory experiments copying the presumed composition of the earth's primitive atmosphere have chemically synthesized, with fair success, the building blocks of life, the organic molecules. In their pioneering investigation of 1953, assuming a methane-ammonia atmosphere, Stanley Miller (1930–   ) and Harold Urey (1893–1981) subjected a gaseous mixture of methane, ammonia, water, and hydrogen to an electrical discharge for about a week. They found that the gas had produced a brown sludge that contained several amino acids, urea, and some other organic

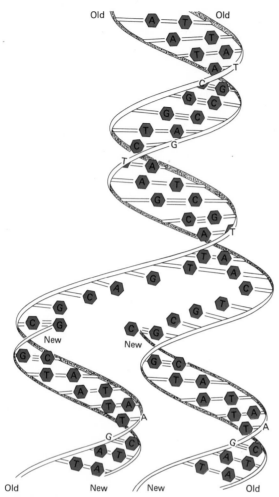

FIGURE 10.11
Unzipping of the DNA molecule to form a new DNA molecule (replication).

FREE OXYGEN, PHOTOSYNTHESIS, AND THE CARBON CYCLE

If free oxygen was not present in the atmosphere in the beginning, how did it become the second most abundant constituent? Some oxygen may have been released from the dissociation of water-vapor molecules into oxygen and hydrogen by the sun's ultraviolet radiation. The lighter, more rapidly moving, hydrogen atoms eventually escaped into space, but earth's gravitational field retained the heavier, slower-moving, oxygen. Oxygen combines so readily with many atoms that most of the early oxygen atoms formed by dissociation must have been rapidly depleted in oxidizing the crustal material of the young earth.

Apparently most of the oxygen now in our atmosphere was produced by *photosynthesis*. This is the name for the process in which plants generate carbohydrates, using carbon dioxide and water as raw materials and sunlight as an energy source, with the release of free oxygen as a by-product. We have evidence for the earliest forms of photosynthesizing organisms (which were not oxygen producers) in the microscopic, one-cell algalike fossils deposited in sedimentary rocks about 3.5 billion years ago.

Eventually, when enough free oxygen had accumulated, a protective atmospheric layer of ozone (O_3) formed from the combination of molecular oxygen (O_2) and atomic oxygen (O). This layer, 12 to 50 kilometers up, absorbed most of the solar ultraviolet light, keeping the radiation from being absorbed in the surface layers of the primordial seas, where it had been the primary source of energy for synthesizing organic compounds.

As the supply of organic compounds was depleted, the earliest organisms, which depended on them for

molecules. Varying the original experiment, other researchers have produced in the laboratory many of the life-associated molecules that are essential components of the nucleic acids. Other experiments have shown that the hydrogen-rich atmosphere of the Miller-Urey experiment is not necessary—just a little hydrogen will do. The essential requirement appears to be that there be no free oxygen in the atmosphere.

More recent experimenters have had some success in the steps that link amino acids to form proteins, imitating the step from building blocks to biological macromolecules (Figure 10.8). The most difficult laboratory experiment, synthesis of the nucleic acids under conditions that are supposed to have predated life, has not yet been achieved.

Life, far from being an aberration on the part of Nature, becomes within the field of our experience, nothing less than the most advanced form of one of the most fundamental currents of the Universe, in process taking shape around us.

Pierre Teilhard de Chardin

food, went into a wholesale decline. Presumably the organisms that survived could utilize the visible-wavelength radiation that penetrated the ozone shield. Green plants and plankton in the oceans, relying on photosynthesis, were favored in this new environment and became dominant in later development. The first organisms using photosynthesis appeared about 3.5 billion years ago. They have had the upper hand ever since.

Plants produce and store energy from sunlight, which animals then consume. By photosynthesis plants and plankton absorb water, carbon dioxide, and solar energy to form the organic compounds they need for growth, and they release oxygen as a waste product. Respiration by plants and animals and oxidation of decaying organic matter in turn consume oxygen and release carbon dioxide as their waste product.

10.4 Biological History of the Earth

DARWINIAN EVOLUTION

Evolution in plants and animals is controlled by two forces: limits that the environment sets for the organisms and changes in their hereditary material. Organisms are constantly modified by chance mutations, sexual selection, and natural selection. Together these produce members of a species that can survive in a changing environment. Mutations occur at random. Unfavorable ones are eliminated because they limit the organism's ability to cope with its environment, to reproduce, or to compete with rival species. Relentless experimenting with cumulative mutation and natural selection has produced the great diversity in life on our planet. Alfred Wallace (1823–1913), an English contemporary of Charles Darwin, hit upon the theory of natural selection at about the same time as Darwin did, but Darwin's *Origin of Species* appeared in 1859 before Wallace had decided to publish his book. Darwinian evolution was the initial step toward our ideas on evolution.

Do we find enough evidence from paleontology and geology to follow biological evolution from the first living organisms to human beings? We can more or less, and a summary of the progress of evolution is shown in Figure 10.12. On the basis of physical evidence there is reason to believe that if the sun's output of radiant energy had varied drastically and rapidly, the gradual evolutionary sequence of biological life would not have been preserved more or less unbroken as we find it.

Contemporary theoretical ideas in stellar evolution postulate that the sun's radiation has steadily increased by 25 percent over the last 4.6 billion years. This means that the early atmosphere of the earth must have been able to provide a more efficient greenhouse effect than it does at present. This in turn requires a different density or composition or both. An atmosphere rich in carbon dioxide that gradually depleted and became nitrogen rich as carbon dioxide combined with crustal rocks could have provided the necessary greenhouse effect, thereby compensating for less solar energy and allowing the development of life.

GEOLOGIC TIME

Earth's geologic history divides into four eras of time: Precambrian, Paleozoic, Mesozoic, and Cenozoic. Each has a dominant biological species, as shown in Figure 10.12. After the primitive oceans formed, the first single-cell organisms of the Precambrian Era evolved in the warm seas. The oldest known (microscopic) fossils that can be interpreted as organisms are those of a simple filamentous bacteria (Figure 10.13) which dates back 3.5 billion years and was found in the Precambrian sedimentary rocks in western Australia.

In the Cambrian Period (about 600 million years ago), the beginning of the Paleozoic Era, there was a remarkable acceleration in the rate of diversification of large complex organisms. Almost 4 billion years passed between the origin of the earth and the appearance of trilobites, which mark the transition to the Cambrian Period. The first vertebrates, primitive fishes, appeared some 500 million years ago; and toward the close of the Paleozoic Era, 220 million years ago, the first reptiles made their appearance. It was only about 70 million years ago, a small fraction of the earth's age, that small mammals evolved. They were followed by other species of mammals, which eventually led to human beings.

HUMAN EVOLUTION

The primitive tree-dwelling mammals such as the lemur that developed stereoscopic vision and high mo-

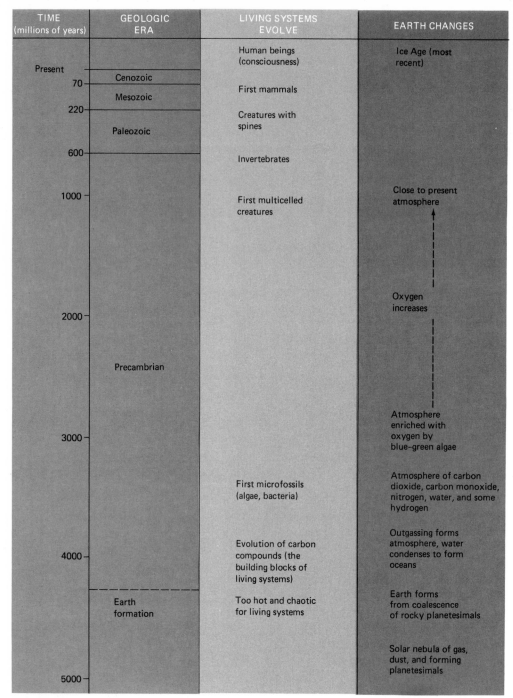

TIME (millions of years)	GEOLOGIC ERA	LIVING SYSTEMS EVOLVE	EARTH CHANGES
Present		Human beings (consciousness)	Ice Age (most recent)
70	Cenozoic	First mammals	
220	Mesozoic	Creatures with spines	
600	Paleozoic	Invertebrates	
1000		First multicelled creatures	Close to present atmosphere
2000			Oxygen increases
	Precambrian		
3000			Atmosphere enriched with oxygen by blue-green algae
		First microfossils (algae, bacteria)	Atmosphere of carbon dioxide, carbon monoxide, nitrogen, water, and some hydrogen
4000		Evolution of carbon compounds (the building blocks of living systems)	Outgassing forms atmosphere, water condenses to form oceans
	Earth formation	Too hot and chaotic for living systems	Earth forms from coalescence of rocky planetesimals
			Solar nebula of gas, dust, and forming planetesimals
5000			

FIGURE 10.12
Evolution of living systems on the earth and atmospheric changes.

bility and could oppose the thumb to the hand were well adapted for survival. From this group, which roamed over eastern and southern Africa, we believe that a precursor of man known as *Australopithecus* evolved. Bones of this creature more than 4 million years old have been found in Tanzania and Ethiopia. It stood 3 to 5 feet in height and weighed from 50 to 120 pounds. Life on the ancient grasslands led to further anatomical changes and to speech, which allowed actions to be coordinated among members of the group.

As far back as 2 million years ago *Australopithecus*'s descendants, *Homo habilis,* could fashion and use tools. As brain capacity increased, the hominid could profit from experience and choose between alternatives, improving the chances of survival. Modern human beings (*Homo sapiens*) are believed to be descendants of the successor to *Homo habilis,* who is known as *Homo erectus.* Skeletal fragments found in Africa and Asia that are around 3 to 4 million years old have produced a raging controversy in anthropological circles over the exact line of human evolution. Regardless of the exact line, 2 million years ago we were not yet "human," but with *Homo erectus* 1 million years ago we were.

For most of humanity's 2 million years people have struggled just to survive. In the 10,000 years since people learned to domesticate animals, grow crops, and live in settlements, our numbers have swelled. The world's population rose from about 10 million in ancient times to about 500 million by the eighteenth century, and since then it has grown to over 4 billion. Grim prospects of overpopulation and overwhelming demands on the world's resources were described as early as 1798 by Thomas Malthus in *An Essay on Population.* It is interesting to note that the germ of the concept of natural selection came to Darwin and Wallace independently, each having read Malthus's treatise early in their work. Our species is not guaranteed perpetual existence on this planet—or even continuation at our present level of intelligence. We can last only by carefully considering what human and natural changes are doing to the earth as a whole. We know that our earth carries only a finite life-support system; we must be particularly careful, therefore, and learn to use its resources wisely.

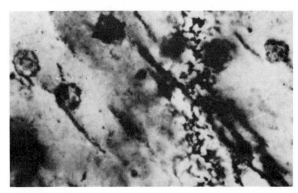

FIGURE 10.13
Preserved microfossils embedded in black Figtree chert of the early Precambrian Era. These specimens are greater than 3.1 billion years old.

chance is lessened when the specimen is found soon after its fall.

Interest in these meteorites was renewed when workers at the NASA Ames Research Center in California discovered amino acids in a fresh chondritic specimen that fell near Murchison, Australia, in September 1969 (Figure 10.14). Extraterrestrial amino acids have since been found in other meteorites, but only a few of these acids are in living cells of earth organisms. Amino acids from meteorites are almost an equal mixture of right- and left-handed molecules. (*Handedness* is the direction in which the plane of polarization rotates when a beam of polarized light passes through the material; see pages 125–126.) Amino acids of biological origin are exclusively left-handed, supporting the idea that the meteoritic amino acids are of extraterrestrial origin. If extraterrestrial life produced amino acids, we believe that they would be exclusively either left- or right-handed; so the mixture of both types in the meteorite specimens suggests a chemical, not a biological, origin.

Along with these acids several other organic molecules have been found. We cannot at once conclude, however, that these organic precursors to life prove that life exists elsewhere in space. Perhaps the one critical step from chemical to biological evolution that took place on earth about 4 billion years ago has been duplicated elsewhere and perhaps it has not; that remains to be determined. What we can conclude is that nature can easily take that first step of biochemical evolution. Of this point we can be relatively certain. If nature has taken additional steps into biological evolution, the signs of such steps may exist somewhere in the solar system besides the earth.

10.5
Life in the Solar System

ORGANIC MOLECULES IN METEORITES
Among meteorites that have been found is a small subgroup of stony meteorites (called carbonaceous chondrites) that contain up to about 5 percent organic molecules. That chondritic meteorites carry organic compounds has been known for more than a century, but only when techniques had been developed for studying lunar rocks could the organic compounds be definitely ascribed to an extraterrestrial origin. Identification is complicated because contamination by terrestrial organic molecules is possible, though the

FIGURE 10.14
A piece of the Murchison meteorite, which fell in Australia in September, 1969. Chemical analysis revealed the presence of amino acids.

LIFE ON THE MOON?

The moon was the first body outside the earth on which searches were made for the presence or remains of living organisms. As mentioned in Chapter 7, the *Apollo* manned landings returned over 800 pounds of lunar surface material. The procedures and the facilities developed here on the earth for their study were as elaborate as any in the history of chemical and organic analyses. Organic chemists had improved old techniques and developed new ones in anticipation of new findings on biochemical evolution and the origin of life that might come from the lunar samples.

But, alas, nowhere in any of the samples was a significant amount of carbon found. Most of the carbon was in the form of traces of carbide, methane, or carbon monoxide. No amino acids, proteins, or nucleic acids were found. Also no water molecules, either free or chemically bound, were found in these surface materials.

These results were not particularly surprising, even if somewhat disappointing, since the surface of the moon is completely exposed to bombardment by solar ultraviolet photons, solar-wind particles, and mete-
oritic matter. Under such an energetic onslaught the likelihood of building long-chain carbon molecules is very small. Thus the chance of finding organic matter on the surface of the moon was extremely small to begin with. However, it is still surprising that so little carbon was found.

MARTIAN LIFE?

The thermally habitable zone in our solar system, the *ecoshell,* where life might flourish on a terrestrial planet, lies between Venus and Mars. The inner limit is roughly the point where water would boil, and the outer limit the point where water would freeze. Since Venus does not seem to be a suitable abode because of its high surface temperature, this leaves Mars as the only realistic place for biological exploration.

The excellent pictures returned from Mars by the *Viking* orbiters neither proved nor disproved that Mars has biological activity. They are good enough to make very remote the possibilities of intelligent life on Mars that could modify the landscape to the extent we have on the earth.

A billion-dollar undertaking, Project *Viking* was designed in part to search for life on Mars. The lander and its instruments were sterilized prior to launch to avoid contaminating the planet's surface with terrestrial organisms. Two cameras periodically scanned the immediate surroundings to look for any large-scale biological life. But the main biological hunt on Mars was for microorganisms in the Martian soil. A motor activated a 10-foot arm that reached out and scooped up a sample of soil. The arm retracted and deposited the soil sample into a hopper on the lander for automated analysis.

LIFE-DETECTION EXPERIMENTS ON MARS

In the various *Viking* biological experiments, depicted in Figure 10.15, all showed an unusually active Martian soil. Water and nutrients added to the soil in the experiments appeared to cause it to imitate biological activity. The answer may lie in an exotic chemistry of the Martian soil. Of great importance in this interpretation of the biology experiments was the results of the gas chromatograph—mass spectrometer experiments. This instrument searched for organic compounds in the soil and failed to find any at either landing site—even under a rock. This is despite the

> Life can only be understood backwards; but it must be lived forwards.
>
> Søren Kierkegaard

ease, as we argued earlier, with which nature can make these molecules even if life is not present. The sensitivity of this instrument was such that many common organic molecules could have been found if they had been present at levels of the order of one part per billion. Even in the most sterile earth environments, organic compounds can be found in abundance when living organisms are absent.

Possibly living organisms developed and evolved in an environment on Mars so different from that of the earth that we did not formulate the experiments correctly, or we are still not interpreting properly the baffling results. At the present juncture it seems unlikely that Mars does have or has had living organisms on its surface.

For the future a robot lander could collect samples on the Martian surface and return them to the earth for analysis in the sophisticated facilities developed

FIGURE 10.15

Viking's life-detection experiments. The pyrolytic release experiment (at left) was designed to detect the carbon cycle. Soil samples were exposed to a simulated Martian atmosphere and then heated to over 600°C. In the labeled release experiment (center) a soil sample was moistened with a nutrient and incubated for about 11 days. Finally, in the gas-exchange experiment (at right) a soil sample was inoculated with an aqueous nutrient in an atmosphere of helium, krypton, and carbon dioxide and was allowed to incubate for several days.

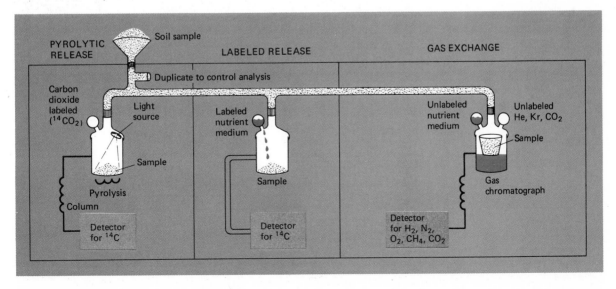

for studying the lunar samples. This is because the best hope for finding evidence for life on Mars seems to be to search for microfossils similar to those found in the 3.5-billion-year-old rocks on earth. Such a project is under study at the present. Of course safeguards would have to be applied to prevent a Martian microbe's escaping into the earth's environment. But it is by no means obvious that, if Martian microbes exist, they can interact with life on the earth and survive.

PROSPECTS FOR LIFE ON MERCURY AND VENUS

Neither Mercury nor Venus is a very promising site for the development of life. With no protective atmosphere to shield the surface from high-energy photons and subatomic particles, it is very hard to believe that any extensive biochemical evolution could have begun on the surface of Mercury. Also, the surface temperature varies from daytime values of 350°C to nighttime temperatures of −170°C.

On the other hand Venus has a very dense atmosphere, which traps solar energy, so that surface temperatures run 450°C or so. In spite of the presence of carbon dioxide and traces of oxygen and water vapor the high temperature suggests a low probability for finding any type of living organism on Venus.

THE JOVIAN PLANETS AND THEIR SATELLITES

The Jovian planets lack the clear demarcation between surface and overlying atmosphere that the terrestrial planets possess. Yet Jupiter and Saturn, with large quantities of hydrogen and helium in their atmospheres, probably have zones in reasonably dense portions of their atmospheres where the thermal conditions are as moderate as those on the earth. These zones are below the clouds, and it is possible that conditions are such that chemical processes that occur on the earth may well be happening in Jupiter's and Saturn's atmospheres. Thus chemical evolution may occur although this does not mean that life is developing. Plans are being developed for a mission that may someday enter Jupiter's atmosphere with an instrumented probe that could search for prebiotic compounds. Finally, there is little to suggest that Uranus and Neptune are abodes for life.

One of the most likely places in the outer solar system for the evolution of organic molecules is Titan, the largest satellite of Saturn. It has been suggested that Titan's surface may be covered with organic molecules and liquid oceans of methane. Missions to Titan are in the planning stage, and at some point in the future we may have definitive evidence for organic compounds one way or the other.

Although evidence has been developed to date that suggests extensive chemical evolution has occurred, whether or not it is heading toward biological evolution is not evident. At this point it is much too early to make anything like a final pronouncement about life in the solar system. In Chapter 20 we shall return to the question of life and its prospects beyond the solar system.

SUMMARY

Origin of the solar system. The architecture of the solar system strongly suggests that the system evolved in an orderly way, shaped by natural forces. Three classes of theories have been advanced to explain this orderly evolution: the nebular hypotheses, the encounter theories, and the protoplanet theories. The protoplanet theory envisions the formation of the solar system as a natural consequence of star formation from interstellar clouds of gas and dust. Small aggregates of matter formed; these planetesimals varied in their chemical composition depending on their distance from the developing sun. The coalescence of rocky planetesimals near the sun produced the terrestrial planets, while icy planetesimals and gaseous matter coalesced to form the Jovian planets. The protoplanet theory, however, fails to account for the observed distribution of angular momentum in the solar system.

Origin of the earth-moon system. Earth formed from coalescing rocky planetesimals some 4.6 billion years ago. Within a few tens of millions of years, the earth became molten and chemical differentiation occurred. Earth began cooling about 4.5 to 4.0 billion years ago, although volcanic activity remained intense. As the earth cooled, a carbon dioxide-rich atmosphere formed. Three theories have been advanced to account for the development of the moon: the fission theory, the condensation theory, and the capture theory.

The origin of life on earth. The prerequisites of life are physical conditions we now associate with the biosphere: a moderate temperature, liquid water, an atmosphere with appropriate chemical composition, carbon from which to build organic molecules, and a source of energy to drive life's forces. The critical starting point of organic evolution was an atmosphere that retained the building blocks of organic molecules—CO_2, CO, N_2, H_2, HCN, and H_2O. Triggered by ultraviolet radiation from the sun, these molecules formed amino acids and simpler building blocks of living systems. Later they gave rise to cells and organisms. Modern experiments have reproduced in the laboratory many of the initial steps of this evolutionary process. The genetic code carried in the DNA molecule located in the chromosomes of the cell nucleus specifically determine the type of organism. The language of heredity is written in a language of four letters and is shared by all the earth's organisms.

Biological history of the earth. Evolution of living organisms is determined by limits set by the environment and changes in hereditary material: chance mutations, sexual selection, and natural selection. Small variations can result in great changes because they persist through immense stretches of geologic time.

Life in the Solar System. Organic molecules have been found in meteorites known as carbonaceous chondrites. Extraterrestrial amino acids found in meteorites are almost certainly of chemical, not biological, origin, however. Studies of the moon and Mars have disclosed no evidence of life. Mercury and Venus, likewise, are unpromising sites for the development of life. The Jovian satellites, on the other hand, especially Titan, may have appropriate conditions for at least the beginnings of the chemical evolution of life.

KEY TERMS

amino acid	fission theory
biochemical evolution	genetic code
biosphere	nebular hypothesis
capture theory	nucleotide
cell	organic molecule
condensation theory	photosynthesis
conservation of angular momentum	planetesimals
deoxyribonucleic acid	protein
ecoshell	protoplanet hypothesis
encounter theory	solar nebula

CLASS DISCUSSION

1. If an interstellar cloud contained only hot gas, should one expect it to contract to form a solar nebula? If not, what must happen for it to contract?

2. When the solar nebula was contracting, why did not all its parts contract radially toward its center? Under what conditions would radial contraction have occurred?

3. How do astronomers account for the difference in composition of the various bodies throughout the solar system?

4. Is it possible that the first steps in chemical evolution, that is, organic molecules, were brought to the earth by meteorlike bodies, asteroidlike bodies, or comets? Explain.

5. Why was the condition of the earth's atmosphere important in the development of life? What role or roles did the atmosphere play?

READING REVIEW

1. Was there a large spread in time between the formation of the various bodies in the solar system?

2. Which planets are most nearly like the sun in chemical composition?

3. What arguments support the proposal that the moon was once part of the earth?

4. Define life as best you can. How does it differ from nonlife?

5. How is the genetic message passed on to both parts resulting from cell division?

6. What role does DNA play in cell activity?

7. How old are the oldest life forms on the earth?

8. When did major diversification in organisms occur on the earth?

9. What test could be used to tell whether organic molecules were produced chemically or biologically?

10. What were the life-detection experiments on the *Viking* landers looking for?

SELECTED READINGS

Breuer, R. A. *Contact with the Stars*. Freeman, 1982.

Dickerson, R. E. "Chemical Evolution and the Origin of Life," *Scientific American,* September 1978.

Gale, G. "The Anthropic Principle," *Scientific American,* December 1981.

Goldsmith, D., and T. Owen. *The Search for Life in the Universe.* Benjamin/Cummings, 1980.

Washburn, S. L. "The Evolution of Man," *Scientific American,* September 1978.

11

The Sun: Our Bridge to the Stars

The sun is a star—a gaseous, self-luminous body—much like stars in the night sky except closer to us. Because other stars are immensely far away, our sun is the only star whose surface we can study in detail. It is, in effect, the guide to an interpretation of regimes of radiation, temperature, motion, and magnetic fields not seen on the earth but common in stars. From a study of the sun we have learned how nuclear fusion—the process of energy production in the stars—has changed and continues to change the chemical composition of the universe. Much of our everyday world (including the existence of life) is the result of the birth, life, and death of stars like the sun. Thus the sun is a vital link in the chain of reasoning that carries us from a Galaxy of stars to a universe of galaxies.

Glorious the sun in mid-career;
Glorious th' assembled fires appear.

Christopher Smart

11.1
The Sun as a Star

THE SUN'S OUTER LAYERS

The sun has roughly 300,000 times the mass of the earth and emits by earth standards an absolutely huge quantity of electromagnetic energy. Only about half a billionth of its radiation is intercepted by the earth: All the power humanly generated on the earth in a year is equivalent to what the earth receives from the sun in about 30 minutes.

Solar astronomers can measure the sun's radiation falling on a unit area of the earth within a certain time. When they correct the measurement for absorption by the earth's atmosphere and some other factors, they obtain what is called the *solar constant* (Table 11.1). How constant it is over periods of hundreds of millions of years is not known now. But for at least decades it has been constant to within about 0.2 percent.

If we multiply the solar constant by the surface area of a sphere whose radius is the earth's mean distance from the sun, we obtain the rate at which radiation is emitted by the sun in all directions at the earth's distance. This number must also be the rate at which the total energy is radiated away from the sun's surface. Thus we find that the sun's rate of emission of radiant

energy over all wavelengths—its *luminosity*—is about 4×10^{33} ergs per second.

This flood of radiant energy comes from what appears to us on earth to be the surface of the sun. It is not in reality a distinct surface but a layer of gas, several hundred kilometers in thickness, called the *photosphere*. Dotted here and there with sunspots, as shown in Figure 11.1, the photosphere is the bottom level of the sun's atmosphere visible to us. Lying above it is a transparent, tenuous layer, the *chromosphere*, which is several thousand kilometers thick. This is topped by an even more rarefied layer, the *corona*, which extends millions of kilometers out from the sun in all directions, as shown in Figure 11.1. These three regions are distinguished from each other by their physical characteristics, but their boundaries are not sharply defined. One region gradually merges into the other.

Why do we see only the photosphere and not the outermost layers? The reason is that light from the chromosphere and the corona is usually too weak to be seen against the brilliance of the photosphere. But they are visible outside the sun's limb (edge) during a total eclipse of the sun, when the moon covers the brighter photosphere. The gases of the chromosphere and corona are transparent in the visible part of the spectrum; and the photons from the photosphere that define our line of sight normally pass through them, and thus our line of sight ends somewhere in the photosphere. Below this level the gases that compose the sun become opaque. However, the chromosphere and corona are observable directly in the short-wavelength regions of the electromagnetic spectrum (the ultraviolet and X-ray regions) or in the longer-wavelength radio region of the spectrum. This is because the photosphere is not very bright in these wavelength regions in comparison to the chromosphere and the corona.

ROTATION OF THE SUN

Rotation is a common characteristic of all astronomical objects. The sun rotates eastward like the planets

◀ Solar eclipse of March 7, 1970, photographed in Mexico.

(except Venus and Uranus). It does not rotate as a solid body, which is not surprising since it is a wholly gaseous body. For the atmosphere, primarily the photosphere, the period of rotation progressively increases from 25 days at the solar equator to about 36 or 37 days at the poles. Astronomers think that this *differential rotation* is caused by an interaction between convection currents below the sun's surface and the overall solar rotation. We see sunspots move across the disk because the sun is rotating, and we can

measure their travel time to find how long it takes the sun to complete one rotation on its axis.

Another method, applicable to all solar latitudes (sunspots rarely appear beyond 40° on either side of the solar equator), uses the Doppler shift of spectral lines from opposite limbs of the sun (Figure 11.2). Since the eastern limb of the sun rotates toward the earth and the western limb away from the earth, the measured difference in the Doppler wavelength shift between the two limbs yields their velocity with re-

TABLE 11.1
Properties of the Sun As a Star

Property	Value	How Determined
Mean distance	1.496×10^8 km	Radar reflection from Venus and Kepler's third law
Angular diameter	0.553°	Direct observation
Radius	6.96×10^5 km	Angular diameter and mean distance
Surface area	6.09×10^{22} cm^2	$A_\odot = 4\pi R_\odot^2$
Volume	1.41×10^{33} cm^3	$V_\odot = \frac{4}{3}\pi R_\odot^3$
Mass	1.99×10^{33} g	Newton's modification of Kepler's third law; $M_\odot = 4\pi^2 a^3 / G P_\oplus^2$
Mean density	1.41 g/cm^3	Mass/volume = $\rho_\odot = M_\odot / V_\odot$
Solar constant	1.36×10^6 erg/cm$^2 \cdot$ s	Ground-based, high-altitude rocket, aircraft measurements
Luminosity	3.83×10^{33} erg/s	Solar constant and mean distance
Mean luminosity per unit mass	1.92 erg/g · s	Luminosity/mass = $L_\odot / M_\odot$
Surface temperature	5770 K	Stefan-Boltzmann law, luminosity, and surface area: $L_\odot = 4\pi R_\odot^2 \sigma T_\odot^4$
Spectral type	G2 V	Classification by absorption spectrum (see Section 12.4)
Apparent visual magnitude	−26.74	(See Section 12.3)
Absolute visual magnitude	+4.83	(See Section 12.3)
Color (B-V)	Yellow; +0.65	(See Section 12.3)
Atmospheric layers	In order outward: photosphere, chromosphere, corona	Direct observation
Rotation period	25 days at equator, 36 days at poles	Motion of sunspots; Doppler shift in photospheric spectrum
Angle between equator and ecliptic	7.2°	Rotation of sun
Magnetic field	≈ 1–2 G averaged over photosphere[a]; hundreds of gauss in active regions; thousands of gauss in sunspots	Zeeman effect
Chemical composition	91% hydrogen, 9% helium, 0.1% heavier elements by numbers of particles	Analysis of photospheric absorption spectrum and mathematical models of the interior

[a] The magnetic field is actually bunched in small bundles of about 300-kilometer size that have magnetic-field intensities of several thousand gauss. These bundles are packed into a network structure in and around active regions. Thus a magnetic-field-intensity average over the entire photosphere does not indicate the actual situation.

FIGURE 11.1

The sun, its atmosphere, and prominent features. The sun is the only star that we can study at close range. It generates energy near its center by fusing hydrogen into helium; this energy, as radiation, flows outward and eventually emerges as sunlight.

The photosphere is the visible surface, or the lowest level, of the sun's atmosphere; it is shown in the high-altitude balloon photograph at the top right. The photosphere is also the region where sunspots occur. A close-up of a large sunspot group that occurred on May 17, 1951 is shown in the middle-left photograph. The dark centers are the umbrae and the surrounding striated areas the penumbra; photospheric granules are also visible outside the sunspot group.

Above the photosphere lies the chromosphere, a layer of transparent gases several thousand kilometers thick.

The outermost layer of the sun's atmosphere is the corona. During a total solar eclipse, we see the corona's pearly white halo extend outward well over a million kilometers.

Extending into the corona and above the chromosphere are prominences. The great eruptive prominence of June 4, 1946 (bottom left) grew to a size almost as large as the sun within an hour; several hours later it disappeared completely. Flares—sudden and short-lived outbursts of highly energetic radiation—emanate from a disturbed region of the sun's surface.

Above: Top, white-light view of the sun, showing the photosphere; two sunspot groups, and bright faculae near limb. Center, enlargement of granulation of the photosphere.
Facing page, top left: The structure of the sun from its energy-generating core to the transient activity in its outer layers.
Middle left: Chromosphere from which prominences extend up into the corona.
Middle right: Large sunspot group whose long axis lies east and west parallel to the solar equator.
Bottom left: Prominence of June 4, 1946, an arch of gas extending into the corona well above the chromosphere (bright line across lower part).
Bottom right, upper: Solar flare of July 16, 1959; a brightening in middle of picture with dark filaments toward edges of picture.
Bottom right, lower: Eclipsed view of the sun, showing the corona extending beyond the photosphere.

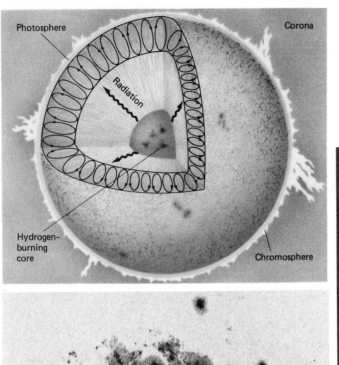

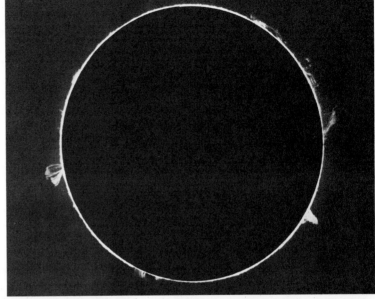

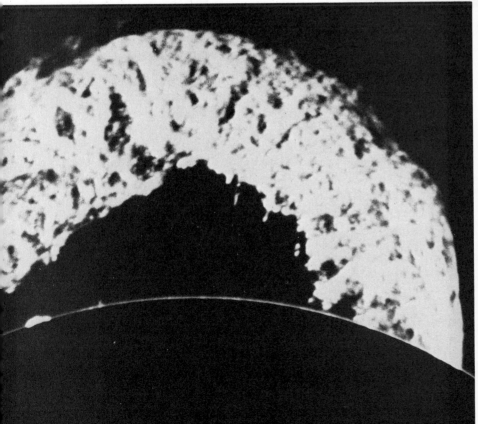

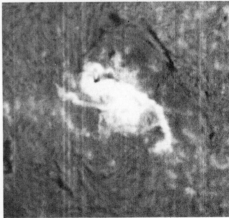

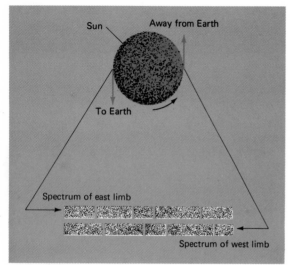

FIGURE 11.2
Doppler shifts of opposite limbs of the sun. The eastern limb rotates toward the earth and the western limb away from the earth. Shown schematically in the spectra below, the Doppler shift between the two limbs is 4 kilometers per second, or 2 kilometers per second relative to the sun's center.

spect to each other. The velocity relative to the sun's center is about 2 kilometers per second at the solar equator. Dividing the distance traveled in one rotation—that is, the circumference—by the velocity gives the time of one complete rotation at the equator: 25 days.

INTERIOR OF THE SUN

Although we cannot see the interior, knowledge of the sun's observable properties (as in Table 11.1) enables us to calculate, from physical laws, a mathematical model of the sun's internal structure (see Table 13.4). To validate the model, we see whether it is consistent with the sun's observable properties. The reasonable success astronomers have had with such modeling for the sun encourages us to believe that theoretical models for other stars can be constructed, as we shall discuss in Chapter 13.

The radiant energy emitted by the sun's photosphere is the result of hydrogen fusion—the conversion of four hydrogen nuclei into one helium nucleus and energy in the deep interior of the sun. This energy is the source of our sunlight; and when first released in the *energy-generating core* in the deep interior, it is chiefly in the form of gamma-ray and

X-ray photons. As the radiation works its way through the solar material toward the sun's surface, atoms and ions absorb and re-emit the energy. After hundreds of thousands of years in a maze of absorption and re-emission, the radiant energy finally reaches the layers about 100,000 kilometers below the sun's surface. There the photons transporting the energy have been so degraded toward longer wavelengths that they are readily absorbed by any neutral hydrogen atoms. As a result, convection (a circulating flow of hot and cold material) takes over as the means of transporting energy. In this region, known as the *hydrogen convection zone*, the movement of energy is like that in a heated room, where cold, heavier air descends to be reheated near the floor and then rises, circulating heat in the room. In the sun hot gases bring energy from the bottom of the convection zone in the interior to the top of the convection zone just below the photosphere, and cool gases return to the interior to start the cycle again. While the convection zone occupies roughly the outer 30 percent or so of the sun's radius, it contains only a few percent of the sun's mass. At the surface most of the energy that left the deep interior is now in the visible spectral region—the sunlight that we observe here on earth about 8 minutes after it leaves the sun.

11.2
The Photosphere

PHOTOSPHERIC TEMPERATURE

Large solar telescopes, such as the one in Figure 11.3, are used by astronomers to study the photosphere of the sun. The temperature of the photosphere is an important property of any star; for it is a measure of the rate at which radiation is emitted by the star. To find the sun's photospheric temperature, we can utilize one of three basic methodologies implied in the Stefan-Boltzmann law, Wien's displacement law, and Planck's radiation law. As discussed in Chapter 4, these laws describe the radiation emitted by blackbodies; but because the sun is not precisely a blackbody, the temperatures derived from these laws differ slightly; they yield an approximate value of 6000 K. One means of visualizing how this value is derived from Planck's law is the degree to which the amount of energy measured at various wavelengths in the solar spectrum approximates a blackbody energy curve. This comparison is shown in Figure 11.4, where we can see that the 6000 K blackbody energy curve is a reasonable approximation to the distribution of energy in the sun's continuous spectrum.

The sun's limb looks darker than the center of the disk. Why? At the sun's edge we are viewing the succession of photospheric layers obliquely, seeing light that comes only from the highest layers of the photosphere. Since the higher layers emit less radiation, they must be cooler, as is evident from the blackbody energy curves in Figure 11.4. Radiation visible to us from the sun's center, however, comes from deeper, hotter layers and is more intense (Figure 11.5). Thus the temperature declines outward through the photosphere. From this fact the astronomer can deduce the decrease of temperature and density through the photosphere and use these data in determining the abundance of the chemical elements. The sun's limb looks sharp-edged because the layers responsible for most of the emitted white light are too thin to be resolved. They are several hundred kilometers thick, while the typical resolution size for 1″ of arc seeing corresponds to about 750 kilometers.

SPECTRUM OF THE PHOTOSPHERE

The spectrum of the visible solar disk displays a continuous band of colors, from red to violet, crossed by many absorption lines, which can be easily seen in Figure 11.6. A German physicist, Joseph von Fraunhofer (1787–1826), in 1814 mapped nearly 600 of the most prominent lines. He designated the strongest absorption lines by capital letters, beginning with A in the red and going to K in the violet. Since the spectrum of the photosphere is an absorption spectrum, we can interpret these lines according to Kirchhoff's third law of spectral analysis (see Section 4.2): The

FIGURE 11.3
McMath Solar Telescope, Kitt Peak National Observatory. The telescope is housed in a 152.4-meter concrete tunnel, partly underground, which is inclined 32° to the horizontal and parallels the earth's axis of rotation. A rotating 2-meter flat mirror called a *heliostat*, mounted on top of the building, tracks the sun. It reflects sunlight down the tunnel onto a 1.5-meter parabolic mirror that focuses light back up the tunnel and down into an underground observing room 91.4 meters away. The image formed by the telescope is nearly a meter in diameter. This is the largest solar telescope now in use.

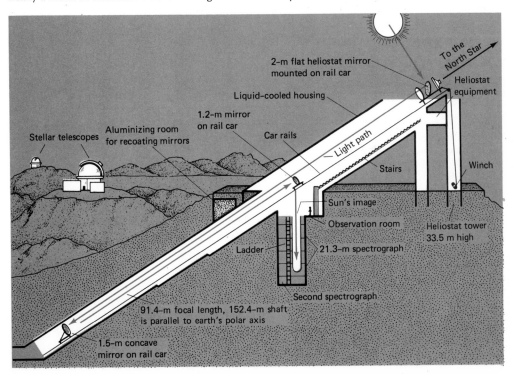

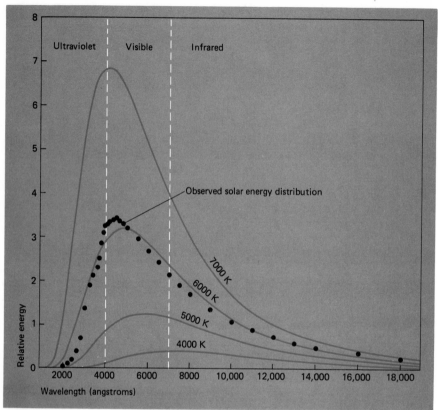

FIGURE 11.4
Blackbody energy curves. The observed distribution of energy in the sun's continuous spectrum (dots) matches the 6000-K theoretical curve best.

FIGURE 11.5
Darkening of the sun's limb. Temperatures at different depths of the photosphere are indicated; the numbers in parentheses are the heights above the lowest level from which solar radiation reaches the earth. Radiation visible to us from the center of the solar disk comes from deeper in the photosphere, where the temperature is higher, than does the radiation from the sun's limb. Drawing is not to scale.

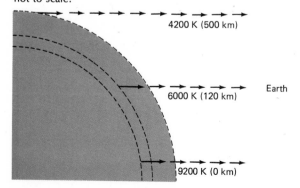

radiation coming up from the interior of the sun has a continuous spectrum; as the radiation passes through the photosphere, certain wavelengths are absorbed by atoms and ions of different chemical species in the photosphere's cooler layers and in the adjoining low chromosphere, causing the observed dark lines. The uninterrupted-wavelength regions, between the absorption lines, are those of continuous radiation that pass into space without being absorbed.

CHEMICAL COMPOSITION

By precisely measuring wavelengths of absorption lines in the solar spectrum, astronomers have identified in the sun nearly 70 of the 92 naturally occurring elements and about 20 molecules. The identifications are made by comparing the wavelengths of lines in the solar spectrum with those obtained from laboratory analyses of the spectra of the elements. The elements missing from the photospheric spectra, mostly the heavier ones, are probably present in the sun's atmosphere, but they are not abundant enough to be detected spectroscopically or their spectral lines are not

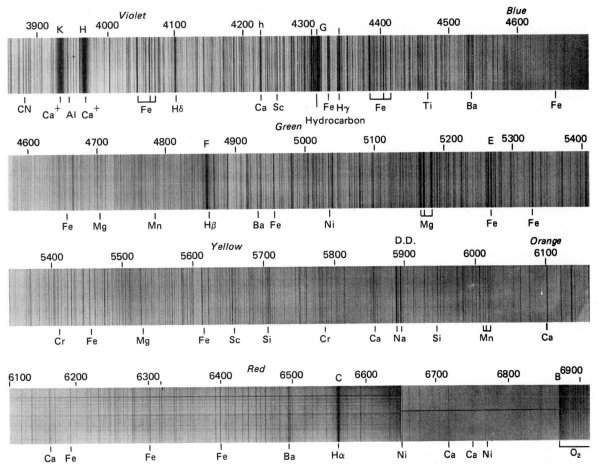

FIGURE 11.6
Solar spectrum from 3900 to 6900 angstroms. The letters (shown above the spectrum) were associated with certain of the strong lines originally assigned by Fraunhofer. The chemical identification of a number of lines is shown below the spectrum. B marks the head of a molecular band, which extends to the right, produced by molecular oxygen in the earth's atmosphere.

in either the visible region or the ultraviolet region, the only regions thoroughly explored. We can estimate the abundances of the photosphere's elements (Table 11.1) by combining our theoretical knowledge of the probability that an atom will absorb radiation at the wavelength in question at a specified temperature with measurements of the line's blackness and width. In Section 12.4 we shall discuss such measurements further.

FINE STRUCTURE OF THE PHOTOSPHERE
In ground-based photographs taken in the white light of the photosphere a number of details are clearly visible. We have already noted that the photosphere darkens toward the limb; near the limb we can also

see bright patches called *faculae*. And with higher resolution, white-light photographs reveal that the entire disk is covered at all times by small, bright features separated by dark lanes called *granules* (see Figure 11.1).

Remarkably clear pictures of the solar surface in narrow wavelength ranges have been made by a telescope mounted in the Skylab station (see Section 5.5). At the altitude that the station was orbiting fine details are not blurred by earth's atmospheric turbulence. These and various high-resolution photographic studies reveal a potpourri of bright solar granules with dark intergranular lanes; these give the surface the salt-and-pepper appearance displayed in Figure 11.1. Sequences of photographs show the granules forming, disappearing, and reforming in cycles lasting

TABLE 11.2
Ten Most Abundant Elements in the Sun's Atmosphere

Atom	Atomic Number	Number (%)	Mass (%)
Hydrogen	1	90.7	69.1
Helium	2	9.1	27.7
Carbon	6	0.036	0.33
Nitrogen	7	0.009	0.10
Oxygen	8	0.072	0.88
Neon	10	0.003 ⎤ ≈ 0.1	0.04 ⎤ ≈ 2
Magnesium	12	0.002	0.05
Silicon	14	0.002	0.06
Sulfur	16	0.001	0.04
Iron	26	0.003 ⎦	0.12 ⎦

the darker intergranular region the difference in brightness corresponds to a difference in temperature of about 200 K. Thus the photosphere is not uniform; it varies not only in height but also laterally across the face of the sun.

The granules in the photosphere are a form of convection, resulting from the upwelling of unstable convective elements from the hydrogen convection zone below the photosphere. As evidence of this convective exchange, spectral lines of the bright centers are Doppler shifted to the violet (moving toward us) and those of the darker intergranular regions are shifted to the red (moving away from us). These shifts between the bright and dark regions indicate that the bright centers are the tops of hot rising gas currents, moving at a few tenths of a kilometer per second, that radiate their excess energy and then form the sinking gas currents of the darker intergranular lanes.

11.3
Sunspots

several minutes. The granules are cells with characteristic diameters of 1000 kilometers and lifetimes of several minutes. At any given time the whole photosphere is broken up into roughly 4 million granules, each occupying about 1 million square kilometers of the surface. From the bright center of the granule to

WHAT ARE SUNSPOTS?

Dark features on the sun have been reported for at least 2000 years. Several sightings per century are con-

TEMPERATURE OF THE SOLAR PHOTOSPHERE

We can find a representative temperature $T_\odot$ for the sun's photospheric layers from the radiation laws of blackbodies since those laws describe very nearly how the sun behaves. One of the three mathematical relations that specify the properties of blackbody radiation is the Stefan-Boltzmann law. It states that the flux of the radiation in ergs emitted per square centimeter each second in all wavelengths is proportional to the fourth power of the temperature. The formula is

$$F = \sigma T^4 = (5.67 \times 10^{-5})T^4,$$

where F is the radiation flux in ergs per square centimeter-second and T is the temperature in kelvins. Dividing the luminosity of the sun ($L_\odot = 3.82 \times 10^{33}$ ergs per second) by its surface area (6.09×10^{22} square centimeters) gives $F_\odot = 6.27 \times 10^{10}$ ergs per square centimeter-second. Hence the surface temperature is

$$T_\odot = \left(\frac{F_\odot}{\sigma}\right)^{\frac{1}{4}} = \left(\frac{6.27 \times 10^{10}}{5.67 \times 10^{-5}}\right)^{\frac{1}{4}} = 5770 \text{ K.}$$

Because temperature varies somewhat over the solar disk and also at different depths in the photosphere, we can take this surface temperature as an average value for the entire solar disk.

tained in ancient Chinese records. Although not the first sightings recorded in Europe, in 1610 Galileo's observations with his telescope gave the first details of sunspots, the most conspicuous of a number of transient phenomena noted in the solar atmosphere in white light.

A typical *sunspot* has a cellular structure with a dark center, the *umbra*, surrounded by a grayish filamentary region, the *penumbra*. The sunspot is intrinsically bright; for the umbra is about one-fourth as bright as the photosphere and the penumbra about three-fourths as bright. The umbra looks dark only because we see it against an even brighter photospheric background, whose temperature is 1800 K higher. A large sunspot group is shown in Figure 11.1

Sunspots develop in a matter of hours as small

LARGE-SCALE MOTIONS OF THE PHOTOSPHERE

In 1960 it was discovered that there was a vertical oscillatory motion in and above the granulation. It has an average period of almost exactly 5 minutes and velocities of about 0.5 kilometer per second. Thus the layers above the convection zone are moving up and down with respect to the mean position of the photosphere and low chromosphere. The typical excursion is on the order of 50 to 100 kilometers. The motion seems to be organized over a few thousand kilometers, and has been reported to cover areas as large as 50,000 kilometers, with roughly two-thirds of the solar surface experiencing oscillations at any given moment. It appears that the 5-minute oscillation may be one extreme in a whole range of solar pulsation modes, with a 160-minute vibration of the whole sun as the other extreme.

Coexisting with the solar granules and the 5-minute oscillations of the solar photosphere is a completely different type of motion detected by Doppler studies of the full disk of the sun. These motions, shown in Figure 11.7, are called *supergranulation cells* because of their resemblance to convective motions and the fact that they are typically an order of magnitude larger than granules (about 30,000 kilometers in diameter). The supergranules show no pattern of bright centers and dark boundaries because the temperature differences are apparently not great enough. Supergranules have a lifetime of about 1 day compared to the several-minute

lifetimes of the granules. At what depth in the sun the supergranulation system begins is unclear. Beyond the name the granules and supergranules may have little in common.

Some solar astronomers believe that an even larger pattern of giant convective cells exists with sizes equal to a substantial fraction of the solar radius. The observational evidence for these giant cells is still weak, and the search continues.

FIGURE 11.7
Supergranulation cells as seen in Doppler-shift picture. Velocities of approach are indicated by lighter areas, and velocities of recession are shown by darker areas. Note the cell-like structure with lighter centers (approach) and darker boundaries (recession). These are the supergranules and are typically 30,000 kilometers across. Compare with Figure 11.14.

pores in the intergranular region of the photosphere. They grow rapidly, and they generally form in adjacent clusters, marking a *sunspot group*, whose orientation is approximately parallel to the solar equator. Each end of the group is often dominated by a large spot surrounded by smaller spots. The very largest groups may cover one-fifth of a solar diameter. Sometimes the sunspot group persists for several months, but the typical lifetime is about 1 week. The typical large spot in a sunspot group is about 10,000 kilometers across; exceptional ones are 50,000 kilometers in diameter, or about four times the diameter of the earth. In a week or so this large spot builds up to its maximum diameter; then its size slowly declines. Individual spots in a sunspot group undergo slow changes from day to day while they maintain their association.

SUNSPOT CYCLE

More than a century ago a German amateur astronomer, Heinrich Schwabe, discovered, during 17 years of observations, that sunspots come and go in a *sunspot cycle* of about 11 years. The plotted sunspot number in Figure 11.8 takes into consideration both the number of sunspot groups and the number of individual spots. The heights of the successive maxima are unequal, and the interval between the peaks and troughs is not constant; the 11-year period is a very rough average. The last minimum occurred in 1976; the last maximum occurred in late 1979. As the cycle progresses, the spots form closer to the equator in each succeeding year, as shown in Figure 11.9. A few sunspots in the higher latitudes, around 35°N or 35°S, herald the beginning of a new cycle as the last spots of the preceding cycle disappear near the equator.

MAGNETIC FIELDS IN SUNSPOTS

Around sunspot groups astronomers observe immense arching or curved features (see Figure 11.13d). They are structures of gas whose shape is determined by curved magnetic lines of force in and around the

FIGURE 11.8
Sunspot cycles for the period 1750 to 1980. The last minimum was in 1976, and the last maximum in late 1979. Note that the 11-year cycle is very approximate and that the peak number is also highly variable. The period since World War II has been a particularly active one for the sun as compared to earlier periods.

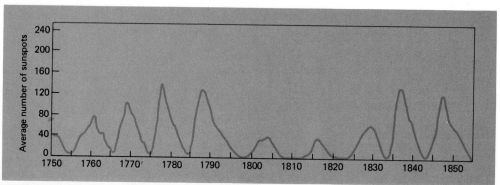

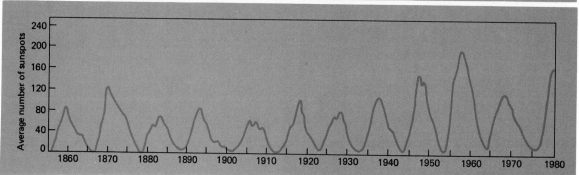

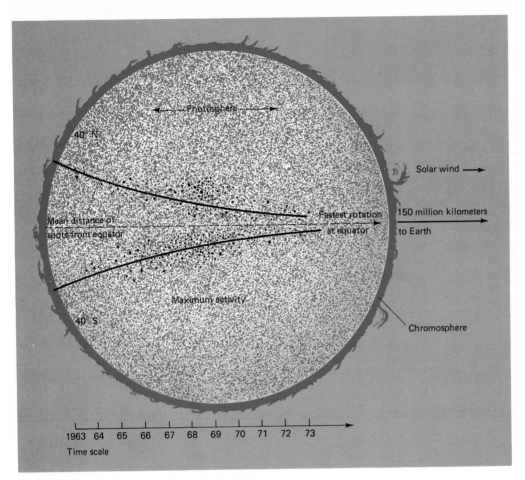

FIGURE 11.9
Equatorial trend of sunspots during the sunspot cycle. As the cycle advances, spots appear progressively closer to the equator.

spot group since sunspots are apparently the centers of intense magnetic fields. Astronomers know this because light from a radiating source that is in a magnetic field has a spectrum in which each absorption line is split into three or more closely spaced components—a phenomenon called the *Zeeman effect* for the Dutch physicist who discovered it in 1896. It was first identified in the absorption lines of the sunspot spectrum by George Ellery Hale (1868–1938) at the Mount Wilson Observatory in 1908. The strength and direction of the magnetic field in a sunspot can be determined from the amount of separation and the degree of polarization of the components of the absorption lines caused by the Zeeman splitting (Figure 11.10).

The leading spots of a spot group (in the forward direction of the sun's rotation) are opposite in polarity from the following spots. Opposite polarities are like

those of the north and south ends of a horseshoe magnet. The measured field strength in sunspots exceeds that of the earth's field by several thousand times. The unit of measure of the intensity or strength of the magnetic field is called a *gauss*; the earth has an average field strength of 0.5 gauss, while the sunspot fields are several thousand gauss. In the northern and southern hemispheres of the sun the polarities of the leading and following spots are also opposite to each other. They reverse in both hemispheres in the following sunspot cycle. That is, the polarity of the leading spot in the northern hemisphere may be north-seeking in one sunspot cycle, while that for the southern hemisphere is south-seeking. Then in the next sunspot cycle the leading spot will be south-seeking in the northern and north-seeking in the southern hemisphere.

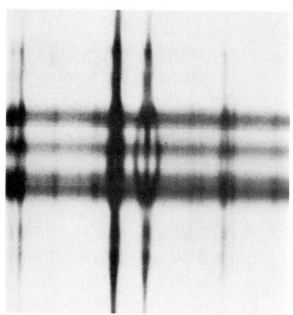 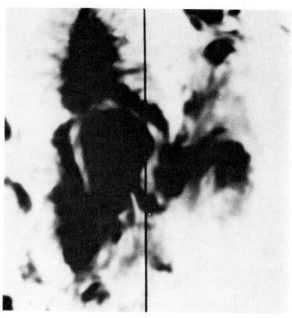

FIGURE 11.10
Zeeman splitting in the spectrum of a sunspot. On the right is a white-light photograph of a complex sunspot group taken July 4, 1974. The vertical black line indicates the position of the spectrograph slit for the spectrum shown to the left. In the center of the spectrum is the line of iron I at a wavelength of 5350 angstroms, which is clearly split into three components by the sunspot magnetic field. The field strength of 4130 gauss is one of the largest ever measured.

As a whole, the sun does not have a general magnetic field like that of the earth. However, by averaging over small localized and intense magnetic fields, one gets the impression of a general field a few times stronger than the earth's field, or several gauss altogether. For the last 25 years astronomers have regularly measured the average of these very localized magnetic fields over areas several thousand kilometers on a side. They find that the magnetic fields in the polar regions of the sun appear to reverse near the time of sunspot maximum. Thus what appears to be a general field is probably the accumulation of surface fields that have drifted to the polar regions. Although astronomers have explanations for sunspot behavior and the magnetic field phenomena that go with it, these explanations are not completely satisfying.

ORIGIN OF SUNSPOTS

Of great importance in understanding sunspots is the question of why they are cooler than their surroundings and how they are able to maintain that condition for so long a period of time. Since energy flows

from hotter to cooler regions, a flood of photons from the high-temperature gas surrounding the sunspot should flow into the cooler spot and eliminate the difference in temperature. The reason that this does not happen must be connected with the very strong magnetic fields that exist in the interiors of spots.

The most acceptable theory of the origin of sunspots suggests that distortions in the magnetic field below the photosphere are created by the sun's differential rotation. As the magnetic lines of force are pulled and stretched unequally at different latitudes by the flow of gas to which they are attached (maximum stretching is at the equator, where rotation is swiftest), the magnetic lines eventually curl into a ropelike structure below the photospheric surface. If a kink forms, the magnetic rope of force may arch up into the photosphere, making a bipolar sunspot group, as shown in Figure 11.11.

As the bipolar group of sunspots breaks up and decays, the field of the following spots of the bipolar group migrates toward the polar regions and eventually neutralizes the polar field, allowing a new one to form, its polarity reversed from the previous cycle.

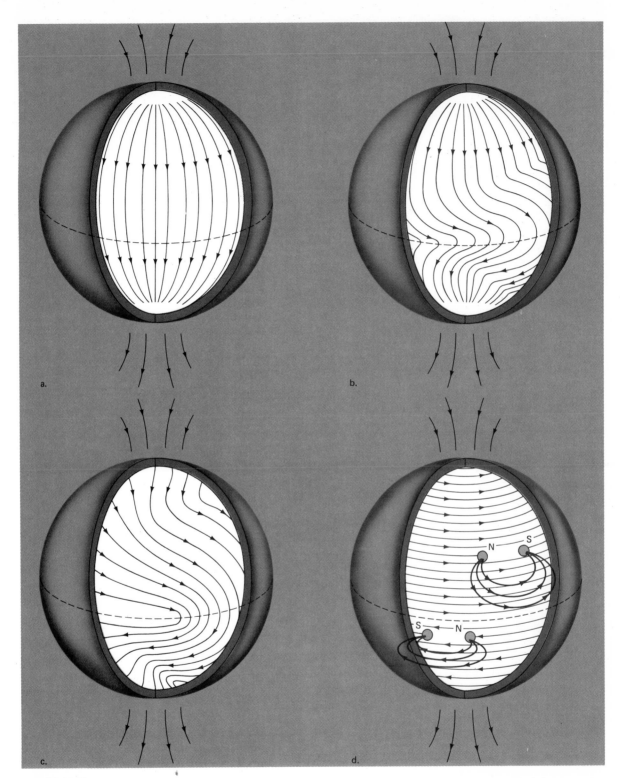

FIGURE 11.11
Origin of sunspots. The magnetic lines of force lying underneath the photosphere (a) are stretched
out along the equator by the sun's differential rotation (b). After many rotations the field lines are
stretched into an intense east-west field under the
photosphere (c). Eventually a loop of magnetic-
field lines bursts through the photosphere (d),

forming pairs of sunspots of opposite polarity
(north-seeking labeled N and south-seeking
labeled S). As shown, the leading spot (westernmost) in the northern hemisphere has an S polarity, and it has an N polarity in the southern hemisphere. In the next cycle the polarities of the
leading spots in each hemisphere will reverse.

The period of complete reversal of the sunspot polarity is thus 22 years. This magnetic cycle is more regular than the 11-year sunspot cycle. That is, if one sunspot cycle were to be 12 years in length, then the following cycle would be close to 10 years so that their sum would be 22 years of the magnetic cycle.

11.4
The Chromosphere

FLASH SPECTRUM

Most of the sun's radiation is from the photosphere and lies in the visible region, where a narrow optical window in earth's atmosphere lets it in (see Figure 5.1). During a total eclipse of the sun, when the moon has just covered the photosphere, a thin (about 2000-kilometer) pinkish fringe of light, called the *chromosphere*, appears beyond the moon's edge (see Figure 11.1). At the same time the corona appears as a much fainter halo surrounding the sun and extending far out into space. Projecting from the chromosphere here and there are rosy arches and loops of gas called *prominences*, which may extend 100,000 kilometers or more into the corona. The chromosphere gets its reddish hue from the large amount of hydrogen gas emitting radiation in the red hydrogen alpha line of the Balmer series.

For the few seconds during an eclipse when only the chromosphere is exposed, the photosphere's normal absorption spectrum is no longer visible, and we see an emission spectrum called the *flash spectrum* (shown in Figure 11.12). It is the spectrum of light originating in the chromosphere. Many of the emission lines match the wavelengths of the absorption lines, but among the exceptions is a bright yellow line produced by helium. Why do helium emission lines appear in the flash spectrum but not in the Fraunhofer absorption spectrum of the photosphere? The reason for this effect is the chromosphere's higher temperature—up to 30,000 K at the highest level—and

lower density. Neutral helium can be excited to emit radiation only when the gas temperature is greater than 10,000 K. And the appearance of ionized helium lines requires temperatures in excess of 20,000 K. Thus the temperature must rise very rapidly from the top of the photosphere up through the chromosphere.

STRUCTURE OF THE CHROMOSPHERE

In the chromosphere astronomers are presented with a very different view of the sun from what they see in the photosphere. Chromospheric events can be monitored, even when there is no eclipse, by photographing the chromosphere in monochromatic light. A single-wavelength photographic device, the spectroheliograph, can make a picture of the solar disk in the residual light of an absorption line, such as the red hydrogen alpha line or the violet K line of singly ionized calcium (see Figure 11.6). The finished picture, a spectroheliogram, is shown for red hydrogen alpha in Figure 11.13b and for the violet calcium K line in Figure 11.13c.

The reason for choosing the residual light of a strong absorption line is that most of the photons from the photosphere have been absorbed and the remaining ones are those coming from the low chromosphere. Photons in the continuous spectrum between the lines come from the bottom of the photosphere, while photons in stronger and stronger absorption lines come from higher and higher up in the photosphere and the low chromosphere. Thus by choosing absorption lines of different strengths, the astronomer can photograph the solar atmosphere at different levels.

Bright patches in the chromosphere, called *plages*, surround the photospheric sunspot groups in spectroheliograms (Figure 11.13d). Hotter and probably more dense than the normal chromosphere, with a spatially averaged magnetic field of a few hundred gauss, plages are typically 10 times larger than the sunspot group lying in the photosphere below them. Plages are nearly always found above regions in the

FIGURE 11.12
Flash spectrum of the chromosphere. The two most intense emission lines are those due to Hα (wavelength 6562 angstroms) and Hβ (λ 4861 angstroms) in the Balmer series of hydrogen. Low-temperature lines due to sodium and magnesium are also present. Note that there are also high-temperature lines of neutral helium, ionized helium, and multiply ionized iron.

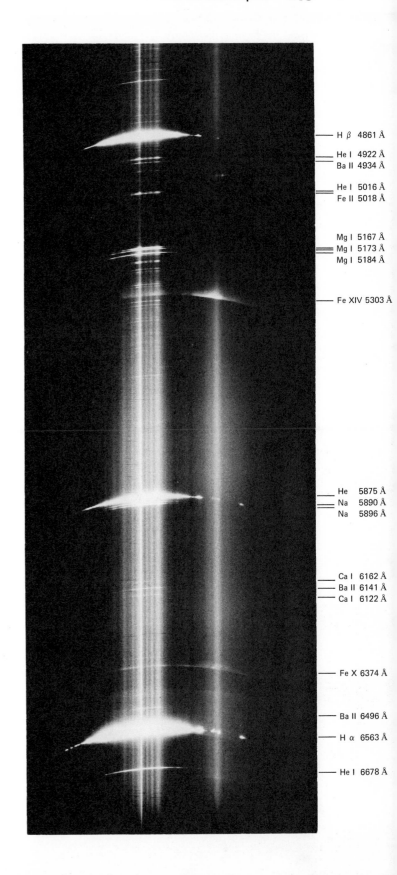

H β 4861 Å

He I 4922 Å
Ba II 4934 Å

He I 5016 Å
Fe II 5018 Å

Mg I 5167 Å
Mg I 5173 Å
Mg I 5184 Å

Fe XIV 5303 Å

He 5875 Å
Na 5890 Å
Na 5896 Å

Ca I 6162 Å
Ba II 6141 Å
Ca I 6122 Å

Fe X 6374 Å

Ba II 6496 Å

H α 6563 Å

He I 6678 Å

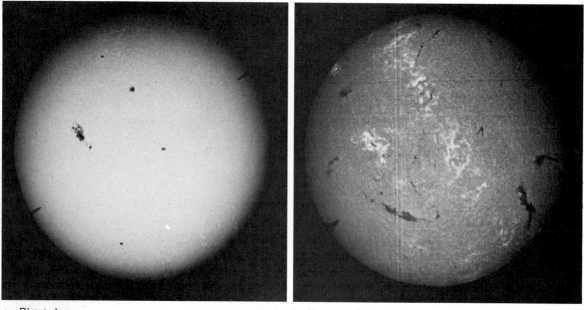

a. Direct view.

b. Hydrogen (alpha line) spectroheliogram.

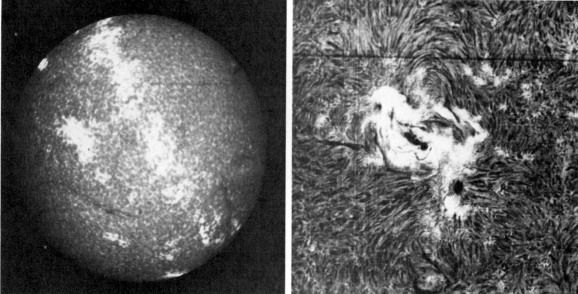

c. Calcium (K line) spectroheliogram.

d. Plage region.

FIGURE 11.13
Four views of the sun. The plage region (d) was photographed in the light of the red hydrogen alpha line. Notice that the sunspot regions observable in the white-light view (a) are also present in the spectroheliograms in the red hydrogen alpha line (b) and the violet calcium K line (c).

photosphere in which a strong magnetic field exists, and they always appear before the spots form. Their usual life span is about 40 to 50 days, during which several spot groups or none at all may form. The plage areas do not look the same in the red hydrogen alpha line (Figure 11.13b) and in the violet calcium K line (Figure 11.13c), but they mark the same general regions of the chromosphere.

In a violet calcium K line spectroheliogram one sees a network of bright gas surrounding dark cells in addition to the larger plages. This *chromospheric network* constitutes the boundaries of the large-scale cells known as convective supergranules and seen as Doppler shifts in the photosphere. Apparently the network is the locus of the very intense and highly localized magnetic-field regions that are concentrated on the boundaries of the supergranule cells by the motions of gas in and around the convective cells.

In short spectroheliogram exposures the chromosphere appears stippled with a myriad of jetlike spikes of gas, or *spicules* (Figure 11.14). Spicules rise rapidly from the chromospheric network, attaining typical heights of almost 10,000 kilometers, and then fade away or collapse in several minutes (Figure 11.15). At any instant 250,000 of them may cover a few percent of the sun's surface. They too outline the boundaries of the supergranule cells. Figure 11.14 shows a magnificent red hydrogen alpha line spectroheliogram with short dark features, like blades of grass, that are thought to be spicules outlining the bright interiors of supergranule cells. Skylab photographs suggest that the chromosphere may have granulelike features much larger than those observed in the photosphere.

SOLAR FLARES

Solar *flares* are perhaps the most complex of the sun's transient phenomena. They vary in size, brightness, and behavior, and they are most common when sunspots are most numerous. A solar flare may suddenly erupt as an intensely bright area in a chromospheric plage (Figure 11.16) when seen in either a red hydrogen alpha line or a violet calcium K line spectroheliogram. Emitting radiation strongly throughout the electromagnetic spectrum; it rises to great brilliance in several minutes and then fades in half an hour to several hours. A flare outburst arises from the sudden release of energy in the chromosphere. The outburst of electromagnetic energy is apparently triggered by the demise of the local magnetic field in the

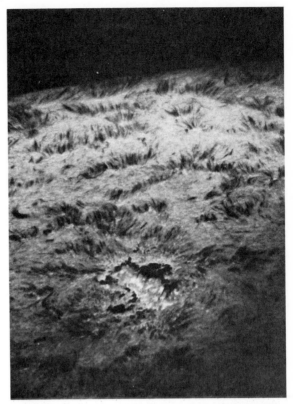

FIGURE 11.14
Hydrogen alpha line spectroheliogram of chromospheric network, with spicules on the boundaries of the supergranule cells. The dark bladelike features are spicules; the bright convective cells that they surround are about 30,000 kilometers across. These cells are the supergranule cells. The area shown is about 100,000 kilometers on a side. This level of the solar atmosphere is several thousand kilometers above the photosphere.

plage. That is, the energy of the flare is first stored in magnetic fields and is then suddenly released.

If we think of magnetic lines of force as being like the rubber band of a toy airplane, then the storage of energy is analogous to winding the rubber band. Twisted magnetic structures are often seen in connection with strong magnetic fields. In a matter of seconds to minutes the energy stored in the magnetic

And thou, O Sun, thou seest all things and hearest all things in thy daily round.
Homer, *The Iliad*

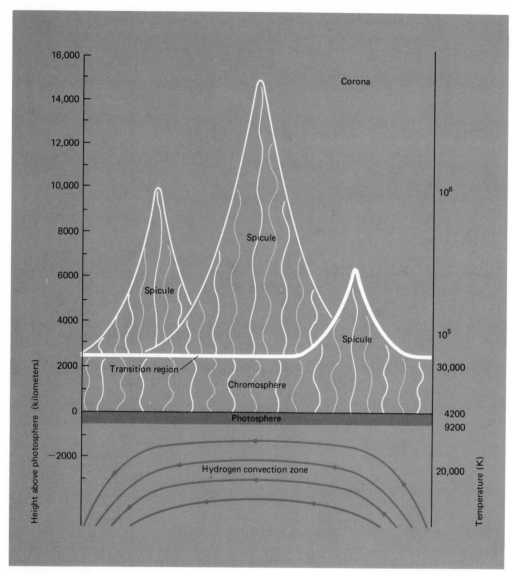

FIGURE 11.15
Spicules extending out of the chormosphere into the corona. Spicules are typically
several thousand kilometers across. They spurt up at a rate of about 20 kilometers per
second to heights of about 5000 to 10,000 kilometers and collapse in several minutes. At
any time there are nearly 1 million spicules in the chromosphere.

field is converted into the kinetic energy of atomic
particles just as releasing the rubber band turns the
propeller of the toy airplane.

In a flare free electrons are accelerated up to velocities of about half the speed of light. As these energetic electrons collide with the ambient gas, they
share their kinetic energy and heat the gas to a few
thousand degrees in the chromosphere and to around
10 million K in the low corona. This heating phase may

last from seconds to minutes and is responsible for
the X-ray, ultraviolet, and visible emission of the flare
(see Figure 11.16). Some high-energy electrons pass
out through the corona, where they excite successive
layers to emit radio frequency radiation.

A major flare is the most energetic of the solar
disturbances, equaling a billion hydrogen bombs in
power. The ultraviolet and X-ray radiation arriving at
the earth can disrupt the ionosphere, persisting some-

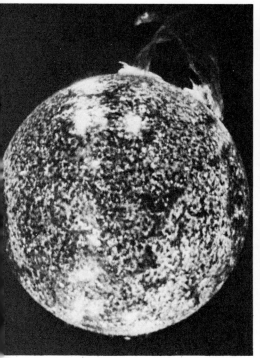 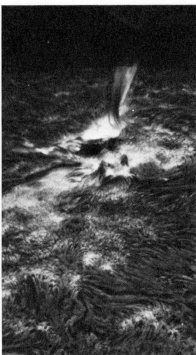 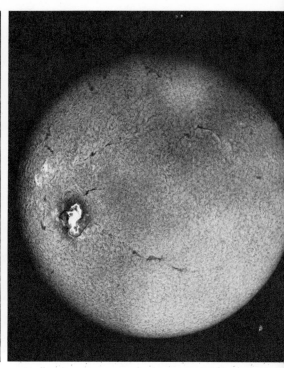

b.   c.

FIGURE 11.16
Solar flares photographed in hydrogen alpha light. (a) A spectacular flare spanning more than 588,000 kilometers across the upper-right edge of the solar disk. (b) A close-up of a surge flare near the limb of the sun. The complexity and activity in the vicinity of a flare is quite evident in this picture. (c) A major flare (August 7, 1972) seen near peak brightness. Note the complexity of the region around the flare.

times for days. Over the few days following the flare outburst subatomic solar particles may spiral into the earth's polar regions, causing brilliant auroral displays and radio blackouts. All these events, which do not necessarily occur with every flare, partially distort the earth's magnetosphere, generate geomagnetic storms, and induce electrical currents that interfere with worldwide communication.

11.5
The Corona

APPEARANCE AND SPECTRUM OF THE CORONA

The corona is the region of the solar atmosphere lying above the chromosphere. As seen during a total eclipse of the sun, it is the large halo of white, glowing gas extending out many solar radii (millions of kilometers) beyond the dark limb of the moon (Figure 11.12 and the chapter opening photograph). A specially designed refracting telescope, called a *coronagraph*, is used to observe the corona from the ground at times when an eclipse is not in progress.

In white-light photographs one can see that the corona is irregular and structured. Beautiful long streamers extend outward from the sun in the equatorial regions. Near sunspot maximum the corona is nearly circular, with streamers radiating out in all directions. Near sunspot minimum it extends farther out in the equatorial region and terminates rather abruptly, with short, thin plumes curving out of the polar areas.

Compared to our knowledge gained from rocket and satellite studies over the last 30 years, eclipse studies have yielded not more than a few hours of observation (an eclipse seldom lasts longer than a few minutes). Even coronagraphic studies, which have greatly increased the observing time, have not matched the best resolution of satellite observations.

> The great globe itself,
> yea all it inherit, shall dissolve
> And, like this insubstantial pageant folded
> Leave not a rack behind.
>
> Shakespeare

In 1973 the launching of the Skylab orbiting observatory, with trained astronauts on board, gave us the opportunity to observe the corona in the X-ray and ultraviolet wavelengths with greater resolution and to record its changing appearance in a virtually continuous fashion for up to 8 months at a time.

Approximately 30 emission lines have been identified in the visible part of the coronal spectrum, and many hundreds of emission lines are known in the ultraviolet and X-ray spectrum. They originate in the peculiarly excited ions of familiar elements, such as iron, from which several to as many as fifteen electrons have been stripped in the corona's extremely hot, tenuous gases. (It takes temperatures from many hundreds of thousands up to several million kelvins to produce such a range of high ionization.) Because of the high temperature, most of the coronal radiation is in the ultraviolet and X-ray regions of the electromagnetic spectrum. This is substantiated in X-ray photographs revealing intense radiation whose source is well outside the photosphere, as is evident in Figure 11.17.

From the millimeter to the meter wavelengths a wide spectral window in the earth's atmosphere lets in radio radiation. The sun, when quiet and undisturbed, normally emits thermal (blackbody) radiation, which is characteristic of the hot corona. When the sun is disturbed, nonthermal radio emission can be quite intense. It comes in short-lived *radio bursts* extending over many wavelengths, long-lived *noise storms* in meter wavelengths, and a slowly varying thermal component in centimeter wavelengths.

STRUCTURE OF THE CORONA

The solar corona appears to be highly inhomogeneous and generally asymmetrical, and it varies with time on both short and long temporal scales. There appear to be three different types of structural region, which collectively characterize the entire solar corona: Coronal holes and coronal active regions are two of them, and they are reasonably well defined in terms of the observational characteristics; the third is the coronal quiet regions, which are not well defined. The *coronal holes* (Figure 11.17) are regions of relatively low temperatures and densities, with magnetic field lines that open out into interplanetary space. Rays and plumes extend out of them essentially away from the sun. Coronal holes are also thought to be the source of most of the subatomic particles in the solar wind (see pages 124–125). The *coronal active regions* are extremely different from the coronal holes. They consist of loop structures, they are the hottest and densest regions of the corona, and their magnetic field lines loop back into the sun instead of extending outward as in the case of the coronal holes. In between these two extremes are the ill-defined *coronal quiet regions*, which appear to be something in between. They are intermediately bright and consist of somewhat amorphous and vaguely looped structures. There are also many small, bright points visible in X-ray pictures like Figure 11.17.

PROMINENCES AND FILAMENTS

With time-lapse photography of the corona we can see spectacular motions of towering masses of luminous gas, the solar *prominences* as shown in Figure 11.1. Projected against the solar disk, they are the dark, threadlike *filaments* seen in Figure 11.13b. Their forms vary from almost stationary, quiescent arches and graceful loops to rapidly moving surges.

The typical prominence is hundreds to thousands of kilometers long and extends several tens of thousands of kilometers above the photosphere. It consists of gas cooler and denser than that in the corona around it. In the more active prominences gas may occasionally rise at rates of hundreds to thousands of kilometers per second, which is sufficient to escape from the sun. Frequently, however, matter appears to rain down from the corona in great luminous masses. Apparently magnetic fields hold up these huge walls

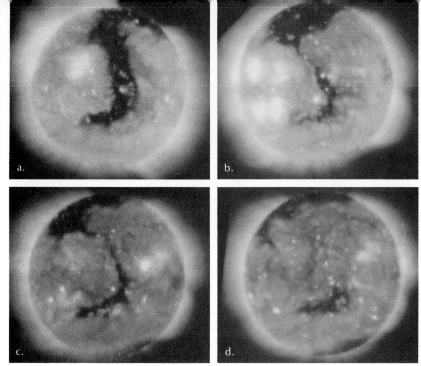

FIGURE 11.17
Boot-shaped coronal hole rotating with the sun. Seen in X rays (magnesium wavelength of 368 angstroms), the sequence of four pictures span 6 days in August 1973. Taken by the crew of *Skylab*, one sees the boot-shaped coronal hole first toward the east limb (a), followed by successive positions toward the west limb as the sun rotates. The coronal holes are the sources of the high-energy subatomic particles of the solar wind.

of gas against the sun's gravitational pull. We can see matter flowing along the body of a prominence, apparently following the curving and looping magnetic lines of force, as in the spectacular eruptive prominence in Figure 11.1.

The mean lifetime of large quiescent prominences is about two to three rotations of the sun. During sunspot maximum 20 filaments may appear on the disk; during sunspot minimum there are typically about 4. Prominences always appear to be associated with a plage or a sunspot group. In fact the large quiescent prominences tend to form along the division between regions of different magnetic polarity in plages. The polarity of each side of the plage region is also the same as that of the sunspots in the photosphere underneath it.

HEATING THE CORONA

There are several bodies of evidence confirming the rise in temperature through the chromosphere into the corona. This temperature rise seemed paradoxical when first recognized over 40 years ago. For a number of years astronomers thought that the corona's high temperature resulted from energy carried into the corona by mechanical waves starting in the turbulent hydrogen convection zone below the photosphere.

As evidence grew that the magnetic fields of the photosphere and chromosphere were highly localized and very intense, it seemed hard to ignore the possibility that these magnetic fields extending up into the corona were not part of the coronal-heating process.

When satellite and *Skylab* X-ray pictures showed that the corona was divided into active regions and hole regions primarily because of the structures of their magnetic fields, it became readily apparent that most, if not all, of the heating involves the magnetic field. The heating is produced by the direct dissipation of the energy in the magnetic field into kinetic energy of motion of the coronal gas.

The high coronal temperature also drives a rapid flow of plasma (composed mostly of protons and electrons), which moves out from the base of the corona as the solar wind. Several radii outside the sun, the velocity becomes supersonic and continues to increase at least up to 1 AU. By then the solar wind is moving at over 400 kilometers per second, eight times the speed of sound in the gas. Because of the sun's rotation, magnetic field lines, which confine the solar-wind particles, spiral outward like water from a rotating sprinkler. Perhaps 600,000 tons of plasma leave the sun every second, which amounts to about 10^{-15} of the sun's mass per year.

11.6
Solar Activity

ACTIVE REGIONS

During the sunspot cycle the general level of all activity in the solar atmosphere follows the number of sunspots. Thus a sunspot group seen in white-light photospheric pictures is just the most visible indicator of a large disturbed region in the solar atmosphere called an *active region*. Such regions can be up to several hundred thousand kilometers in extent. The

DEFLECTION OF STARLIGHT: TESTING RELATIVITY THEORY

A total eclipse of the sun provides a chance to test relativistic against classical physics. Newtonian physics says that a stream of photons from a star grazing the limb of the sun will be deflected inward 0.875 arc second (see Figure 11.18). General relativity theory predicts that since space-time in the sun's vicinity is warped, a ray of light traveling along the shortest path in this curved space is deflected by 1.75 arc seconds. This is exactly twice the amount predicted by Newtonian theory.

How can we test these predictions? During a total solar eclipse bright stars can be seen on a photograph of the darkened sky around the eclipsed sun. Another photograph of the same area can be made at night (a few months earlier or later) with the same telescope when the sun is in a different place in the sky. Star posi-

tions on the two photographs are then compared to show the stars around the eclipsed sun shifted away from the sun. The amount of deflection decreases with distance from the sun's limb. This test favors the answer given by general relativity theory although the difficulties of measuring keep this from being a definitive test. This observational test has become almost a standard procedure at every total eclipse of the sun.

Bending of radio waves near the sun has also been checked in recent years by radio interferometry. The apparent position of a pointlike radio source about to be occulted or grazed by the sun can be found relative to nearby sources. Bending of the radio waves passing through the solar corona (caused by refraction by the corona) must be separated from that

caused by the curvature of space-time in the sun's vicinity. The measured gravitational deflection agrees within 1.5 percent of the values predicted by Einstein's general theory.

When Mars was near conjunction with the sun from late November to mid-December, 1976, radio signals from the *Viking* orbiters grazing the sun on their way to the earth were bent and slightly delayed due to the warping of space around the sun. Mars was then about 321 million kilometers from the earth, and radio signals took about 42 minutes for the round trip. The difference in the travel time of the signals when Mars was near the sun compared with that when Mars was well separated from the sun weeks later, 0.0002 second, was in exact agreement with that predicted by the general theory of relativity.

FIGURE 11.18
Bending of starlight grazing the sun's limb that is the result of the curvature of space-time near the sun. The deflection angle α is 0.875" of arc according to Newtonian calculations but 1.75" of arc according to Einstein's general relativity. The starlight, which is deflected when passing the sun, appears to an observer on the earth to be coming from a point farther from the edge of the sun than is actually the case. The deflection decreases for stars farther from the solar limb.

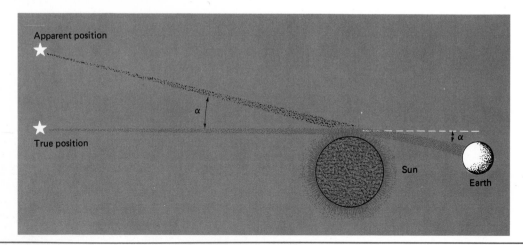

various transient phenomena that are part of the active region are summarized in Table 11.3 and are illustrated in Figure 11.19. The common bond among these visible features is the magnetic field. It appears first, followed by the faculae in the photosphere and the plage in the chromosphere. This is followed by the sunspot group, flare activity, and prominences. The precise behavior is somewhat different for each active region; but there is little doubt that the phenomena in different regions are related.

The solar cycle of activity (the succession of active regions over a 22-year period) is fundamentally a mag-

TABLE 11.3
Summary of Solar Active Regions

Atmospheric Region	Visible Activity	Description
Photosphere	Faculae	Observed in white light, denser and hotter than normal photosphere, granular and irregular, average life 15 days, can exist for 80–90 days, temperature excess up to 1000 K, magnetic-field strengths up to several thousand gauss
	Sunspots	Observed in white light, cooler than photosphere by 1800 K, average life 6 days, strong magnetic fields up to several thousand gauss, size about 10,000 kilometers
Chromosphere	Plages	Seen in monochromatic light, lifetime about 40 days, hotter and denser than normal chromosphere, magnetic-field strengths up to several hundred gauss, size about 50,000 kilometers
	Flares	Brief brightening in plages, life about 20 minutes, size about 30,000 kilometers, enhanced particle emission in solar wind and solar cosmic rays
Corona	Holes	Best seen in X rays, relatively low-temperature and -density regions with magnetic-field lines opening out into interplanetary space, source of rays, plumes, and solar-wind particles, size up to hundreds of thousands of kilometers, changeable in days and weeks.
	Active regions	Best seen in X rays, relatively hot and dense regions consisting of magnetic loop structures, source of coronal streamers, size up to hundreds of thousands of kilometers, changeable in days and weeks, occur over chromospheric plages
	Quiet regions	Something in between coronal active regions and coronal holes, consist of somewhat amorphous and vaguely looped structures, size up to hundreds of thousands of kilometers, changeable in days and weeks
	Prominences (filaments)	Chromospheric material in corona, much cooler than surrounding corona, lifetimes up to 60–90 days if quiescent, height 30,000 kilometers, length 200,000 kilometers, and thickness 5000 kilometers, exhibit motions associated with magnetic-field strengths up to several hundred gauss
	Radio bursts	Motions of fast electrons in upper corona

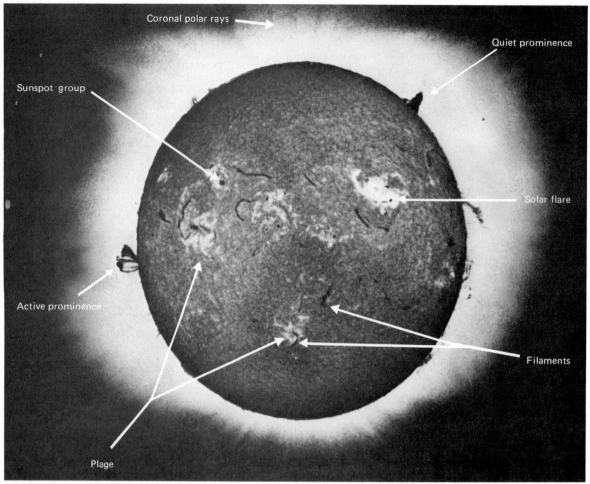

FIGURE 11.19
Composite spectroheliogram of the solar disk and the chromosphere, photographed in the light of the red hydrogen alpha line and superimposed on the inner corona. Various transient solar features are marked.

netic cycle. It is a many-year variation in the amount of magnetic field that erupts onto the solar surface. The average latitude of the eruption varies as well, producing the great complexity of observed events. The cause of the eruption is apparently some oscillatory or pulsational motion deep inside the sun, which in turn causes a dynamolike action in the subphotospheric magnetic fields that burst through to the surface.

SOLAR ACTIVITY AND THE EARTH

As one might guess, transient activity on the sun produces a variety of effects here on the earth. This transient behavior on the earth also follows a rough cycle in unison with the sun's cycle. An example of this relation is the increase in auroral activity in the earth's atmosphere during sunspot maximum and its decrease during minimum. Changes in solar activity also affect the earth's weather and climate, but how and over what time scales these effects occur we do not clearly understand. In recent years some interesting discoveries have been made about the constancy of the sun's activity and its relation to the earth.

IS SOLAR ACTIVITY CONSTANT?

As we stated at the beginning of this chapter, approximately 1.4 million ergs of radiant energy fall on each square centimeter of the earth every second. The the-

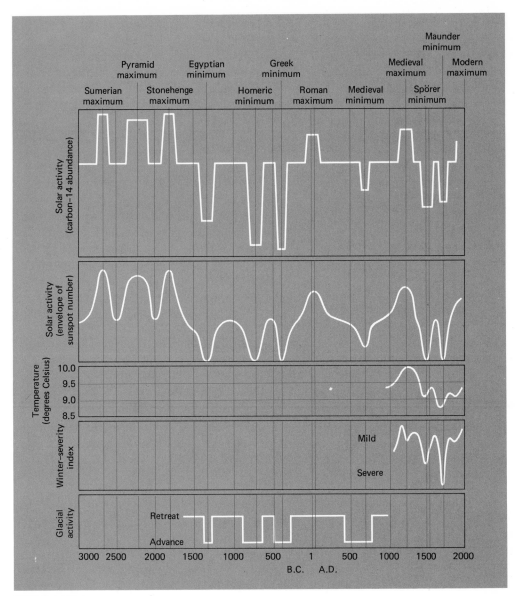

FIGURE 11.20
The sun's variation in activity since the Bronze Age is shown in the top two panels. Below are shown the corresponding estimates of the temperature in Europe, the winter-severity index from Paris and London, and Alpine glacier activity. It appears that long-term climatic activity rises and falls in response to long-term solar activity.

ory of stellar evolution, discussed in Chapters 13 through 15, suggests that the sun's luminosity must have increased by perhaps 30 percent since it began its existence some 4.6 billion years ago. However, as recently as 1975, paleoclimatic evidence concerning long-term temperature variations on the earth showed that any change has been less than 3 percent in the last 1 million years. One can argue that, if the solar luminosity had been as much as 25 percent less than the present value, the oceans would have frozen. On a shorter time scale, observations at Kitt Peak National Observatory between 1975 and 1979 have suggested a decrease in the photospheric temperature corresponding to a 0.6 percent decrease in the solar constant. Nevertheless, questions regarding the long-term constancy of the sun and the earth's probable response to any change are still unanswered.

Another question is the constancy of solar transient activity. Does the 22-year cycle of activity repeat itself, period after period, for millions of years? In 1893 E. Walter Maunder (1851–1928), a British astronomer, found from historical records from Europe that very few sunspots were seen in the period from 1645 to 1715, now known as the Maunder minimum. Within the last several years Maunder's work has been confirmed and extended by the American astronomer John A. Eddy (1931–   ).

Eddy has shown that in addition to the absence of sunspots very few aurora were observed in Europe and that during eclipses the corona was absent or very weak. Virtually no sunspots were reported from Asia during the Maunder minimum even though naked-eye sunspot-sighting reports exist there from as early as 28 B.C. Finally, measurements of the amount of carbon 14 ($^{14}C_6$), a radioactive isotope of carbon, in tree rings show that during the Maunder minimum there was an excess of this isotope in the earth's atmosphere. We believe that the high-energy subatomic particles called *cosmic rays*, moving randomly through the Galaxy, collide with the nucleus of nitrogen atoms ($^{14}N_7$) in our atmosphere, converting it to carbon 14. When the sun is very active, the interplanetary magnetic field is strong, and Galactic cosmic rays are deflected away from the earth. Thus high levels of carbon 14 in the earth's atmosphere correspond to low levels of solar activity.

Historical research by Eddy has shown a correlation among carbon 14 abundance in tree rings, winter severity, Galactic cosmic-ray activity, and solar activity, as shown in Figure 11.20. Periods of colder climate appear to coincide with low levels of solar activity, as evidenced by unseasonably cold weather in Europe between the sixteenth and eighteenth centuries. There is evidence that at least a dozen similar periods of minimal solar activity, lasting from 50 to 200 years, have occurred since 3000 B.C. Thus after so much effort devoted to trying to understand the 22-year solar magnetic cycle, we have discovered that it may be only a transient feature itself. Changes in solar magnetic activity are probably caused by changes in the pattern of convection underneath the photosphere. The interior of the sun may not be so constant in its behavior as once thought by astronomers.

SUMMARY

The sun as a star. The sun's radiant energy is emitted by the outer, gaseous region of the sun, known as the photosphere. Surrounding the photosphere is the transparent chromosphere and surrounding it, the corona. Although boundaries between these regions are not sharply defined, they differ in their physical characteristics. Nuclear processes in the sun's interior are responsible for the immense output of radiant energy in various parts of the electromagnetic spectrum. Like other stars, the sun rotates on its axis. Because the interior of the sun is a gas, the outer layers can undergo differential rotation. Beneath the photosphere, hidden from our view, is the hydrogen convection zone.

The photosphere. The temperature of the photosphere of any star is a measure of the rate at which energy is radiated by the star. Three ways of measuring the sun's photospheric temperature are based on the blackbody radiation laws: the Stefan-Boltzmann law, Wien's displacement law, and Planck's radiation law. The approximate value derived for the photospheric temperature is 6000° K. The absorption spectrum of the photosphere is a continuous spectrum of colors crossed by many absorption lines. By studying the solar spectrum we can identify about 70 of the 92 naturally occurring elements and about 20 molecules. Hydrogen accounts for about 91 percent, helium 9 percent, and everything else less than 1 percent of the composition. The photosphere has a fine structure of solar granules. The bright centers are hot, rising gas currents, while the cooler, sinking gas currents are the dark intergranular regions.

Sunspots. The typical large sunspot has a cellular structure with a dark center (umbra) surrounded by a grayish, filamentary region (penumbra). The umbra appears dark because it is cooler than the surrounding photosphere. Sunspots form in groups and occur in approximately 11-year cycles. As the cycle progresses, spot groups form closer and closer to the equator. The magnetic activity of sunspots is apparently based on a

22-year period, which is more regular than the 11-year cycle.

The chromosphere. During a total eclipse of the sun, the photosphere is obscured and the chromosphere may be observed. At this time an emission spectrum of the chromosphere (the flash spectrum) may be observed. The emission spectrum shows that the temperature rises through the chromosphere to as high as 30,000° K toward the corona. The bright plages in the chromosphere are found above regions of intense magnetic fields and sunspot groups in the photosphere. A bright chromospheric network exists in addition to the larger plages. Jetlike spikes (spicules) rise from the chromospheric network. Solar flares, the most complex and energetic of the sun's transient phenomena, may erupt as bright areas in plages, as seen in spectroheliograms.

The corona. From spectroscopic evidence, it is apparent that the temperature of the corona is on the order of several million degrees K. Because of the low density and high temperature of the corona, it is a source of X-ray, ultraviolet, and radio emission. The structure of the corona is highly inhomogeneous and asymmetrical and changes with time. Coronal holes, coronal active regions, and quiet regions characterize the structure of the corona. In addition, prominences, cooler and denser masses of gas, are observed extending above the photosphere into the corona.

Solar activity. Transient activity on the sun in the photosphere, chromosphere, or corona are related to each other and collectively form active regions.

KEY TERMS

active region
chromosphere
corona
coronal active region
coronal hole
facula
flare
flash spectrum
hydrogen convection zone
luminosity

photosphere
plage
prominence
solar granules
spicule
sunspot
sunspot cycle
supergranulation cells
surface temperature
Zeeman effect

CLASS DISCUSSION

1. How does the determination of the solar constant lead to finding the total radiant energy output of the sun? What type of spectrum does the solar luminosity have? How does it compare with the radiation of a blackbody?

2. If an X-ray photon created in the sun's energy-generating core has no chance of penetrating the overlying matter to reach the solar atmosphere, how is it possible to make X-ray pictures of the sun?

3. What are sunspots? What layers of the solar atmosphere are they to be found in? Why are they darker than surrounding regions?

4. Why are there dark absorption lines in the spectrum of the photosphere and bright emission lines in the spectrum of the chromosphere and of the corona? Why does the ultraviolet spectrum of the sun contain bright emission lines?

5. How do astronomers determine the chemical composition of the sun? Does it contain all the 92 naturally occurring elements? Is there direct evidence of all 92?

6. What is meant by an active region? What do we observe to be associated with an active region? Describe its development and demise.

READING REVIEW

1. What is the name astronomers give to the rate at which the sun radiates energy over all wavelengths from the photosphere? What is its numerical value in metric units?

2. What are the names given to the three regions of the solar atmosphere?

3. How is the radiant energy produced that is radiated into space? Where is it produced?

4. Is the "surface temperature" of the sun actually a surface temperature? Explain.

5. How many of the chemical elements have been identified in spectra of the photosphere? How many molecules? Which element is most abundant? Which element is second, third, fourth, fifth?

6. How do we know that the solar granules are circulating gas?

7. What is a sunspot group? How large is it? Does it change?

8. How do astronomers detect the presence of magnetic fields in gases like the photosphere?

9. How long is the sunspot cycle? How long is the magnetic cycle? Which is more regular in length?

10. How do we know that chromospheric temperatures can reach 30,000 K?

11. What would one expect to see in the corona overlying the plage when observed in the X-ray wavelengths?

12. Name all the visible features that are part of an active region. In what regions of the solar atmosphere do they occur?

PROBLEMS

1. If the sun has emitted energy at its present rate for its 4.6-billion-year life, what is the total amount of energy emitted? On the average how much energy has each gram of solar material emitted over the life of the sun? How meaningful is the amount-per-gram figure in understanding the sun?

2. If the sun's surface temperature were 9000 K instead of roughly 6000 K, how much greater would be the sun's luminosity? What change would this make in the wavelength of the point of maximum intensity? If one could compensate for this increase in surface temperature by changing the solar radius, how much of a change would be required? Is it an increase or decrease of the radius?

3. Using the information contained in Table 11.1, find the amount of radiant energy in all wavelengths emitted by the photosphere per square centimeter per second.

SELECTED READINGS

Eddy, J. A. "The Case of the Missing Sunspots," *Scientific American*, May 1977.

Frazier, K. *Our Turbulent Sun.* Prentice-Hall, 1982.

O'Leary, B. "The Stormy Sun," *Sky and Telescope*, September 1980.

Pasachoff, J. M. "Our Sun," in M. A. Seeds, ed., *Astronomy: Selected Readings.* Benjamin Cummings, 1980.

Washburn, M. *In the Light of the Sun.* Harcourt Brace Jovanovich, 1981.

12
Stars: "The Pale Populace of Heaven"

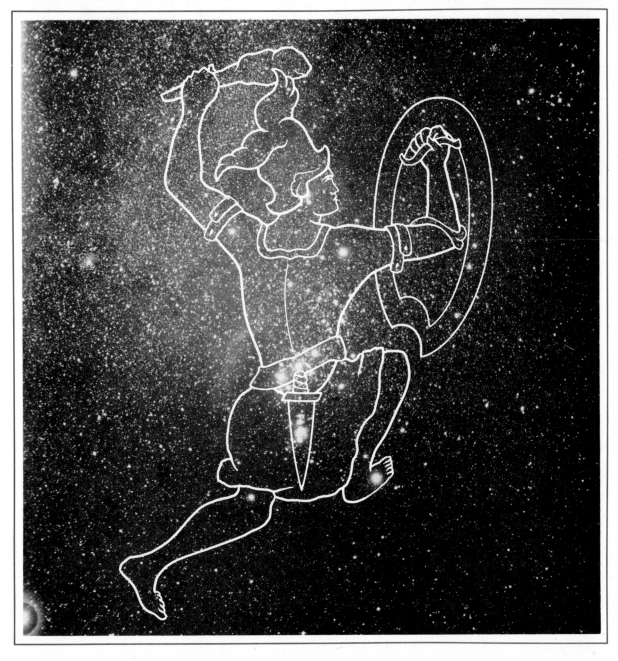

To help you learn something about the organization of the sky, star maps have been included in Appendix 2. Having spent a night looking at the stars, you will not find it hard to understand the fascination they have held for generations of human beings. But what are they? It is only in the last 100 years or so that we have been able to say with any certainty that the stars are great spheres of glowing gas like the sun. For decades astronomers have been seeking answers to these questions and others that tell us about the nature of the stars in our Galaxy and presumably the universe. We have compiled a great body of knowledge about stars—a census of different stellar population groups within the Galaxy. This chapter is devoted to surveying those data.

12.1
Distances of Stars

TRIGONOMETRIC PARALLAX
AND FINDING DISTANCE

Unless we know the distance of a star, it is impossible to determine most stellar properties. The basic technique for determining distance is *trigonometric parallax*. You can understand how it works by looking first with one eye and then with the other at a pencil held at arm's length; as you do so, the pencil seems to shift first one way and then the other against a distant background, the shift diminishing as the distance of the pencil from your eyes is increased. Applied to stars (Figure 12.1), the parallactic shift that we can see is very tiny. It was not until the years from 1837 to 1839 that astronomers were first able to measure the parallax for three fairly nearby stars: α Centauri, 61 Cygni, and Vega.

The tiny apparent displacements of a star against background stars in a photograph are compared, in principle, every 6 months because the parallactic shift is a maximum when the earth is at the opposite ends of the diameter (whose length is 2 AU) of its orbit.

◀ Winter constellation of Orion and the surrounding region. The three prominent stars in a line form the mighty hunter Orion's belt. The Orion Nebula, the fuzzy object below the belt, is his sword. The chapter title is from Robert Browning's "Balaustion's Adventure."

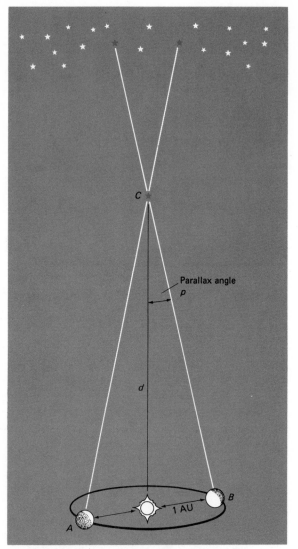

FIGURE 12.1
Parallax diagram. If we look at star C from points A and B on the earth's orbit 6 months apart, the star will seem to have shifted slightly compared with the more distant stars in the background. Measurement of the parallax angle p, in arc seconds, combined with knowledge of the earth's mean distance from the sun (1 AU) yields the distance d of the star by trigonometry: distance in astronomical units, $d = 206,265/p''$; distance in light years, $d = 3.26/p''$; distance in parsecs, $d = 1/p''$; 1 ly = 9.46×10^{12} km, 1 pc = 3.26 ly = 206,265 AU. (See Appendix 1 for abbreviations of units of measure.)

Even in a telescope of 15-meter focal length the star closest to us, α Centauri, has a parallactic displacement on a photographic plate of only 0.01 centimeter;

so many photographs must be made over a period of many years in order to measure such small shifts in position.

To calculate the distance of a star from the sun in astronomical units, we need to insert the parallax angle *p* in the first formula of Figure 12.1 (derived by simple trigonometry) and solve for *d*. Because the distance in astronomical units may be an extremely large number that is difficult to remember, astronomers prefer two larger units of distance. One is the *light year*, which is the distance that light travels in 1 year (the velocity of light times the number of seconds in 1 year). The second is a still larger unit known as the *parsec*, the distance of a star whose parallax is 1″ of arc. The name comes from the first three letters in the words *par*allax and *sec*ond.

Table 12.1 lists distances and other data on the nearest stars to the earth, those besides the sun within 15 light years. As one can see in Table 12.1, the parsec (equal to 3.26 light years) and the light year are units of distance characteristic of the separation between stars in the solar neighborhood.

The limit for measuring trigonometric parallax is about 300 light years, or 100 parsecs. As many as 500,000 stars may be within this range, and 6000 of these have been observed for trigonometric parallax. For stars whose parallaxes are too small to be measured astronomers fortunately have other ways of measuring distance.

FINDING DISTANCES OF MORE REMOTE STARS

If we know a star's true brightness, we can use the inverse-square law of light (page 74), which is a relationship between apparent brightness and distance, to find how far the star is from us. For example, if one star is twice as distant as another of the same intrinsic brightness, its apparent brightness is reduced by a factor of 4 (inversely as 2^2, the square of twice the distance); if three times as distant, the reduction is $3^2 = 9$ times; if four times as distant, the reduction is $4^2 = 16$ times; and so on. Hence, if we know the true brightness of the particular class of stars to which our subject star belongs, we can calculate its distance by the inverse-square law from its apparent brightness. The critical point in this method is recognizing the subject star as a member of a class whose luminosity is known so that we also know the intrinsic brightness of the subject star.

Astronomers have used variations of the inverse-square technique to build step by step a set of overlapping distance scales. We begin with the nearby stars, whose distances are known from trigonometric measurements of parallax, and move toward the more distant stars, applying the appropriate variation of the inverse-square law. Through the rest of the book we shall describe these various distance-determination techniques.

And thus we die,
Still searching, like poor old astronomers
Who totter off to bed and go to sleep
To dream of untriangulated stars.

Edwin Arlington Robinson

12.2
Brightnesses of Stars

APPARENT MAGNITUDES

Before photography the only way of determining a star's apparent brightness was by visual estimates. Ancient Greek astronomers were the first to use a scale of apparent magnitude *m* as a measure of the brightness of a star as it appeared in the sky. They picked a natural scale for assigning magnitude that is the one the eye follows: Equal ratios of brightnesses correspond to equal differences of magnitude. Hipparchus and Ptolemy roughly graded the apparent brightnesses of the stars into six magnitude classes: the brightest stars assigned to magnitude class 1, the next brightest to 2, and so on to 6 for the faintest stars. Experimenting with telescopes of various apertures, William Herschel concluded that a bright star of first magnitude was about 100 times brighter than a faint star of sixth magnitude.

In 1856 British astronomer Norman Pogson (1829–1891) quantified the system of magnitudes so that the ratio of apparent brightness of first-magnitude stars to sixth-magnitude stars is 100 to 1. This ratio corresponds to *any* five-magnitude difference, such as between ninth- and fourteenth-magnitude stars. From this rule and a logarithmic formula one can determine the brightness ratio corresponding to any magnitude difference, as can be

TABLE 12.1
Stars within Fifteen Light Years of the Earth[a]

Star Designation	Distance (ly)	Spectral Type A	Spectral Type B[b]	B-V Color A	B-V Color B[b]	Visual Apparent Magnitude A	Visual Apparent Magnitude B[b]	Visual Absolute Magnitude A	Visual Absolute Magnitude B[b]	Luminosity (sun = 1) A	Luminosity (sun = 1) B[b]	Mass[g] (sun = 1) A	Mass[g] (sun = 1) B[b]	Radius[h] (sun = 1) A	Radius[h] (sun = 1) B[b]
Sun		G2 V		+0.65		−26.8		+ 4.8		1.0		1.00		1.00	
α Centauri[c,f]	4.3	G2 V	K0 V	+0.68	+0.88	− 0.1	+ 1.5	+ 4.4	+ 5.7	1.5	0.44	1.14	0.93	1.27	0.94
Barnard's star[d,f]	6.0	M5 V		+1.74		+ 9.5		+13.2		4.4×10^{-4}		0.15		≈0.07	
Wolf 359[e]	7.5	M8 V		+2.01		+13.5		+16.7		2.0×10^{-5}		≈0.06		≈0.02	
BD + 36° 2147	8.2	M2 V		+1.51		+ 7.5		+10.5		5.2×10^{-3}		≈0.25		≈0.18	
Luyten 726-8[f]	8.4	M6 V	M6 V	+1.85		+12.5	+13.0	+15.4	+15.9	6.0×10^{-5}	4.0×10^{-5}	0.11	0.11	0.16	0.15
Sirius[e,f]	8.6	A1 V	wd	0.00	−0.12	− 1.5	+ 8.5	+ 1.4	+11.2	29.0	2.8×10^{-3}	2.20	0.94	1.68	0.01
Ross 154	9.4	M5 V		+1.70		+10.6		+13.3		4.0×10^{-4}		≈0.13		≈0.06	
Ross 248	10.2	M6 V		+1.91		+12.3		+14.8		1.0×10^{-4}		≈0.09		≈0.04	
ε Eridani[c,f]	10.7	K2 V		+0.88		+ 3.7		+ 6.1		0.30		0.74		≈0.73	
Luyten 789-6	10.8	M6 V		+1.76		+12.2		+13.5		1.2×10^{-4}		≈0.12		≈0.06	
Ross 128	10.8	M5 V		+1.96		+11.1		+14.6		3.3×10^{-4}		≈0.09		≈0.04	
61 Cygni[c,f]	11.2	K5 V	K7 V	+1.17	+1.37	+ 5.2	+ 6.0	+ 7.5	+ 8.3	8.3×10^{-2}	4.0×10^{-2}	0.58	0.57	≈0.53	≈0.43
ε Indi	11.2	K5 V		+1.05		+ 4.7		+ 7.0		0.13		≈0.58		≈0.59	
τ Ceti	11.3	G8 V		+0.72		+ 3.5		+ 5.9		0.39		≈0.78		≈0.68	
Procyon	11.4	F5 IV-V	wd	+0.42		+ 0.4	+10.7	+ 2.7	+13.0	7.0	5.0×10^{-4}	1.77	0.65	2.06	0.01
Σ 2398[e]	11.5	M4 V	M5 V	+1.54	+1.59	+ 8.9	+ 9.7	+11.2	+12.0	2.8×10^{-3}	1.3×10^{-3}	0.41	0.41	≈0.14	≈0.10
BD + 43° 44[e,f]	11.6	M1 V	M6 V	+1.56	+1.80	+ 8.1	+11.0	+10.4	+13.3	5.8×10^{-3}	4.0×10^{-4}	≈0.26	≈0.13	≈0.20	≈0.07
CD − 36° 15693	11.7	M2 V		+1.48		+ 7.4		+ 9.6		1.2×10^{-2}		≈0.31		≈0.26	
G51-15	11.9	M? V		+2.06		+14.8		+17.0		1.0×10^{-5}		≈0.05		≈0.05	
Luyten 725-32	12.3	M5 V		+1.83		+11.5		+13.6		3.0×10^{-4}		≈0.12		≈0.06	
BD + 5° 1668[e]	12.3	M4 V		+1.56		+ 9.8		+12.0		1.4×10^{-3}		≈0.18		≈0.10	
CD − 39° 14192	12.5	M0 V		+1.40		+ 6.7		+ 8.8		2.5×10^{-2}		≈0.38		≈0.35	
Kapteyn's star	12.7	M0 V		+1.56		+ 8.8		+10.8		4.0×10^{-3}		≈0.23		≈0.17	
Krüger 60[f]	12.8	M4 V	M6 V	+1.62	+1.80	+ 9.7	+11.2	+11.7	+13.2	1.7×10^{-3}	4.4×10^{-4}	0.28	0.16	0.35	0.28
Ross 614	13.4	M5 V	?	+1.71		+11.3	+14.8	+13.3	+16.8	4.0×10^{-4}	2.0×10^{-5}	0.11	0.06	≈0.06	
BD − 12° 4523	13.7	M5 V		+1.60		+10.0		+11.9		1.4×10^{-3}		≈0.18		≈0.10	
Wolf 424	13.9	M6 V	M6 V	+1.80	+1.80	+13.2	+13.4	+15.0	+15.2	8.0×10^{-5}	7.0×10^{-5}	≈0.08	≈0.08	≈0.03	
Van Maanen's star	14.0	wd		+0.56		+12.4		+14.2		1.7×10^{-4}		0.60		≈0.01	
CD − 37° 15492	14.5	M3 V		+1.46		+ 8.6		+10.4		5.8×10^{-4}		≈0.14		≈0.06	
Luyten 1159-16	14.7	M8 V		+1.82		+12.3		+14.0		2.0×10^{-4}		≈0.11		≈0.05	
BD + 50° 1725	15.0	K7 V		+1.36		+ 6.6		+ 8.3		4.0×10^{-2}		≈0.43		≈0.43	

[a] Adapted from Lippincott, S. L., Space Science Review, Vol. 22, page 153, 1978.

[b] A is the brighter component and B is the fainter component of the star listed in the first column.

[c] Additional companion in system.

[d] Nonvisible companion or companions.

[e] Suspected nonvisible companion or companions.

[f] Source of low-energy X rays.

[g] ≈ means mass derived by authors from mass-luminosity relation (Figure 12.15).

[h] ≈ means radius derived by authors from temperature for spectral type and luminosity.

seen in Table 12.2. With the apparent magnitude the great range in apparent brightness among celestial bodies is illustrated by the values listed in Table 12.3. (It is essential to remember that large negative numerical values of magnitude mean bright objects, while large positive values designate faint objects.)

From the table, for example, we see that the difference in visual apparent magnitude between Sirius and Venus at its brightest is $-1.5 - (-4.4) \simeq 3$, which corresponds to a brightness ratio of about 16 to 1; hence Venus can appear 16 times brighter than Sirius in the night sky. In fact Sirius is intrinsically much brighter than Venus because Sirius is a large star and Venus a tiny planet. The only reason Venus appears brighter is that it is very close and Sirius is very far away. An important question to ask is whether the stars appear bright in our night skies because they are nearby or because they are intrinsically bright. The answer to this question involves what is termed a star's absolute magnitude.

ABSOLUTE MAGNITUDE AND DISTANCE MODULUS

To find the intrinsic brightness of a star, we must know its distance and apparent magnitude. Suppose that we calculate (using the inverse-square law of light) what the apparent magnitude of stars would be if they were all placed at the same distance from us. Comparing the stars' brightnesses at the same distance is, of course, a comparison of their intrinsic luminosities. The reference distance selected by astronomers is 10 parsecs (32.6 light years). The magnitude a star would have if it were 10 parsecs from us is called the star's *absolute magnitude*.

The difference between a star's apparent magnitude and its absolute magnitude, $m - M$, is called its *distance modulus* since it is proportional to the ratio of the star's distance to 10 parsecs. From its numerical value we can determine how many times brighter or fainter the object appears compared with its brightness at the reference distance of 10 parsecs.

To measure the apparent magnitude of a star, astronomers have set up stellar-magnitude sequences in various parts of the sky similar to the one shown in Figure 12.2. These are stars whose apparent magnitudes are known and against which the magnitudes of other stars can be determined.

As we discussed in Section 5.2, the color of radiation affects our perception of its brightness. (At this point you might wish to review the discussion on radi-

TABLE 12.2
Brightness Ratios Corresponding to Magnitude Differences

Brightness Ratio	Magnitude Difference[a]
1:1	0.0
1.6:1	0.5
2.5:1	1.0
4:1	1.5
6.3:1	2.0
10:1	2.5
16:1	3.0
40:1	4.0
100:1	5.0
1000:1	7.5
10,000:1	10.0
1,000,000:1	15.0
100,000,000:1	20.0

[a] The mathematical relation between the magnitude difference of two stars, 1 and 2, and their brightness ratio is $b_1/b_2 = 2.512^{(m_2 - m_1)}$, or in logarithmic form $\log(b_1/b_2) = 0.4(m_2 - m_1)$, where b_1/b_2 is the ratio of brightness.

TABLE 12.3
Visual Magnitudes of Selected Objects

Object	Apparent Magnitude (m)	Absolute Magnitude[a] (M)
Sun	−26.7	+4.8
Full moon	−12.5	
Venus (at brightest)	−4.4	
Sirius (brightest star)	−1.5	+1.4
α Centauri	0.0	+4.4
Vega	0.0	+0.5
Antares	+0.9	−4.3
Andromeda galaxy	+3.5	−21.2
Faintest naked-eye star	+6.0	
Faintest star photographed by 5.1-m Hale telescope	+24.0	

[a] The difference $m - M$ is the distance modulus and from it a star's distance d can be determined by the equation

$$d = 10^{0.2(m-M)+1},$$

or in logarithmic form

$$\log d = 0.2(m - M) + 1.$$

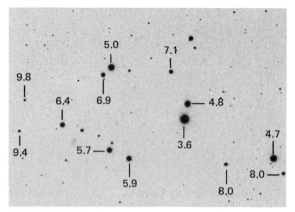

FIGURE 12.2
Photographic negative of a star field with several apparent magnitudes marked. An image's size and darkness are used to determine the apparent magnitude by comparison with known standards, such as those labeled in this photograph.

ation detectors on page 103.) So before proceeding further, let us talk about the color of stellar radiation.

COLORS AND TEMPERATURES

The numerical value of a star's magnitude depends on the spectral region that we are looking at when we measure the magnitude. The distribution of radiant energy with wavelength is certainly not constant, as Figure 11.4 for the sun shows. Because photographic emulsions and photoelectric devices do not respond in the same fashion to different color regions of the spectrum, they are not measuring the same wavelength range when used to determine the apparent magnitude of a star. One means of compensating for this effect is to use filters that transmit a narrow range of wavelengths. This technique ensures that the results, obtained by using the same color filter with different radiation detectors, are comparable. Some common filter designations are ultraviolet U, blue B, green or visual V (which approximates the human eye's color sensitivity), red R, near infrared I, and so forth. Hence we should qualify the apparent magnitude with the color region under consideration, such as the blue apparent magnitude or the visual apparent magnitude, when making comparisons. One magnitude scheme, known as the UBV photometric system, covers the spectral range from approximately 3000 to 6000 angstroms in three segments, as shown in Figure 12.3.

The difference between the magnitudes measured in two different color regions of the spectrum is called a *color index*, or simply a *color*. A frequently used color index is the difference between the blue and visual magnitudes, designated as B–V. A blue star, for example, has a brighter blue magnitude B than its visual magnitude V. Since brighter means algebraically smaller values on the magnitude scale, B–V is negative. The opposite is true for an orange or red star, where B–V is positive. The zero point of the B–V scale is arbitrarily assigned to a blue-white star whose spectrum is classified AO V (see Section 12.3).

Does the color index tell us anything? Yes indeed; for magnitudes for different color regions of the same object differ because the distribution of radiant energy varies with wavelength in accordance with

FIGURE 12.3
UBV photometric system. Color filters light only through the narrow spectral regions shown. In the UBV system these spectral regions are centered around 3600 angstroms for the ultraviolet, 4350 angstroms for the blue, and 5550 angstroms for the visual. The percentage of light transmitted through the filter drops off fairly rapidly on either edge of the spectral region. Notice that the B–V color index (the difference between the blue and the visual magnitudes) is positive for the 3000- and 5000-K blackbody energy curve. The B–V color index can be calibrated as a measure of the temperature of the blackbody energy curve, which approximates the energy curve for a star. Such a calibration is given in Table 12.5.

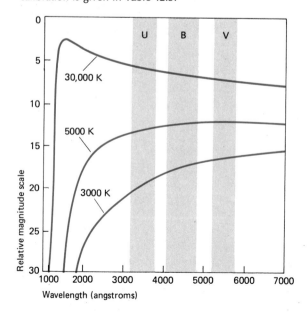

Planck's radiation law for blackbodies (see Section 4.2). We can therefore use the color index to determine the star's temperature. (In Table 12.5 the experimentally determined relationship is given between the B–V color index and the star's surface temperature; for most stars it is an average of the high temperatures at the bottom of the star's photosphere and the lower temperatures found in the uppermost layers; the surface temperature ranges from about 3000 to 40,000 K.)

BOLOMETRIC MAGNITUDES AND STELLAR LUMINOSITY

A star's *luminosity* is the total amount of energy it radiates per second from its surface. But the magnitudes we have considered so far correspond to the amount of radiation in a specific wavelength interval, such as the visual apparent magnitude. A more fundamental measurement, the *bolometric magnitude,* represents all of a star's radiation integrated over all wavelengths received outside the earth's atmosphere; that is, it is a measure of a star's luminosity. Unless this magnitude can be observed directly from an instrumented satellite, a theoretical correction using the star's surface temperature must be applied to the ground-observed visual magnitude to compensate for the radiation that does not pass through the earth's atmosphere. The correction is largest where the peak of the energy curve lies outside the visual range, as it does for the hottest and coolest stars. In summary a star's luminosity is derived by first measuring its distance and apparent magnitude, from which its absolute magnitude is calculated; a bolometric correction is added to the absolute magnitude to give the bolometric absolute magnitude, which can then be converted to the star's luminosity.

Values for luminosity are given in Table 12.4 for the 30 stars with the largest apparent brightnesses. As is evident from the table, most stars that appear bright are intrinsically bright and do not appear bright merely because they are nearby.

RADIUS OF STARS

Knowing a star's temperature and luminosity allows us to calculate another property of a star: its radius. From the temperature we can determine the energy emitted per second from each square centimeter of the star's surface. The total amount of energy radiated per second by the entire star, which is its luminosity, depends not only on its temperature but also on its surface area. We can thus calculate how large the star must be to have that luminosity. The radii of stars are found to vary from one-hundredth of the sun's radius (or about the size of the earth) to about 1000 times that of the sun, which is about 5 AU, or five times the radius of the earth's orbit.

12.3
Spectra of Stars

SPECTRAL CLASSES

In the latter part of the last century and the first half of this one, astronomers worked to classify the spectra of the stars into an orderly sequence. The first large study was a photographic survey of the spectra of nearly a quarter of a million stars, which was begun in 1884 at the Harvard College Observatory.

As that survey ended in 1924, when the relationship between atomic structure and emission of radiation was better understood, the Harvard astronomers realized that the great diversity in spectral appearance, shown in Figure 12.4, resulted primarily from the stars' differing surface temperatures and not from differences in the abundance of elements in their atmospheres. Most stellar atmospheres have a chemical

The number is certainly the cause. The apparent disorder augments the grandeur, for the appearance of care is highly contrary to our idea of magnificence. Besides, the stars lie in such apparent confusion, as makes it impossible on ordinary occasions to reckon them. This gives them the advantage of a sort of infinity.

Edmund Burke

TABLE 12.4
Thirty Brightest Stars[a]

Star Name	Constellation Designation	Distance[b] (ly)	Spectral Type A	Spectral Type B[c]	B-V Color A	B-V Color B[c]	Visual Apparent Magnitude A	Visual Apparent Magnitude B[c]	Visual Absolute Magnitude A	Visual Absolute Magnitude B[c]	Luminosity (sun=1) A	Luminosity (sun=1) B[c]	Radius (sun=1) A	Radius (sun=1) B[c]
Sirius[d,e,f]	α CMa	8.6	A1 V	wd	+0.00	−0.12	−1.5	+ 8.5	+1.4	+11.2	29	2.8×10^{-3}	1.7	≈0.01
Canopus[e]	α Car	120?	F0 II		+0.15		−0.7		−3.5		2.2×10^{3}		25	
Arcturus	α Boo	35	K1 III		+1.28		+0.0		−1.1		3.7×10^{2}		26	
Rigel Kent[d,e,f]	α Cen	4.3	G2 V	K0 V	+0.68	+0.88	+0.0	+ 1.5	+4.4	+5.9	1.5	0.44	1.3	0.9
Vega[e]	α Lyr	25	A0 V		+0.00		+0.0	+10.4	+0.7	+11.1	63	2.5×10^{-2}	2.6	
Capella[d,e,f]	α Aur	40	G5 III	G0 III	+0.80		+0.1	+ 0.6	−0.4	+ 0.1	1.4×10^{2}	75	6.7	5.4
Rigel[d]	β Ori	850?	B8 I	B5 V	−0.03		+0.1	+ 6.6	−7.0	− 0.8	1.2×10^{5}	6.7×10^{2}	70	
Procyon[d,f]	α CMi	11.4	F5 IV-V	wd	+0.42		+0.4	+10.7	+2.7	+13.0	6.6	5.0×10^{-4}	2.1	
Achernar	α Eri	125	B3 V		−0.16		+0.5		−2.5		6.0×10^{3}		7.7	
Betelgeuse[d]	α Ori	650?	M1 I		+1.85		+0.5	+10.4	−6.0	+ 3.9	1.2×10^{5}	2	1.0×10^{3}	
Hadar[d]	β Cen	360?	B1 III		−0.23		+0.6	+ 3.8	−4.6	− 1.4	6.1×10^{4}	1.2×10^{3}		
Altair	α Aql	16.1	A7 V		+0.22		+0.8	+ 9.4	+2.3	+10.9	10.6	2.9×10^{-2}	1.5	
Aldebaran[d]	α Tau	60	K5 III	M2 V	+1.54		+0.9	+10.7	−0.5	+ 9.3	3.0×10^{2}	0.13		
Acrux[d,f]	α Cru	400?	B1 IV	B3 V	−0.26		+0.9	+ 2.0	−4.6	− 3.5	7.1×10^{4}	1.3×10^{4}		
Antares[d,f]	α Sco	300?	M2 I	B4 V	+1.83		+1.0	+ 5.2	−3.8	+ 0.4	1.6×10^{4}	3.0×10^{2}	4.0×10^{2}	
Spica[d]	α Vir	260?	B1 III	B2 V	−0.23		+1.0	+ 2.5	−3.5	− 2.0	2.3×10^{4}	4.0×10^{3}	7.4	
Pollux	β Gem	35	K0 III		+1.00		+1.1	+11.3	+1.0	+11.2	50	2.2×10^{-2}		
Fomalhaut	α PsA	22	A3 V	K4 V	+0.09	+1.10	+1.2	+ 6.5	+2.0	+ 7.1	15.0	0.20	1.5	
Deneb[d]	α Cyg	1600	A2 I		+0.09		+1.3	+11.3	−7.2	+ 2.8	9.0×10^{4}	7		
Mimosa[d]	β Cru	500?	B0 III		−0.23		+1.3	+ 7.3	−4.6	+ 1.4	8.2×10^{4}	30	12	
Regulus[d]	α Leo	70	B7 V	K1 V	−0.11	+0.86	+1.4	+ 8.2	−0.4	+ 6.4	3.1×10^{2}	0.26	3.2	
Adhara	ε CMa	650?	B2 II		−0.21		+1.5	+ 7.9	−5.0	+ 1.4	7.2×10^{4}	25	17	
Castor[d]	α Gem	50	A1 V	A5 V	+0.03		+1.6	+ 2.5	+0.7	+ 1.6	55	20		
Shaula[d]	λ Sco	330?	B2 IV	B?	−0.22		+1.6	+11.9	−3.4	+ 6.9	1.3×10^{4}	0.16		
Gacrux	γ Cru	230?	M4 III		+1.59		+1.6	+ 6.7	−2.6	+ 2.5	8.7×10^{3}	8.4		
Bellatrix[d]	γ Ori	300?	B2 III		−0.22		+1.6	+12.1	−3.2	+ 7.3	1.7×10^{4}	0.17	7.1	
El Nath	β Tau	120	B7 III		−0.13		+1.7		−1.1		2.4×10^{2}			
Miaplacidus	β Car	160	A2 IV		+0.00		+1.7		−1.7		5.5×10^{2}			
Alnilam[d,e]	ε Ori	1500?	B0 I		−0.19		+1.7	+10.4	−6.6	+ 2.1	5.5×10^{5}	13	35	
Al Nair	α Gru	55	B7 IV		−0.13		+1.7	+11.8	−0.5	+10.6	1.5×10^{2}	3.7×10^{-2}	1.9	

[a] Adapted from D. Hoffleit, *The Bright Star Catalogue,* 4th revised edition, Yale University Observatory, 1982.
[b] Distances of more distant stars are estimated from spectroscopic parallaxes while others are from trigonometric parallax.
[c] A is the brighter component and B is generally the fainter component of the star listed in the first column.
[d] Spectroscopic binary.
[e] Source of low-energy X rays.
[f] Mass determinations in solar masses: Sirius, A-2.20, B-0.94; Rigel Kent, A-1.14, B-0.93; Capella, A-2.67, B-2.55; Procyon, A-1.77, B-0.65; Acrux, A-14, B-10; Antares, A-15.5, B-7.0.

composition very nearly the same as that of the sun: an overwhelming amount of hydrogen, a little helium, and traces of the other elements (see Table 11.2).

Each of the seven *spectral classes* — O, B, A, F, G, K, M — in the common one-dimensional classification scheme is divided into 10 parts, or *spectral types*, from 0 to 9. The sun's type is G2, 0.2 beyond G0 toward the next class, K. In Table 12.5 we list the spectral classes and describe their most distinguishing features, including the surface temperature and B–V color index for the hottest spectral type in each spectral class (that is, A0 in A, G0 in G, and so on). Often stars of spectral classes O, B, and A are referred to as "early-type stars," while stars of classes G, K, M are known as "late-type stars." "Early" and "late" are also used to denote direction along the spectral sequence from O to M.

Data, including the spectral type, are given in Table 12.4 for the 30 brightest stars. Spectral types for the nearest known stars are in Table 12.1. The point to remember is that the spectral classification scheme is a sorting of stars by their surface temperatures. The significance of the Roman numerals in the spectral column is discussed in the next subsection.

LUMINOSITY CLASSES

A second dimension to spectral classification was added to the Harvard scheme by astronomers at Yerkes Observatory in the early 1940s; it subclassified stars of similar surface temperatures into *luminosity classes*. The scheme is based on subtle but observable differences in the strength of certain absorption lines in a star's spectrum. The luminosity classes are designated by Roman numerals: I is for very luminous *supergiants*; II, for *bright giants*; III, for *giants*; IV, for *subgiants*; and V, for *main-sequence* or *dwarf stars*. You can see in Figure 12.5 that, although the spectra of the A0 supergiant, the A0 giant, and the A0 main-sequence stars are much alike (and similarly for the B2

FINDING A STAR'S RADIUS

Recall that the Stefan-Boltzmann radiation law (see Section 4.2) states that the radiation emitted from each square centimeter of the surface in 1 second of time is proportional to the fourth power of the temperature, or

$$F = \sigma T^4 = (5.67 \times 10^{-5})T^4,$$

where F is the surface emission in ergs per square centimeter per second. The luminosity L is equal to the surface area of the star multiplied by the unit surface emission; that is, $L = 4\pi R^2 F$, where R is the star's radius. Hence $L \propto R^2 T^4$, and $R \propto \sqrt{L/T^4}$. The angular diameters of some of the largest stars have been measured directly with either an optical or an intensity interferometer and by lunar occultation studies. Direct measurement is difficult at best for even large stars and is thus of limited applicability.

We can express the star's radius in terms of the sun's, eliminating the constant of proportionality:

$$\frac{R_*}{R_\odot} = \sqrt{\frac{L_*}{L_\odot}} \left(\frac{T_\odot}{T_*}\right)^2,$$

where $*$ is the value for the star and $\odot$ is that for the sun.

Example: Barnard's star (second closest to the sun) has a bolometric absolute magnitude of +10.3, compared with the sun's +4.8. From the magnitude difference of 5.6 between the two bodies we find (from Table 12.1) that the luminosity of Barnard's star is 0.00044 that of the sun. The temperature of Barnard's star is $T_* = 3200$ K; that of the sun is $T_\odot = 5800$ K. Then $L_*/L_\odot = 0.00044$ and $T_\odot/T_* = 2$. Substituting the numerical values in the formula, we get

$$\frac{R_*}{R_\odot} = \left(\frac{5800}{3200}\right)^2 \sqrt{0.00044} = (3.3)(0.021) = 0.07.$$

The diameter of Barnard's star is less than one-tenth that of the sun.

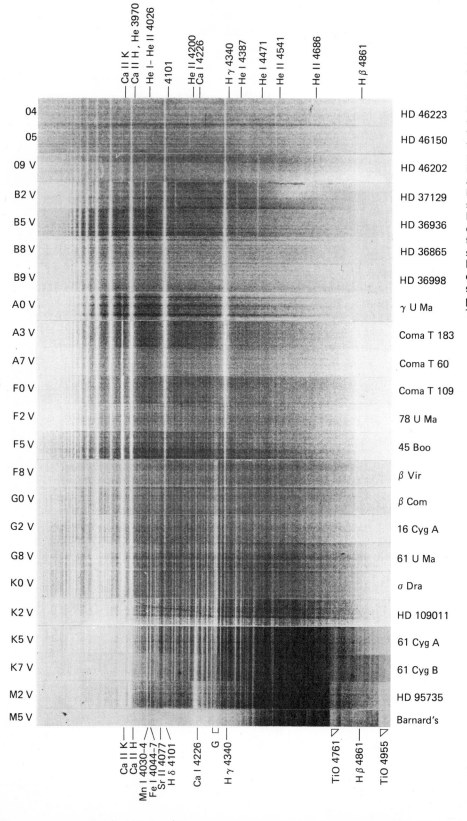

FIGURE 12.4
Stellar spectral sequence. Shown as negative prints rather than positive, the absorpiton lines (that is, the absence of light) are white, while the presence of light (the continuum) is dark. The spectral sequence shows examples of each spectral class and is designated at the left. Prominent absorption lines of several elements and their wavelengths in angstroms are identified at the top and bottom. Note, for example, how the strength of the hydrogen lines is greatest in spectral class A and then diminishes towards hotter and cooler stars. The rise and fall of line intensities of several elements throughout the spectral sequence is displayed graphically in Figure 12.8.

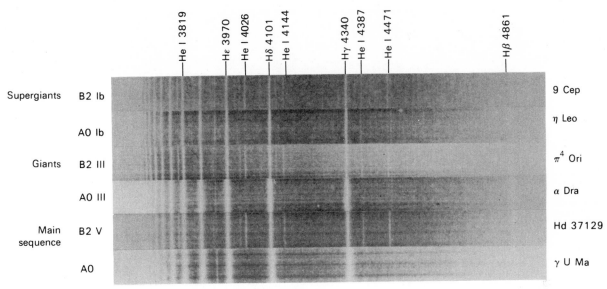

FIGURE 12.5
Luminosity effects shown as a negative print rather than a positive. Note the difference in the appearance of the Balmer series of hydrogen in the spectrum of a supergiant, giant, and main-sequence star. The three B-type stars are of the same spectral type since the neutral helium lines have the same intensity in the spectrum of all three stars. The three A stars are the same spectral type since the intensity of the CA II K and Mg II 4481 lines are about the same in the three stars.

stars), differences are visible in the lines of the Balmer series. The differences in the Balmer lines are not so pronounced in cooler stars as they are in the hot stars.

Luminosity differences in stellar spectra are more subtle effects than temperature differences are. Thus with low-quality spectrograms it is not always possible to determine a star's luminosity class even though its spectral type can be roughly estimated. Once a star's luminosity class is known, though, we can find its distance from its apparent and absolute magnitudes, that is, its distance modulus $(m - M)$.

The names for each of the luminosity classes are more than poetic. As we have already seen, two stars of the same spectral type (or temperature) must differ in radius if they are to have different luminosities. The more luminous the star, the larger will be its radius. Supergiants are the largest stars, the bright giants next largest, and so forth. The significance of the name *main sequence* is that approximately 90 percent of all stars are dwarf or main-sequence stars.

In addition to stars of the same spectral type that are brighter than main-sequence stars, there are stars that are less luminous than main-sequence stars. These are the *subdwarfs* and the *white dwarfs*. We shall not discuss the subdwarfs again, but we shall have a great deal to say later about the white dwarfs, which are fascinating members of the world of stars.

12.4
Messages in the Spectrum of a Star

The differences observed in stellar spectra stem from the manner in which atoms of various elements absorb and reradiate energy under different temperature and density conditions in stellar atmospheres. By far the most important of these variables affecting stellar spectra is temperature, the primary factor in Figure 12.6, which explains why the strengths of the absorption lines of the neutral and ionized atoms vary as they do from one spectral type to another.

TABLE 12.5
Spectral Classes of Stars

Spectral Class	Intrinsic Color	B–V Color Index[a]	Surface Temperature[a] (K)	Prominent Absorption Lines
O	Blue	−0.32	41,000	Ionized helium; multiply ionized oxygen, nitrogen, silicon; neutral helium; hydrogen weak
B	Blue	−0.22	30,000	Neutral helium strongest at B2; ionized helium weak or absent; hydrogen stronger; ionized oxygen, carbon, nitrogen, silicon
A	Blue white	0.00	9,500	Hydrogen strongest at A0; ionized calcium, magnesium, iron, and titanium strong
F	White	+0.33	7,240	Hydrogen weakening; more ionized and neutral metals
G	Yellow white	+0.60	5,920	Hydrogen weaker; ionized calcium very strong; other ionized metals weaker; neutral metals stronger; CH present
K	Orange	+0.81	5,300	Neutral atoms very strong; ionized calcium still strong; few ionized metals; hydrogen still weaker; CN usually strong
M	Red	+1.41	3,850	Strong neutral atoms; hydrogen very weak; titanium oxide bands prominent

[a] For the hottest spectral type in the class, such as A0 in A, with the exception of spectral class O, where it is O4, the hottest known O-type stars.

Atomic processes, large-scale motions, and chemical composition also affect stellar spectra, but they do it in a much more subtle fashion than temperature does.

Astronomers can learn a surprising amount about a star's physical and chemical properties from the shape, width, and strength of absorption lines in its spectrum. You can see how the line intensity varies across a part of a typical spectrogram in Figure 12.6b. The absorption lines are the dips below the continuous background spectrum, where the tiny wiggles are made by the photographic emulsion's graininess.

Absorption lines are broadened by several atomic processes. As an example of these, the absorption lines in the spectra of supergiants and bright giants are narrow because of a lower density in their photospheres, with consequent fewer collisions between atomic particles, than that in the photospheres of main-sequence stars. And large-scale magnetic fields in a few stars are found to broaden absorption lines by a Zeeman effect of several thousand gauss compared with a few gauss for the sun. The Zeeman broadening of stellar absorption lines is illustrated in Figure 12.7a.

ATOMIC PROCESSES IN STELLAR ATMOSPHERES

In actual practice it is not easy to sort out all the processes that affect the shape of an absorption line.

LARGE-SCALE MOTIONS IN THE OUTER LAYERS

Large-scale motions in the outer layers of a star—such as random thermal motions, streams of gas analogous

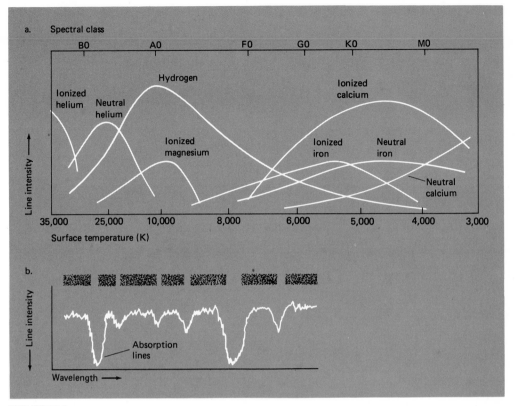

FIGURE 12.6

(a) Relative strengths of absorption lines of various elements as a function of spectral class. The principal factor responsible for the variation in line intensities is the star's surface temperature. It determines the relative number of atoms of a particular element having an outermost electron excited to a particular energy level and hence the strength of the absorption lines. Higher temperature also favors a higher degree of ionization. Corresponding spectral classes are shown on the horizontal scale above. The behavior shown here is described in Table 12.5. (b) The line intensity of several absorption lines in the spectrum of one star. The spectrogram (shown schematically) is scanned with a narrow beam of light and recorded on a strip of moving paper. More light passes through where the absorption line is than elsewhere, so the recording pen shows a larger dip where the line is. The lower the dip, the stronger the absorption line.

to solar granules and supergranules, and rotation of the star—all broaden absorption lines through the Doppler effect. Rates of rotation for the O and B stars are several hundred kilometers per second compared with the few kilometers per second for stars like the sun. This difference is shown schematically in the widths of the absorption lines in Figure 12.7b.

CHEMICAL COMPOSITION OF STARS
Analyzing absorption lines to determine the chemical composition of a star's atmosphere is an important field of research in astronomy. The intensity of an absorption line depends on the fraction of neutral or ionized atoms of an element capable of absorbing that particular wavelength of radiation. This fraction depends on the star's temperature, the atmospheric-gas density, and the abundance of the element. The as-

tronomer generally knows something about the first two factors from the star's luminosity and spectral type; the uncertain factor is the abundance of the element. From judicious estimates for the abundance, temperature, and density the astronomer can calculate theoretical absorption-line strengths that agree with the observed line intensities (Figure 12.6), element by element. By so doing, he determines abundances for the elements in the star's photosphere and refines the value of its atmospheric temperature and density.

Abundance studies for several hundred stars give values much like those for the sun (see Table 11.2). But some exceptions appear in the abundances both of all elements heavier than hydrogen and helium and of certain elements in some stars (Figure 12.8). The heavy-element abundances appear to vary from 0.01 percent to about 4 percent by mass among all stars.

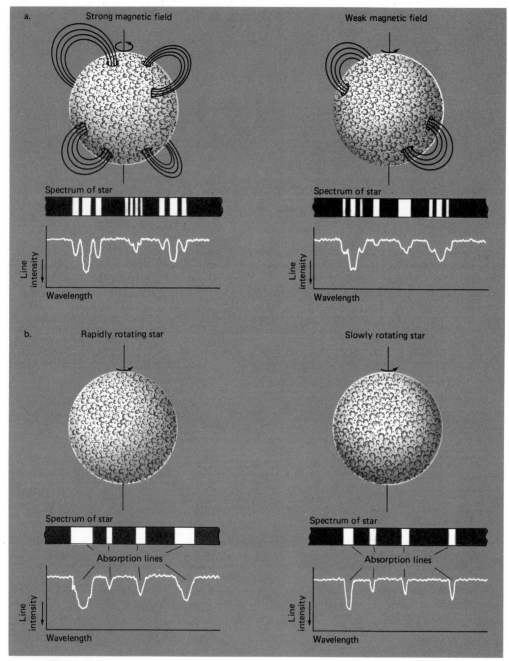

FIGURE 12.7
Messages in stellar spectra, examples of what astronomers can learn from the spectra of stars. (a) The Zeeman effect produced by magnetic fields on the surface of the star. (b) The Doppler broadening of absorption lines caused by the rotation of the star.

ACTIVITY IN OUTER LAYERS OF STARS

In recent years astronomers have made strong arguments for the hypothesis that the hydrogen convection zone underneath the sun's photosphere is re-sponsible for its differential rotation. In turn the differential rotation produces the 22-year magnetic cycle and all the associated activity seen in the photosphere, chromosphere, and corona. Evidence for the

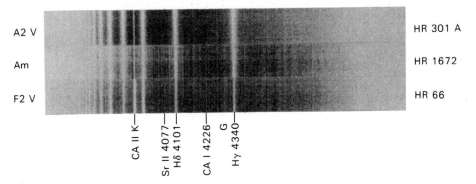

A2 V

Am

F2 V

HR 301 A

HR 1672

HR 66

CA II K
Sr II 4077
Hδ 4101
CA I 4226
G
Hγ 4340

FIGURE 12.8
Suspected chemical-composition effects in the
spectrum of a star. The star labeled Am (on the
left) has a Ca II K line (shown as a negative print)
whose strength is similar to that of the A2 V star
above it. However, the weak absorption lines due
to a variety of heavy elements are more nearly like
those of the F2 V star below it. The spectra of a
number of stars display conflicting information,
which most astronomers believe to be due to
chemical-composition effects.

existence of such properties in other stars would be to
find activity similar to that of the sun in their outer
layers.

Spectroscopic evidence has been found for chromospheres in all main-sequence stars from spectral
class F stars to the coolest red dwarfs of spectral class
M. Giant stars in this same range (F through M) also
show the spectroscopic characteristics that indicate
the presence of a chromosphere. Among the supergiants, however, chromospheres appear to be present
primarily in those of spectral class G and redder. For
the stars of earlier spectral classes (O, B, and A) no
such spectroscopic evidence for chromospheres has
been found. There is dramatic evidence for hot coronae in the early-type stars as well as in spectral classes
F through M. The evidence comes from X-ray emission
detected by the *Einstein Orbiting Observatory.*

The stars whose luminosity includes significant
amounts of energy emitted in the X-ray region of the
spectrum cover a range of spectral type and absolute
magnitude. They occur among all the main sequence
spectral classes and many of the red giants and include many of the blue and red supergiants. For example, about 30 stars in the Hyades open cluster in Taurus (see Section 12.7) have X-ray emission. Other
examples are the red supergiants Betelgeuse (M2 I)
and Antares (M1 I) and the bright giant Canopus (F0
II), which have X-ray emission 100 to 1000 times the
solar luminosity. Final examples are the three main-

sequence A stars Sirius (both the A star and the white
dwarf companion), Vega, and Altair.

From spectral classes O to A the X-ray emission
seems to decrease, while there is no recognizable systematic behavior for spectral classes G to M (class F
appears to be a transition group).

ACTIVITY CYCLES IN STARS
Among stars only the sun is close enough for astronomers to observe activity on its surface directly. The
technique used to detect stellar activity cycles depends on estimating the fraction of a star's observable
disk covered by chromospheric plages. When the sun
is more active, a greater fraction of its face is covered
by plages (about 20 percent at maximum activity) and
a smaller fraction when less active (0 percent at minimum activity).

Measurements for stellar activity have been made
at the Mount Wilson Observatory for about 400 main-
sequence stars within 80 light years of the sun, as
shown in Figure 12.9. In the study 91 stars were followed in time, some as long as 14 years, to see
whether there is a cycle of activity similar to the
11-year cycle of sunspot activity. A sizable fraction of
the stars studied show a cyclic behavior analogous to
that of the sun. The time periods, where fully measured, vary from about 7 years to almost twice that
period. Hence the sun does not appear to be unusual
with its 11-year cycle of activity. Although these results
are far from the final answer, they suggest that rotation, convection, and magnetic fields occur in the
outer layers of at least the main-sequence stars of
spectral classes F through M.

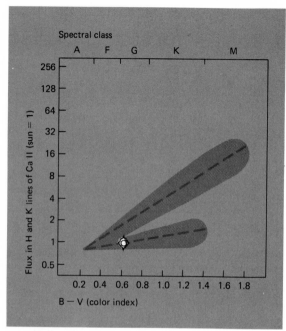

FIGURE 12.9
Active regions on stars as measured by the residual radiation in the H and K lines of singly ionized calcium coming from chromospheric plages. The two teardrop-shaped regions are based on measurements for about 400 main-sequence stars that are within about 80 light years of the sun. The dashed lines are rough averages that seem to suggest two almost distinct groups. Evidence suggests that the upper group consists of younger stars than does the bottom group; so surface activity apparently declines as a star ages.

12.5
Variable Stars

Astronomers recognize several tens of thousands of stars that vary in brightness. To analyze the variability, astronomers usually employ photometric observations, spectrum analysis, and Doppler measurements. From the photometric observations they can plot the observed change in apparent magnitude over a specific interval of time, which is called a *light curve*; a plot of the Doppler line shift with time is known as a *velocity curve* (Figure 12.10).

Many of the variable stars belong to the group of *pulsating variables*, which we believe owe their variability to a regular expansion and contraction (Table 12.6). For this reason most of the supergiant stars are

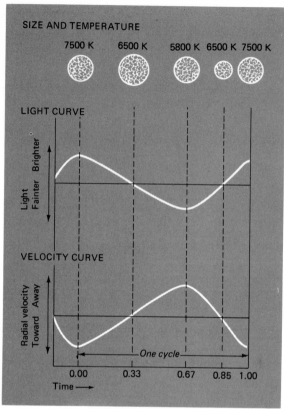

FIGURE 12.10
Typical Cepheid variations during one light cycle. Stellar diameter changes are exaggerated in this drawing.

variable to some extent. In fact within a range of surface temperature from about 7000 to 3500 K there are no known supergiants that are not variable in brightness to some degree.

The simple theory of stellar pulsation predicts that the period of oscillation is inversely proportional to the square root of a star's mean density. A star whose mean density is four times greater than another will pulsate in half the time. Thus stars of the lowest mean density have the longest periods, and these are also the largest and brightest stars—the red supergiants. This dependence of the period on mean density is borne out by observational data for Cepheids.

CEPHEID VARIABLES

The *classical Cepheids*, numbering about 700, form one important subdivision of the pulsating variables. Named after the first star of their kind discovered, δ Cephei, they have the following characteristics:

TABLE 12.6
Pulsating Variable Stars

Variable Name	Spectral Range	Luminosity Class	Period Range (days)	Median Absolute Magnitude	Maximum Range in Brightness (magnitudes)
Cepheids:					
Classical Cepheids	F to G	I	2 to 40	−1.5 to −5	Up to 2
W Virginis Cepheids	F to G	I	10 to 30	0.0 to −3.5	Up to 2
RR Lyrae variables	A to F	III	≤1	+1 to 0.0	≤2
δ Scuti stars	F	IV	≤1	+2 to 0.0	<0.25
β Canis Majoris or β Cephei stars	B	III	0.1 to 0.3	−2 to −4	0.1
Long-period variables:					
RV Tauri	G to K	II	30 to 150	−2 to −3	Up to 3
Long-period Mira-type variables	M	III	80 to 600	+2 to −2	>2.5
Semiregular	G to M	III, I	30 to 2000	0 to −3	Up to 2
Irregular	K to M	III	Irregular	<0	Up to several

They are yellow supergiants of great brilliance and spectral types late F to early K; their brightness varies periodically over about one magnitude during an interval of 2 to 40 days; their spectra show Doppler shifts synchronized with the periodic changes in brightness (see Figure 12.10).

The classical Cepheids are the most luminous of the Cepheids. Sparsely sprinkled within the Galaxy's disk, they are a very small segment of the population of the spiral arms. Another group of Cepheids, somewhat less luminous than classical Cepheids with periods generally between 10 and 30 days, is the *W Virginis stars* (named after their prototype). They are found in the population of stars in the halo portion of the Galaxy and in several globular star clusters. Still another type of pulsating variable, present by the thousands in the Galaxy's central and halo regions and in globular clusters, constitutes a third class of Cepheids. They are known as *cluster variables,* or more often as *RR Lyrae variables* (also named for their prototype). These are bluish-white giant stars fainter than the other groups of Cepheids but up to a hundred times brighter than the sun. Their periods of brightness variation average about a half day.

The more luminous the Cepheid, the longer the period of variation. This extremely important correlation between period and intrinsic brightness is known as the *period-luminosity relation.* It is repre-

sented graphically by a plot of the median absolute magnitude against period (Figure 12.11). (The median magnitude is halfway between the maximum and minimum magnitudes.) Calibrating the absolute-

FIGURE 12.11
Period-luminosity relations for Cepheids. The very important relationship between absolute magnitude and period helps us to determine the distance of individual Cepheids or of stellar systems that contain Cepheids. Basically, the longer is the period of light variation, the intrinsically brighter is the Cepheid.

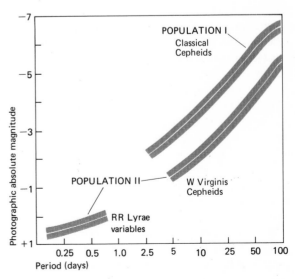

magnitude scale is difficult because no Cepheid is close enough to have its parallax measured; astronomers have had to resort to less direct methods to set the zero point of the absolute magnitude scale.

LONG-PERIOD AND IRREGULAR VARIABLES

Nearly 4000 luminous red giants are observed to have cyclic brightness changes of several magnitudes with a period of about a year. Known as the *long-period variables*, these form another large group of the pulsating variable stars. Variations in the spectra and velocity curves of the long-period variables are complex, with a poorly understood cause that is apparently not identical with what makes Cepheids pulsate.

Hundreds of other variable stars, covering a wide range of luminosities and colors, go through such irregular and often baffling changes that the cause of their behavior remains obscure. More will be said about the reasons for a star's varying brightness in Chapters 14 and 15. But the important points to remember are that a small percentage of all stars vary in brightness and that their variability attracts attention to them and to their location in the Galaxy.

12.6
Stars in Binary Systems

Half or more of the stars may be stars that are in orbit about another star. They are thus stars whose fates are permanently linked by gravity. Among the sun's neighbors out to 15 light years at least half are in multiple systems (see Table 12.1). Most of the multiple-star systems are *double*, or *binary*, *stars*, whose components may be separated by a large fraction of a light year or almost touching. In a binary the stars orbit in elliptical orbits around the system's center of mass. The more massive component (the primary) has the smaller orbit; the relative size of each star's orbit is inversely proportional to its mass, as shown in Figure 12.12. One can visualize the system's relative motions by imagining how a dumbbell with two unequal spheres would move if rotated around its center of balance, which is its center of mass.

Binary systems are classified according to the means of detection, which in turn depends on the separation between the components and the distance

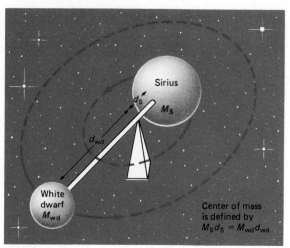

FIGURE 12.12
Finding the masses of binary stars from Newton's modified version of Kepler's third law (see page 53): $M_1 + M_2 = 4\pi^2 a^3/GP^2$. If we measure the masses M_1 and M_2 in units of the sun's mass, the semimajor axis a in astronomical units, and the orbital period P in years, the relation becomes: $M_1 + M_2 = a^3/P^2$. For the bright star Sirius and its white dwarf companion, $a = 20$ AU and $P = 50$ years so that $M_S + M_{wd} = 20^3/50^2 = 3.2 M_\odot$. From the definition of the center of mass it is observed that $d_{wd}/d_S = 2.3$ so that $M_S = 2.3 M_{wd}$. Combining these two results, we find that $M_S \simeq 2.2 M_\odot$ and $M_{wd} \simeq 1.0 M_\odot$.

of the system from the earth. In practice such considerations lead to three classes known as visual, spectroscopic, and eclipsing binaries.

VISUAL BINARIES

Double-star systems that have separations large enough for us to see both companions are called *visual binaries*. In these systems the secondary's motion around the primary is often obvious. The observed path is a projection of the true orbit on the plane of the sky. Yearly measurements of the apparent separations and motions of the two stars around each other must be made and the system's distance from the earth must be known to calculate their individual orbits. These vary in size from a few astronomical units to thousands of astronomical units, with periods of revolution from several years to many thousands of years. An example is shown in Figure 12.13 for the visual binary Xi Boötis.

Astrometric binaries form a subgroup of the visual binaries and are defined by the fact that with today's

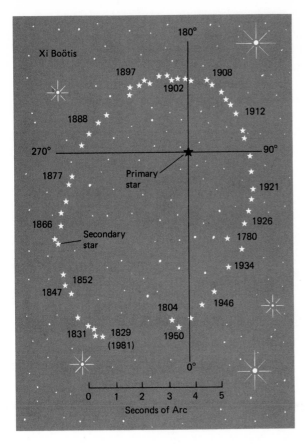

technology the secondary star in the system cannot be seen because it is much fainter than the primary is. What we observe are tiny periodic wiggles in the visible star's apparent motion across the celestial sphere (called *proper motion*); from these we infer that an invisible companion exists. The variation arises from the visible star's orbital motion around the center of mass of the system combined with the system's motion relative to the sun, as pictured in Figure 12.14. As the visible star (in white) orbits around the center of mass, which moves, let us say, upward toward the right-hand side of the page (its proper motion), it follows a wavy path across the sky. In a few astrometric binaries the invisible companion seems to have a planetlike mass.

◀ **FIGURE 12.13**
Orbit plotted for a visual binary. Successive positions are shown for the faint secondary star in the visual binary Xi Boötis as it orbited the primary from 1780 to 1950. The apparent elliptical orbit is actually the projection of the true elliptical orbit on the plane of the sky (which is tangent to the celestial sphere).

▼ **FIGURE 12.14**
Path of astrometric binary across the celestial sphere.

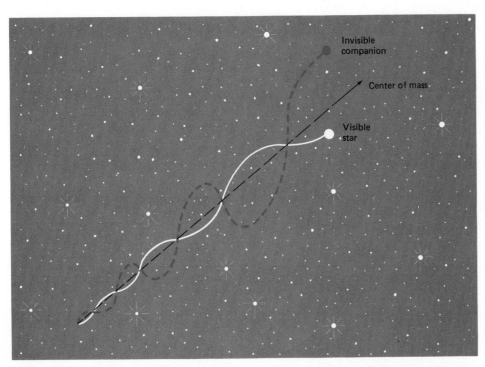

SPECTROSCOPIC BINARIES

Systems whose components cannot be resolved visually may nevertheless be shown to be binaries by periodic Doppler shifts of the spectral lines in the composite spectrum produced by one or both stars. These systems are the *spectroscopic binaries,* with orbits varying from a fraction of an astronomical unit in size to several tens of astronomical units. The corresponding orbital periods range from hours to several years. When the absorption lines of both stars of such systems are visible in the composite spectrum, two sets of absorption lines shift periodically back and forth relative to each other. The maximum shift occurs when the two stars are perpendicular to the line of

sight, and there is no Doppler shift (the two sets of absorption lines merge) when the two stars are moving across the line of sight a quarter of a revolution later.

Most often only the brighter component's spectrum is discernible on the spectrogram, and its absorption lines shift periodically back and forth. We can determine whether or not the system is a binary system by analyzing the velocity curve, that is, a plot of the change in radial velocity during the spectroscopic binary's period of revolution (Figure 12.15).

ECLIPSING BINARIES

Eclipses of closely paired stars whose orbits are seen more or less edgewise can occur as one star passes in front of the other, producing a decrease in the brightness of the system. These double-star systems are called *eclipsing binaries.* First one star and then the other passes periodically between its companion and

FIGURE 12.15
Spectroscopic binary system in which the spectral lines of both components appear. The black component is about 25 percent more massive than its companion. Thus both stars will be of approximately the same spectral type.

ELLIPTICAL ORBITS

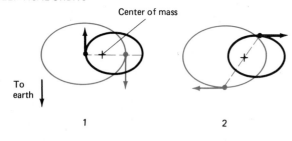
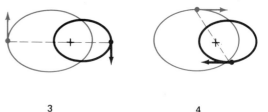

DOPPLER LINE SHIFTS IN THE COMPOSITE SPECTRUM

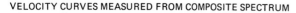

VELOCITY CURVES MEASURED FROM COMPOSITE SPECTRUM

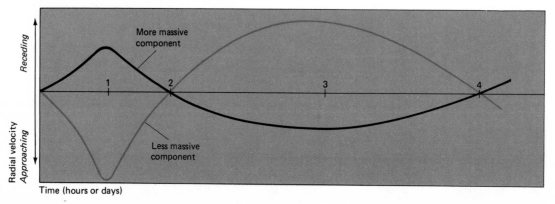

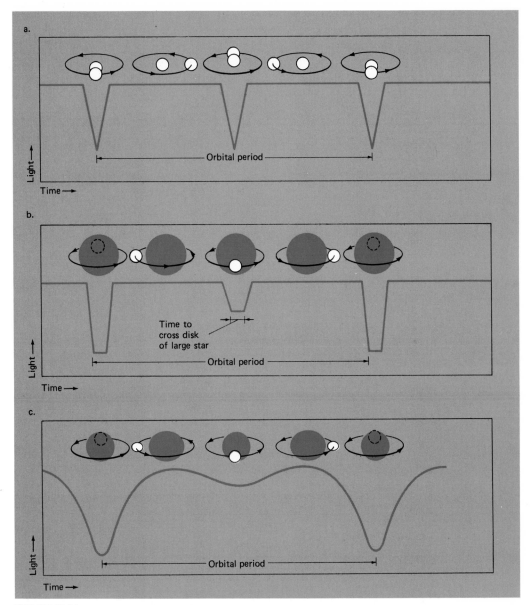

FIGURE 12.16
Representative eclipsing binary systems: (a) partial eclipses; (b) alternating total and annular eclipses; (c) total and annular eclipses of tidally distorted companions.

us, temporarily cutting off all or part of the eclipsed star's light (Figure 12.16).

When the two eclipsing stars are equal in brightness and size (Figure 12.16a) the minima in the light curve are V-shaped and of equal depth, whether the eclipses are partial or total. The more nearly total the eclipse is, the deeper are the minima. In Figure 12.16b, if most of the light comes from the smaller, brighter component, we see the deeper minimum when the larger, fainter star totally eclipses the smaller one. Halfway around the orbit the eclipse is annular and the minimum is much shallower when the brighter star passes in front of the fainter one. Both minima are flat-bottomed since that represents the time it takes the small star to cross the disk of the large star. In Figure 12.20c the stars are tidally distorted into an ellipsoidal shape. The depths of the two minima differ, as in Figure 12.16b, because the two components are unequal in luminosity and size.

By analyzing the light curve's general shape and measuring how long the eclipses diminish the light and by how much, we can derive a model of the system. If we also observe the eclipsing binary as a spectroscopic binary, we can combine data from the light curve and the velocity curve to determine such properties of the double stars as radii, masses, densities, temperatures, and true orbital sizes.

HOW MANY STARS ARE BINARIES?

Observational evidence supports the contention that at least half of all stars are members of binary systems. We must qualify that remark by saying that astronomers are not able to sample at will stars across the Galaxy or in neighboring galaxies; our statement extends results found for nearby stars to a generality about all stars. Also, if the two components of a binary system are of approximately the same mass, their identification as a binary is fairly easy; but the greater the difference in mass, the more difficult the binary identification becomes. As an illustration, the sun does not appear to have a stellar companion although we cannot entirely rule out the possibility that a companion star of very small mass (and consequently faint luminosity) exists far beyond the orbit of Pluto.

In the case of the sun does the existence of a planetary system take the place of a stellar companion? This is a very difficult question to answer. A study at Kitt Peak National Observatory of 123 nearby stars that are similar in mass to the sun found that 57 definitely have one companion, 11 have two companions, and 3 have three companions. Thus 58 percent of those studied definitely have companions. Using a variety of arguments, the astronomers doing the study concluded that in the 123-star sample probably 67 percent have normal stellar companions, 15 percent black dwarf companions (a nonluminous stellar companion), and 20 percent planets. Or in other words there is at least inferential evidence that sunlike stars have some kind of companion. From this conclusion it is probably going too far to infer that *all* stars occur in some kind of companion relationship.

MASS-LUMINOSITY RELATION

We cannot directly obtain one very important quantity, the stellar mass, from isolated single stars be-

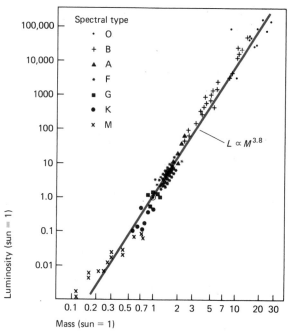

FIGURE 12.17
Mass-luminosity relation for main-sequence, or luminosity class V, stars. The plot shows that, the more massive the star, the greater its luminosity, or intrinsic brightness. Not all stars conform to this relation, such as the extremely luminous stars of luminosity classes I, II, and III or the faint white dwarf stars.

cause their gravitational effects on other stars are insignificant. When stars are close enough, as double stars are, they orbit each other because of their mutual gravitational attraction, and as a result we can find the masses of these stars relative to the sun's mass as shown in Figure 12.10.

In this way an important correlation, the *mass-luminosity relation*, has been found to exist between the mass and luminosity of main-sequence stars (Figure 12.17). The relationship has also been derived theoretically from the fundamental laws governing matter and radiation in stellar interiors, to be discussed in Chapter 13. Notice in Figure 12.17 that a star's luminosity is approximately proportional to the fourth power of the mass for stars brighter than the sun and to less than that for stars fainter than the sun. A reasonable compromise is that the luminosity is proportional to the 3.8 power of the mass. For example a star of two solar masses is 14 times more luminous than the sun.

The correspondence between mass and luminosity

in Figure 12.17 does not apply to the stars that are not main-sequence stars, which suggests that their internal structures differ from those of main-sequence stars. The very luminous supergiants, bright giants, and red giants lie somewhat above the curve and the white dwarfs considerably below it.

Although luminosities for main-sequence stars vary greatly—from a few ten-thousandths of the sun's brightness to almost 100,000 times the sun's brightness—the range of masses is very modest, going from several hundredths of the sun's mass to about 50 times it.

MULTIPLE-STAR SYSTEMS

Some gravitationally linked systems have more than two stars. For example there may be a distant third star revolving around a close pair or possibly two close pairs (usually short-period spectroscopic binaries) revolving around each other in a longer period. Other combinations might have more than four stars. Take, for example, the second-magnitude star Castor in the constellation Gemini: With a small telescope one can easily pick out two visual components, which have a period of revolution of about 400 years. But when we observe the stars' spectra, we find that the primary is itself a spectroscopic binary, with a period of 9.2 days (Figure 12.18), and the secondary is also a spectroscopic binary, with a period of 2.9 days. A ninth-magnitude star called Castor C, which is only 1.2' of arc away from the visual pair and takes many thousands of years to orbit them, is an eclipsing binary, with a period of 0.8 day. Thus all three classes of binary are represented in Castor's sextuple system.

12.7
Stars in Clusters

The existence of stars in groups seems to be a rather common occurrence. As astronomers have searched the solar neighborhood, they have found, in addition to the binary and multiple-star systems, larger aggregates of stars, called *open clusters*. Far across the Galaxy we can see even larger groups of stars dotting the Galactic landscape, the *globular clusters*. Star clusters exist because their member stars are close enough to each other to be bound into a physical group by the stars' mutual gravitational attraction.

Whereas typical separations between the stars in the solar neighborhood that are not members of a star cluster (known as *field stars*) are about 5 or 6 light years, the typical separation for the members of an open cluster is probably 2 to 3 light years, and for a globular cluster it is a few tenths of a light year. Another way of characterizing the separation between stars is to state it in terms of star size: Using the solar radius as a typical star size, we find that field stars are separated by several tens of millions of solar radii, the stars of an open cluster by a few tens of millions, and the stars of a globular cluster by a few million. Even though cluster members are not much closer to each other compared with field stars, their gravitational attraction is capable of holding a cluster together for up to tens of billions of years in the case of globular clusters and up to a few billion for open clusters.

A number of clusters have proper names, such as

FIGURE 12.18
Spectrum of the single-line spectroscopic binary α^1 Geminorum. The spectrum, photographed at different times, shows the change in Doppler shift that results from orbital motion. α^1 Geminorum is the brighter visual companion of the double star Castor in the constellation of Gemini. The bright-line comparison spectrum is that of titanium.

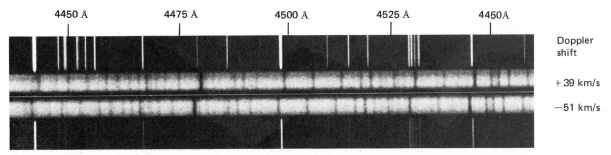

4450 Å	4475 Å	4500 Å	4525 Å	4450Å

Doppler shift

+ 39 km/s

−51 km/s

the Hyades, the Pleiades, and Praesepe, in addition to various catalog designations, but most clusters are known by a number in some catalog. For example many of the entries in the Messier catalog (see inside front cover) are clusters, the Pleiades being M45 and Praesepe M44.

OPEN CLUSTERS

Close to the band of the Milky Way that runs around the sky, astronomers have identified some 1100 open clusters. They contain anywhere from a score of stars to many hundreds of stars distributed over volumes that are several tens of light years across. Open clusters have a somewhat loose appearance and only approximate a spherical shape. Information on some typical open clusters is given in Table 12.7, with a picture of an open cluster shown in Figure 12.19.

Several of the open clusters are easily visible to the naked eye. Two of the best known are both in the constellation of Taurus—the Hyades and the Pleiades (see Plate 19). To the eye the Hyades is a V-shaped group of stars marking the face of the bull. With tele-

scopes astronomers have found several hundred stars in the Hyades cluster, which is some 15 light years across. The Pleiades contains about the same number of stars and is approximately the same size as the Hyades. But because it is almost three times farther away than the Hyades cluster is, the Pleiades, looking like a tiny dipper of six stars in the shoulder of the bull, occupies a much smaller area on the sky than does the Hyades.

In the 1930s open clusters were instrumental in demonstrating that there was matter lying between the stars—the interstellar medium (see Chapter 14). Astronomers now know that interstellar matter, which is composed of gas and a fine dust, hides from our view most of the open clusters in the Galaxy. Undoubtedly there are many more than 1100 open clusters; estimates suggest that as many as 18,000 open clusters are buried in the Galaxy's spiral arms and disk.

For open clusters the stellar composition varies from cluster to cluster. There are some whose members are predominantly or exclusively main-sequence stars; the brightest in these clusters are brilliant, blue,

TABLE 12.7
Selected Open Clusters, O Associations, and Globular Clusters

Name	Distance (ly)	Diameter (ly)	Estimated Mass (sun = 1)	Estimated Age (yr)
Open clusters:				
Ursa Major	70	25	300	2×10^8
Hyades	140	15	300	6×10^8
Pleiades	430	15	350	5×10^7
NGC 752	1,200	15	—	1×10^9
M67	2,700	15	150	4×10^9
NGC 188	4,600	20	—	1×10^{10}
NGC 2362	4,900	10	—	5×10^6
M11	5,600	20	250	8×10^7
h Persei	7,300	50	1,000	1×10^7
χ Persei	7,800	50	900	1×10^7
O associations:				
I Orionis	1,600	—	3,000	—
IV Sagittarii	5,500	—	—	—
I Persei	6,200	—	180	—
Globular clusters:				
M4	9,100	60	60,000	—
ω Centauri	16,000	300	—	$\approx 2 \times 10^{10}$
M13	25,000	95	300,000	$\approx 1 \times 10^{10}$
M5	28,000	85	60,000	$\approx 1 \times 10^{10}$
M92	33,000	120	140,000	$\approx 1-2 \times 10^{10}$
M3	42,000	115	210,000	$\approx 1-2 \times 10^{10}$
M15	46,000	125	6,000,000	$\approx 1 \times 10^{10}$
NGC 7006	160,000	140	—	—

main-sequence, giant, or supergiant stars. Contrasting with these open clusters are those whose membership does not include any bright, blue stars but whose brightest stars are red giants and supergiants. In such clusters there is a total absence of the blue stars of spectral classes O, B, and A. What we have just described seems to be opposite ends of a reasonably continuous gradation of stellar composition in open clusters. That is, there are also open clusters with a mixed stellar composition, containing some bright, blue stars and some bright, red stars. In nearby open clusters faint, yellow and red dwarfs of the main-sequence luminosity class are also observable. And some even fainter white dwarf stars have been found in those nearby open clusters that contain bright, red stars. We shall have more to say about the stellar composition of open clusters in Chapter 13.

STARS IN ASSOCIATION

A Russian astronomer, V. A. Ambartsumian (1908–   ) was the first to note that far from the solar system there exists a considerable number of loose stellar groupings. These sparsely populated *stellar associations* contain highly luminous O and B stars (and no bright, red stars) and are mixed with interstellar gas and dust in the spiral arms that lie in the Galactic disk. They contain up to 100 or so stars and are up to several hundred light years in diameter. Estimates place the number of associations at a few hundred. Three representative O and B associations are listed in Table 12.7. To the naked eye one of them appears to be the middle star in the sword of Orion, but it is actually an association of O stars known as the Trapezium. These hot stars ionize the hydrogen gas surrounding the association and produce the emission nebula shown in Plate 15. This is the kind of Galactic environment in which many associations are found.

The individual stars are separating rapidly from each other because the association has too little mass to bind them permanently. Hence associations are highly unstable, with a maximum life expectancy of only a few million years before they completely disperse into the mainstream of stars in the Milky Way.

GLOBULAR CLUSTERS

The largest and most densely concentrated of all stellar groups are globular clusters. A photograph of the magnificent globular cluster that is just visible to the

FIGURE 12.19
Open star cluster, NGC 4755, in the constellation of the Southern Cross. It has been titled the "Jewel Box" because a bright red star (Kappa Crucis) is situated in the middle of a cluster of white stars.

FIGURE 12.20
Globular star cluster, M13, in Hercules. It is just visible to the naked eye.

naked eye in the constellation of Hercules appears in Figure 12.20. Astronomers have discovered a couple of hundred globular clusters surrounding the center of our Galaxy, somewhat like an immense spherical frame of reference in which the Galaxy rotates. Other clusters undoubtedly are hidden from us by dust in or near the Galactic plane. The largest globular cluster may have over 0.5 million stars packed into a diameter of 100 or so light years. The nearest clusters are about 7000 to 8000 light years from the sun, while the farthest are over 100,000 light years from us.

Red giants dominate photographs of globular clusters because they are more luminous, even though they may exist in fewer numbers, than main-sequence stars. RR Lyrae variable stars are also commonly found in globular clusters and are useful for determining the distance of the cluster. Because they vary in brightness in periods of less than 1 day, the RR Lyrae stars are readily identifiable. Also, all the RR Lyrae stars are of very nearly the same visual absolute magnitude (≈ 0.0). Thus they are good candidates on which to apply the inverse-square law of light so that we can determine a cluster's distance. For clusters too distant to resolve into individual stars astronomers can use (with precautions) angular diameters and integrated light to estimate distance. Totally unexpected and as yet unexplained was the detection by the *Orbiting Astronomical Observatory* (*OAO-2*) of a few faint but extremely hot blue stars in several clusters. Their light had not registered on earth-based photographs because our atmosphere absorbs ultraviolet light.

The photographs of globular clusters look crowded; but the individual stars, though more closely spaced than are those in the solar neighborhood (up to a thousand times, or several stars per cubic light year), are sufficiently far apart to escape collision with each other. They move in extremely elliptical orbits from one side through the center to the other side of the cluster. If the earth orbited a star in a globular cluster, the nearest stars would be light months from us instead of light years; and although there would be many more bright stars in the night sky than we see around the sun, the night sky would still not be close to being as bright as the daytime sky.

Having surveyed the stellar census, we shall wish to bring these seemingly disparate data for stars into a coherent picture. The picture that has been revealed through the work of many astronomers is that of the life story of stars: They are not eternal; they are born, live out their lives, and die as part of a changing universe. Chapters 13 to 15 will cover the life of stars.

SUMMARY

Distances of the stars. The basic technique for determining distances is trigonometric parallax. Distances are measured in light years and parsecs. The limit of measuring distance by parallax is 300 light years. Distances of more remote stars can be determined from the apparent brightness and the inverse-square law of light. The critical factor is to recognize the intrinsic brightness of the class of stars of which the subject star is a member.

Brightness of stars. The method of describing the relative brightness of stars is apparent magnitude, brightest stars being assigned to magnitude class 1 down to magnitude 6 for those stars just visible to the naked eye. The assumption is that equal ratios of brightness correspond to equal differences of magnitude. To find the intrinsic brightness of a star, we must know its distance and apparent magnitude. The magnitude of a star would have if it were 10 parsecs distant is called the absolute magnitude. Because brightness and magnitude depend on what part of the spectrum is being observed, color indexes can be used to determine a star's surface temperature. The bolometric magnitude is a measure of the star's luminosity, or the radiated energy per second in all wavelengths.

Spectra of stars. Stars have been classified on the basis of spectral appearance. Because most stars have similar chemical compositions, differences in spectra are due principally to differences in their surface temperatures. Stars of similar surface temperatures can be divided into luminosity classes based on the relative strengths of certain absorption lines in a star's spectrum. A star's luminosity depends on both its surface temperature and radius.

Messages in the spectrum of a star. Astronomers can learn a great deal from the shape, width, and strength of absorption lines in a star's spectrum: temperature, atomic processes, large-scale motions, and chemical composition. It is difficult, however, to sort out the factors influencing an absorption line. Atomic processes include the Zeeman effect produced by magnetic fields. Large-scale motions in a star's atmosphere include thermal motions, streams of gas, and rotation. Relative abundance of elements in a star's atmosphere is determined by studying the intensity of absorption lines in the star's spectra.

Variable stars. Many stars vary in brightness. Photometric observations allow astronomers to plot the light curve and the velocity curve, by which to describe the apparent changes in a star's brightness over time. A major constituent of the pulsating variables are the Cepheids: classical Cepheids, W Virginis stars, and cluster variables (or RR Lyrae variables). The period-luminosity relation for Cepheids can be used as a means of determining their distances.

Stars in binary systems. At least half the stars in the Galaxy are gravitationally bound in binary systems. Double-star systems that have separations large enough to be seen are called visual binaries. Astrometric binaries are ones in which only the bright primary star can be seen, while the faint secondary is not observable. Spectroscopic binaries may be resolved by spectroscopic study of variations in the combined absorption spectrum of the two stars. Eclipsing binaries are detected when one star and then another eclipse each other in a binary system. More than two stars may be related in some multiple star systems.

Stars in clusters. Large aggregates of stars—called open clusters—and even larger groups—called globular clusters—exist within the Galaxy. Open clusters may contain from scores to hundreds of stars. The stellar composition of open clusters varies widely. Stellar associations of highly luminous 0 and B stars lie in the Galaxy's spiral arms. The largest and most densely concentrated stellar groups are globular clusters whose brightest stars are red giants.

KEY TERMS

absolute magnitude	mass-luminosity relation
apparent magnitude	open cluster
bolometric magnitude	parallax
Cepheid variable stars	parsec
color index	spectral class
eclipsing binary	spectroscopic binary
giant stars	stellar association
globular cluster	subgiant stars
luminosity class	supergiant stars
main-sequence stars	visual binary

CLASS DISCUSSION

1. In their study of stars what kinds of data do astronomers consider vital to their understanding?

2. What is the relationship(s) among the following quantities, if any: apparent magnitude, luminosity, bolometric magnitude, absolute magnitude, visual apparent magnitude?

3. What is the relationship(s) among the following, if any: spectral type, temperature, color index, luminosity?

4. How do astronomers distinguish among dwarfs, subgiants, giants, bright giants, and supergiants? What is the distinguishing characteristic among these classes of stars? What type of classification scheme do these classes represent?

5. Why is the period-luminosity relation for Cepheids important for astronomers?

6. What kind of relation, if any, exists between the mass and luminosity of a star? If such a relation exists, are there any limitations to it, and can you guess why such a relationship might exist?

READING REVIEW

1. What is the limit of applicability in stellar parallax?

2. Does an apparent-magnitude measurement cover all wavelengths? If not, what does it cover?

3. How is absolute magnitude defined?

4. How do astronomers define color index and distance modulus, and do they differ from each other?

5. How many of the 30 brightest stars are also nearby stars? How are you defining "nearby"?

6. In which spectral class are the lines of the Balmer series the strongest?

7. What is the range in luminosities as compared with the range of masses of stars?

8. What are the popular names for luminosity classes III, I, V, II, and IV?

9. Do stars all have the same chemical composition? What is their composition generally? What is the range of difference, if any?

10 Are most of the stars in our Galaxy variable stars? If not, how many are variables?

11. What decides whether a binary system is a visual, astrometric, spectroscopic, or eclipsing binary?

12. Are the properties of associations, open clusters, and globular clusters similar or quite different? Explain.

PROBLEMS

1. The parallax measured for the star 61 Cygni is 0.291" of arc. What is the distance of this star in astronomical units, light years, and parsecs?

2. If the visual apparent magnitude of Sirius is −1.5 and Barnard's star is +9.5, what is the ratio in visual apparent brightness of Sirius to Barnard's star? (Hint: By the use of Table 12.2 magnitude differences can be added or subtracted, while the corresponding operation with brightness ratios is multiplication and division, respectively). What is the distance in parsecs and light years of the Andromeda galaxy for the apparent and absolute magnitudes given in Table 12.3?

3. If the radius of the sun is approximately 7×10^{10} centimeters and its surface temperature is approximately 6000 K, how much greater is its luminosity than that of the K5 V star ϵ Indi, whose radius is 3.5×10^{10} centimeters and whose surface temperature is close to 5000 K? What is Epsilon Indi's luminosity in units of the sun's luminosity?

SELECTED READINGS

Abt, H. A. "The Companions of Sunlike Stars," *Scientific American*, April 1977.

Cohen, H. L., and J. P. Oliver. "Star Colors," *Sky & Telescope*, February 1981.

Kaufmann, W. J. *Stars and Nebulae*. Freeman, 1978.

Percy, J. "Pulsating Stars," *Scientific American*, June 1975.

Wilson, O. C., A. H. Vaughn, and D. Mihalas. "The Activity Cycles of Stars," *Scientific American*, February 1981.

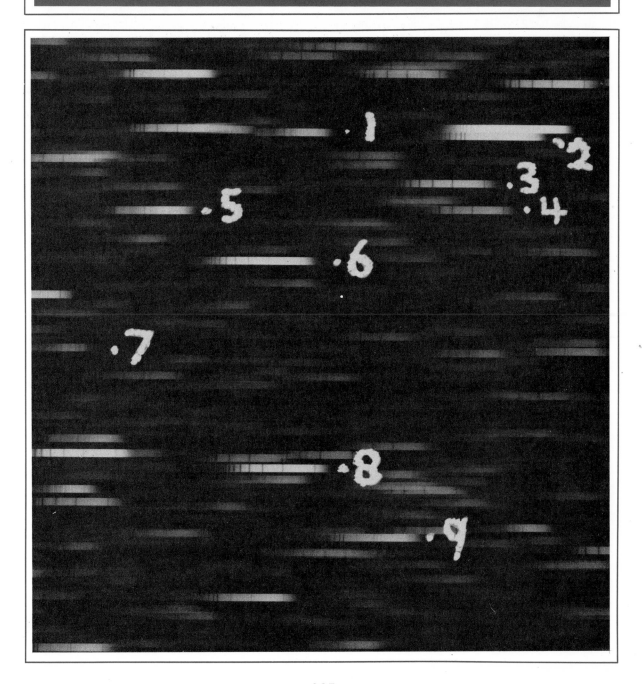

In Chapter 12 we presented some general physical characteristics of stars. As a prelude to this chapter the range of properties such as radius, mass, mean density, surface temperature, and luminosity for as wide a sample of stars as possible is reviewed in Table 13.1.

One characteristic that appears in Table 13.1 for which there has as yet been no discussion is the age of stars. Stars have definite ages that cover an immense range. In this chapter and Chapters 14 and 15 we shall consider such questions as how stars are created, what keeps them shining, and how they end their lives.

In brief a star evolves from a large cloud of gas and dust when it contracts under the force of gravity; unchecked, the gravitational force would squeeze the star down to almost nothing. Opposing forces, principally gas pressure, counterbalance this contraction. The pressure would be decreased by the loss of energy as a star radiates if it did not replace the energy drained by the luminosity through nuclear burning.[1] During the star's life, the ashes of one nuclear-burning phase provide the fuel for the next burning phase. Eventually the star exhausts its nuclear fuels

[1]By *nuclear burning* we mean those nuclear reactions in which low-mass nuclei are fused together at *very* high temperatures to produce heavy nuclei plus energy; we do not mean ordinary combustion, which involves atomic and molecular reactions.

TABLE 13.1
Range of Stellar Properties

Parameter	Approximate Limits[a]
Mass	10^{-2}–$10^{2}M_\odot$
Radius	10^{-2}–$10^{3}R_\odot$
Mean density	10^{-7}–$10^{7}\rho_\odot$
Luminosity	10^{-5}–$10^{5}L_\odot$
Surface temperature	10^{3}–10^{5}K
Heavy-element mass abundance	0.05–$2Z_\odot$
Age	10^{4}–10^{10} years

[a]Solar units are used: $M_\odot = 1.99 \times 10^{33}$ g; $R_\odot = 6.96 \times 10^{5}$ km; $\rho_\odot = 1.41$ g/cm³; $L_\odot = 3.83 \times 10^{33}$ erg/s; $Z_\odot = 0.02$, fraction of all elements heavier than hydrogen and helium. The rare neutron stars (Section 15.5) are not included.

◄ Examples of spectral classification. Representative spectrum-luminosity classes are indicated for nine stars in the southern hemisphere, photographed with the Curtis-Schmidt telescope of the University of Michigan. 1. K2 III; 2. F2 IV; 3. AO V; 4. A6 IV; 5. KO III; 6. B2 IV; 7. M1 (III); 8. B9 V; 9. F5 V.

and ends its life as a white dwarf, a neutron star, or possibly a black hole.

In its simplest form that is the life story of most stars. But before we can fill in details, we must be able to produce a coherent picture from stellar data. We do this by using the Hertzsprung-Russell diagram, an astronomer's most important tool for studying stars. It allows us to organize stellar data and to look for various relationships among stars.

13.1 The Hertzsprung-Russell Diagram

CORRELATING SPECTRAL CLASS AND LUMINOSITY
Between 1911 and 1913 the Danish astronomer Ejnar Hertzsprung and the American astronomer Henry Norris Russell independently developed the diagram that we now call the *Hertzsprung-Russell,* or *H-R, diagram.* Plotting the spectral type (or, equivalently, the color index or surface temperature) for many stars on the horizontal axis across the diagram against the absolute magnitude (or luminosity) on the vertical axis, they found that the resulting points were not scattered at random over the diagram. Instead, the points lie in well-defined regions, as illustrated in Figure 13.1, which suggests a continuous relationship among the stars of each region.

MAIN SEQUENCE, RED GIANTS, AND WHITE DWARFS
The most conspicuous region of the H-R diagram is the sequence of stars running from the bright, hot stars in the upper left-hand corner to the faint, cool stars in the lower right-hand corner. This sequence is called the *main sequence,* and it contains most of the stars that could be plotted on the diagram. The sun is a G2 main-sequence star and lies in roughly the middle of the H-R diagram among the yellow dwarfs. In Chapter 12 these stars were said to form one class in the luminosity classification scheme (luminosity class V). Clearly, the main-sequence stars vary from extremely luminous O stars to very faint M dwarfs—a range of about a billion in luminosity. Nevertheless, because of their number and the common features of their internal structure, the main-sequence stars are considered to belong to a single class.

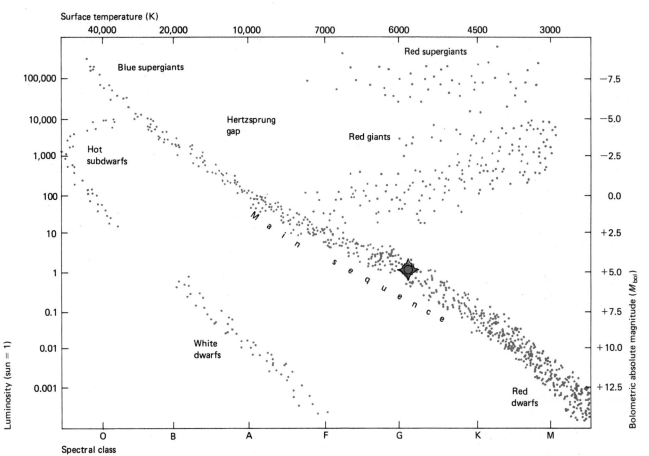

FIGURE 13.1
Schematic representation of the Hertzsprung-Russell (H-R) diagram. When we plot a star's temperature (or spectral class) against its luminosity, we find that the points for a large number of stars are not scattered at random but are confined to fairly well defined regions. The most prominent regions are these:

— *Blue supergiants:* bluest, most luminous, hottest; moderately large stars; low densities and large masses; rare. Example: Rigel.

— *Red supergiants:* orange to red in color; the largest stars and among the brightest; large masses and extremely low densities; few in number. Example: Betelgeuse.

— *Red giants:* actually yellow, orange, and red; considerably larger and brighter than the sun; average to larger-than-average masses and low densities; fairly scarce. Example: Arcturus.

— *Main-sequence stars:* blue, white, and yellow stars higher than the sun on the main sequence are somewhat larger, hotter, more massive, and less dense than the sun; plentiful in number. Example: Sirius. Orange and red stars below the sun on the main sequence are somewhat smaller,

cooler, fainter, less massive, and denser than the sun; very plentiful in number. Example: Epsilon Eridani. An important subgroup is the red dwarfs, which are the coolest and reddest stars on the lower end of the main sequence; considerably fainter and smaller than the sun; small masses and high densities; the most abundant stars. Example: Barnard's star.

— *Hot subdwarfs.* quiescent novas, central stars of planetary nebulae, and other subluminous hot blue stars; masses in the solar range and very high densities; fairly rare. Example: central star of the Ring Nebula in Lyra.

— *White dwarfs:* mostly white and yellow; extremely faint and tiny by solar standards; enormously high densities; terminal evolutionary development; quite plentiful. Example: the companion of Sirius.

HENRY NORRIS RUSSELL   (1877–1957)

Henry Norris Russell, director of the Princeton observatory, was one of America's most beloved and distinguished astronomers of this century. His grasp of all phases of astronomy was truly awesome.

In 1912 Russell developed a method of calculating the orbital elements of an eclipsing binary, a method that is still used. He showed how to derive the relative dimensions of the two components and their densities. Beginning in 1903, in collaboration with the British astronomer Hinks, of Cambridge, he undertook a program of deriving stellar parallaxes photographically. The work, which was completed in 1910, was instrumental in his discovery of a relationship between the absolute magnitudes and spectral types of the stars. A plot of this relationship showed the existence of two types of red star: one highly luminous, the other quite faint. By 1913 he had refined this correlation, now known as the Hertzsprung-Russell diagram, and was using the terms *giants* and *dwarfs* to distinguish between the two groups.

In the late 1920s Russell applied the newly developed quantum theory to determining the abundances of the elements in stars. From an analysis of the intensity profiles of the solar absorption lines, he derived the relative abundance of some 50 different elements in the solar atmosphere. He also applied this technique to a number of stars. The research revealed the very high abundance of hydrogen in the sun and stars, a result of great importance in our understanding of the role of hydrogen in astrophysical processes. Russell also made important contributions to the theory of stellar structure. He showed that the physical properties of a star can be found solely from its mass and chemical composition (the Vogt-Russell theorem).

The second most prominent region in the H-R diagram is the region broadly labeled as *red giants*. These are the luminous stars in spectral classes F, G, K, and M lying above the main sequence in a region that angles up toward the upper right-hand corner of the diagram. Despite their being in the same luminosity class (III), the red giants vary by at least a factor of 100 in luminosity. On the average they are 100 times more luminous than the sun, and they vary in surface temperature from 3000 to 7000 K.

The stars of luminosity classes I and II are called *red supergiants* if they lie on the cool side of the diagram in spectral classes G, K, and M, and *blue supergiants* if they are early-type stars of classes O and B. The red supergiants and the blue supergiants can be hundreds of thousands of times more luminous than our sun.

The last region of importance, which spans the spectral classes B, A, and F, contains the faint stars lying below the main sequence; these are called *white dwarfs*. (When we refer to a star as being on or off the main sequence, we refer to its position in the H-R diagram and not to its actual position in space). White dwarfs are typically a few thousandths of the luminosity of the sun even though their outer layers are hotter than those of the sun.

STELLAR RADIUS ON THE H-R DIAGRAM

Stars of similar spectral type or surface temperature can be vastly different in size (recall that their luminosity is proportional to their radius squared and the fourth power of their surface temperature). You can see in Tables 12.1 and 12.4 how dwarfs, giants, and supergiants differ in size. These figures show (as you would expect) that, because of its larger surface, a more luminous star radiates more energy than does a less luminous star of the same spectral type (temperature). Figure 13.2 is an H-R diagram for the bright stars listed in Table 12.4. On it are plotted lines along which all stars have the same radius. As is evident, the radii of stars increases from the lower left-hand corner toward the upper right-hand corner of the diagram.

Although there is no unique position on the H-R diagram for stars of a given mass, except as was discussed for the main-sequence stars in their mass-luminosity relation, the more massive stars are, very

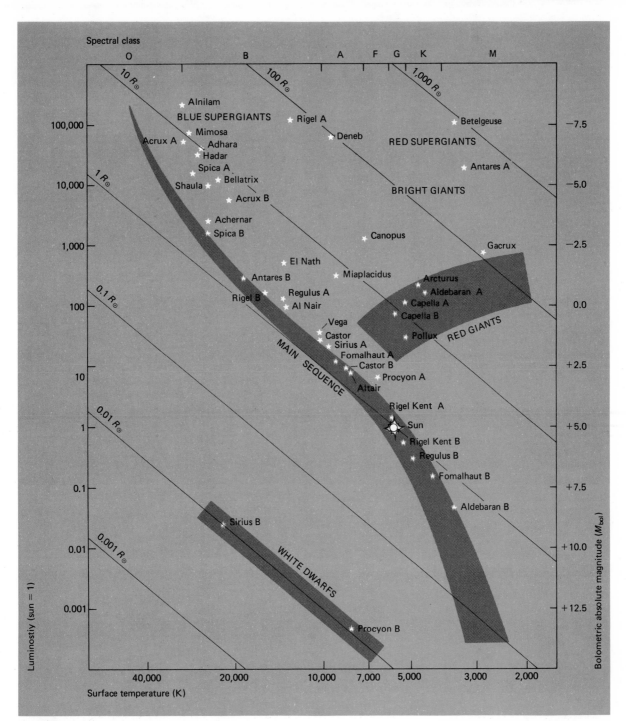

FIGURE 13.2
H-R diagram for the 30 brightest stars, their faint
companions, and the sun. Note that the majority
of the bright stars are intrinsically more luminous
than the sun. Two of the five companions are
white dwarfs, Sirius B and Procyon B. Compare
this diagram with the H-R diagram for the nearest
stars in Figure 13.3

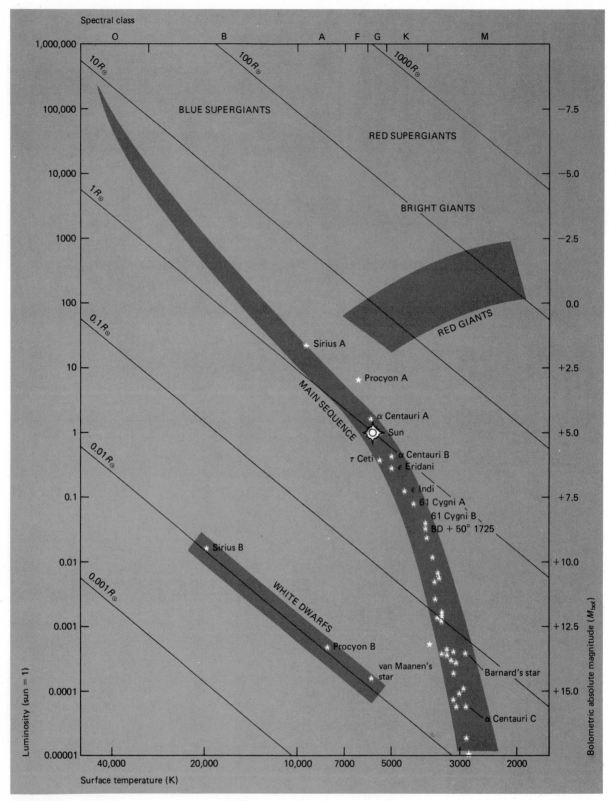

FIGURE 13.3
H-R diagram for stars within 15 light years of the sun. A and B refer to the brighter and
fainter components of a multiple system. Three white dwarfs appear at the lower left.
Compare this diagram with the 30 brightest stars in Figure 13.2. Clearly the nearby stars
are small and faint ones, while the bright stars are large and luminous and are not in
general nearby.

roughly, in the upper part of the diagram. From Table 13.1 the range of mass is about 10,000 between the least and the most massive stars, while radii vary by almost 100,000. For typical white dwarf stars, whose mass and radius are about 1 and 0.01 in units of the sun's values, the mean density is about 10^6 grams per cubic centimeter. The mass and radius of typical red giants are about 1 and 100 in solar units so that the mean density is about 10^{-6} gram per cubic centimeter. Thus the mean density of the stars roughly decreases in the same direction as the radius increases.

BRIGHT STARS VERSUS NEARBY STARS

The H-R diagram for the brightest stars in the sky in Figure 13.2 is striking in that the majority of the stars are not main sequence. Of the 30 brightest stars plotted, roughly 70 percent are stars above the main sequence (either subgiants, giants, bright giants, or supergiants), while roughly 30 percent are main-sequence stars. This again emphasizes that the apparently bright stars are bright because they are intrinsically bright and not because they are nearby.

In the H-R diagram for stars in the sun's immediate neighborhood (see Table 12.1), out to 15 light years, it is striking that no giants or supergiants appear (see Figure 13.3). The sun outranks all but three stars in size and luminosity. This sample also contains three white dwarfs, but most of the stars are red dwarfs that are located on the lower end of the main sequence.

For the nearby stars of Table 12.1 there are approximately 50 stars in a sphere of radius 15 light years, or about 0.004 star per cubic light year (one star in almost 300 cubic light years). Figure 13.4 displays the number of main-sequence stars from Table 12.1 by spectral type and approximate mass. As is evident, there are many more faint, small-mass red dwarfs than anything else. For every 1000 small, red, class M stars there are not more than 400 class A to K main-sequence stars and 1 class B or O giant or supergiant star. The evolutionary process that forms stars (Chapter 14) appears to favor the formation of faint, small-mass stars.

What we see when we scan the Galaxy are the intrinsically bright stars, not necessarily the most numerous type of star. Thus it is not really possible to do a survey of the Galaxy by type of star. However, using the results for the nearby stars and some educated guesswork, astronomers have devised the estimate for a Galactic census given in Table 13.2. On the assumption that the total number of stars in the Galaxy is about 400 billion, then 90 percent are main-

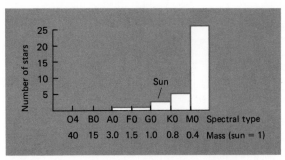

FIGURE 13.4
Number of stars for each spectral class among nearby main-sequence stars. Within 15 light years of the sun are 27 main-sequence M stars, 11 A to K main-sequence stars, and 3 white dwarfs, and the rest are nonvisible ones (Table 12.1). The number of stars per cubic light year is about 0.004, or one star in a volume of almost 300 cubic light years. The total mass of these stars is approximately $17.3M_\odot$ so that the mean density of matter is about $0.001M_\odot$ per cubic light year. The typical mass of the stars in this sample is about $0.4M_\odot$, corresponding to roughly an M0 V star. The total luminosity of the stars in this sample volume is about $34L_\odot$, or the typical star is about 0.8 as luminous as the sun. A volume of 1000 cubic light years contains about $1M_\odot$ of stellar matter, which is emitting about $2.4L_\odot$ of radiant energy.

sequence F through M stars, 9 percent white dwarfs, 0.5 percent red giants, and 0.5 percent everything else. All the stars intrinsically brighter than the red dwarfs (M V stars) and white dwarfs account for only 24 percent of all stars and only 42 percent of the mass of the stars in the Galaxy. However, they provide 99 percent of the luminosity emitted by stars in the Galaxy. In other words 58 percent of the Galaxy's mass in the form of stars provides 1 percent of its luminosity. We might anticipate the same result for other galaxies beyond the Milky Way. As an example, it would take about 1000 stars intrinsically as bright as Vega (A0 V) to equal the brightness of Rigel (B8 Ia), about 50,000 as bright as the sun (G2 V), and almost 2.5 billion as bright as Wolf 359 (M8 V). Red dwarfs in the solar neighborhood are not as numerous as that; we conclude that it is the intrinsically bright stars that are responsible for the light emitted by a galaxy, while much of the mass of a galaxy is tied up in stars that do not contribute appreciably to its brightness.

PULSATING VARIABLE STARS
ON THE H-R DIAGRAM

In Chapter 12 we saw that a number of stars vary in brightness because they pulsate. Although the pul-

TABLE 13.2
Census of Stars in the Milky Way Galaxy

Luminosity Class	Spectral Class	Typical Mass ($M_\odot$)	Typical Luminosity ($L_\odot$)	Number of Stars[a]	Cumulative Percent		
					Number	Mass[b]	Luminosity[c]
Supergiants (I, II)	O–M	?	50,000	$\approx 10^5$	≈ 0	≈ 0	≈ 3
Red giants (III)	F–M	≈ 1.2	40	$\approx 2 \times 10^9$	0.5	0.6	≈ 41
Main sequence (V)	O	≈ 25	80,000	$\approx 10^4$	0.5	0.6	≈ 42
	B	5	200	300×10^6	0.6	1.6	≈ 70
	A	1.7	6	3×10^9	1.2	4.6	≈ 79
	F	1.2	1.4	12×10^9	4.2	13.6	≈ 87
	G	0.9	0.6	26×10^9	11.2	$\simeq 27$	≈ 94
	K	0.5	0.2	52×10^9	$\simeq 24$	$\simeq 42$	≈ 99
	M	0.25	0.005	270×10^9	$\simeq 91$	$\simeq 80$	≈ 100
White dwarfs	B–F	≈ 1.0	0.005	$\approx 35 \times 10^9$	≈ 100	≈ 100	≈ 100
Total				$\approx 400 \times 10^9$			

[a] Estimating the total number of stars in the Galaxy to be 400 billion.

[b] Estimated contribution by stars to the mass of the Galaxy is $\approx 175 \times 10^9 M_\odot$.

[c] Estimated contribution by stars to the luminosity of the Galaxy is $\approx 20 \times 10^9 L_\odot$, or $\approx 8 \times 10^{43}$ ergs/s.

[d] Dashed line shows at what point 50 percent of the total has been accumulated.

sating stars that we considered, the Cepheids and the red variables, are intrinsically bright stars, there are less luminous stars that also appear to pulsate. If in the H-R diagram of Figure 13.5 we mark the boundaries of the regions in which the Cepheids and the red variable stars are located, we find that Cepheids of all types form an *instability strip*. It lies between the upper end of the main sequence, containing bright, blue stars, and the region where red giants and supergiants are located. On the other hand, long-period and irregular variable stars are red giants and supergiants, and they lie to the cool side of the Cepheid instability strip. Why this might be the case will be discussed in Section 15.4 (see also Figure 15.1).

H-R DIAGRAMS FOR CLUSTERS

The H-R diagrams in Figures 13.1, 13.2, and 13.3 are for field stars or the brightest stars of the night sky or the nearby stars. Other than their common location (they are relatively near the sun), the stars of these diagrams need not have any significant relationship with each other. On the other hand, the stars in a cluster have been with each other since birth, and they should be more comparable than stars selected at random.

Therefore it is important to consider H-R diagrams for clusters. As the story of the evolution of stars unfolded in the 1950s and 1960s, the H-R diagram of clusters played a vital part in unlocking the secret of how stars evolve.

Figure 13.6 is an H-R diagram for two nearby open clusters: the Hyades, the nearer one, and the Pleiades, some three times farther away. Both clusters have well-delineated main sequences but not the substantial number of red giants one sees in the H-R diagram of the bright stars. The Pleiades cluster has no red giants, and the Hyades has four. When astronomers compare the H-R diagram for a number of open clusters (see Figure 15.3), a gradation is found from those with a well-developed upper main sequence and no red giants to those with no massive blue stars on the upper main sequence and a large number of red giants.

In contrast to H-R diagrams for open clusters those for the globular clusters are fairly similar to each other but distinctly different from those of the open clusters. One such H-R diagram is shown in Figure 13.7 for the globular cluster known as M3 in the constellation Canes Venatici. In it one sees that there are no massive blue stars on the upper end of the main se-

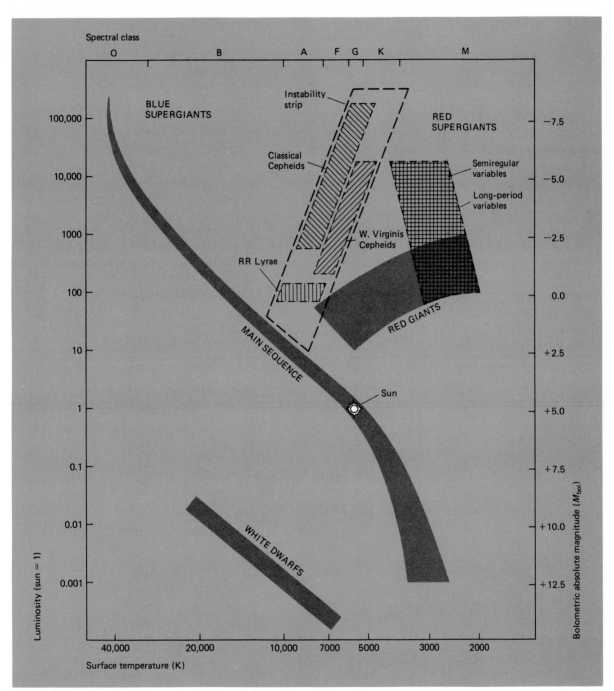

FIGURE 13.5
H-R diagram for pulsating variable stars. The Cepheids are yellow stars that range from giants up to supergiants, while the long-period and semiregular variables are M stars.

quence. The striking feature is the extremely well-developed red giant branch and the existence of a horizontal branch, which is not present in the H-R diagram of an open cluster.

INTERPRETING THE H-R DIAGRAM

The significance of the H-R diagram became clear several decades ago, when astronomers realized that it is a panorama portraying how stars evolve. A star's position in the diagram is a function of its mass, radius, luminosity, chemical composition, and current age. There is a physical reason why the points in the diagram are not scattered at random: The natural forces that guide the star's evolution confine them to stable portions of the diagram while they are converting matter into radiant energy. For example the Hertzsprung gap (see Figure 13.1) is a region in which

there are known to be very few stars; it must then be a region through which evolving stars pass quickly. (The full story of stellar evolution is the subject of Chapters 14 and 15.)

13.2
Stellar Populations

Among its billions of stars our Galaxy contains a vast assortment of stars of varying size, mass, temperature, color, age, and chemical composition. At least nine-tenths are main-sequence stars, most of the rest being white dwarfs. Giants are rare, and supergiants are still more scarce. Less than half the Galaxy's total population of stars is made up of single stars; the rest is

FIGURE 13.6
H-R diagram for two open clusters. The B-V color is used instead of the surface temperature and the visual apparent magnitude instead of luminosity since all the stars in a cluster are at approximately the same distance. With these coordinates the diagram is called a *color-magnitude diagram*. The Pleiades are shown as dots and the Hyades as plus

signs. Notice that the Hyades cluster has four giants; the Pleiades has none. The absolute-magnitude scale on the right applies only to the Hyades cluster. It is evident from the apparent magnitudes that the Pleiades cluster is considerably more distant than the Hyades.

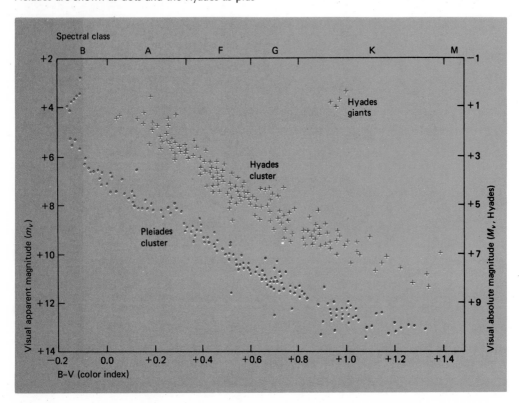

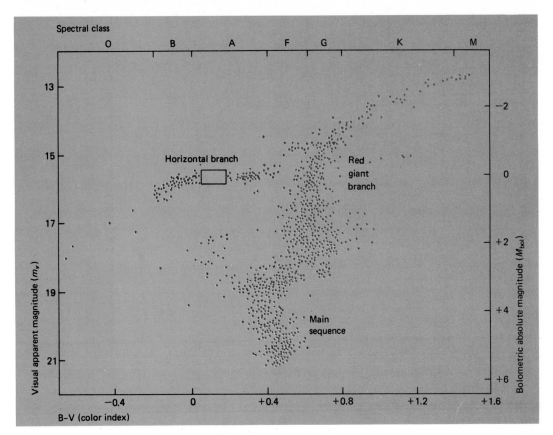

FIGURE 13.7
H-R diagram for the globular cluster M3 in the constellation Canes Venatici. Note the absence of any blue stars on the main sequence and the presence of the horizontal branch. The box represents the location of the RR Lyrae variables.

DISTANCES OF CLUSTERS BY MAIN-SEQUENCE FITTING

An excellent method of finding out how far away open star clusters are involves yet another application of the inverse-square law of light. To use this technique, we must have absolute magnitudes and colors (or spectral types) to construct an H-R diagram for at least one cluster, such as the Hyades, which is 144 light years away. The distance of the Hyades has been determined by several methods.

The most direct means of determining the distance of the Hyades is to measure parallaxes of individual stars in the cluster; this procedure works for the Hyades, but most clusters are too distant to apply it. An-

other method is that of spectroscopic parallaxes, which are based on a determination of the intrinsic brightness of stars from luminosity criteria seen in their spectra.

Having an H-R diagram for the Hyades cluster, let us see how a distance can be determined for the Pleiades by main-sequence fitting: First we have to plot an H-R diagram for the stars of the Pleiades cluster, using color indexes (as a measure of surface temperature) and visual apparent magnitudes (since all the stars of the cluster are at the same distance). Next we read on the vertical scale the differences between the ap-

parent magnitude of the Pleiades main-sequence stars and the apparent magnitude of the Hyades main-sequence stars at several points along the main sequence and average them. This averaged difference, when added to the distance modulus ($m - M$) of the Hyades, gives the distance modulus of the Pleiades cluster. The distance modulus of the Hyades is about $+3.0$, and we find that the difference between m (Pleiades) and m (Hyades) is about $+2.6$, so that for the Pleiades $m - M = 5.6$. A distance modulus of 5.6 corresponds to a distance of 430 light years, which is the value listed in Table 12.7.

physically related doubles and multiples or is in clusters and groups. Some unusual stars, such as the pulsating and eruptive ones, vary in brightness, which draws special attention to them. Even at great distances their variability is easy to recognize. But variable stars are actually very rare, probably less than 0.1 percent of the stars in the Galaxy.

Is there a pattern beneath the Galaxy's rich variety? Very much so, as we shall try to illustrate in this chapter and the next three. In the early 1940s Walter Baade (1893–1960) found clues identifying what appeared to be two distinct populations of stars in the Andromeda galaxy. Separating the never-before-resolved central region of that galaxy into its constituent stars, he found multitudes of red giants like those in the globular clusters of our Galaxy. These he named Population II stars to distinguish them from what he called Population I stars, the highly luminous blue supergiants in the spiral arms of the Andromeda galaxy. To Baade it was clear that these population groups were like those in our Galaxy. The massive Population I stars within our Galaxy, which are associated with the bright gaseous nebulae in the spiral arms, are younger than are the Population II stars, which are found in the globular clusters, the Galaxy's halo, and its central bulge. A definite relationship exists between a star's physical characteristics and its location in the Galaxy. In many regions the pattern is not pronounced, but in general astronomers have found that there are two distinct groups of stars analogous conceptually to Baade's population groups but with somewhat different characteristics. They are known as the *spheroidal population* and the *disk population;* they are differentiated by their age, physical properties, chemical composition, motions, and location.

Astronomers now realize, because of an accumulation of observations and theories on stellar and Galactic evolution, that there are gradations between the two populations. As we discuss in Chapter 16, the two population groups can be subdivided by their evolutionary history, by the location, and by the motions of their stars.

13.3 Physical Structure of Stars

The sun is the astronomer's primary model for the stars. We feel confident that we know the physical laws operating inside the sun and presumably in other main-sequence stars. For example, if the sun emitted radiant energy simply because it is a hot body without any internal source of energy, it should cool at a rate fast enough for us to measure the decrease. But the evidence suggests that the sun is not cooling significantly; thus the sun—and other main-sequence stars—must have a source of energy deep in the interior that replaces the energy radiated away from the surface.

HYDROSTATIC EQUILIBRIUM AND IDEAL GASES

For a star to exist without either expanding or contracting, the upward forces on each layer from center to surface must equal the downward force. Such a condition is called *hydrostatic equilibrium,* and at any distance from the center the weight of the overlying layers is balanced by the upward pressure of gas in the layers closer to the center. Through most of the star's existence the gases inside it conform to what is called the *perfect-gas law* so that the pressure of the gas (which results from the forces imparted by colliding gas particles) is proportional to the density and temperature of the gas.

At the high temperatures inside stars collisions between gas particles will strip electrons from atoms, leaving bare nuclei and free electrons—a *plasma.* Because most of an atom is just empty space, free electrons and bare nuclei can crowd much closer together in a plasma than they can in a gas composed of atoms, allowing the material inside stars to remain gaseous even at very high density. In the high-temperature and high-density regions near stars' centers gas particles increase their speed, and so they collide more often and more violently. Consequently the pressure exerted by the gas is greater in the center and declines outward. If the gas pressure exceeds the force of gravity, the star must expand; if gravity is greater, the star should contract.

So may we read, and little find them cold;
Not frosty lamps illuminating dead space,
Not distant aliens, not senseless Powers.
The fire is in them whereof we are born;
The music of their motion may be ours.

George Meredith

For the hottest stars on the main sequence the high-energy photons of the very intense radiation fields inside these stars can add to the pressure, a phenomenon known as *radiation pressure,* which helps balance the weight of the outer layers.

THERMAL EQUILIBRIUM

During most of a star's existence the amount of energy generated inside the star equals the amount radiated away into space at its surface; the star maintains *thermal equilibrium*. As a result, the temperature and pressure inside the star do not change during these periods.

Thermal equilibrium is self-regulating in that, if more energy is released in the center of a star than is radiated away at the surface, the temperature inside the star should rise. Because the gas pressure depends directly on the temperature, the pressure should increase, and the star should expand with heat energy being transformed into *gravitational potential energy*. This change in turn cools the gas, decreasing the gas pressure, and hydrostatic equilibrium is achieved at a larger radius. If, on the other hand, too little energy is produced, the star should contract and heat up, which increases the gas pressure and stops further contraction. Hydrostatic equilibrium is now found at a smaller radius.

What would happen if the pressure of the gas did not depend on temperature? Heating or cooling from changes in the amount of energy generated would not be checked by increasing or decreasing the gas pressure followed by subsequent changes in the star's size. This situation can exist and is called a *degenerate state*. It may apply to all the constituents of the gas or to only one of them, such as the free-electron component, which would then be called a *degenerate electron gas*. We shall see in Section 15.2 how degeneracy develops in stars.

TRANSPORTING ENERGY

Energy may be transported in one or more of three distinct ways although usually one means is more efficient than the others under the prevailing conditions. One of these means is *conduction*, which is the way heat is transferred along a metal rod placed in a fire. In the interior of main-sequence stars obeying the perfect-gas law this is a very ineffective means of energy transfer. But in the cores of stars where the material has become degenerate conduction becomes the primary way of transporting energy.

A more frequently occurring way for transporting energy in stars is *convection*, in which gas circulates between hot and cool regions, transferring thermal energy to the cool region (see Figure 13.8 and Section

DEGENERATE GASES

In Section 4.4 it was pointed out that there is a limit to the number of electrons that may occupy a particular orbit, or energy level, in the atom. This *exclusion principle*, originally proposed by the theoretical physicist Wolfgang Pauli (1900–1958), also applies to electrons that are not bound to a nucleus but merely confined to a fixed volume, such as the deep interior of a star. In such a volume there are only certain discrete energy states available to the electrons. When the density of a gas is quite low, such as in the air in a room, there are always enough unoccupied energy states available so that electrons may readily gain or lose energy in order to move

from one energy state to another. However, when the density of a gas is quite large, such as in the interior of a star (see Table 13.1), the lowest energy states may all be filled, with only the highest still unoccupied. In this case electrons cannot readily gain or lose energy, and they will become highly incompressible. Even though thermal energy may be extracted from the gas, the gas still may not cool down since the electrons cannot give up energy by moving to a lower energy state, and they exert a dominant influence on the behavior of the overall gas.

In a degenerate gas the average pressure is high enough to keep the

material from being compressed by gravity. Also, because the kinetic energies of the electrons are quite high and the rate of collision between the electrons and other particles is quite low, the degenerate electrons can travel great distances at velocities that can approach the speed of light. And unlike the pressure in a perfect gas the pressure that degenerate electrons exert has little to do with their temperature. When electron degeneracy occurs within stellar matter, major changes in the thermal balance of energy can lead to mechanical instabilities.

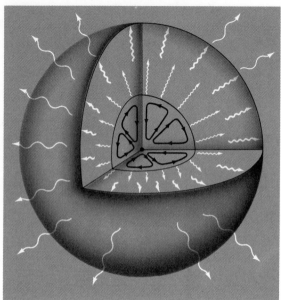

FIGURE 13.8
Schematic illustration of the interior of a star, showing a convective zone in the center and a radiative zone as an envelope. Energy is carried out of the center by mass motions in the convective zone. Through the radiative zone energy moves outward by successive emission and absorption of photons. The greater the opacity of the material is, the greater the number of photon absorptions and re-emissions is.

11.1). Once a pattern of circulation is established, convection can be a very efficient mechanism. For layers in which temperature changes quite rapidly with depth, convection develops as the principal means of carrying energy. These layers are referred to as a *convective zone,* and in them the atomic constituents are well mixed by the continual stirring.

Radiation is the third way of moving energy. Inside a star photons resulting from energy generation diffuse outward through the intervening material (Figure 13.8). After a photon is emitted, it very soon is either absorbed by an atom (or ion) or scattered by free electrons after traveling a characteristic distance that ranges from small fractions of a centimeter deep in the star's interior to several kilometers in the photosphere. Although the direction in which a reemitted or scattered photon can move is generally arbitrary, there will be a net drift of photons outward from the center to the surface of the star. Where energy is transported by radiation, the layers of material are known as a *radiative zone.* In a radiative zone chem-

ical elements go through very little mixing; any chemical inhomogeneity developing here should persist.

OPACITY

Radiation and matter continually interact by absorption, reemission, and scattering of photons, processes that impede the outward flow of radiant energy (Figure 13.8). Matter's resistance to the flow of radiation through it is called *opacity.* In regions of large opacity the temperature drops rapidly outward, and convection will take over as the primary way of transporting energy. When the density is low, radiation travels more freely through a star.

Energy is liberated in the deep interior of stars as a comparatively small number of high-energy gamma-ray photons. As these photons work their way out of the star, the absorption and reemission by overlying matter degrades the gamma-ray photons into millions of lower-energy ones by the time they reach the photosphere. Radiant energy takes hundreds of thousands of years to diffuse through the sun from its energy-generating core.

THERMONUCLEAR FUSION

Over a century ago astronomers understood that the energy already radiated by the sun could never have been supplied by ordinary combustion (for example, by the burning of wood or coal). Another way of producing energy is by the conversion of gravitational potential energy into heat by contraction. In the nineteenth century this was thought to be the only source of the sun's energy. We now know that contraction is a vital source of energy, on which a star can draw at various stages in its life. But at its current luminosity our sun could not survive on gravitational contraction alone for more than about 15 million years.

For stars like the sun a source of energy must keep the luminosity approximately constant for billions (not just millions) of years. The question plaguing astronomers in the early part of this century was what that source is. The answer is the fusion of small-mass nuclei to form more massive nuclei. Sir Arthur Eddington (1882–1944) suggested in 1920 that fusion of hydrogen could form helium and that this could be the long-sought fuel. After it was found that stars have vast quantities of hydrogen, the physicist Hans Bethe (1906–   ) proposed in 1938 a way in which four hydrogen nuclei (four protons) could be converted into

ARTHUR STANLEY EDDINGTON (1882–1944)

Eddington was a brilliant scholar. His first position after graduating from Cambridge University (1906) was that of chief assistant at the Royal Observatory in Greenwich, where he ex-

celled in practical astronomy. In 1913 he was appointed Plumian Professor of Astronomy at Cambridge and a year later made director of the observatory.

In the years following, his intuitive insight, bold imagination, and mastery of mathematics led him to important discoveries over a wide range of problems. Eddington was the first to model the interior of a star under radiative equilibrium, pointing out that the condition for stellar equilibrium involved three forces: gravity, gas pressure, and radiation pressure. Recognizing the importance of ionization in stellar interiors, he boldly assumed that, because of the high ionization of the internal gases, the perfect-gas condition prevailed within the interiors of the stars, except for the white dwarfs. This hypothesis was later accepted. He demonstrated that energy could be transported by radiation as well as by convection and that the centers of stars must be at very high temperatures—in the millions of

degrees. An extremely important result that emerged from his research was his theoretical formulation of the mass-luminosity relation, verified later by stellar data.

Eddington suspected that the chief source of stellar energy was subatomic and that hydrogen played a dominant role in supplying this energy. Later, in 1938 and 1939, Bethe introduced the theory for the fusion of hydrogen into helium, which clarified the picture of stellar energy generation and substantiated Eddington's speculations.

In 1919 Eddington organized a solar-eclipse expedition to Brazil to photograph the stars in the neighborhood of the eclipsed sun. This was to test whether or not a beam of starlight would bend when going by the sun, as predicted by Einstein's general theory of relativity. Although difficult to measure, the observed deflection was in rough agreement with Einstein's predicted value—the first observational test of relativity theory.

a helium nucleus, releasing energy. If many hydrogen nuclei are converted, they will release sufficient energy through this process (known as *thermonuclear fusion*) to keep stars shining for billions of years.

What determines whether hydrogen can be fused to form helium? The answer is the temperature and density of the gas; the higher the temperature and density, the more readily the process will proceed. Thermonuclear reactions will therefore be most numerous in a star's central region, where the temperature and density are highest. The reactions will gradually decline to zero somewhere out from the center, where temperature and density are too low to sustain them. This distance from the center, then, defines the *energy-generating core* of the star.

THE *p-p* CHAIN AND CNO CYCLE

Hydrogen burning proceeds by two principal schemes: the *proton-proton chain* (*p-p* chain) and the

carbon-nitrogen-oxygen cycle (CNO cycle; see Figure 13.9). In each process four protons are fused into one helium nucleus with a slight loss in mass, which is converted into energy. Which thermonuclear process produces more energy depends on the temperature. Up to about 16 million K the *p-p* chain dominates. Beyond that temperature, however, the CNO cycle takes over as the most important thermonuclear process. The average rate of energy generation for the entire sun, which depends primarily on the *p-p* chain, is about 2 ergs per gram per second. For a star of 10 solar masses the average rate of energy generation, supplied principally by the CNO cycle, is about 1000 times greater than that of the sun.

The mass of the end product of hydrogen burning, ^{4}He, is 0.71 percent less than the combined masses of four reacting protons ($4 ^1$H). What has happened to the rest of the mass? Early in this century Einstein pointed out that there is an *equivalence between mass and energy*. Mass is just one more manifestation of en-

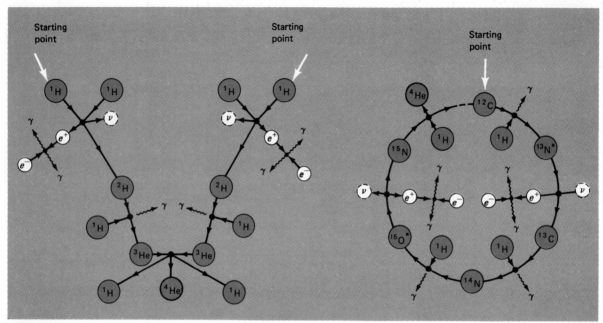

FIGURE 13.9

The two principal fusion processes for hydrogen burning. The *p-p* chain predominates in the sun and less massive stars. The CNO cycle is predominant in stars hotter than the sun. *Key to symbols:* ^{1}H = hydrogen 1 nucleus (proton); ^{2}H = hydrogen 2 nucleus (deuteron); ^{3}He = helium 3 nucleus; ^{4}He = helium 4 nucleus; e^- = electron; e^+ = positron; ν = neutrino; γ = gamma-ray photon; ^{12}C = carbon 12 nucleus; ^{13}N* = unstable nitrogen 13 nucleus; ^{13}C = carbon 13 nucleus; ^{14}N = nitrogen 14 nucleus; ^{15}O* = unstable oxygen 15 nucleus; ^{15}N = nitrogen 15 nucleus.

ergy, and what is conserved in any type of interaction between particles of matter is the total energy, including the energy equivalent of the mass. The equivalence is symbolized in Einstein's equation $E = mc^2$, where E is the energy, m is the mass, and c is the velocity of light.

In hydrogen burning 1 gram of hydrogen is converted into 0.9929 gram of helium plus 6.4×10^{18} ergs of energy—exactly 0.71 percent of the original 1 gram of hydrogen times c^2. In what form does the energy appear? In the various steps several gamma-ray photons are created and degraded by absorption and re-emission into many photons having the same total energy. Also, some of the material particles created have large kinetic energies, which they will soon redistribute to other particles by collisions. Thus both radiant energy and heat energy come from the mass that is lost in these thermonuclear fusion processes.

Translated into practical units, every second the sun converts 600 million metric tons of hydrogen into 596 million metric tons of helium and 4 million metric tons of mass into energy. This energy will diffuse to the surface, where it will supply the 3.83×10^{33} ergs of energy radiated away into space each second. In its core the sun has enough hydrogen to keep it shining for about 10 billion years. So far in 4.6 billion years the sun has used up about half of its core's hydrogen supply and lost about 0.043 percent of its mass.

13.4
Mathematical Models of Stars

Each of the physical processes described above depends on several physical quantities, among them temperature, density, pressure, and the mass and luminosity of the star. We can use a symbol to represent the numerical value of each quantity and combine these symbols into mathematical equations embodying their relationships. These equations, called *equations of stellar structure,* describe how mass, pressure, temperature, and luminosity vary outward from the center of the star. Within the equations are additional quantities, such as density, chemical composition, opacity, and rate at which energy is generated. To construct models of stars, we take their observed properties—such as mass, radius, luminosity, sur-

face temperature, and estimated chemical composition—as constraints in solving the equations of stellar structure at a discrete number of points along the radius. The solution is a mathematical model, or a *stellar model.* The computer makes it possible to develop these models in reasonable lengths of time for several hundred points. Without the high-speed digital computer much of the progress we have made in the last 20 years in this field would not have been possible.

An example of such a mathematical model for the

STEPS IN HYDROGEN BURNING

Step 1 in the *p-p* chain is fusion of two colliding protons (^{1}H) to form a *deuteron* (^{2}H), which is the nucleus of the hydrogen isotope deuterium, resulting in the emission of a positron (e^+) and a neutrino (ν). This reaction happens, on an average, once every 14 billion years for each isolated pair of protons. The time for the entire thermonuclear process is determined by this first step, and it is only the enormous quantity of hydrogen in the cores of stars that makes this process a significant source of energy.

The positron is a positively charged particle with the mass and other characteristics of an electron; it is the antiparticle for the electron. The collision of a positron with an electron destroys them as matter and creates two gamma-ray photons. The *neutrino,* on the other hand, is a massless, chargeless particle traveling at the speed of light and has a low probability of interacting with matter. It immediately escapes from the star, carrying away about 2 percent of the energy released in the *p-p* chain of reactions.

Step 2 in the *p-p* chain is the collision within a few seconds of another proton with the deuteron to fuse and form the light isotope of helium, resulting in the emission of a gamma-ray photon. Finally, in step 3 two ^{3}He

nuclei collide every few million years and fuse to form the heavy isotope of helium (^{4}He), accompanied by the return of two protons. (We should point out that there are other branches of these reactions leading to the same end.)

All told, six protons have taken part in producing two ^{3}He nuclei, from which one ^{4}He nucleus is produced and two protons returned to the reservoir of fusionable matter.

The other hydrogen-burning reaction, the CNO cycle, has six steps occurring at rates between 80 seconds and 300 million years but leading to the same result as the *p-p* chain; that is, the conversion of four protons to produce one helium nucleus and to liberate energy. The cycle begins with ^{12}C and closes with the return of ^{12}C so that carbon is only a catalyst that makes the reaction go.

TABLE 13.3
Thermonuclear Reactions to Hydrogen Burning

Hydrogen Burning	Nuclear Reaction	Energy Released (ergs)	Mean Reaction Times
p-p chain	1. $^1H_1 + {}^1H_1 \rightarrow {}^2H_1 + e^+ + \nu$	1.9×10^{-6}	14×10^9 yr
	2. $^2H_1 + {}^1H_1 \rightarrow {}^3He_2 + \gamma$	8.8×10^{-6}	6 s
	3. $^3He_2 + {}^3He_2 \rightarrow {}^4He_2 + 2{}^1H_1$	$\underline{2.1 \times 10^{-5}}$	1×10^6 yr
		4.2×10^{-5}	
CNO cycle	1. $^{12}C_6 + {}^1H_1 \rightarrow {}^{13}N_7^* + \gamma$	3.1×10^{-6}	13×10^6 yr
	2. $^{13}N_7^* \rightarrow {}^{13}C_6 + e^+ + \nu$	2.4×10^{-6}	420 s
	3. $^{13}C_6 + {}^1H_1 \rightarrow {}^{14}N_7 + \gamma$	1.2×10^{-5}	2.7×10^6 yr
	4. $^{14}N_7 + {}^1H_1 \rightarrow {}^{15}O_8^* + \gamma$	1.1×10^{-5}	3.2×10^8 yr
	5. $^{15}O_8^* \rightarrow {}^{15}N_7 + e^+ + \nu$	2.8×10^{-6}	82 s
	6. $^{15}N_7 + {}^1H_1 \rightarrow {}^{12}C_6 + {}^4He_2$	$\underline{7.9 \times 10^{-6}}$	1.1×10^5 yr
		4.0×10^{-5}	

*Denotes an unstable nucleus that will spontaneously change into a different nucleus.

TABLE 13.4
Mathematical Model for the Sun

Fraction of the Radius	Radius (10^3 km)	Temperature (10^6 K)	Density (g/cm^3)	Fraction of Central Pressure[a]	Fraction of Mass	Fraction of Luminosity
0.0	0	15.5	160	1.00	0.0	0.0
0.04	28	15.0	141	0.84	0.008	0.08
0.1	70	13.0	89	0.46	0.07	0.42
0.2	139	9.5	41	0.15	0.35	0.94
0.3	209	6.7	13.3	0.035	0.64	0.998
0.4	278	4.8	3.6	0.007	0.85	1.00
0.5	348	3.4	1.0	0.0014	0.94	1.00
0.6	418	2.2	0.35	0.0003	0.982	1.00
0.7	487	1.2	0.08	0.00004	0.994	1.00
0.8	557	0.7	0.018	0.000005	0.999	1.00
0.9	627	0.31	0.002	0.0000003	1.000	1.00
1.0	696	0.006	3×10^{-7}	4×10^{-13}	1.000	1.00

[a] $P_c = 3.4 \times 10^{17}$ dyne/cm^2.

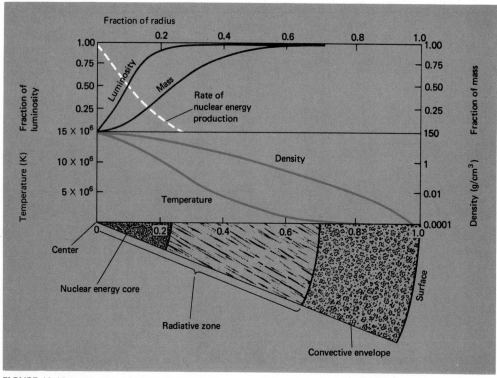

FIGURE 13.10
Plot of a mathematical model of the solar interior along a radius from the center to the surface, based on Table 13.3. The wedge through the sun at the bottom shows the primary zones of interest: energy-generating core (also a radiative zone), radiative zone, and convective envelope. Temperature and density variations along the radius are shown in the bottom panel; the fraction of the mass and luminosity at a given point along the radius is shown in the top panel.

> There is one glory of the sun, and another glory of the moon, and another glory of the stars: for one star differeth from another star in glory.
>
> I Corinthians, 15

present sun is given in Table 13.4 and is shown schematically in Figure 13.10. As the star evolves, it alters the structure of its layers from center to surface. In the layers responsible for nuclear burning the chemical composition is changing as well.

We can calculate a sequence of mathematical models simulating the restructuring that the real star presumably undergoes as it ages during its lifespan of millions of human generations. For each model in the sequence the surface temperature and luminosity represent a fixed point in the H-R diagram at a given time. How do we test the validity of our model? We see how well this time sequence of points, or *evolutionary path,* for one star helps us predict the distribution of real stars in the H-R diagram. Both open and globular clusters are important observational keys in checking our results. The color-magnitude diagrams of clusters, such as Figures 13.6 and 13.7, give the best evidence on stellar aging because a cluster is a group of stars with a definite range of masses that began their existence at about the same time and in the same place. They also formed from the same material and so at first were reasonably similar chemically. From the distribution of cluster stars in the H-R diagram, we can then deduce details of how individual stars age.

13.5
Main-Sequence Stars

RESULTS FROM STUDIES OF STELLAR EVOLUTION

If our knowledge of the physical processes going on inside the star is correct, we should be able to predict the star's position in the H-R diagram for different ages during its lifetime. We can thus explain the existence of the various regions of the diagram and trace the evolutionary sequence that carries stars from one region to another.

Also, the fraction of stars in any region of the H-R diagram should equal the fraction of a star's existence spent in that region of the diagram. If the red supergiant phase is a short part of the star's life, we should see relatively few red supergiants—and we do. If the main sequence is a long phase in the star's existence, then many stars should be main-sequence stars. We actually observe the main sequence to be the most densely populated region in the H-R diagram (as shown in Table 13.2).

Let us begin the study of the evolution of stars with the most common, the main-sequence stars. On this sequence the hot stars of spectral classes O and B are the most massive ones and the cool, red dwarfs of class M the least massive. The main sequences for the various population groups differ just slightly from each other, a difference traceable to the variation among them of the heavy-element abundance.

MEANING OF THE MAIN SEQUENCE

Stars spend most of their lives on or near the main sequence for two reasons: the large yield of energy per gram from hydrogen fusion as compared with other sources of nuclear energy and the vast amount of hydrogen available. The H-R diagram for the stars in the sun's immediate neighborhood (Figure 13.3) and the color-magnitude diagrams for open clusters (Figure 13.6) are illustrations of this fact.

The approximate time a star spends on the main sequence is proportional to its mass divided by its luminosity, which can be derived with the help of Einstein's mass-energy equivalence. The more massive a star, the greater is its emission of radiant energy per gram of matter and the shorter its time on the main sequence. This point is illustrated by the numerical estimates in Table 13.5 for luminosity per gram and duration on the main sequence, given that the sun will be a main-sequence star for about 10 billion years.

Now let us consider the reason for the shorter hydrogen-burning phase for more massive stars. After contraction onto the main sequence, the central temperature must be higher than for lower-mass stars because a higher gas pressure is needed in the center of the star to balance gravity. Consequently the temperature difference between the center of the star and its photosphere will be larger the more massive the

TABLE 13.5
Approximate Time Stars Are on the Main Sequence

Spectral Class	Surface Temperature (K)	Mass (sun = 1)	Luminosity (sun = 1)	Luminosity/ Mass (erg/s · g)	Time on Main Sequence (yr)
O7	37,500	25	80,000	6140	3×10^6
B0	30,000	15	10,000	1280	15×10^6
A0	9,500	3	60	38.6	500×10^6
F0	7,240	1.5	6	7.68	2.5×10^9
G0	5,920	1.0	1	1.92	10×10^9
K0	5,240	0.8	0.6	1.44	13×10^9
M0	3,850	0.4	0.02	0.09	200×10^9

star is. For a $15M_\odot$ star (15 times the mass of the sun) the central temperature is around 35 million K compared with 7 million K for a star of $0.25M_\odot$. More thermal energy flows from the interior of massive stars, making them more luminous than less massive stars. The more massive a star, the more rapidly it must burn hydrogen to supply the energy loss from its surface. The CNO cycle in the massive stars also depends more strongly on temperature and consumes the hydrogen faster than does the p-p chain that operates in stars of low mass. The dividing line between the two is about $2M_\odot$.

SOLAR NEUTRINO EXPERIMENT

How certain are astronomers about the thermonuclear processes going on inside stars? So far the only experiment designed to test the theory directly is one that tries to detect the neutrinos created in the thermonuclear processes. Neutrinos pass freely out of the sun into space because of the extremely low probability that they have for interacting with matter. They carry away, at the speed of light, a small fraction of the energy generated in the sun. By detecting solar neutrinos, astronomers can have first-hand information on the average temperature in the hydrogen-burning core.

One scheme for detecting solar neutrinos uses a huge tank filled with 400,000 liters of dry-cleaning fluid (Figure 13.11). It is located deep in a South Dakota gold mine to shield the chlorine atoms in the solvent from cosmic-ray particles; nothing is allowed to reach them but neutrinos. When the nucleus of a chlorine atom (^{37}Cl) captures a neutrino, it is transformed into a radioactive argon (^{37}Ar) nucleus, which has a half-life of about 35 days; the argon nucleus is recovered, and its decay is monitored. We find that the observed flux of solar neutrinos is about four times smaller than the rate predicted from standard mathematical models of the sun. By manipulating the solar model, which is subject to some uncertainties anyway, the discrepancy can be reduced but not eliminated. Other explanations for the discrepancy have been proposed, but none has received widespread endorsement. Plans are being made for a new experiment that will measure the lower-energy neutrinos that are not counted in the experiment described above.

Recently there has been another explanation, which arises from two studies of the historical measurements of the sun's diameter. Both studies seem to find that the sun has been shrinking for the last 100 or so years. The question is by how much. The contraction, if real, amounts to about 0.01 to 0.1 percent per century. Further speculation prompted by this finding suggests that the sun may undergo a long-term cycle of expansion and contraction. During contraction the sun derives heat energy from gravitational potential energy and lowers its rate of hydrogen burning. This would account for the emission of fewer neutrinos. These results are a long way from being thoroughly verified.

The solar neutrino problem is serious; for it casts doubt on our knowledge of the details of the structure and/or energy generation in main-sequence stars. Thus we are forced to look more carefully at the details of these processes in the sun if the experiment is completely correct. But it is unlikely that the solar

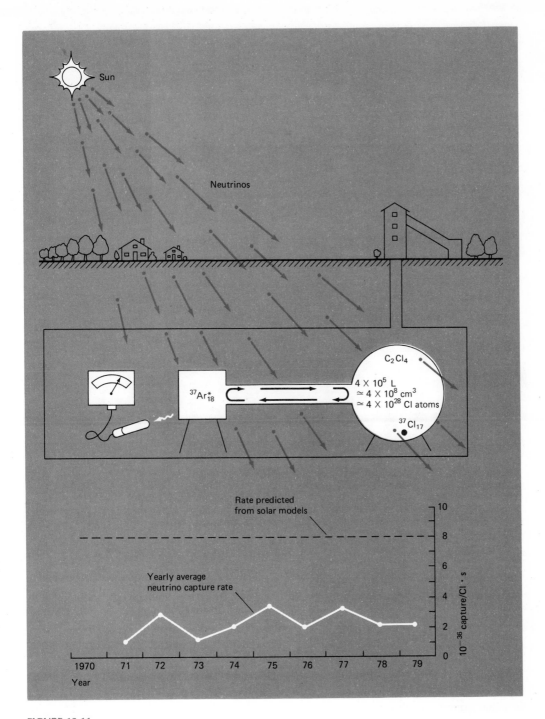

FIGURE 13.11

Solar neutrino experiment. The experiment is designed to detect the high-energy neutrinos in hydrogen burning; the detection depends on the capture of a neutrino by a chlorine nucleus, which transforms it to radioactive argon, which in turn can be identified and counted. The experiment permits solar neutrinos to be counted as a function of time. As shown by the graph, the observed capture rate is below that predicted by solar models.

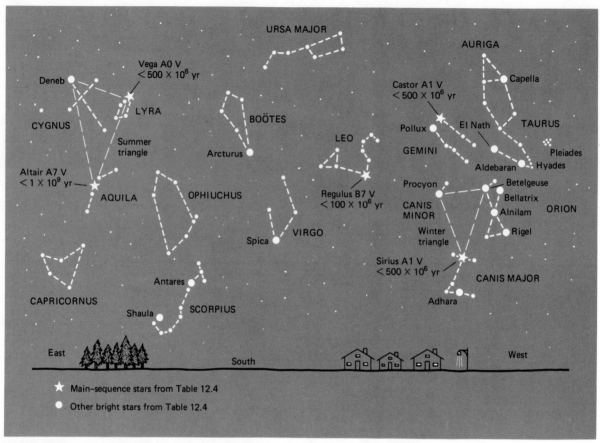

FIGURE 13.12
Main-sequence stars among the 30 brightest stars. The view of the night sky is more than one could see in one season of the year. Nevertheless, it should help you find the four main-sequence stars among the 30 brightest stars of Table 12.4 that are visible to the naked eye: Vega (Lyra), Castor (Gemini), Regulus (Leo), and Sirius (Canis Major). These four stars are all less than 500 million years old.

neutrino problem forecasts the failure of the present theory of stellar evolution. The details of the theory will change in time as a result of the solar neutrino problem and others, but the general outline seems likely to endure.

MAIN-SEQUENCE STARS OF THE NIGHT SKY

How does the time a star spends on the main sequence affect the stars in our night skies? A star like Spica, in the constellation of Virgo, is a B1 star, which may be just above the main sequence but is at most about 15 or 20 million years old (see Table 13.4). Stars like Sirius in Canis Major and Vega in Lyra are early A stars and are less than 500 million years old (Figure 13.12). The main-sequence life of nearby Barnard's

star, an M5 red dwarf in the constellation of Ophiuchus, is greater than 200 billion years, and it will still be fusing hydrogen in its core long after the sun has ceased to shine. During the sun's 4.6 billion years on the main sequence, thousands of massive O and B stars have come into existence and left the main-sequence phase, but all the small-mass M stars born in that period have hardly begun their stay on the main sequence.

LIMITS OF THE MAIN SEQUENCE

If we say that a star is an object held together by gravity that is or has been self-luminous, then the smallest mass for a star that derives energy from thermonuclear burning is about $0.1 M_\odot$. This limit is set by the mass needed to produce a central temperature that is sufficient to initiate hydrogen burning. Yet objects even smaller than $0.01M_\odot$ can survive a very long time on just the energy from gravitational contraction. These "stars" would, however, be so faint that they could easily escape detection. The only exception would be those that are companions to low-mass M dwarfs that are very nearby. These might be detectable as astrometric binaries.

At the upper end of the main sequence the mass of a star is probably determined more by the amount of interstellar matter available when it was formed than by its internal structure. Stars of about $60M_\odot$ and greater have such a delicate balance between gravitational and pressure forces that equilibrium could be prevented by any irregularities of motion inside the star. There are a few O stars for which mass estimates suggest that they are near $100M_\odot$. Limits on the masses of main-sequence stars determined from binaries are consistent with these figures.

As hydrogen burning progresses in a star on the main sequence, the central core is slowly depleting hydrogen and converting it to helium. Because the gas pressure depends on the density, or number of particles, converting four hydrogen nuclei to one helium nucleus must reduce the gas pressure. Hydrogen burning will therefore be accompanied by a very slight contraction of the energy-generating core and a heating up of the material. Because of this the star brightens slightly; it also increases the temperature difference between the center and surface to cause a greater outflow of radiation, and the outer portion of the star expands, increasing the radius. This is part of the reason for some of the width of the main sequence evident in H-R diagrams. As noted earlier, estimates from mathematical models for the sun suggest that it has increased its luminosity by 20 to 30 percent during its 4.6 billion-year existence as a main-sequence star.

In Chapter 14 we shall examine the birth of stars or the process that brings them to the main sequence, while in Chapter 15 we explore the aging processes that carry stars away from the main sequence toward their deaths.

SUMMARY

The Hertzsprung-Russell diagram. The H-R diagram is an astronomer's most important tool for studying stars. Plotting surface temperature (spectral type) against luminosity (absolute magnitude) outlines definite regions that reveal a great deal about the structure of stars. The most conspicuous region is the sequence of stars running from bright, hot stars (upper left) to faint, cool stars (lower right)—the so-called main sequence. Lying above the main-sequence stars are the regions occupied by red giants, bright giants, and red and blue supergiants. Below the main sequence is the region of white dwarfs and subdwarfs. The significance of the H-R diagram is that it portrays how stars evolve. The clusters of points on the diagram represent stable periods in the lives of stars. The star's position in the diagram is a function of the star's mass, radius, luminosity, chemical composition, and current age.

Stellar populations. At least 90 percent of all stars in our Galaxy are main-sequence stars. Most of the rest are white dwarfs. Astronomers identify two populations of stars, although there are gradations between them: the spheroidal population and the disk population. They are subdivided by their evolution, locations, and motions in the Galaxy.

Physical structure of stars. The sun and other main-sequence stars have a source of energy within them that replaces energy radiated from the surface. Most stars are in hydrostatic equilibrium; they neither expand nor contract. Toward the center of the star, high temperatures heat the gases ionized, creating a plasma that is capable of achieving a high density while remaining a gas. In the hottest stars, radiation pressure of high-energy photons can help balance the weight of the outer layers. During most of a star's existence, it maintains thermal equilibrium so that as much energy is released at the center as is radiated from its surface. Energy may be transported by one of three means: conduction, convection, or radiation. The sources of energy in main-sequence stars is thermonuclear fusion in which four protons fuse to form a helium nucleus, with the liberation of energy. Hydrogen burning proceeds by the proton-proton chain and the carbon-nitrogen-oxygen cycle. In low-mass stars, the *p-p* chain predominates; in high-mass stars the CNO cycle predominates.

Mathematical models of stars. Equations of stellar structure describe how mass, pressure, temperature, and luminosity vary outward from the center of a star. The solution of these equations is a mathematical stellar model. From these models we can infer how stars evolve.

Main sequence stars. Stars remain on or near the main sequence for two reasons: first, the large yield of energy per gram from hydrogen fusion; second, the vast amount of hydrogen available. The approximate time a star spends on the main sequence is proportional to its mass divided by its luminosity. The solar neutrino experiment is an effort to determine the temperature at which these thermonuclear processes proceed rather than to deduce them theoretically.

KEY TERMS

disk population stars
early-type stars
energy-generating core
gravitational potential energy
H-R diagram
hydrogen burning
hydrostatic equilibrium

late-type stars
main-sequence lifetime
opacity
perfect gas
spheroidal population stars
thermal equilibrium
thermonuclear fusion

CLASS DISCUSSION

1. Discuss how the various physical properties of stars, such as luminosity, mass, radius, surface temperature, and density, vary from one part of the H-R diagram to another.

2. Discuss, with regard to similarities and differences, the distinctive aspects of the H-R diagrams for the field stars, the bright stars of the night sky, the nearby stars, variable stars, open clusters, and globular clusters.

3. Where does the energy in a ray of sunlight striking the surface of the earth begin the journey leading to its encounter with the earth? Trace the entire path step by step, including the approximate length of time spent at each step in the journey.

4. Why are the more-massive main-sequence stars even more luminous than one would expect if they were only proportionally brighter than the less-massive stars? Why is the main-sequence lifetime of the massive stars shorter than that of the low-mass stars?

5. Explain how stellar models (mathematical models) can be used to understand the evolution of stars. Is it possible to follow the evolution of an individual star by watching it? If not, how do we know stars really evolve?

READING REVIEW

1. What is the range of masses for stars? What interval do their luminosities span?

2. Name the various regions of the H-R diagram.

3. Are most of the bright stars in the night sky bright because they are the nearby stars?

4. Do all open clusters contain red giant or supergiant stars?

5. What percentage of all the stars in our Galaxy are main-sequence stars? What percentage are white dwarfs?

6.   Is there any relation between the location of a star in the Galaxy and its physical properties?

7.   What does it mean for a star to be in hydrostatic equilibrium or in thermal equilibrium?

8.   How does the sun generate the energy that it radiates away as its luminosity? Is its supply of energy inexhaustible?

9.   Where does Einstein's equivalence between mass and energy play a role in stars? What is thermonuclear fusion?

10. What is the significance of the main-sequence stars? Even though main-sequence stars vary widely in their physical properties, what do they have in common?

SELECTED READINGS

Jastrow, R. *Red Giants and White Dwarfs,* 2nd ed. Norton, 1974.

Marshall, L. A., L. G. Chin, and W. F. van Altena. "Star Cluster Membership," *Sky & Telescope,* August 1981.

Payne-Gaposchkin, C. *Stars and Clusters.* Harvard University Press, 1979.

14
Interstellar Matter: Birthplace of Stars

Before we can discuss the evolution of stars, a topic begun in Chapter 13, we need to know something about the material and the environment in which stars are born. Therefore the first three sections of this chapter are on the nature of the interstellar medium, some of the physical processes going on in it, and how it has changed during the history of the Galaxy. With these points as a basis the current theory for the birth of stars should be more understandable.

Interstellar matter is primarily a gas, in which hydrogen is the chief component. In regions near very luminous and very hot stars the gas is ionized, while in other regions it is a fairly cold gas that allows molecules containing up to 10 or so atoms to exist. Mixed in with interstellar gas is a very fine dust, whose grains are about the size of the smoke particles that can be seen as flashes in a shaft of light coming through a window. Interstellar dust, however, appears to have a very different chemical composition from smoke particles.

Interstellar matter is not uniformly spread throughout the Galaxy but is clumped together in interstellar clouds that vary from small features in the disk of our Galaxy to giant cloud complexes in the Galaxy's central bulge. The stars of our Galaxy—and presumably the stars in all the billions of galaxies in the universe—are born in interstellar clouds. And when they come to the end of their lives, many stars throw off matter that mixes with the rest of the interstellar medium, where it forms new interstellar clouds, and finally becomes the matter composing new generations of stars. Thus new stars form from matter that was once or maybe several times part of an earlier generation of stars.

14.1
Gas, Dust, and Molecules among Stars

Early in this century astronomers believed that interstellar space in our Galaxy was fairly transparent and that any dimming of starlight could be ignored. Then in the 1930s astronomers discovered that distant stars, compared with nearer ones, are diminished in brightness and reddened in color by an *interstellar medium* that absorbs and scatters starlight. The interstellar matter of our Galaxy, and presumably other galaxies, is a mixture of atomic and molecular gases, mostly hydrogen, along with small solid particles, called *grains* or *dust*, concentrated primarily in the plane of the Galaxy.

If you exhale your breath once and let it expand into an evacuated cubical enclosure 1 kilometer on a side, the resulting density of your breath will exceed the density in most parts of the interstellar medium. Although this suggests that interstellar space is nearly a vacuum, there is a significant amount of matter lying between the stars because of the vast volume of space. In space the interstellar medium is found in clumps that are referred to as *interstellar clouds*; they come in a variety of sizes, compositions, and forms.

INTERSTELLAR GAS

The gaseous component of interstellar matter is confined almost entirely to a thin disk in the plane of the Galaxy. In the vicinity of the sun the disk is only about 1000 light years in thickness. About 90 percent of it is hydrogen, of which perhaps half is in molecular form and half in atomic form. Atomic hydrogen occurs in both neutral and ionized forms. Molecular hydrogen and ionized hydrogen are found in only a small fraction of interstellar space compared with the vast volume in which neutral hydrogen atoms are located. And where hydrogen molecules or ions are located, they are the most prevalent form of hydrogen, the other being absent. Because hydrogen is the main ingredient in the interstellar gas, astronomers generally designate a region in which hydrogen is predominantly ionized as an *H II region* and a region where hydrogen is predominantly neutral as an *H I region*.

Starlight passing through cool interstellar gas is selectively absorbed, producing a few absorption lines superimposed on the normal spectra of stars. These *interstellar lines* can be differentiated from the spectral lines of the O and B stars because they are usually narrow; they are not characteristic of a hot star's photosphere, and they have different Doppler shifts from the stellar absorption lines. Interstellar lines are more difficult to identify in stars of later spectral classes, which have many absorption lines. Frequently we see several sets of Doppler-shifted interstellar lines; this means the starlight has passed through several intervening clouds moving at different speeds along the line of sight, as illustrated in Figure 14.1.

In the visible region of the electromagnetic spectrum, astronomers have identified absorption lines belonging to such elements as sodium, calcium, and iron and molecules like cyanogen (CN) and methylidine (CH). In the ultraviolet part of the spectrum molecular hydrogen, atomic hydrogen, carbon, nitrogen, oxygen, iron, and other absorption lines have been identified in observations with the *Copernicus* satellite. A surprising find is a large amount of deuterium, the heavy isotope of hydrogen, compared with its abundance on earth. Since 1964, in the radio portion of the spectrum, lines of hydrogen, helium, and carbon have been observed, lines resulting from electron transitions between upper energy levels near

FIGURE 14.1
Multiple interstellar lines of ionized calcium and neutral sodium in the spectrum of the B0 supergiant ε Orionis. Clouds of gas in space absorb small amounts of energy from the starlight as it passes through them, producing interstellar absorption lines in the spectra of distant stars. (ε Orionis is about 1500 light years away; see Table 12.4). The strength of such lines depends on the number of absorbing atoms lying along the line of sight. The five different sets of Doppler-shifted lines of singly ionized calcium (H and K lines) and neutral sodium (D lines) give the following velocities for the five absorbing clouds through which the starlight has passed: +3.9, +11.3, +17.6, +24.8, and +27.6 kilometers per second, respectively. Above is an idealized representation of the clouds with light from ε Orionis passing through three of the five clouds. The arrows indicate the velocities of the clouds and ε Orionis relative to the sun.

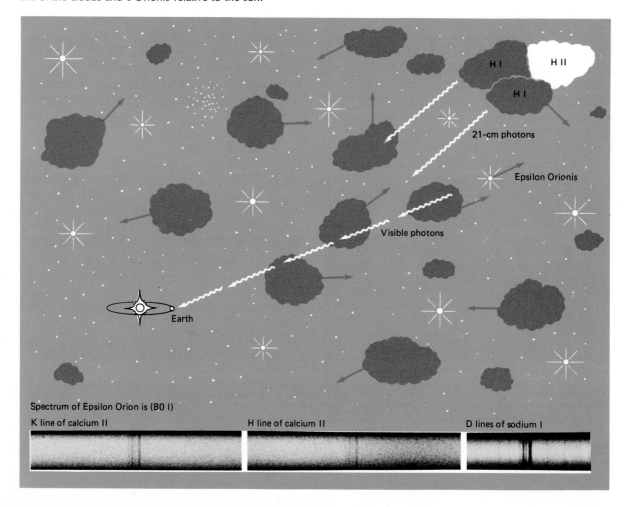

Spectrum of Epsilon Orion is (B0 I)

K line of calcium II   H line of calcium II   D lines of sodium I

KARL G. JANSKY   (1905–1950)

After graduation from the University of Wisconsin, Jansky joined the Bell Telephone Laboratories in New Jersey in 1928. His work there dealt with the problems of the short-wave radio telephone. In 1931 he was assigned the task of tracking down the crackling static noises that plagued overseas telephone reception. At the Holmdel station he constructed a large directional antenna system.

Jansky recorded two well-known kinds of atmospheric static: crashes from local thunderstorms and noise from distant thunderstorms reflected from the ionosphere. From his records he later singled out a weak, third kind of static that could hardly be distinguished from the internal receiver noise. Through headphones the weak noise sounded like a steady hissing. At first Jansky thought that the interference came from the sun, but after a year of careful measurement he concluded that the radio waves came from a specific region on the sky every 23 hours and 56 minutes. Suspecting that the radiation was coming from an astronomical source, he attempted to trace its origin. He knew from his study of astronomy that the period of the earth's rotation relative to the stars was 4 minutes less than the 24-hour period relative to the sun, and this was the clue that the radio noise originated in space beyond the solar system. He found that its direction coincided with the constellation of Sagittarius toward the center of the Milky Way.

At the age of 26 Jansky had made a historic discovery—that celestial bodies could emit radio waves as well as light waves. But his results, published first in 1933, received little attention. Not until the end of World War II was the significance of his achievement widely appreciated.

Jansky's serendipitous discovery gave birth to a new branch of astronomy, radio astronomy. In Jansky's honor astronomers named the unit of radio flux the jansky (10^{-26} watts per square meter per hertz).

---

the ionization limit. For example a free electron may be captured into level 110 of a hydrogen atom, from which it can cascade down to level 109 and emit a radio photon with a wavelength of 6 centimeters. Although in the overall interstellar medium the abundances of the chemical elements that are heavier than helium are similar to what they are in the sun and other stars, the chemical elements are not spread out uniformly. This makes it hard to decide what typical element abundances in the interstellar medium are. It may be that there are no really typical values.

THE 21-CENTIMETER LINE

A new means of exploring the interstellar medium and its structure became available to astronomers in 1951, when there was discovered a spectral line at 21 centimeters (1420 megahertz) produced by neutral hydrogen. How are photons formed that have a 21-centimeter wavelength, which is in the radio region of the electromagnetic spectrum?

As an electron revolves about the proton, it and the proton also spin like tiny rotating tops (Figure 14.2). Once in 11 million years (on the average) an electron, if spinning in the same sense as the proton to which it is bound and not disturbed by collisions with other atomic particles, will spontaneously change its spin to the opposite sense. This change drops the atom into a lower energy state, creating a photon, whose wavelength is 21 centimeters, to carry away the difference in energy. Within an interstellar cloud an electron may actually, reverse its spin much sooner, as often as once every several hundred years during collision with a passing atom. Random collisions between particles in the interstellar medium can transfer kinetic energy to a bound electron and cause it to flip over and align its spin with that of the proton.

Even though the time lag for producing a 21-centimeter photon is inordinately long, a ready supply of 21-centimeter radiation is always available because of the enormous number of hydrogen atoms along a line of sight through the Galaxy.

The emission of 21-centimeter photons not only confirms the importance of hydrogen as the primary

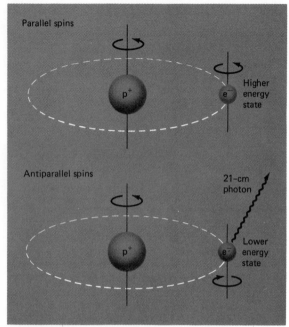

FIGURE 14.2
In the upper neutral hydrogen atom the spins of the proton and the electron are parallel; that is, they have the same sense of rotation, such as counterclockwise. In the lower atom they are anti-parallel, or have opposite senses of rotation, such as the proton counterclockwise and the electron clockwise. When they are parallel, the electron is in a higher energy state than it is when they are antiparallel. The transition from parallel to anti-parallel produces a 21-centimeter photon as shown.

constituent of the interstellar medium but also provides radio astronomers with a valuable tool for studying the structure of the Galaxy. Because of its long wavelength, a 21-centimeter photon can travel greater distances through interstellar space than photons of visible light can. This is because electromagnetic waves are more likely to interact with bits of matter the closer their wavelengths are in size to the characteristic size of the matter. Thus electromagnetic waves with visible wavelengths more readily interact with atoms, molecules, or very small solid particles than do waves with very long radio wavelengths.

INTERSTELLAR MOLECULES

Since 1963 radio astronomers have found a surprising number of *interstellar molecules*, including many organic ones (those containing carbon), by searching for their spectral fingerprints, which occur in the centimeter and millimeter region of the electromagnetic spectrum. From approximately 150 radio spectral lines some 50 or more molecules (Table 14.1) — containing mostly combinations of hydrogen, carbon, nitrogen, and oxygen — have been identified; the number of different molecules discovered is increasing yearly. Some of the newly discovered molecules are familiar inorganic compounds, such as ammonia (NH_3), water (H_2O), and several containing sulfur.

A number of interesting organic molecules have also been found, such as formaldehyde, methyl alcohol and ethyl alcohol. The latter has nine atoms in the molecule; it and the heaviest, cyano-octatetra-yne, indicate that, however they are produced, the mechanism can form rather complex molecules. Some of the interstellar organic molecules have not yet been produced in a chemistry laboratory so that study of the interstellar medium is adding a new dimension to organic chemistry.

Compared with hydrogen, the amounts of other molecules that occur in interstellar space are small — less than one-thousandth that of hydrogen. A few of their spectral lines are observed as absorption lines instead of emission lines whenever enough molecules lie along the line of sight toward a Galactic or extragalactic radio source that emits continuous radiation. In addition to being found in our Galaxy, the hydroxyl radical, water, formaldehyde, hydrogen cyanide, ammonia, and carbon monoxide have also been detected in several nearby galaxies. Thus their presence in the interstellar medium of our Galaxy is not an isolated event in the universe.

Molecules are primarily found in dark cloud complexes such as those in the constellations of Orion, Taurus, and Ophiuchus. Toward the Galactic center there is a ring of cloud complexes on the outskirts of the Galactic bulge, and there are some large molecular clouds closer to the center. Other locations for molecules appear to be widely distributed across the Galaxy in localized regions containing interstellar clouds. Some are even concentrated in tiny regions comparable in size to the solar system. When we discuss the structure of our Galaxy in Chapter 16, we shall have more to say about the locations of molecules.

How these molecules were formed is not well understood. A two-atom collision can produce a di-atomic molecule, but it is very difficult to imagine a sequence of successive collisions that can produce a polyatomic molecule with nine atoms. But before we

TABLE 14.1
Molecules Detected in the Interstellar Medium

Number of Atoms in the Molecule

2	3	4	5	6	7	8	9	11
H_2[a]	H_2O	NH_3	CH_4	CH_3OH	CH_3NH_2	$CHOOCH_3$	CH_3CH_2OH	HC_9N
CH[b]	C_2H	H_2C_2	CH_2NH	CH_3CN	CH_3C_2H	CH_3C_2CN	CH_3CH_2CN	
CH^+[b]	HCN	H_2CO	NH_2CN	CH_3SH	CH_3CHO		CH_3OCH_3	
OH	HNC	$HNCO$	$HCOOH$	NH_2CHO	CH_2CHCN		HC_7N	
CN[b]	HNO	$HNCS$	HC_4		HC_5N			
CO[a]	HCO	H_2CS	HC_3N					
C_2[c]	HCO^+	C_3N	H_2C_2O					
CS	HCS^+							
SiO	O_3							
NO	HN_2^+							
NS	H_2S							
SO	OCS							
SiS	SO_2							

[a] Identified by ultraviolet lines.
[b] Identified by visible lines.
[c] Identified by infrared lines.

discuss a possible formation for the molecules, we should introduce the interstellar dust.

INTERSTELLAR DUST

Interstellar dust consists of solid grains of microscopic size whose composition and properties are not like the dust on earth. Photographs of regions along the Milky Way are laced with dark patches that are large clouds containing dust as well as gas. The dimming of starlight is due almost entirely to interstellar dust; for the gaseous component of the interstellar medium is known to be fairly transparent to starlight. In fact the interstellar gas is billions of times more transparent to visible light than is air at sea level on the earth. Clinging close to the Galactic plane, interstellar dust completely shuts off our view of the Galactic center in the visible wavelengths, and it keeps us from seeing extragalactic objects near the Galactic plane. In other spiral galaxies seen edgewise this dust is the dark lane that passes centrally across the galaxy (see Figure 17.5).

Dimming by interstellar dust is greatest for ultraviolet light, less for visible light, and least in the infrared wavelengths. For visible light the loss can be as much as 0.7 magnitude per 1000 light years (the average is about half this value) near the Galactic plane. This means that for a star at the center of the Galaxy,

about 30,000 light years away, only about 1 photon out of every 100 billion reaches us. If we do not correct the observed apparent magnitude of a distant star for the loss of light, its distance calculated from the distance modulus is too large. In the X-ray, infrared, and radio spectral regions, however, we can observe all the way to the Galactic center since the dust is transparent to those wavelengths. Because blue light is affected twice as much as red light by interstellar dust, light from a distant star not only looks dimmer but is also redder than it should be for the spectral type of the star (Figure 14.3). Astronomers refer to this effect as *interstellar reddening*. Because of this effect, color indices measured for distant stars are in error and must be corrected before they can be used as a measure of the star's temperature.

About 1 percent of the mass of interstellar clouds is due to dust and 99 percent to gas. The average density of the dust is about one grain per 10^{13} cubic centimeters, or one grain in a cube 200 meters on a side. This is a very low density when compared to the typical interstellar gas density of one atom per cubic centimeter. The density of the dust grains can be much larger in small localized regions, such as the heart of an interstellar cloud. The density of gas will also be larger; and it appears that the ratio of dust to gas in most of the interstellar medium is constant to within a factor of about two.

What is the size and composition of the dust grains? Since the scattering of photons is strongly dependent upon the size of the scattering particle, the strong scattering of visible light by interstellar grains suggests that a grain is comparable in size to the wavelength of light, say, a few hundred-thousandths of a centimeter or smaller. At this size the typical grain should contain about 100 million atoms. The grains are most likely composed primarily of elements heavier than hydrogen and helium. From the reddening, dimming, and polarizing of starlight astronomers conclude that dust grains are probably assorted graphite, iron particles, silicon carbide, silicates, and ices.

It seems possible that most of the interstellar dust can come from material that is being blown out of stellar atmospheres. There are several phases in a star's life, starting from birth and ending with its death, when the star can lose matter. As an example, among the brightest sources of infrared radiation are the glowing dust shells around some stars, called *circumstellar shells* (Figure 14.4). Apparently the grains intercept short-wavelength radiation from the central star, heat up, and reradiate the energy as long-wavelength infrared photons.

Interstellar dust grains may be the means of forming interstellar molecules. It is thought that hydrogen and other types of atoms can accrete on the cold surfaces of the grains, where they bond together to form molecules. These molecules can escape from the grain surface by absorbing a low-energy photon of starlight (or by some other means). Apparently the enveloping dust cloud prevents ultraviolet starlight or other energetic photons from reaching the interstellar molecules and dissociating them.

14.2 Interstellar Clouds and Nebulae

INTERSTELLAR CLOUDS

In photographs of the Milky Way our view of the starry background is partly or wholly blocked by *dark nebulae*. They contain denser concentrations of interstellar dust than occur generally in the Galactic plane. One such dark region is a long, chainlike complex, composed of dozens of isolated and connected dark *interstellar clouds*, that stretches about halfway around the Milky Way from the constellation Cygnus to Crux. This obscuring strip forms the Great Rift dividing the Milky Way into two branches, as shown in Figure 16.1. In many regions along its length this dark nebulosity separates into tangled lanes of absorbing material that partially cover bright, glowing, gaseous nebulae.

FIGURE 14.3
Scattering and reddening by interstellar dust. Since interstellar reddening does not cause a shift in wavelength, be careful not to confuse this phenomena with "redshifts" due to motion (Doppler effect).

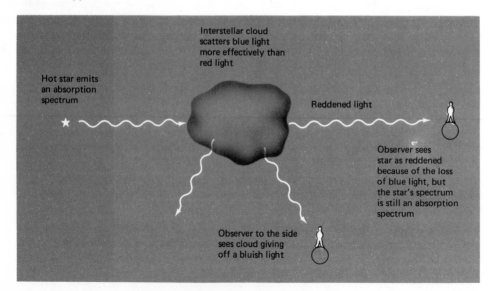

Hot star emits an absorption spectrum

Interstellar cloud scatters blue light more effectively than red light

Reddened light

Observer sees star as reddened because of the loss of blue light, but the star's spectrum is still an absorption spectrum

Observer to the side sees cloud giving off a bluish light

probably accounts for some of the wide variation in the properties quoted for interstellar clouds. Both types of clouds are irregularly shaped and are from 0.1 to 50 light years in diameter. The giant molecular clouds can be as much as several hundred light years across. Their temperatures go from about 100 K for the diffuse clouds down to 10 to 20 K for the dark clouds. Interstellar clouds may take up about 4 percent of space in the Galactic plane, with typical masses of several solar masses up to 10^4 solar masses for diffuse clouds and up to 5×10^5 solar masses for the giant molecular clouds. Their densities, which may vary from 100 particles per cubic centimeter for diffuse clouds to more than 1 million particles per cubic centimeter for dark clouds, are low compared with the 10^{19} air molecules per cubic centimeter in the air which we breathe. Even so, dark clouds can be remarkably opaque because of the accumulated effect of light extinction on starlight as it passes through an enormous length of interstellar dust in the clouds.

Typical separations between clouds appear to be on the order of hundreds of light years. The total number of the giant molecular clouds in our Galaxy may run up to several thousand, representing a couple of billion solar masses of interstellar matter. The largest single concentration of giant molecular clouds is the ring mentioned earlier, lying some 15,000 light years from the center of the Galaxy (Figure 14.5). It has been suggested that this ring may contain as much as 90 percent of all the interstellar matter in the Galaxy.

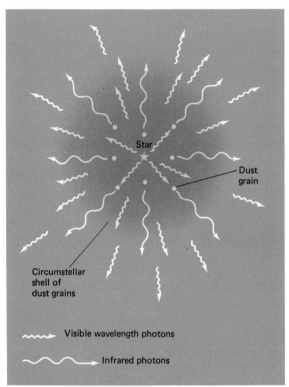

FIGURE 14.4
Circumstellar shell. Radiation emitted by the central star is absorbed by dust grains in the circumstellar shell, warming them and causing them to radiate energy in the infrared spectral region. The thicker the dust shell, the greater the amount of energy emitted by the central star that will be converted to infrared radiation.

Even though ground-based observations have provided us with important information about the properties of interstellar clouds, much of our understanding of them has come from ultraviolet studies with the *Copernicus* and *IUE* satellites. We now find that the clouds can be divided into *diffuse clouds*, which are thin enough for us to observe stars behind them, and *dark clouds*, which are so opaque that stars behind them cannot be seen. Some of the dark clouds are of immense extent and are the locations for many different types of molecules; these are known to astronomers as *giant molecular clouds*. The intercloud region (between clouds) contains a high-temperature and low-density gas, much of it ionized hydrogen, and some dust grains.

For astronomers it is still not clear in all the examples under study where individual clouds leave off and groups of clouds (or cloud complexes) begin. This

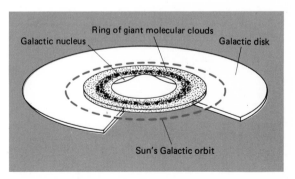

FIGURE 14.5
Ring of giant molecular clouds about the center of our Galaxy. Lying between roughly 12,000 and 26,000 light years with strongest concentration at about 17,000 light years, these giant molecular clouds contain about 90 percent of all the interstellar matter in the Galaxy. The dashed line represents the orbit of the sun about the Galactic center.

CARBON MONOXIDE IN DARK INTERSTELLAR CLOUDS

In dark interstellar clouds hydrogen is primarily in the form of molecules rather than atoms so that clouds are not sources of 21-centimeter radiation. In addition molecular hydrogen has no spectral features in the visible or near-infrared part of the spectrum. With the discovery of a strong emission feature at 2.6 millimeters due to carbon monoxide (CO) radio astronomers have acquired a marker of molecular hydrogen's location and a new probe for investigating dark clouds. Dark clouds are the primary locations for the interstellar molecules, and the CO molecule is much more abundant in them than in the general interstellar medium. CO has proved to be an invaluable aid to the study of dark interstellar clouds, as suggested by Figure 14.6.

FIGURE 14.6
CO contour lines superimposed on an interstellar cloud complex. This negative reproduction shows emission from H II regions as dark spots near upper center. The molecular clouds are the white spaces outlined by the black contour lines that show emission by the carbon monoxide molecule. The pluses are the locations of strong infrared sources. And, of course, stars appear as black dots.

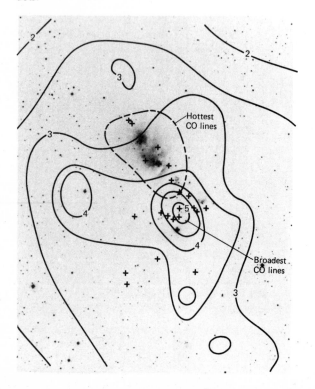

DARK CLOUDS WITH ENERGY SOURCES: INTERSTELLAR MASERS

Some of the dark interstellar clouds are detectable from their continuous emission of radio waves and infrared radiation. These clouds obviously have sources of energy within them. In 1965 astronomers accidentally found microwave emission produced by the hydroxyl radical (OH) that is coming from dark clouds. The character of the emission was peculiar, and the region from which it came was very small. They found that these small regions were also bright sources of infrared radiation, but they emit virtually no visible light.

The radio emission was much stronger than could be accounted for by random thermal collisions. Clearly there had to be a mechanism that was selectively exciting the OH molecule. It is thought that infrared radiation from nearby stars excites OH molecules and they are stimulated to deexcite by interaction with stellar photons of the right wavelength. The emitted radiation in turn stimulates other molecules to radiate in the same fashion, producing an avalanche of emission. Thus the radiation in a normally weak line is greatly amplified. The word *maser* used to describe this phenomenon is an acronym for *m*icrowave *a*mplification by *s*timulated *e*mission of *r*adiation.

Astronomers know of several hundred OH masers and several dozen H$_2$O masers operating in dark interstellar clouds. They are also found in the atmospheres of the red giants that are variable stars. In general the masers in molecular clouds are brighter than those in luminous red stars, but those associated with stars seem to be more numerous. Our interest here, however, is the significance of the presence of masers in clouds where astronomers believe stars are forming. Clearly the masers in clouds signify that energetic events are occurring at specific points in molecular clouds. Such events are most likely star formation.

H II EMISSION NEBULAE

Even far from hot O and B stars their emission of ultraviolet photons is still so great that the ultraviolet photons can ionize the hydrogen gas of an interstellar cloud (H I region). With the ionization of hydrogen the region becomes an H II region, or an *emission nebula* (Figure 14.7). Stars of spectral type O5 emit enough ultraviolet photons to ionize hydrogen out to

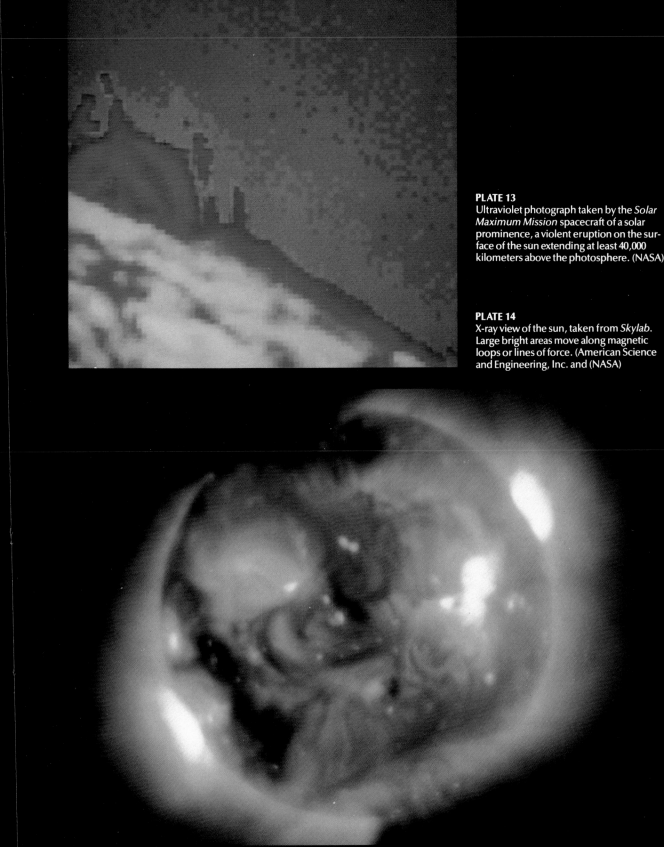

PLATE 13
Ultraviolet photograph taken by the *Solar Maximum Mission* spacecraft of a solar prominence, a violent eruption on the surface of the sun extending at least 40,000 kilometers above the photosphere. (NASA)

PLATE 14
X-ray view of the sun, taken from *Skylab*. Large bright areas move along magnetic loops or lines of force. (American Science and Engineering, Inc. and (NASA)

PLATE 15

Great Nebula in Orion (M42) photographed with the Hale 5.1-meter reflector. The Orion Nebula is the birthplace of new stars some 1500 light years from earth. Its overall diameter is about 20 light years. The nebula consists mostly of glowing hydrogen gas with some helium and lesser amounts of heavier atoms, all stimulated to fluorescence by the ultraviolet light of its recently created stars. This phenomenon is similar to the mechanism causing certain mineral substances to glow when irradiated with ultraviolet ("black") light. In the nebula stellar ultraviolet light striking the atoms is absorbed and energy is reradiated in the visible colors. (Copyright by the California Institute of Technology and Carnegie Institution of Washington. Reproduced by permission from Palomar Observatory, California Institute of Technology.)

PLATE 16

Using data obtained from the *Kuiper Airborne Observatory,* scientists created this image of the Orion Nebula to show how it would appear through a large telescope sensitive to infrared. The infrared image was superimposed on a black-and-white photograph provided by the Lick Observatory. The colors represent infrared wavelengths between 0.03 and 0.1 mm. (NASA)

LATE 17

North America Nebula in Cygnus (NGC 7000) photographed with the 1.2-
meter telescope of the Hale Observatories. The pinkish glow arises from
hydrogen gas shining by the same mechanism that causes gas in an emission
nebula to radiate: the ultraviolet light of nearby hot stars, which is absorbed
by the gaseous medium, is reradiated in the visible region of the spectrum.
Copyright by the California Institute of Technology and Carnegie Institution
of Washington. Reproduced by permission from Palomar Observatory,
California Institute of Technology.)

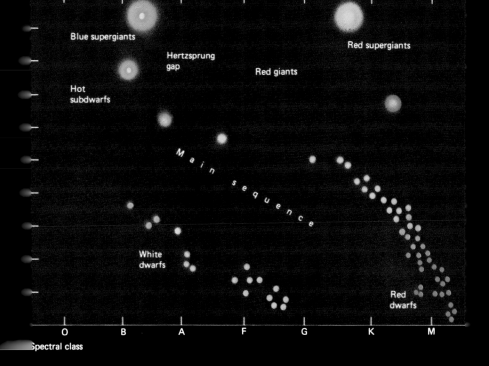

Blue supergiants

Hertzsprung
gap

Red supergiants

Red giants

Hot
subdwarfs

M a i n

s e q u e n c e

White
dwarfs

Red
dwarfs

O   B   A   F   G   K   M

Spectral class

g-Russell diagram showing the integrated or composite color for stars on the main
the red giant region, and the white dwarfs. The differing diameters of the stars are
each of the regions in the diagram.

associated nebulosity in Taurus (M45) photographed with the 1.2-meter telescope on
ar. The nebulosity around the brighter stars of the Pleiades star cluster contains dusty
scatters starlight. The phenomenon is analogous to the color glow observed around
on a foggy night, which results from the scattering of the lamplight by moisture par-
bulosity is bluer than stars because the shorter waves are scattered more strongly
er waves. (Copyright by the California Institute of Technology and Carnegie Institu-
ngton. Reproduced by permission from Palomar Observatory, California Institute
y.)

PLATE 20
Globular cluster M13 in the constellation of
Hercules. (NASA)

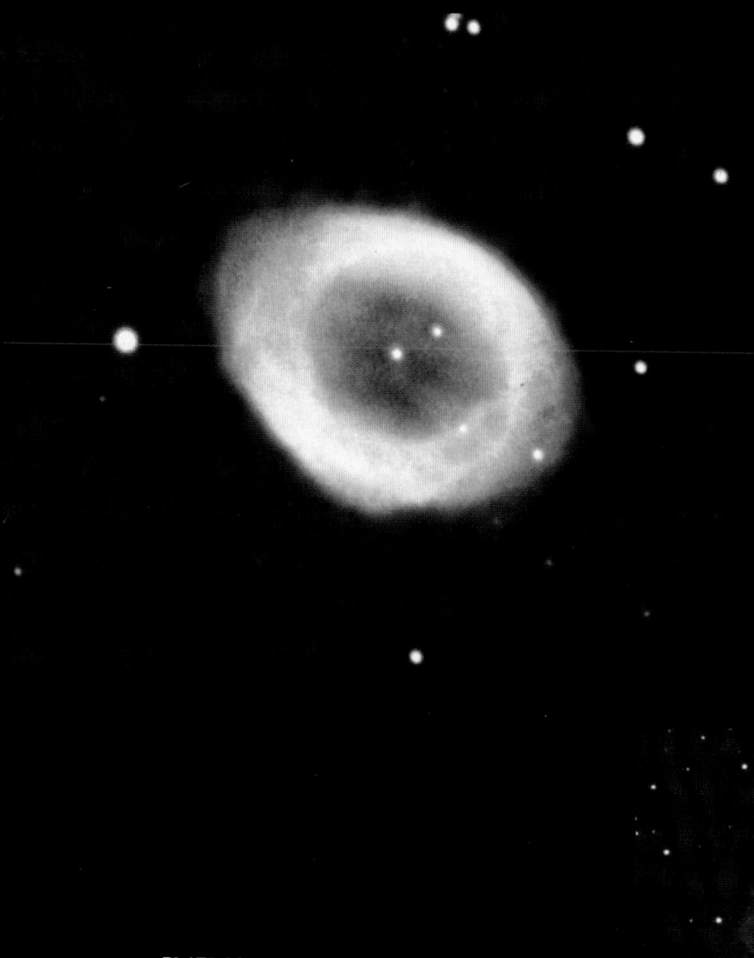

PLATE 21
Ring Nebula in Lyra (M57) photographed with
the 5.1-meter Hale reflector. The central star is a
hot dwarf whose strong ultraviolet light has stimulated its ejected "smoke rings" into lumines-
cence. (Copyright by the California Institute of
Technology and Carnegie Institution of Washington. Reproduced by permission from Palomar
Observatory, California Institute of Technology.)

PLATE 22
Crab Nebula in Taurus (M1) photographed with
the 5.1-meter Hale reflector. Its distance is about
5000 light years and its diameter, which is still
growing, is 6 light years. (Copyright by the California Institute of Technology and Carnegie
Institution of Washington. Reproduced by permission from Palomar Observatory, California
Institute of Technology.)

PLATE 23
NGC 6946, a type Sc galaxy in Cepheus. (NASA)

PLATE 24
NGC 4594 or M104, a type Sa spiral galaxy in Virgo. (© McDonald Observatory–J.D. Wray)

PLATE 25
Great galaxy in Andromeda (M31) photographed with the 1.2-meter Schmidt telescope of the Palomar Observatory. The Andromeda galaxy and its two elliptical satellites are about 2,200,000 light years distant. The light by which we see these objects began its journey about the time humans first appeared. The closely wound dust-streaked spiral arms glow blue with the light of bright young Population I stars. The pinkish-white of the central portion comes from old red Population II stars. This major galaxy, about 130,000 light years in diameter, is the nearest of the spirals. We view it at a rather sharp angle through a field of foreground stars inside our own Galaxy. (Copyright by the California Institute of Technology and Carnegie Institution of Washington. Reproduced by permission from Palomar Observatory, California Institute of Technology.)

PLATE 26
X-ray photograph of quasars. This picture, taken by NASA's *High Energy Observatory* Satellite 2 in March 1979, shows the bright quasistellar object 3C273 at the lower right. The X-ray emission in the upper left is from a never-before-seen quasistellar object that is probably the brightest and most distant one yet discovered. Most of the white and blue dots are not other X-ray sources, but are artifacts of the equipment and experimental procedures. (NASA)

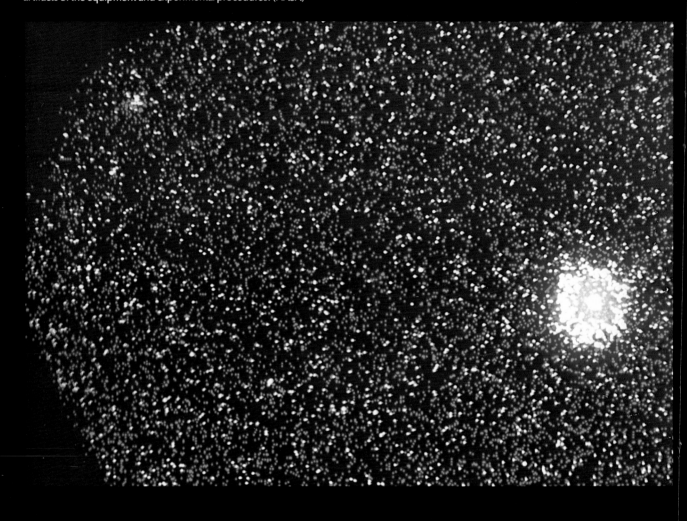

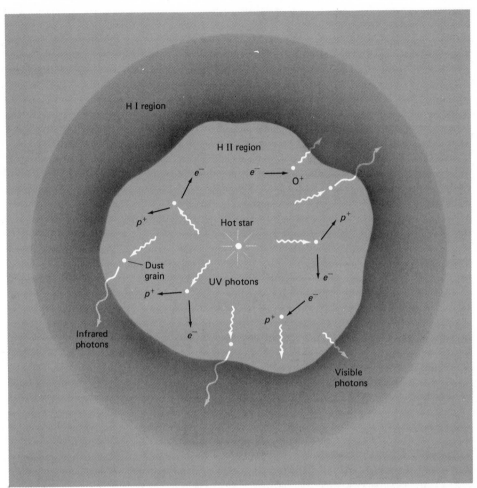

FIGURE 14.7
H II region about a hot star. Ultraviolet photons are absorbed by neutral hydrogen atoms, ionizing the atom and creating free electrons that excite the surrounding gas. The dust near the hot star also absorbs some radiation, is heated, and emits infrared radiation. The wavelength of most of the infrared radiation emitted lies between 30 and 100 microns.

distances of 300 light years from the star. For cooler spectral types the surrounding H II region is smaller; an A0 star creates an ionized region about it that is less than 1 light year in radius. Table 14.2 lists some of the properties of a few emission nebulae.

In panoramic photographs of the plane of the Milky Way, such as Figure 16.1, one sees many bright glowing regions whose spectrum is an emission spectrum. The Balmer alpha line of hydrogen is responsible for the vivid red color of many H II regions. These H II regions are produced by hot stars and are associated with interstellar clouds either by being surrounded by them or by being on the edge of a cloud complex.

H II regions occur in about six distinct categories,

depending upon their size and the density of free electrons resulting from the ionization of hydrogen. Astronomers refer to the smallest as ultracompact ones and the largest as supergiant H II regions. The smallest ones are from a few tenths to a few tens of light years in diameter and their masses range from a few tenths to a few solar masses. These categories of H II regions are generally buried in dark molecular clouds so that in the visible part of the spectrum they are almost totally obscured form view or are heavily reddened if seen (the larger ones generally can be seen).

In the three categories of large H II regions, the regions are from a few to 1000 light years in diameter,

TABLE 14.2
Properties of Selected Emission Nebulae

Name	Messier Number	Distance (ly)	Diameter (ly)	Mass (sun = 1)	Density (particles/cm³)	Temperature (K)	Spectral Type of Exciting Stars
Orion	M42	1600	20	300	600	9000	O
Trifid	M20	3200	15	150	100	8200	O
Lagoon	M8	4000	30	1000	80	7500	O
Eagle	M16	5600	20	500	90	8000	O
Omega	M17	5200	30	1500	120	8700	B

and they contain anywhere from tens to hundreds of millions of solar masses of ionized matter. These three categories are the types of emission nebulae that are seen most readily, with the supergiant H II regions being by far the brightest objects in the spiral arms of our Galaxy and other spiral galaxies. They occur sometimes in groups and sometimes isolated from each other. As for the shapes of these H II regions, they range from readily definable shapes to large complex ill-defined regions. Altogether it is estimated that about 1 percent of the mass of our Galaxy is tied up in the form of H II emission nebulae.

WHAT TYPE OF INTERSTELLAR MEDIUM SURROUNDS THE SUN?

Before leaving a discussion of the interstellar medium, we should ask about the nature of the interstellar matter that surrounds our solar system. From what we have said about giant molecular clouds, the sun is obviously not sitting in the middle of one of them. Observations with the ultraviolet satellites place the sun in the low-density (about 0.1 particle per cubic centimeter) and high-temperature gas of the intercloud region. Although the sun in its motion relative to the stars of the solar neighborhood (to be discussed in Chapter 16) could encounter a dense cloud (greater than 100 particles per cubic centimeter), it is not likely to happen soon.

14.3
Mass Loss by Stars

There are many stars that, for one reason or another, lose mass at one or several times during their lives. Such mass loss is important to the star for its effect upon the star's evolution. It is also important because it alters both the dynamics and the chemical composition of the interstellar medium.

The evidence for the loss of matter by stars comes from direct telescopic observations and from indirect studies of the spectra of stars. For some stars matter is ejected in one gigantic explosion, such as for the supernova and to a lesser extent the nova. The planetary nebula is another example of a single expulsion of material from the star. However, the planetary nebula, compared to the supernova, is a very gentle event. There are also stars for which the evidence points to an almost continuous loss of matter over a substantial period of their lives.

STELLAR WINDS

The continuous loss of matter is called a *stellar wind*, and it is thought to be roughly analogous to the loss of mass by the sun in the solar wind. As discussed

They cannot scare me with their empty spaces
Between stars—on stars where no human race is.

Robert Frost

◄ **FIGURE 14.8**
Halo of gas around the red supergiant Betelgeuse due to its stellar wind is seen in this computer-enhanced image.

FIGURE 14.9 ▼
Interstellar bubble blown by stellar wind. Known as NGC 2359 in Canis Major, this bright filamentary bubble is created by a strong stellar wind's colliding with a reasonably dense interstellar medium. The dark patches surrounding the bubble are dark molecular clouds. The central star is a relatively rare Wolf-Rayet star, which is as hot as or hotter than an O star. It is apparently losing 10 to 100 times more matter in the form of a stellar wind than does an O star. Without the dense interstellar surroundings the shell of the bubble would not glow.

earlier, the solar corona is a tenuous, spherical halo of very hot gas, whose temperature is 1 to 2 million K. These very high temperatures arise from the deposition of energy in the coronal gases from the dissipation of energy stored in coronal magnetic fields (see Section 11.5). The pressure of the gases in the hot corona exceeds that of the interstellar medium surrounding the sun, and thus the solar corona continuously expands, being replaced by material from the photosphere. Such a wind is said to be thermally driven.

Do other stars have thermally driven stellar winds? There is no reason to believe that the sun is unique in this respect. The amount of mass lost by the sun in this way is so small that it would take literally trillions of years for it to lose a significant fraction of its total mass. It is therefore unlikely that we can detect such a low rate of mass loss in other stars. Even though there are large numbers of low-mass stars like the sun, it seems unlikely that stellar winds by these stars could have appreciably altered the nature of the interstellar medium.

Red giants and supergiants are stars with very large radii and very low mean densities. Atomic and molecular particles in the outer layers of their atmospheres are not held so tightly by the gravity of the star as is the case for main-sequence stars. More substantial stellar winds occur in the red giants and supergiants, but with smaller velocities than the solar wind (astronomers have actually photographed a faint halo of scattered light from enormous envelopes of gas surrounding the red supergiant Betelgeuse; see Figure 14.8).

The recognition that some hot stars of spectral classes O and B continuously lose mass dates from about 1929. Surveys using the *Copernicus* ultraviolet satellite have now shown that all stars brighter than bolometric magnitude −6 are losing appreciable amounts of mass by stellar winds.

Estimates are that O stars contribute as much as 30 percent of the matter being lost by stars even though as a class they contribute only about 1 to 2 percent of the Galaxy's luminosity. Their winds may be driven as much by the pressure of the immense numbers of photons they emit (radiation pressure) as by thermal effects (Figure 14.9).

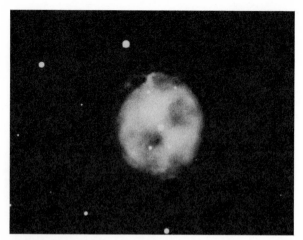

FIGURE 14.10
Planetary nebula known as the Owl Nebula (M97)
in the constellation of Ursa Major.

PLANETARY NEBULAE

In contrast to a continuous loss of matter is the expulsion at one time of the outer layers of a star, such as the mild and slow ejection of the surface layers that forms planetary nebulae.[1] In photographs, such as Plate 21 of the Ring Nebula in Lyra (M57) or Figure 14.10 of the Owl Nebula in Ursa Major (M97), one can see a small, hot, subluminous central star surrounded by a nebulous shell of ionized gas. The shell is expanding slowly outward at speeds of about 30 kilometers per second.

The nebulous shell emits an emission spectrum produced by rarefied common gases like hydrogen, helium, oxygen, neon, and sulfur. The source of energy to produce the spectrum is the ultraviolet light of the hot central star, whose surface temperature is around 100,000 K. At such temperatures most of the star's luminosity is composed of ultraviolet photons. The luminosity of the star is typically 1000 times that of the sun, while the radius is only a few tenths that of the sun. Such conditions provide the nebula surrounding the star with sufficient energy to give the gas a kinetic temperature of about 10,000 K, a density of several thousand particles per cubic centimeter, a diameter of several tenths of a light year, and a mass of a few tenths that of the sun. The degree of ionization in the surrounding nebula is somewhat higher than that in an H II region, but otherwise they are very

---

[1]The name was given by the eighteenth-century astronomer William Herschel, who noted the resemblance to the disk of a planet. The planetary nebulae certainly have nothing to do with the planets of our solar system.

---

similar. Infrared observations indicate that a great deal of dust accompanies the gases in the nebulous shell. Heated by the absorption of ultraviolet and visible photons, the dust in turn radiates in the infrared region of the spectrum.

The planetary nebula's precursor may be a red giant star of moderate mass located in the Galactic disk. Many stars may pass through the planetary nebula stage although we have identified only about 1000. This transition is brief—lasting a few tens of thousands of years—reducing our chance of seeing it. Astronomers estimate that there are somewhere between 20,000 and 50,000 planetaries in our Galaxy and that a few form each year.

NOVAE

The occurrence of a *nova* is announced by a rapid rise in a star's brightness, amounting to tens of thousands of times its original brightness in a few hours. This is followed by a slow decline in light that may persist for a year before the star settles down to its former obscurity.

We think we understand the cause of this sudden outburst of light. The nova's spectrum shows that matter has been expelled from the star since, to begin with, there is a large Doppler shift of the absorption lines toward the blue, indicating a velocity of approach. Soon after the outburst very broad emission lines appear, indicating that a transparent gaseous shell has been ejected at a high speed, usually a few hundred to 2000 or 3000 kilometers per second. The front part of the shell has the largest blue shift, since it is approaching us, and the back part has the largest red shift, indicating that it is receding. All the other parts of the shell have velocities lying between these two extremes. The amount of matter expelled is estimated to be somewhere between a few hundred-thousandths and a few tenths of a solar mass. In several recent novae radio radiation has been detected, which arises from the thermal energy in their expanding envelopes of ionized gas. And with the X-ray satellites several novae have also been found to be X-ray emitters. About 30 novae may occur in our Galaxy each year; a few even go through recurrent outbursts.

There is strong spectroscopic and photometric evidence that many—and perhaps all—novae are the hotter members of close binaries. In Chapter 15 we shall discuss the evolution of binary stars and shall see in it a natural explanation for the nova outburst.

SUPERNOVAE

A *supernova* is the explosion of a star of such immense proportions that it can be observed in an external galaxy even when the rest of the galaxy cannot be seen. These exploding stars suddenly attain luminosities up to several billion times that of the sun. As many as five supernova outbursts may occur in our Galaxy each century, according to present estimates. Most of them in our Galaxy probably escape detection because of the heavy obscuration by interstellar dust in the plane of the Galaxy. In other galaxies their occurrence varies from several times a century in the brightest and largest spiral galaxies to one every few centuries in the faintest spirals. So far approximately 400 supernovae have been identified in other galaxies.

Of the 100 or so supernova remnants that radio astronomers have found in our Galaxy at least 8 are known X-ray objects and 13 have also been identified optically. Table 14.3 lists some examples of supernova remnants in the Milky Way. They are formed by the expanding shell of gas resulting from the explosion of the star. Somewhat like the sonic boom of a jet airplane, the shell creates a shock front that compresses and pushes interstellar gas ahead of it. This process heats the gas to temperatures of millions of degrees, causing it in turn to emit X rays. This seems to be what has happened to the Loop Nebula in Cygnus, shown in Figure 14.11. Identifying old supernova remnants optically is very difficult because the expanding gas shell thins out so rapidly that it can no longer push

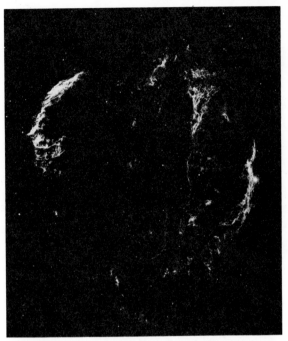

FIGURE 14.11
Loop Nebula in Cygnus, photographed in red light with the 1.2-meter Schmidt telescope, Hale Observatories. The photograph reveals the filamentary remnants of a supernova that exploded about 20,000 years ago and has since reached a diameter of 120 light years. X-ray emission has been detected coming from a point source at the center of the loop. It is possibly from the remaining corpse of the star responsible for the outburst.

TABLE 14.3
Some Supernova Remnants

Object	Age (yr)	Distance (ly)	Diameter (ly)
Cassiopeia A	300	10,000	15
Kepler's SN	370	20,000	13
Brahe's SN	400	9,800	20
Crab Nebula[a]	925	6,500	8
Lupus	975	4,000	35
Puppis A	4,000	7,200	55
Vela X[a]	10,000	1,600	130
Cygnus Loop[b]	20,000	2,500	120
IC 443[c]	60,000	5,000	70

[a] Pulsar inside remnant.
[b] Suspected neutron star at center.
[c] Pulsar near remnant.

interstellar matter ahead of it. Presumably many supernova remnants exist throughout the Galaxy, but we cannot identify them. Ages for the supernova remnants that we observe in our Galaxy are probably less than 100,000 years.

At least two kinds of supernovae are recognized. The major difference between them is in their spectra and maximum luminosity; but the general behavior of both is pretty much the same. Type I supernovae have been observed in all types of galaxies, but they occur most often in the disk of the spiral galaxies. Their maximum luminosity is about 4 billion times that of the sun. For Type I supernovae there is a rapid decline in brightness after maximum luminosity, which is followed by a slowing of the decline with time.

Type II supernovae reach a maximum luminosity of up to 600 million times that of the sun and exhibit a greater variety of light-curve shapes and spectral changes than do Type I supernovae. They appear most

That nova was a moderate star like our good sun; it stored no doubt a little more than it spent
Of heat and energy until increasing tension came to the trigger point
Of a new chemistry; then what was already flaming found a new manner of flaming ten-thousand fold
More brightly for a brief time; what was a pin-point fleck on a sensitive plate at the great telescope's
Eyepiece now shouts down the steep night to the naked eye, a nine-day super-star.

Robinson Jeffers

often in the arms of spiral galaxies but rarely appear in elliptical galaxies.

Although both types of supernovae have very complex and variable spectra, which are not yet fully understood, they both show spectroscopic evidence for very high expansion velocities, which are on the order of 15,000 to 20,000 kilometers per second. How much matter is blown off to return to the interstellar medium? The amounts of mass ejected are not known for certain, but the best estimates suggest that the Type I supernovae belong to the spheroidal population and lose about half a solar mass of material; the Type II supernovae, on the other hand, are stars belonging to the disk population and may eject more than five solar masses (in some cases $50M_\odot$ of stellar matter). In Chapter 15 we shall learn more about what leads a star to become a supernova.

CRAB NEBULA: THE MOST FAMOUS SUPERNOVA
In A.D. 1054, Chinese chronicles say, a "guest star" suddenly appeared in the constellation Taurus; for many months it was visible to the naked eye. This date corresponds very closely to the estimated birth of the Crab Nebula shown in Figure 14.12, the remnant of the most celebrated and most studied supernova. By assuming that the velocity of the nebula across the line of sight is the same as the velocity from the Doppler shift of its spectral lines, astronomers have measured the rate of increase in angular size and

hence estimated its date of birth. From this they have also estimated that the nebula is about 6500 light years away.

In Figure 14.12 you can see the Crab Nebula's outer filamentary structure and an inner smoothed-out amorphous region. Two kinds of spectra have been observed in the light from the Crab: The spectrum of the expanding filamentary network is an emission-line spectrum similar to that of planetary nebulae, while that of the amorphous region is a continuous spectrum. The emission lines and part of the continuous radiation are produced by normal thermal processes. But part of the continuous radiation is nonthermal synchrotron radiation emitted by high-energy electrons spiraling around the magnetic lines of force that permeate the nebula.

Where does the energy come from that has sustained the emission by the Crab Nebula for so many centuries? It was a mystery until the discovery of the pulsar, which we will discuss in Chapter 15. We recognize the pulsar to be the remains of the star whose explosion caused the outburst. And it is this bizarre little stellar remnant, a neutron star, that supplies the energy to power the nebula.

FIGURE 14.12
Crab Nebula, photographed in the red spectral region. Its visible diameter is 8 light years. The filamentary network and associated amorphous region are clearly visible. The arrow points to the pulsar that is thought to be the star responsible for producing the nebula.

14.4
Birth of Stars

How and where are stars born? Observational evidence points to the interstellar gas and dust clouds along the Galaxy's spiral arms as being the birthplaces of stars. As some stars—like those responsible for planetary nebulae, novae, and supernovae—approach the end of their lives, they return some of their mass to the interstellar medium. New generations of stars are thus forming from the "ashes" of previous generations.

COLLAPSE OF INTERSTELLAR CLOUDS

Apparently the particles composing interstellar matter are not subjected to a net force by which an excess gravitational attraction from their neighbors pulls them together. If they were, within several hundred million years all the matter in the interstellar medium should collapse and fragment into stars. Thus all the interstellar matter would have been used up early in the Galaxy's history, and no more stars could form.

The very existence of interstellar matter and its organization into clouds of up to several hundred thousand solar masses argue that gas pressure in the clouds is sufficient to balance the effects of gravity. The first step in making new stars is to compress a cloud to strengthen gravity's effect so that the material in the cloud can contract and fragment into smaller clouds that eventually collapse to form stars.

A promising way of getting this process going is the traveling compressional wave, or density wave, which astronomers believe is responsible for the Galaxy's spiral-arm structure (see Section 16.3). As the wave moves past a cool molecular cloud, it compresses the cloud, driving the particles closer together. Their mutual gravitational attraction is now greater than the gas pressure. If the compressed cloud has no other force that can halt contraction, the collapse continues until the matter heats up, raising the gas pressure sufficiently to resist further contraction.

Another possible mechanism for compressing a molecular cloud is a supernova outburst. An expanding shock front on the leading edge of the gas shell expelled by a supernova explosion impinges on a cloud and compresses it by factors of 10 or more,

triggering gravitational collapse. The discovery that some young stellar associations are located inside the expanding shells of old supernova remnants certainly makes this a believable mechanism.

Finally the collapse of a cloud could begin if the cloud could be cooled so that the gas pressure would go down. There are several possible ways of cooling the cloud, such as dust grains' radiating away energy as infrared radiation.

Regardless of what starts the process, the fragmenting molecular cloud breaks into smaller units; the fragments attract more matter and grow in mass. The rate at which stars are created from the fragments of an interstellar cloud and the number of stars of different masses formed probably depend on several factors: total mass, density, temperature, magnetic fields, and the amount of internal motion stirring the material. The mechanism forming stars favors small-mass stars (see Table 13.2) since we observe many more small-mass stars than we do large-mass ones. Finally it appears that only a small fraction, 1 or 2 percent, of the matter in dark clouds actually forms stars.

Small, dark blobs have been photographed against many of the bright star-filled regions and luminous H II nebulosities of the Milky Way. They are called *globules* (see Figures 14.13 and 14.14) and may be an early stage in the coalescence of matter on the way to forming stars. Their diameters are hundreds to thousands of times that of the solar system, and they contain several tens of solar masses of material.

Matter in collapsing fragments converts its gravitational potential energy into thermal energy, some of which is radiated away into space as infrared radiation. At some point, however, a significant amount of energy goes into dissociating molecular hydrogen to form atomic hydrogen; later more energy is needed to ionize all chemical species. Because this energy used to dissociate and ionize is not available as thermal energy, the collapsing fragment is prevented from even approaching hydrostatic equilibrium. Consequently in a very short time (hundreds to tens of thousands of years) the fragment collapses from a small fraction of a light year in diameter (several million solar radii) to a few thousand solar radii (see Figure 10.3).

During collapse of the fragment, matter in the star has been growing hotter, emitting more visible light and less infrared radiation. Because it is cooler, however, dust in the surrounding stellar nebula out of

FIGURE 14.13
Unusual gaseous nebula in Serpens, photographed in red light with the 5.1-meter Hale reflector. The tiny, dark spots (globules) projected against the bright background are believed to be condensations of gas and dust that may some day begin to radiate as stars.

which the star is forming absorbs visible photons, heats up, and reemits the energy in the infrared. Thus the stellar nebula conceals the developing star until most of the surrounding gas and dust is either attracted to it or blown out of the system by it.

There are several examples in which astronomers have apparently witnessed interstellar dust rearranging itself over a period of years to reveal, if not a developing star, the place where one or more stars will be eventually.

PROTOSTARS: FIRST APPEARANCE ON THE H-R DIAGRAM

Eventually the central regions of a forming star become opaque and slow the outward flow of radiation. The effect of this is to stem the loss of energy so that the temperature rises and the gas pressure increases. This causes the central region to slow from a free-fall collapse to a gradual contraction as it approaches a balance between gas pressure, which is pushing outward, and weight resulting from gravity, which is pushing inward. Now the embryo star can appear on the H-R diagram for the first time; it begins its evolution on the coolest fringes of the diagram on tracks determined from stellar models; see the right-hand side of Figure 14.15.

Once the forming star has stabilized somewhat, it is in the red giant region of the H-R diagram, although

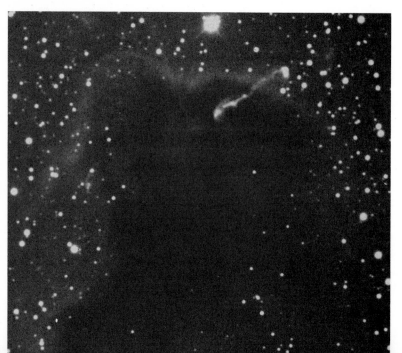

FIGURE 14.14
Dark globule in the constellation of Vela, photographed in very red light with the 4-meter telescope at Cerro Tololo, Chile. Note the two bright objects connected by an "umbilical cord." They are known as Herbig-Haro objects and probably contain protostars (see Figure 14.15). The luminous rim of the dark globule is due to reflection of starlight. From all indications dark globules are the birthplaces of stars.

it is not called a red giant; it is a *protostar*. The temperature of a protostar's surface is about 4000 K, and energy in its deep interior is transported to the surface entirely by convection, which extends from center to surface. In this slower contraction phase the protostar decreases its luminosity but keeps about the same surface temperature; most of the process of accreting matter has been completed.

Gravitational contraction eventually raises the temperature in the protostar's core to 1 million K or so, which is hot enough to destroy by nuclear reactions such light nuclei as deuterium, lithium, beryllium, and

FIGURE 14.15
Theoretical pre-main-sequence evolutionary tracks for stellar models ranging from 0.5M☉ to 15M☉. The points along each path are labeled with the approximate time (in years) elapsed during contraction to the *zero-age main sequence* (when hydrogen burning supplies all of a star's luminosity).

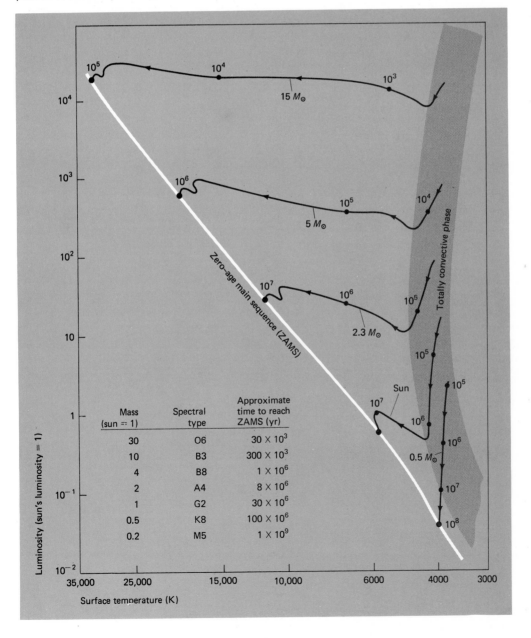

Mass (sun = 1)	Spectral type	Approximate time to reach ZAMS (yr)
30	O6	30×10^3
10	B3	300×10^3
4	B8	1×10^6
2	A4	8×10^6
1	G2	30×10^6
0.5	K8	100×10^6
0.2	M5	1×10^9

boron (initially present in small quantities). These are the first stages of the star's thermonuclear existence, from which it derives very little energy; the next stage is to initiate hydrogen burning.

By the time the protostar's central temperature has risen to several million degrees, the *p-p* chain can be ignited, and hydrogen burning begins to supply some of the luminosity, at first in small amounts. Several million years later the protostar arrives on the zero-age main sequence in the H-R diagram, where hydrogen burning supplies 100 percent of the luminosity and contraction has virtually ceased. Astronomers define the *zero-age main sequence* (Figure 14.15) as the position in the H-R diagram along which protostars of different masses cease to contract (and thus are stable configurations) and derive all their luminosity by burning hydrogen. The protostar is now a full-fledged star.

How long does it take to go through the protostar stage to reach the hydrogen-burning phase? For the sun it was about 30 million years; approximate times are listed for other stellar masses in Figure 14.15. As the figure shows, stars exceeding the sun's mass evolve quite rapidly, while for less massive stars the protostar phase is longer than that of the sun. It is usual among astronomers to date the star's age from the zero-age main sequence onward, since the time it has spent contracting out of the interstellar medium to the main sequence is only a small fraction of its life span, typically a few tenths of a percent.

Some stellar models indicate that protostars with masses of less than about $0.1 M_\odot$ never become hot enough at their centers to fuse hydrogen. They pass the lower end of the main sequence and continue contracting toward extremely high densities. These very-low-mass stars apparently bypass normal stellar evolution and proceed slowly to becoming degenerate red dwarfs, where electron degeneracy dictates the physics of their deep interior. With less than a hundredth the mass of the sun such objects may well end up as Jupiterlike planets (Jupiter's mass is $0.001 M_\odot$).

FIGURE 14.16
Star formation on the edge of a giant molecular cloud.

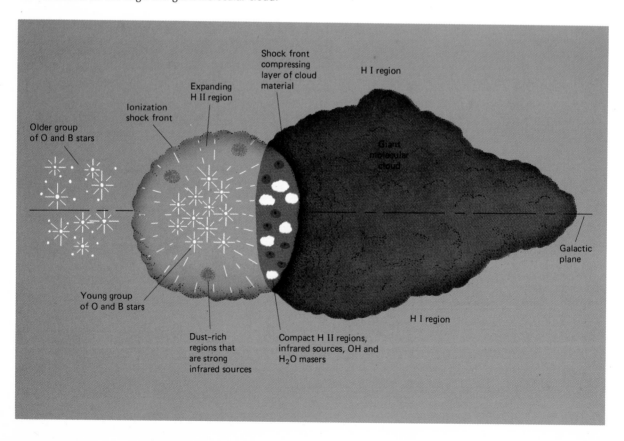

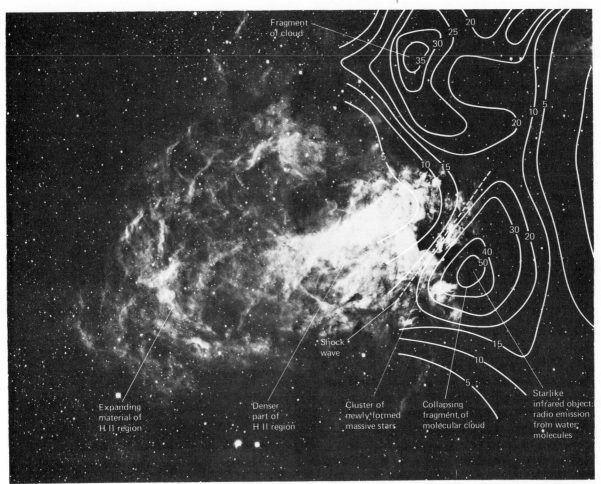

FIGURE 14.17
Birthplace of stars, the Omega Nebula, M17. The bright H II region shows an asymmetrical pattern of development. Note that to the left the expanding H II region seems to advance unopposed. To the right the H II region is adjacent to the dark molecular cloud shown by the radio contour lines for the CO molecule. The numbers on the contour lines indicate the CO temperature in kelvins.

There is little doubt among astronomers that rotation is also a crucial factor during the collapse of interstellar clouds and the contraction of the resulting fragments through the protostar stage into main-sequence stars. Rotation probably determines whether the results will be multiple-star systems (high rotation), binaries, stars with planets, or just single stars (low rotation). At present the attempts to include the effects of rotation have been crude and not satisfactory.

STELLAR NURSERIES

By the time several O stars arrive on the zero-age main sequence from the collapse of a giant molecular cloud, they will produce enough ultraviolet radiation to evaporate the dust grains and ionize the gas in their vicinity; this process forms an H II region. Figure 14.16 is an attempt to depict the formation of an O association on one end of a giant molecular cloud. Star formation advances across the cloud by the formation of new O stars. This is because the O stars' emission of ultraviolet radiation (which forms the H II region) in conjunction with their stellar wind creates a shock front that compresses the cloud. This in turn initiates new fragmentation and collapse, forming more stars. When the massive O stars reach the end of their lives, they also undergo a supernova outburst that adds to the compression of the cloud and furthers star formation. Finding H II regions in dark-cloud complexes definitely demonstrates that stars are forming, as the example of the Omega Nebula shows in Figure 14.17.

349

Canst thou bind the sweet influences of Pleiades, or loose the bands of Orion?

Job, 38

The white contour lines show the position of radio emission by the CO molecule and mark the location of the remaining dark molecular cloud.

When a large, dark molecular cloud begins to fragment in selected regions into a cluster of protostars of differing masses, the evolving stars will reach the main sequence at different times. The more massive stars begin burning hydrogen first, and in beadlike progression the others arrive along the zero-age main sequence from the upper left down to the lower right in the diagram. Stars of lower mass will lie progressively farther above and to the right of the zero-age main sequence at any instant of time after contraction starts.

The open cluster NGC 2264 shows this progression quite well in the H-R diagram of Figure 14.18. The less-massive cluster stars should eventually arrive on the lower portions of the main sequence in order of their mass. Many of the cluster's stars even have gas and dust shells. The cocoons, or shells, around these stars contain large quantities of dust apparently inherited from the original giant molecular cloud.

Additional evidence that can be used to identify recently formed stars comes from the coexistence of T Tauri variable stars in open clusters and veiling clouds, out of which they seem to have originated. The T Tauri variables have characteristics that we might expect for objects going through pre-main-sequence evolution; in particular they lie above the main sequence in the H-R diagram.

Sometimes the very luminous O and B stars, which are definitely quite young, are intermingled with T Tauri stars. Because the O and B stars are massive and already on the main sequence, it is presumed that the T Tauri stars are in the $0.2M_\odot$-to-$3M_\odot$ range, contracting toward the main sequence. Typical radii measured for them are about five times that of the sun. We do not definitely know whether all stars that will originate in a giant molecular cloud form simultaneously. The formation of individual stars may spread over a period as long as 10 million years.

About 4.6 billion years ago, after a protostar stage of about 30 million years, the sun settled on the main sequence for a long, uninterrupted period of hydrogen burning. This stable phase in its life should continue for another 5 billion years. As stated earlier, a star's life on the main sequence is the longest and most quiescent phase in its life. After a star exhausts its supply of hydrogen fuel, the star must undergo some relatively rapid changes that lead eventually to its death.

FIGURE 14.18
Color-magnitude diagram of the young cluster NGC 2264. Crosses represent T Tauri stars.

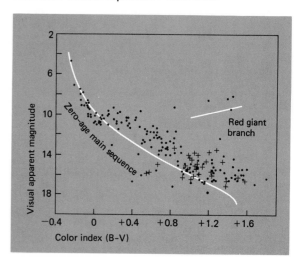

SUMMARY

Gas, dust, and molecules. The interstellar medium is a very low-density mixture of atomic and molecular gases and small dust particles. Interstellar gas — most of it hydrogen — is almost all found within a thin disk in the plane of the Galaxy. Starlight passing through cool interstellar gas is absorbed, forming interstellar absorption lines. The 21-centimeter photon, emitted as the electron in atomic hydrogen spontaneously changes the orientation of its spin, gives astronomers a useful tool for studying the structure of the Galaxy because 21-centimeter radiation can travel great distances through interstellar space. A number of interesting interstellar molecules have been discovered in cool localized regions of interstellar space. Some of these 50 or so molecules are organic ones containing as many as 9 atoms. Although far less abundant than interstellar gas, the interstellar dust is almost entirely responsible for the dimming and reddening of starlight. The dust grains probably contain 100 million or so atoms of varying chemical species.

Interstellar clouds and nebulae. Dark nebulae containing unusually dense concentrations of interstellar dust block our view of the disk of the Galaxy. These interstellar clouds can be divided into diffuse clouds and dark clouds on the basis of their opacity. Immense dark clouds containing interstellar molecules are known as giant molecular clouds. Although the most plentiful constituent of dark interstellar clouds is molecular hydrogen, it does not emit 21-centimeter radiation. Carbon monoxide, also present, however, strongly emits radio radiation and has proved to be an invaluable tool in studying dark interstellar clouds. Several lines of evidence indicate that energetic events are occurring at specific points in dark molecular clouds which are most likely stars forming. Ultraviolet radiation from hot O and B stars ionize hydrogen in interstellar clouds forming H II regions. These emission nebulae are the brightest objects in the spiral arms of galaxies.

Mass loss by stars. The loss of mass by a star affects the dynamics and chemical composition of the interstellar medium. Matter may be ejected in a single explosion or lost more gradually by a star. The continuous loss is called a stellar wind. Red giants and supergiants have large stellar winds (while hot O and B stars are losing large quantities of matter through stellar winds). There may also be explosive loss of matter as in the formation of a planetary nebula. The most spectacular eruptive stars are supernovae, in which their brightness may increase by a factor of several billion in a few hours. Accompanying the outburst of light is probably the ejection of vast amounts of matter into the interstellar medium.

Birth of stars. The first step in the formation of a star is the gravitational loss of an interstellar cloud so that interstellar material fragments to form stars. In collapsing, gravitational potential energy is converted to thermal energy. Some energy is radiated into space. As the star collapses, it grows hotter; eventually the central region of the star becomes opaque and stems the loss of energy, and the gas pressure rises. When equilibrium is reached, the body is a protostar. When the central temperature has risen to several million degrees, hydrogen burning is initiated. When all luminosity is supplied by the fusion of hydrogen and contraction has ceased, the protostar is a zero-age main-sequence star. In a cluster of protostars, the more massive ones begin hydrogen burning first, and in beadlike progression smaller and smaller-mass stars arrive on the zero-age main sequence.

KEY TERMS

dark cloud
diffuse cloud
emission nebula
giant molecule cloud
globule
H I region
H II region
interstellar dust
interstellar gas
interstellar line

interstellar reddening
mass loss
nova
planetary nebula
protostar
stellar wind
supernova
T Tauri variable stars
21-centimeter line
zero-age main sequence

CLASS DISCUSSION

1. If, as stated in this chapter, the interstellar gas is extremely transparent, how do astronomers know that there really is gas throughout interstellar space?

2. What are interstellar dust grains? If the average density of interstellar dust is one grain per 10^{13} cubic centimeters (that is, about the volume of a large office building), how can it be responsible for obscuring vast reaches of space from our view? Is it possible that astronomers are wrong on the density of interstellar dust?

3. Describe interstellar clouds. What should space in the disk of the Galaxy be like with interstellar clouds in it? Do you visualize the clouds as touching each other or isolated and far apart? What is the intercloud region like?

4. What is so significant about the loss of mass by stars other than that they happen to be doing it? Describe the various ways stars are losing matter. Could a star waste away by throwing off mass until there was nothing left of the star?

5. Describe the process by which stars form. Where do they form? How many at a time? Are they all large, or small, or a mixture of sizes? What signs should one look for in order to find new stars?

READING REVIEW

1. What is the composition of the interstellar gas? In what forms is the element hydrogen found?

2. What is the density of interstellar gas in interstellar clouds and in the intercloud region and the average over very large volumes of space?

3. What is the 21-centimeter line? How is it produced?

4. How does the interstellar dust affect light of many different wavelengths passing through it?

5. How much matter do the giant molecular clouds contain? What is its composition?

6. How are H II regions created? Are they very significant as constituents of the Galactic disk? Name one.

7. What types of stars possess stellar winds? Should other stars like the sun have stellar winds?

8. What is a supernova, and is it very bright? Name an example of a supernova.

9. Where on the H-R diagram does a forming star first appear?

10. What is the relationship, if any, between H II regions and young O stars?

SELECTED READINGS

Bok, B. J. "Early Phases of Star Formation," *Sky & Telescope*, April 1981.

Chaisson, E. J. "Gaseous Nebulas," *Scientific American*, December 1976.

Lada, C. J. "Energetic Outflows from Young Stars," *Scientific American*, July 1982.

Malin, D. F. "Dust Clouds of Sagittarius," *Sky & Telescope*, March 1978.

Wynn-Williams, G. "The Newest Stars in Orion," *Scientific American*, August 1981.

15
Death of Stars

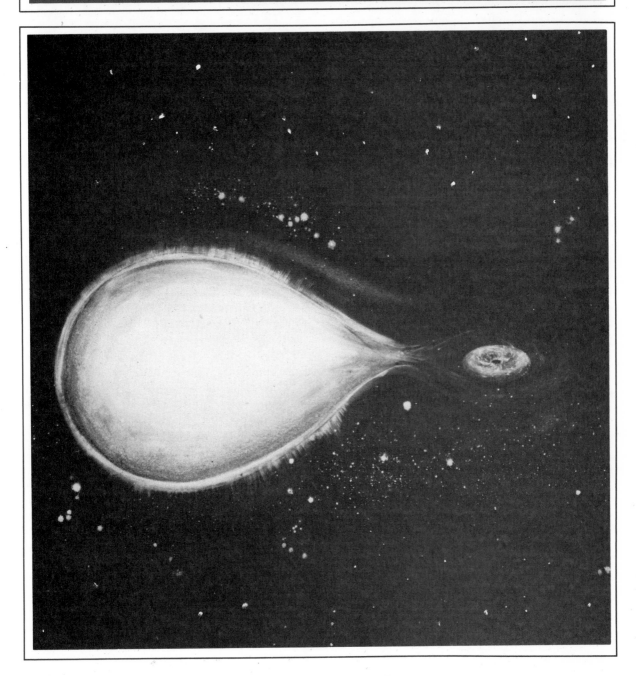

In Chapters 13 and 14 we have discussed in the life of stars the hydrogen-burning, or main-sequence, phase and then the birth of stars. After stars have exhausted their hydrogen fuel, while on the main sequence, they are unable to support the weight of their outer layers with their internal gas pressure. This crisis means that they must renew their battle against gravitational compression through a sequence of contracting, heating, and igniting new sources of nuclear fuel in order to restore and maintain thermal equilibrium. This chapter discusses these advanced stages of stellar evolution, which ultimately lead to the death of stars.

15.1
End of Main-Sequence Life:
Hydrogen Exhaustion
in the Core

During hydrogen burning the core of a star contracts very slowly as the star tries to maintain hydrostatic equilibrium. This contraction converts gravitational potential energy to thermal energy, raising the core temperature and the star's overall luminosity. The star moves up from the zero-age main sequence toward the time when it will exhaust its hydrogen fuel.

The main sequence consists of a broad avenue of stars of different ages that evolve from the zero-age main sequence to their respective hydrogen-exhaustion points (for example, Figure 15.1). Unlike what happens in pre-main-sequence contraction the star's overall radius does not decrease but increases as the outer layers absorb some of the outflowing energy. Enough energy is absorbed to lift these layers against the gravitational pull of the underlying layers. The H-R diagram records increases in luminosity and radius at the surface of the star; it does not show the core's slow contraction. For a star like the sun the radius will roughly double during its main-sequence life as about 12 percent of its hydrogen is depleted.

As hydrogen burning ends, hydrostatic equilibrium shifts in favor of gravity, and the core contracts more rapidly. The contraction releases substantial amounts

◀ Artist's conception of an accretion disk around a black hole fed by matter pulled from a blue supergiant.

of gravitational potential energy, further heating the deep interior. The layers just outside the former energy-generating core are now hotter too, and hydrogen burning migrates to a relatively thin shell surrounding an inactive helium core. Although the core continues to contract and heat, it is the burning of hydrogen in the surrounding shell that is responsible for supplying the luminosity at the star's surface. The evolution of the star now carries it along a path leading to the red giant region in the H-R diagram.

The *red giant branch* in Figure 15.1 is the portion of the evolutionary path that extends steeply upward at the extreme right. When the contracting helium-rich core reaches about 120 million K (Table 15.1), the second major thermonuclear reaction, *helium burning*, begins. In this reaction three ^{4}He (helium) nuclei fuse to form a ^{12}C (carbon) nucleus and two gamma-ray photons. Stars burn helium (and hydrogen in a shell) for about 5 to 20 percent of the time spent burning hydrogen on the main sequence.

When helium begins to burn, rapid contraction in the core ends. And just as hydrogen burning on the main sequence leads to a slow contraction of the core, helium burning also produces a slow contraction of the energy-generating core. On the main sequence the core is about 20 percent of the star's inner radius; but when helium starts to burn, the core is more like 0.1 percent of the radius. The star's structure has thus changed dramatically from what it was on the main sequence.

When will contraction raise the temperature in the core to 120 million K so that helium may begin burning? This event is determined by the star's main-sequence mass. To see this fact, we can summarize the approach to helium burning and subsequent evolution by dividing the stars into two groups: stars whose masses are less than about $2M_\odot$ and those whose masses are greater than $2M_\odot$. As a means of distinguishing these groups, we shall refer to those less than $2M_\odot$ as *low-mass stars* and those greater than $2M_\odot$ as *high-mass stars*.

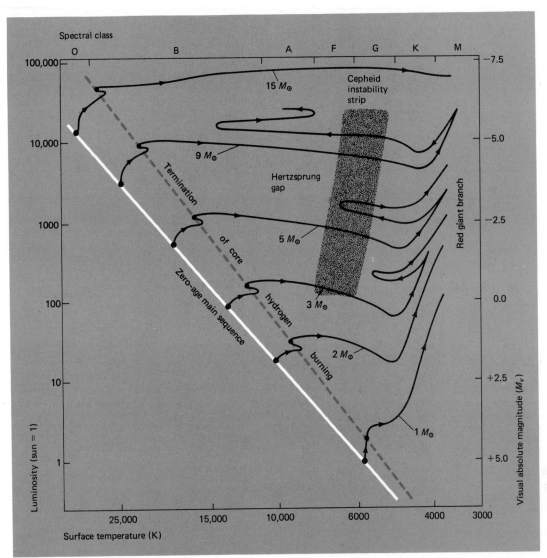

FIGURE 15.1
Theoretical evolutionary tracks in the H-R diagram for disk population stars from $1M_\odot$ to $15M_\odot$. Stars evolving away from the zero-age main sequence eventually leave the main sequence at the point (dashed line) when they have consumed the hydrogen in their cores. Up to that time their evolution has been very slow because of hydrogen burning; when the hydrogen fuel is spent, evolutionary changes become extremely rapid.

15.2
Red Giant Evolution
for Low-Mass Stars

HELIUM FLASH

Let us consider first the post-main-sequence evolution of the low-mass stars. When hydrogen is exhausted, they begin an internal restructuring that eventually ends when they are red giants. The result of this action is, first, to decrease their surface temperature but to keep about the same luminosity (Figure 15.1). Second, before helium burning actually begins, these stars will brighten appreciably and follow an evolutionary path upward into the red giant region (Figure 15.2). For the sun this entire change takes about 1 billion years. For stars less massive than the sun the time is even longer.

By the time a star like the sun reaches the red giant region, the helium-rich core (containing about 30 per-

TABLE 15.1
Thermonuclear Burning and Electron Degeneracy

Thermonuclear Process	Mass on Main Sequence Needed to Burn (sun = 1)	Ignition Temperature (K)	Approximate Density (g/cm³)	Electrons Degenerate at Densities Greater Than
Hydrogen burning: $H \rightarrow He$	0.1	4×10^6	$10^1 – 10^2$	$\approx 10^3$
Helium burning: $He \rightarrow C, O$	0.4	120×10^6	$10^3 – 10^6$	$\approx 10^5$
Carbon burning: $C \rightarrow Ne, Na, Mg, O$	4.0	600×10^6	$10^5 – 10^8$	$\approx 10^7$
Oxygen, neon, and silicon burning: $Ne \rightarrow O, Mg$; $O \rightarrow Si, S, P$; $Si \rightarrow Ni \rightarrow Fe$	8.0	$1 \times 10^9 – 3 \times 10^9$	$> 10^7$	$\approx 10^9$

cent of the star's mass) has been compressed to a volume about twice that of the earth, with a density of about 10^6 grams per cubic centimeter. At this density the gas no longer obeys the perfect-gas law; for the electrons have become degenerate (page 317).

For low-mass stars the presence of degenerate electrons in their cores causes helium to ignite very differently from the way it would in a perfect gas. In a perfect gas any rise in temperature is followed by an increase in pressure, and the contraction then underway is halted. The increase in pressure expands the core slightly, cools it, and reduces the thermonuclear burning rate until thermal equilibrium is restored.

But when it is a degenerate electron gas, matter behaves more like a solid than like a gas. The core does not readily expand with an increase in temperature, and if it cannot expand, it cannot cool; the rate of helium burning increases with temperature. As more energy is generated, the temperature continues to rise; in turn this increases the rate at which energy is generated, which pushes the temperature still

higher; and a runaway condition follows. In a few hours temperatures leap to hundreds of millions of degrees, causing the generation of as much energy as 100 billion stars of solar luminosity—about that of a whole galaxy. This explosive flash of energy is called a *helium flash*. Even as huge as the energy generation is, the helium flash is not enough to blow the star apart. Instead all the energy goes into removing electron degeneracy, and this so alters the structure that the star may now burn helium in a core that acts as a perfect gas.

The helium flash just described is generated in computer models of stars—astronomers have not seen and never will see the flash in a real star. This is because the helium flash occurs in the deep interior of the star hidden by millions of kilometers of gas.

OPEN-CLUSTER H-R DIAGRAMS
Astronomers have found observational evidence that confirms much of this evolutionary sequence in the

I seem to have stood a long time and watched the stars pass.
They shall also perish I believe.
Here today, gone tomorrow, desperate wee galaxies
Scattering themselves and shining their substance away
Like a passionate thought. It is very well ordered.

Robinson Jeffers

H-R diagrams for open clusters. As individual stars in a cluster evolve, their rate of aging is determined by their mass. The more massive stars, because they are more luminous, age faster than do stars of lower mass, as illustrated in Figure 15.3. It is from the relative aging of the cluster's members that astronomers can find the age of a cluster. The critical time when a star reaches the point at which it exhausts its hydrogen depends on the mass of the star. This in turn is related to its luminosity and hence its age. From model star calculations we can place ages on the vertical axis of the H-R diagram, corresponding to the luminosity of

the observed *turnoff point* of the smallest star to have reached hydrogen exhaustion, like that shown in Figure 15.4.

In that figure the youngest cluster is NGC 2362, whose age we estimate to be less than 2 million years. None of its blue supergiants has yet crossed the Hertzsprung gap. In the next youngest cluster, h and χ Persei, some of the stars that were originally massive blue supergiants have evolved across the diagram and are now in the red giant region. Next youngest is the Pleiades, followed, in order, by M41, M11, the Hyades, NGC 752, M67, and NGC 188.

The oldest open clusters shown in Figure 15.4 are M67 and NGC 188. The most obvious difference between these two is that all subgiant and giant stars in M67 are more luminous than comparable ones in NGC 188. This is because stars leaving the main sequence in M67 are more massive and therefore younger than

FIGURE 15.2
Helium core and hydrogen-burning shell in low-mass stars prior to helium flash. As star radius (7×10^6 kilometers) and helium-core radius (5×10^3 kilometers) show, the drawing is not to scale. The hydrogen-burning core while a low-mass star is a main-sequence star is approximately to scale.

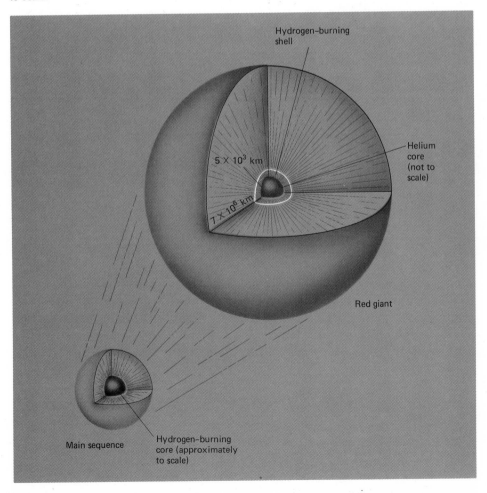

Hydrogen–burning shell

5×10^3 km

7×10^6 km

Helium core (not to scale)

Red giant

Main sequence

Hydrogen–burning core (approximately to scale)

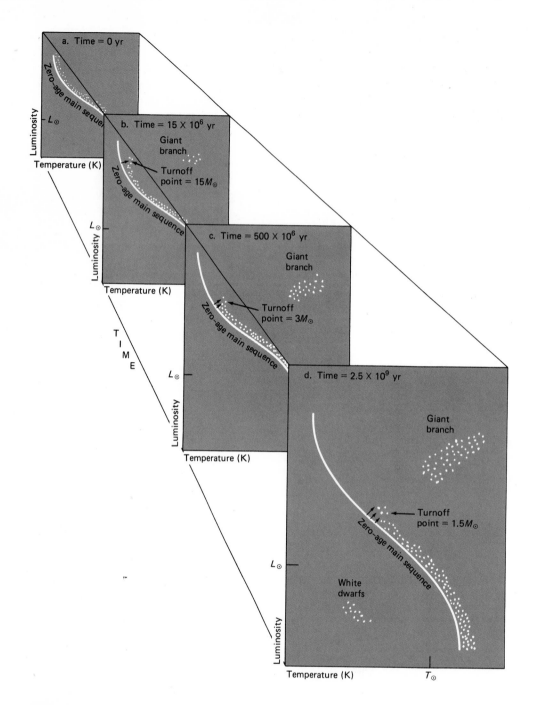

FIGURE 15.3

Evolution of an open cluster. In schematic form the evolution of an open cluster is shown at four periods: (a) time equals zero, when most of the cluster stars are burning hydrogen on the zero-age main sequence; (b) 15×10^6 years have passed, when a few massive stars on the upper main sequence have exhausted hydrogen and are red giants; (c) 500×10^6 years have passed, when the most massive star on the main sequence is $3M_\odot$; (d) 2.5×10^9 years have passed, when some white dwarfs are present.

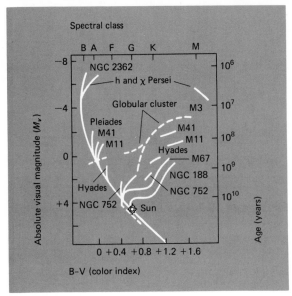

FIGURE 15.4
Composite H-R diagram for several open clusters and one globular cluster. A cluster's age can be estimated from the location of the turnoff point, which is the point locating the most massive star still on the main sequence.

those in NGC 188. The subgiant branches in M67 and NGC 188 strongly resemble the evolutionary tracks of stars of about $1.25M_\odot$ and $1M_\odot$, respectively. Estimated ages are about 3 billion years for M67 and about 5 billion years for NGC 188.

HORIZONTAL-BRANCH STARS

One conspicuous difference between an H-R diagram for open clusters and one for globular clusters is the *horizontal branch* in that for the globular clusters (Figure 15.4). This region of stars runs almost horizontally across the diagram from the red giant branch to the very blue stars at a visual absolute magnitude of about zero. What does the horizontal branch represent in the evolution of stars?

To answer that question, first remember that the helium flash in the solar-mass stars reduces the stars' luminosity over a very short time interval, say, about 10,000 years. After the flash the drop in core temperature reduces the star's luminosity while the surface temperature increases. This causes the star to move down and to the left of its original position in the red giant region and onto the horizontal branch. From stellar models we are led to believe that the

horizontal-branch stars are stars between about 1 and $0.5M_\odot$ that are burning helium.

The horizontal branch is a relatively stable phase in a star's life. Yet there is a period when the star can become temporarily unstable and pulsate as an RR Lyrae variable (Section 12.4). Even though the comparison in Figure 15.5 suggests reasonable agreement between the H-R diagrams for the stars of the globular cluster M92 and a theoretical cluster made up of stellar models of different masses, astronomers still have a great deal to learn about the horizontal-branch stars.

As the time approaches when helium will be exhausted at the star's center, the battle resumes between gravity and gas pressure. The conversion of helium to carbon causes the core to contract slowly and heat, and the star moves upward from the horizontal branch. But most important for low-mass stars compression does not sufficiently heat the core to start carbon burning, as shown in Table 15.1. This end to nuclear burning is the beginning of the end for low-mass stars.

FIGURE 15.5
H-R diagram for stars in the globular cluster M92. Solid lines indicate the average theoretical positions of various stages in the evolution of model stars of different masses.

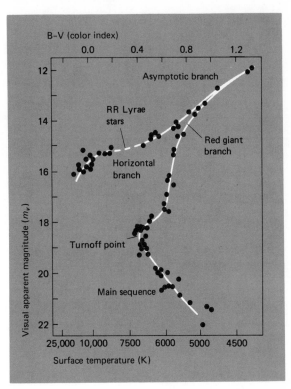

In the H-R diagram, a star evolves from the horizontal branch along a path which brings it into the red giant region for a second time. The outer layers are bound so loosely by the gravitational attraction of the inner layers that the star may expel large amounts of material in the form of a stellar wind. Current observational evidence suggests that the more luminous red giants and supergiants are indeed losing significant amounts of matter to the interstellar medium (Section 14.3).

FIGURE 15.6

Low-mass star undergoing planetary-shell ejection. Outer layers are expelled as a shell at velocities of about 30 kilometers per second. Ultraviolet photons from the central star ionize hydrogen and other species in the ejected shell. The recombination of electrons and protons or ions produces the visible radiation we see coming from the planetary shell.

15.3 Death of Stars like the Sun

STARS IN PLANETARY NEBULAE

A most obvious loss of mass occurs when a star ejects its hydrogen-rich envelope to form a planetary nebula (Section 14.3). How does this happen? While a star evolves toward the red giant branch, burning helium and hydrogen in two separate shells, it apparently pulsates slightly. Eventually the outer layers separate from the core and expand outward at several tens of kilometers per second to become the shell of the planetary nebula. The path in the H-R diagram, along which the star now evolves very rapidly, is across the top of the diagram toward the high-temperature side.

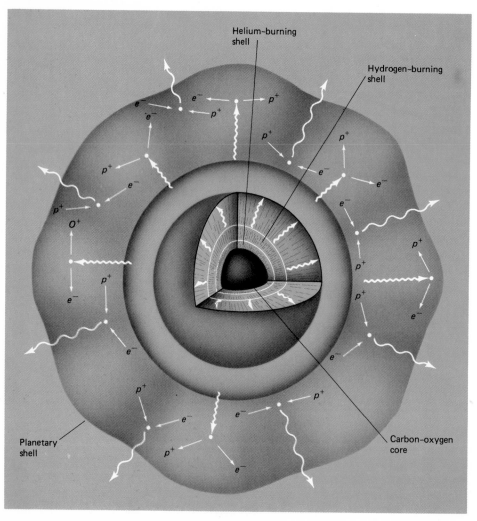

Helium-burning shell

Hydrogen-burning shell

Planetary shell

Carbon-oxygen core

The ejected shell holds most of the original outer layers that missed thermonuclear burning because they were never hot enough (Figure 15.6). Near the red giant branch the luminosity and temperature of the star's original surface appear in the H-R diagram. But as the shell expands, the former core is exposed to view so that its luminosity and temperature become the quantities plotted in the H-R diagram. Since the former core is hot, the star's position is now on the blue side of the diagram. By the time the central star moves into the white dwarf region, the expanding nebula has mixed with the interstellar medium and is no longer recognizable.

Whether the sun will eject its outer layers as a planetary nebula is somewhat uncertain. Most, if not all, of the disk-population stars with main-sequence masses between about 1 and $4M_{\odot}$, as well as some low-mass spheroidal-population stars, may form planetary nebulae after being red giants and before becoming white dwarfs.

WHITE DWARFS

By the time a star has exhausted its nuclear fuel supply, gravitational contraction has compressed the material to extremely high densities, and the electrons are once again degenerate. Electron degeneracy in the core prevents further contraction, and the only source of energy for the star is its store of thermal energy, which supplies the luminosity, cooling its interior. The star is now a *white dwarf*.

Mathematical modeling of white dwarfs leads to a surprising result: The larger the mass, the smaller will be its radius. For example, a white dwarf whose mass is $0.4M_{\odot}$ has a radius equal to about 1.5 percent of the sun's radius, or about 10,000 kilometers. But a white dwarf whose mass is $0.8M_{\odot}$ has a radius that is about 1 percent of the sun's radius, or about 7000 kilometers (about the size of the earth). A theoretical limit, called the *Chandrasekhar limit* after S. Chandrasekhar (1910–   ), is reached for a white dwarf of mass $1.4M_{\odot}$ in that it has a radius of zero. We can infer from this that a star whose initial mass is less than $1.4M_{\odot}$ can evolve into a stable white dwarf. For an initial mass exceeding $1.4M_{\odot}$ the star must lose mass at some time during its life if it is to become a white dwarf. We think it most probably does so during its red giant evolution or afterward.

A star of solar mass enters the white dwarf stage as a small, hot, blue object. As energy is radiated away,

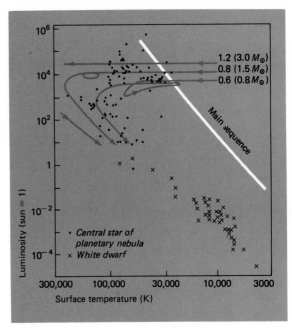

FIGURE 15.7
Theoretical evolutionary tracks of three red giants that have evolved out of the red giant region (not shown). Numbers in parentheses give the mass before the envelope was ejected. Evolutionary tracks are shown for the central star into the white dwarf region of the H-R diagram.

thermal energy is depleted, and the temperature declines. This cools the white dwarf, changing it from blue to white to yellow and eventually to red. Since the star can no longer contract, the evolutionary tracks of white dwarfs roughly follow lines of constant radius. The tracks are parallel to the main sequence but well over to the left and below it in the H-R diagram (Figure 15.7).

Most of the white dwarfs that have been discovered are relatively close to the sun. Their very low luminosity makes them too faint to be seen at greater distances even in large telescopes. Of the 50 or so nearest stars three are white dwarfs; the companions of Sirius and Procyon and van Maanen's star (Figure 13.3). The total number of known white dwarfs is several thousand. The conservative estimate of Table 13.2 places their total number in the Galaxy around 35 billion. Most of them are presumably descendants of the oldest disk-population stars that have evolved into white dwarfs during the Galaxy's 15-billion-year life span.

After billions of years a white dwarf's thermal energy will be exhausted. As the star cools, its rate of

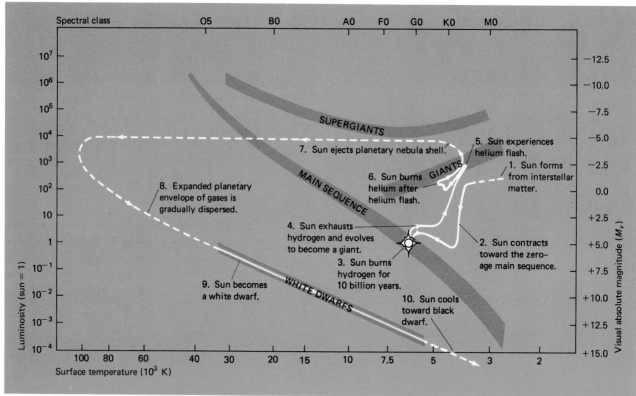

Spectral class   O5   B0   A0   F0   G0   K0   M0

5. Sun experiences helium flash.

7. Sun ejects planetary nebula shell.

6. Sun burns helium after helium flash.

1. Sun forms from interstellar matter.

8. Expanded planetary envelope of gases is gradually dispersed.

4. Sun exhausts hydrogen and evolves to become a giant.

2. Sun contracts toward the zero-age main sequence.

3. Sun burns hydrogen for 10 billion years.

9. Sun becomes a white dwarf.

10. Sun cools toward black dwarf.

SUPERGIANTS

MAIN SEQUENCE

GIANTS

WHITE DWARFS

Luminosity (sun = 1)

Surface temperature (10³ K)

Visual absolute magnitude (M_v)

FIGURE 15.8
An H-R diagram of the evolutionary history of the sun from birth to death, showing a number of critical stages in its life. The sun is now about halfway through its life as a main-sequence star, burning hydrogen. All the other phases of its life either were or will be much shorter in years than that of the main sequence.

cooling slows, making its approach to its final non-luminous state as a *black dwarf* quite long. The Galaxy is probably not old enough for very many white dwarfs to have cooled sufficiently to become black dwarfs. This approach to obscurity, though, is definitely a one-way track, from which nothing can save the white dwarf.

Our knowledge of the evolution of low-mass stars is not so complete that we can state precisely what evolutionary path a particular star, such as the sun, will follow. What has been described in the preceding paragraphs is a general evolution for all low-mass stars. If the sun does indeed proceed along that general evolutionary path, then Figure 15.8 traces out its entire evolution from formation in the interstellar medium to final state as a white dwarf.

15.4
What Happens to Massive Stars after the Main Sequence?

Just as for low-mass stars, the day comes in the lives of high-mass stars (greater than about $2M_\odot$), like Regulus (B7 V), when they also exhaust the hydrogen fuel in their cores. The evolutionary path in the H-R diagram followed by high-mass stars, such as Betelgeuse (M1 I) and Antares (M2 I), is described in this section and the next two. Evolution for high-mass stars is very different in many ways from what we have just considered for low-mass stars.

HELIUM BURNING IN HIGH-MASS STARS
With hydrogen gone in the core of the massive main-sequence star, the core shrinks drastically, driving up

the density and temperature and exaggerating the differences between the star's central regions and its outer layers.

For stars above $2M_\odot$ in Figure 15.1 evolution away from the main sequence is quite rapid, almost horizontal, across the H-R diagram toward the red giant region. These stars expand their radii by a factor of about 10, which is a hundredfold increase in their surface area. Still the temperature is reduced enough to keep the luminosity roughly constant. The region across which the stars are evolving in the H-R diagram is the *Hertzsprung gap*. Very few stars have been found in this gap; so the transition must be rapid—occurring in 50 million years to less than 1 million years.

Just when will contraction raise the temperature in the core to 120 million K so that helium burning may begin? Again this is decided by the star's mass. For stars between 2 and $9M_\odot$ or so the time comes when the star is in the red giant region. When helium burning begins, the star's evolution upward along the red giant branch in the H-R diagram stops. Following this, the star, still burning helium in the core and hydrogen in a surrounding shell, contracts its radius and moves to the high-temperature side of the red giant branch and then back to the low-temperature side a second time, as shown in Figure 15.1.

The path these stars follow in the H-R diagram apparently leads some of them across the region known as the *Cepheid instability strip* (Figure 15.1). A Cepheid variable pulsates only in its outermost layers, where small rhythmic expansions and contractions of these layers cause the cyclic variation in the Cepheid's light and velocity described in Chapter 12. A Cepheid is apt to pulse only for a brief time late in its evolution. Once pulsation begins, it continues for a considerable time until evolution carries the star toward higher temperatures in the H-R diagram. The classical Cepheids appear to be a spiral-arm component of the disk-population stars in their helium-burning phase whose masses range from 3 to $9M_\odot$.

Stars more massive than about $9M_\odot$ begin and complete core helium burning after hydrogen exhaustion before they cross the Hertzsprung gap leading to the red giant branch.

CARBON BURNING

After core helium burning ceases, the parts of the star that have burned helium are rich in carbon and oxygen (Figure 15.9). The next step in the thermonuclear evolution is carbon burning, which a star more massive than about $25M_\odot$ initiates even before crossing the Hertzsprung gap. *Carbon burning*, which takes

SIRIUS B: A WHITE DWARF

In the constellation of Canis Major, the Great Dog, we find the brightest star in the sky, Sirius (α Canis Majoris). Sirius, also called the Dog Star, is one of the three members of the winter triangle of very bright stars, the other two being Procyon (α Canis Minoris) and Betelgeuse (α Orionis). Sirius B, the faint companion of Sirius, was one of the first white dwarf stars discovered. Procyon also has a white dwarf companion. Table 15.2 contrasts some of the physical properties of Sirius B with the sun, our stellar yardstick, and the earth. Sirius B is about 10,000 times less luminous than Sirius and is very difficult to photograph.

TABLE 15.2
Some Physical Properties of Sirius B, the Sun, and the Earth

Quantity	Earth	Sirius B	Sun
Mass (sun $= 1.99 \times 10^{33}$ g)	$3 \times 10^{-6}M_\odot$	$0.94M_\odot$	$1.00M_\odot$
Radius (sun $= 6.96 \times 10^5$ km)	$0.009R_\odot$	$0.008R_\odot$	$1.00R_\odot$
Luminosity (sun $= 3.83 \times 10^{33}$ erg/s)	$\approx 0.0L_\odot$	$0.003L_\odot$	$1.00L_\odot$
Surface temperature (K)	287	27,000	5780
Gravitational redshift (km/s)	0.0	89 ± 16	0.6
Mean density (g/cm³)	5.5	2.8×10^6	1.41
Central density (g/cm³)	9.6	3.3×10^7	1.6×10^2
Central temperature (K)	4200	2.2×10^7	1.6×10^7

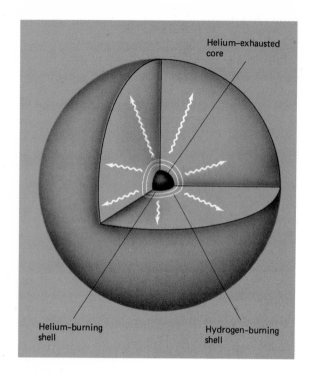
Helium-exhausted core

Helium-burning shell

Hydrogen-burning shell

FIGURE 15.9
Red giant phase for high-mass star of about $9M_\odot$ or greater: inert core and helium- and hydrogen-burning shells. Core and thermonuclear fusion shells not to scale.

3-to-$9M_\odot$ stars, they have returned to the red giant branch for a second time (Figure 15.1). The core is slowly contracting and heating. The increased density of an already-dense core squeezes the electrons and ions even closer together, imposing degeneracy on the electrons.

Once the central temperature in the core has reached about 600 million K, carbon nuclei initiate the thermonuclear fusion reactions in carbon burning. Carbon burning in an electron degenerate gas can start explosively, just as helium burning does. Thus the core can have a *carbon flash*, which removes the degeneracy and allows carbon burning to restore thermal equilibrium.

Studies of stellar models suggest, however, that carbon burning in some stars may start so explosively as to blow the star completely apart, as shown in Table 15.3, where we summarize evolution of all masses of stars. Astronomers are not yet absolutely certain that this explosion actually occurs and is one type of supernova outburst, but it seems probable.

place at temperatures above 600 million K, is the fusion of two carbon nuclei to produce primarily neon (^{20}Ne) or sodium (^{23}Na) or magnesium (^{24}Mg).

By the time helium is exhausted in the core of the

TABLE 15.3
Summary of Stellar Evolution by Mass of Star

Range of Mass While Main-Sequence Star	Thermonuclear-Burning Sequence	Evolution After Main-Sequence Stage	Final State of Star
$<0.1M_\odot$	None	None	Black dwarf
0.1–$0.5M_\odot$	Hydrogen	Red giant	White dwarf
0.5–$4M_\odot$	Hydrogen Helium	Red giant Horizontal branch Small mass loss or planetary nebula (?)	White dwarf
4–$8M_\odot$	Hydrogen Helium Carbon	Red giant Horizontal branch (?) Large mass loss Pulsation Explosive supernova (?)	White dwarf or neutron star or (?)
9–$60M_\odot$	Hydrogen Helium Carbon Oxygen Neon Silicon (?)	Red giant Large mass loss Explosive or implosive supernova (?)	Neutron star or black hole

GRAVITATIONAL REDSHIFT: TESTING RELATIVITY THEORY

In his general theory of relativity Einstein showed that for a distant observer an object in an intense gravitational field would be contracted, have gained mass, and have slowed down its clock time. Consequently, if the object is an atom, time dilation should also play a role when it emits a photon. Relativity predicts that the wavelength of the photon is lengthened, or shifted to the red, by an amount that depends on the strength of the gravitational field. The relative change in wavelength ($\Delta\lambda/\lambda$) is proportional to the mass of the attracting body divided by its radius (Figure 15.10).

This effect, known as the *gravitational redshift*, has practical astronomical interest because it occurs when a photon of light escapes from a star. If the star's gravitational field is sufficiently intense, we can measure the change in wavelength. We cannot easily measure this effect for the sun, but for a white dwarf of solar mass and small size (where the mass divided by the radius is large) the gravitational redshift is of measurable size. It has been observed for several white dwarfs in binary systems; this is made possible because, by using the companion's spectrum, we can differentiate between the gravitational redshift of the spectral lines and the Doppler shift produced by the system's radial velocity. The measured redshifts agree satisfactorily with those predicted by theory. And the gravitational redshift has been verified with even greater accuracy in a laboratory experiment.

FIGURE 15.10
Photon escaping from the strong gravitational field of a white dwarf with mass M_2 and radius R_2. The wavelength of the escaping photon is longer than an identical photon escaping from a main-sequence star of mass M_1 and radius R_1 if $M_1 = M_2$ and $R_1 > R_2$. Hence the absorption lines in the white dwarf's spectrum should exhibit a small redshift.

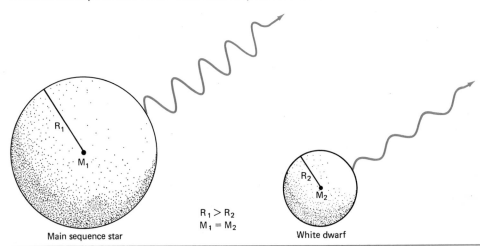

$R_1 > R_2$
$M_1 = M_2$

Main sequence star   White dwarf

For stars with main-sequence masses greater than about $9M_\odot$, carbon burning begins in a gaseous core that still obeys the perfect-gas law. There is no carbon flash, just as there was no helium flash for any of the high-mass stars (greater than $2M_\odot$).

OXYGEN, NEON, AND SILICON BURNING

In stars more massive than about $9M_\odot$ compression of their cores makes them hot enough to initiate the next round of thermonuclear reactions, known as *oxygen*, *neon*, and *silicon burning*, at 1 to 3 billion K (Table 15.1). Nuclear burning for these massive stars ends with the nuclear reactions that produce the iron nucleus. Because the iron nucleus is the nucleus most resistant to any type of structural change, thermonuclear burning beyond iron does not release energy but takes up thermal energy from its surroundings. From here on the star cannot go through the alternate contracting, heating, and nuclear-burning sequence that has sustained the star against gravity through most of its existence.

The evolution of the most massive stars ($15M_\odot$ and up) is not at all clear (Table 15.3). Nevertheless, several important predictions can be made from stellar models. As was pointed out in the preceding subsections, stars more massive than $9M_\odot$ initiate helium burning as blue supergiants, and those more massive than $25M_\odot$ also burn carbon on the blue side of the H-R diagram. While most massive stars probably become red supergiants, there is reason to believe that a vigorous mass loss by the very rare O stars of 50 to $100M_\odot$ may prevent them from ever crossing the H-R diagram to become red supergiants. That is, their entire life span is spent as extremely hot, luminous stars. This point is still unsettled because they are so rare.

The massive star has evolved through nuclear burning into a very small, extremely dense core with shells like those of an onion, which have different chemical compositions because of the nuclear burning. Various thermonuclear reactions can take place simultaneously: silicon burning at the center and neon, oxygen, carbon, helium, and hydrogen burning in successive shells outward, as depicted in Figure 15.11. Surrounding these shells is a highly distended, hydrogen-rich envelope. The star is very large, very bright, and quite red. Thermonuclear burning is a viable energy source for shorter periods as synthesis proceeds toward heavier elements. The $25M_\odot$ star has spent about 5 to 10 million years in hydrogen burning, 0.5 to 1.0 million years in helium burning, 500 to 1000 years in carbon burning, 6 to 12 months in oxygen burning, and a mere day or so in silicon burning.

15.5
Supernovae, Pulsars, and Neutron Stars

SUPERNOVA OUTBURSTS

For massive stars the final phases of the nuclear burning of oxygen and silicon can be violently explosive if they occur in degenerate or nearly degenerate matter. The star may undergo a bomblike detonation or, after exhausting all its nuclear fuels, suffer a catastrophic collapse and explode. Supernova outbursts (Section 14.3) are probably these violent events, predicted by astronomers' computer studies of model stars.

What causes supernova outbursts? Astronomers do not agree on the exact cause. There has been no case

in which a star that underwent a supernova outburst was identified and studied before the outburst. However, a supernova is believed to start with the gravitational collapse of a massive star having a degenerate iron core, as sketched in Figure 15.12.

With the explosion of the star its gravitational potential energy is released both as radiant energy and as kinetic energy of ejected matter, liberating as much energy as the sun expends in 10^8 years. What is left is a rapidly spinning compressed remnant of the former star.

As the outer layers of the star are blown off in the catastrophic explosion, a great flux of high-energy radiation and high-energy particles is released along with an avalanche of neutrinos. Stellar-model calculations indicate that large numbers of free neutrons should exist during a supernova outburst. When a star blows up, the neutrons can rapidly react with different nuclei to create the heavy elements beyond iron in the

FIGURE 15.11
Successive burning shells about a silicon-burning core (not to scale) in a spiral-arm-population star with a mass greater than about $15M_\odot$.

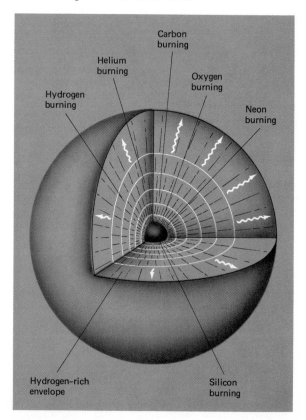

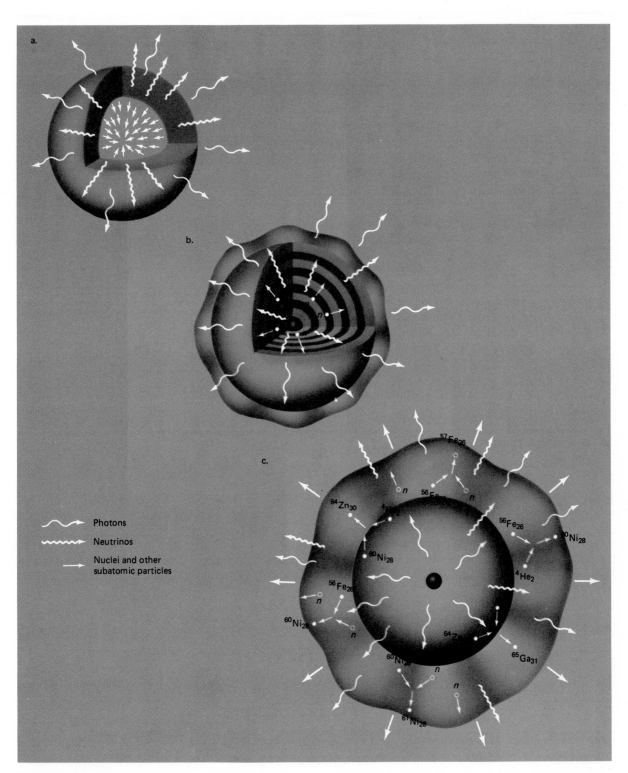

FIGURE 15.12
Supernova outburst: (a) The cessation of thermonuclear reactions leads to gravitational collapse; (b) shock waves moving out and expelling material; (c) the shell that nuclear reactions have encircled and are enriching with elements heavier than iron.

periodic table. By doing so, the expanding cloud, when it disperses into space, increases the heavy-element composition of the interstellar medium. Thus successive generations of stars formed from interstellar clouds, enriched with supernova debris, will have more of the heavy elements than the older stars do. If this is so, then all the familiar elements of our everyday world that are heavier than iron, such as silver in bracelets or gold in watches or mercury in thermometers or tin in cans, were synthesized during a supernova outburst before the solar system formed 5 billion years ago.

Although it might seem at first that the remains of a star after a supernova outburst should be less than $1.4M_\odot$ and thus be a white dwarf, this is apparently not what happens. Even as far back as the 1930s a few astronomers felt that the remaining star was likely to be too massive to be a white dwarf. We now believe that these massive stars end in one or the other of the strangest objects in all the cosmos: The remaining star will be either a neutron star or a black hole.

The collapse will halt at approximately the density of the atomic nucleus if the collapsing mass is less than about $3M_\odot$ with a radius of the order of several tens of kilometers. This object is a *neutron star*. If the collapsing mass is much greater, the end is presumably a *black hole* with a radius of a few kilometers.

PULSARS: COSMIC LIGHTHOUSES

In 1967 a strange object in the constellation Vulpecula was discovered to be emitting pulses of radio radiation with a very short period of 1.337 seconds between pulses. Later the object was named a *pulsar*, an acronym for *pul*sating radio *s*tar. It is now believed by astronomers to be a neutron star. Within a few weeks after the discovery three more pulsars were found,

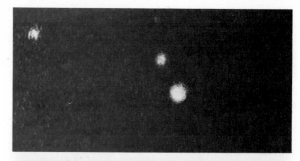
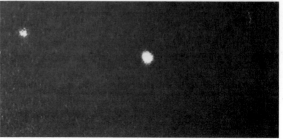

FIGURE 15.13
Crab pulsar at maximum and minimum light. This body flashes on and off about 30 times per second. Rapid atmospheric scintillation is responsible for the difference in the appearance of nonvarying stellar images.

and the count now stands at over 300. The periods between pulses range from 0.002 to 3.8 seconds, the majority having periods from 0.5 to 1.0 second. Of all the pulsars discovered only two have been conclusively identified with an optical source: the Crab Nebula pulsar and Vela supernova-remnant pulsar.

In 1968 radio astronomers found that a source within the Crab Nebula was emitting pulses of radio radiation at a rate of 30 times a second. The source was soon pinpointed as one of the two central stars in the nebula (Figure 14.12). Since then the star has been observed to pulse optically at the same frequency as the radio pulses (Figure 15.13). Pulses of X-ray and gamma-ray radiation were also found whose rate matched the rates for the pulses in optical and radio radiation. The rate at which the Crab Nebula is emitting radiant energy in all wavelengths is on the order of 10^{38} ergs per second, which is 100,000 times greater than the 10^{33} ergs per second emitted by the sun. Thus the Crab Nebula is comparable in luminosity to the most luminous supergiant although the way in which the emitted energy is distributed in wavelength is significantly different.

The Crab pulsar, as well as others, is not at the geometrical center of the surrounding nebula. If the supernova outburst was not symmetrical, the neutron star may have been forced out in one direction and the expanding nebulous shell in the opposite direction, an example of Newton's third law. If enough time has elapsed since the supernova outburst, the neutron star may no longer even lie inside the nebula, and this may in part explain why no pulsar has been found associated with most supernova remnants.

The time interval between pulses is very constant—for some pulsars about one part in a billion. For their distances the amount of energy in each pulse from a pulsar is enormous. The very short duration of the pulses also indicates that the emitting region is quite small. For example, if the pulse duration is a 100 microseconds (0.0001 second), the size of the emitting region is less than about 30 kilometers, or the distance that light travels in 100 microseconds.

Electrons in interstellar space affect the velocities of radio waves: the longer the wavelength, the slower the wave's velocity. Radio waves coming from a sharp pulse are therefore spread out along the way into a train of waves. The shorter wavelengths arrive on the earth before the longer ones. We can estimate the distance to the pulsar (the path length) from this delay time. Calculations of this kind indicate that most pulsars are relatively nearby Galactic objects, at an average distance of about 3000 light years, most of them being in or near the Galactic plane.

NEUTRON STARS

Astronomers now seem agreed that pulsars are rapidly rotating neutron stars, the possible remnants of supernovae. During the collapse of a star's core, leading to the supernova outburst, electrons and protons are jammed together to form neutrons. Neutrons crowded together in the core of a star are subject to the exclusion principle (page 317), just as are electrons; so they can become degenerate also. As such they can exert a pressure outward that balances the weight of the overlying layers if it is not too great. Thus if the small, dense object left after the supernova outburst is between about 1.4 and $3M_\odot$, pressure produced by degenerate neutrons can balance the weight of matter and stop any further collapse. Such an object is a *neutron star* and is, astronomers believe, a permanently stable configuration of matter.

Conserving whatever rotational motion the presupernova star had, the neutron star spins rapidly since it is so much smaller than the original star. Only

a body 10 to 30 kilometers in diameter with a density approaching nuclear densities (approximately 10^{14} grams per cubic centimeter) could survive the disruptive force of such rapid rotation. Also, if it conserves its original magnetic field of even a few tens of gauss, the enormous reduction in size amplifies the magnetic field to incredible intensities. The magnetic field in neutron stars may run as high as 10^{12} gauss. By comparison the earth's magnetic field intensity is 0.5 gauss, and the largest magnetic field produced in a laboratory is about 300 thousand.

The neutron star's high rotational speed provides a reservoir of energy that can power a continuous flow of charged particles streaming from the star's magnetic poles, as shown in Figure 15.14. Such a flow of charged particles emits coherent electromagnetic waves in a highly directed cone of radiation that spins with the neutron star. When this searchlight beam sweeps across our line of sight, we see a pulse of radiation from the neutron star every fraction of a

FIGURE 15.14

Model of a pulsar. The rotating neutron star is typically about $2M_\odot$, with a radius of about 10 kilometers. The charged particles are accelerated by the magnetic field of the neutron star (up to about 10^{12} gauss in intensity) and flow out along the magnetic axis, producing radio radiation. The magnetic axis of the rapidly rotating neutron star must be properly oriented for us to catch the flash of radiation when one of the rotating beams sweeps past our line of sight. Otherwise we detect no pulses.

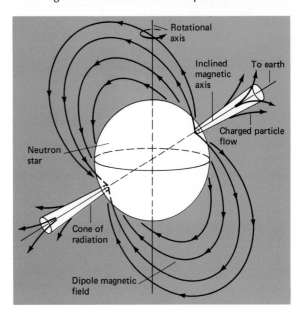

second or so, or we witness a pulsar. And if not, the neutron star is not otherwise observable since it is so very small.

The magnetic field should slow the pulse rate by ultimately converting the neutron star's rotational energy to electromagnetic radiation. This prediction has been confirmed by observations for over 90 pulsars that show a decrease in the pulse rate. Pulsars with the shortest periods are thus apparently the youngest ones.

It also follows that the Crab pulsar, with the shortest period yet discovered (0.033 second), must be the youngest, with an estimated age of 1000 years, or the age of the Crab Nebula.

The average age of the remaining pulsars is about 2 million years; the oldest pulsars are about 10 million years old. Our failure to detect optical pulsations in all but the Crab and Vela pulsars may mean that the light flashes are a transient phenomenon, occurring only during a pulsar's youth.

The neutron star is unlike anything in human experience. Yet it is not so bizarre as a black hole.

15.6
Stellar Black Holes

CORPSES OF DEAD STARS

In Chapter 13 we pointed out that throughout a star's existence there is a continuing struggle between gravity and various countering forces that prevent contraction of the star. During most of a star's life this force is the kinetic pressure that is maintained through the thermal energy supplied by different thermonuclear reactions.

Once the nuclear fuels have been used up, kinetic pressure can no longer be the mainstay against the forces of gravity. For stars whose masses are less than $1.4M_\odot$ contraction leads, as we have seen, to a degenerate electron gas, and the degenerate electrons can indefinitely provide a pressure that balances gravity.

These stars are the white dwarfs, and they are stable and will suffer no further contraction. This is the most common form of stellar demise since the vast majority of stars are low-mass stars.

For high-mass stars, if mass loss during the stars' lives does not reduce them to less than $1.4M_\odot$ by the time they exhaust their nuclear fuels, they cannot become stable white dwarfs. Such stars may go through a supernova outburst. And if the core remaining after the outburst is less than about $3M_\odot$, gravitational collapse cannot be halted by degenerate electrons but can be halted by the degenerate neutrons that form. The object is now a neutron star. The pressure provided by the degenerate neutrons is sufficient to maintain the balance against gravity, and the neutron star is a stable object at a radius considerably smaller than that of the white dwarf. This is the second form for a dying star.

A BLACK HOLE: WARPING OF SPACE-TIME

What happens if the remaining core mass after the supernova outburst is larger than $3M_\odot$? Is there a third form for the dying star? The pressure of the degenerate neutrons cannot halt the collapse, and as far as we know, there are no forces that can halt it.

There is thus nothing to balance gravity and the collapse continues. As the star becomes more compact and its size decreases, the intensity of its gravitational field increases dramatically. In our normal experience the gravitational force is the weakest of all forces in nature. But in this case gravity overpowers everything, and this contracting, superdense mass of matter causes the space-time geometry in its vicinity to warp about it. Eventually this warping becomes so great that space-time folds in over itself as the star passes through the event horizon (Figure 15.15), the point of no return, and the star disappears into a *black hole*. The enormous forces of gravity have so modified the geometry of local space-time for the black hole that there are no paths by which photons or particles of matter may escape the black hole. There are only paths into it, none away from it.

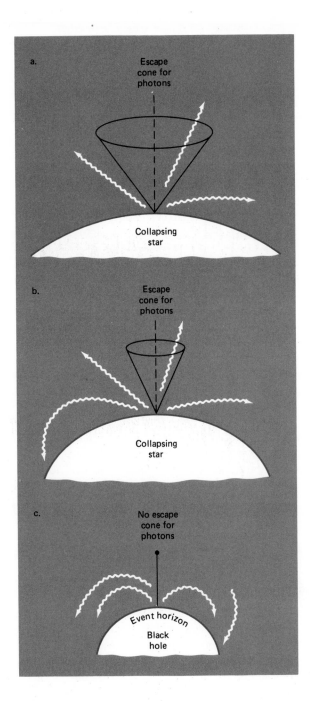

FIGURE 15.15
Collapsing star and its shrinking escape cone for photons. The escape cone is a useful way of visualizing the warping of space about a dense collapsing star that is becoming a black hole; it contains the paths traversed by photons that lead away from the collapsing star. As the collapse proceeds from (a) to (b) to (c), the apex angle of the cone shrinks so that, as the collapsing body just reaches the event horizon, only photons moving radially away can still escape; on and inside the event horizon the cone completely shuts, and escape is no longer possible.

events can be seen by the outside world. Neither light nor matter can escape from the powerful gravitational field of the black hole, and all communication from inside the event horizon is lost with the outside world.

The radius of the event horizon is the *Schwarzschild radius*, named after the German astronomer, Karl Schwarzschild (1873–1916), who first explored the geometry of space around a point mass, using Einstein's theory of general relativity. The calculated Schwarzschild radius for the sun is about 3 kilometers although the sun is not the kind of star that we should expect to experience such a collapse leading to a black hole.

As the contracting matter approaches the Schwarzschild radius, a distant observer will see the first stages of collapse followed by a slowing down of the action. The observer cannot witness the final stage of collapse because time slows down in an intense gravitational field. The greater the intensity of the gravitational field, the more time slows down. In other words at the event horizon time stands still for the outside observer, as we try to illustrate in Figure 15.16. An observer inside the Schwarzschild radius would see matter crushed to a stupendously high density in a relatively short period of time.

This means that when we observe light from a collapsing star, we see two effects: The first is that the star grows dimmer because of the warping of space-time over itself such that it is harder for photons to find exit paths to the outside; second, as we explained in the gravitational redshift, the atom can be thought of as being a small clock, and as photons are emitted, their frequencies decrease as the collapse proceeds. That is, the entire spectrum of emitted radiation is gravitationally shifted further and further to the red, as depicted in Figure 15.16. As we watch the collapse, light from the star grows fainter and redder.

THE EVENT HORIZON

The distance from the black hole's center to the boundary of the *event horizon* marks the region of curved space within which the collapsed body becomes invisible to an external observer. The name is very appropriate since it is the last point from which

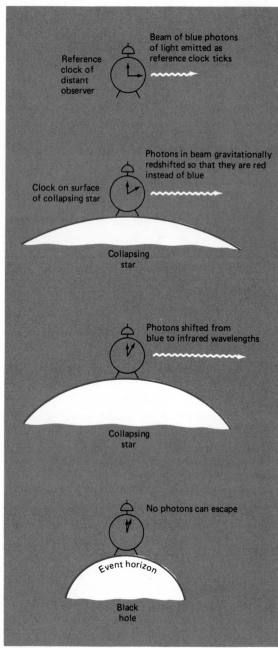

FIGURE 15.16
Slowing of time in a warped space-time geometry for three different identical collapsing stars. The four clocks, a distant reference clock and one on each of the three collapsing stars, are identical. The distant observer sees each clock on the collapsing stars tick slower compared with his clock, depending on how far the collapse has advanced. Also the beams of light emitted by clocks as they tick are gravitationally redshifted compared to his.

GEOMETRY OF SPACE-TIME NEAR A BLACK HOLE

The geometry of space near a Schwarzschild black hole (Figure 15.17) results from the simplest idea: a spherical, nonrotating, gravitationally collapsed body. The nearer we come to the collapsed body, the more space is warped.

Once past the event horizon, the photons are trapped inside the black hole and cannot escape. At the very center, or the heart, of the black hole there can develop eventually a *singularity*. That is a point where the mass of collapsed body is concentrated into zero volume, producing an infinite density! No one knows whether nature so abhors a singularity that none actually exists, but astronomers feel extremely uncomfortable whenever a mathematical formulation of a physical phenomenon suggests the formation of a singularity. We are still not sure what is happening physically at the center of a black hole.

Different objects falling into a black hole lose all their identity except for mass, angular momentum, and electric charge. This meager set of descriptive properties for a Schwarzschild black hole is in stark contrast with the extensive set of descriptive properties for the object when it was, say, an active star. Then astronomers described it with such properties as temperature, pressure, density, and energy-generation rates as they vary with the radial distance from the center.

DETECTING A BLACK HOLE

Even though the remains of a star after going through a supernova outburst may become a black hole, they are not isolated gravitationally from the rest of the universe. One place to look for black holes is in binary systems, where we might detect a black hole (which should have a mass greater than about $3M_\odot$) from the orbital motion of the binary system's visible component. Another possibility is that gas streaming from the visible component of the binary onto its black-hole companion might emit detectable X-ray radiation. The X-ray source Cygnus X-1 may be a black hole in orbit about a blue supergiant star, a possibility we shall discuss in the next section.

ROTATING BLACK HOLES

Since all bodies in the universe rotate, it seems logical to consider the possibilities of rotating black holes. As

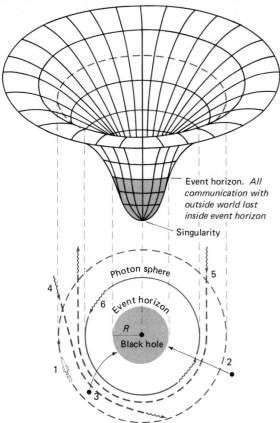

FIGURE 15.17
Space geometry of a black hole and possible trajectories. *R* is the Schwarzschild radius. The increased warping of space as we come closer to the black hole results in several possible trajectories for objects in its vicinity. The trajectory will depend on three variables, the object's speed, its direction, and its distance.

1. An object (spaceship) moving at one-half the speed of light in the last stable orbit.

2. An object approaching head on is sucked into the black hole.

3. An object on a nonradial trajectory falls into the black hole on a curved path.

4. A photon passing outside the photon sphere is deflected at a large angle.

5. A photon outside the photon sphere returns in a direction opposite to its original path.

6. At the boundary of the photon sphere a photon moves continuously in a circular orbit.

in the case of a neutron star the enormous reduction in size from the original star to the black hole should cause the black hole to spin at immense speeds.

Based on the general theory of relativity, the rotating black hole (called a *Kerr black hole*) has two kinds of event horizons: an outer event horizon produced by the rotation and, as in the nonrotating black hole, an inner one. Both event horizons merge at the poles of rotation, as shown in Figure 15.18. The space-time region between the two surfaces is called the *ergosphere*. It should shrink in size the faster the rotation because the outer event horizon grows smaller and the inner one grows larger.

Two very interesting phenomena occur for rotating black holes: The first is that, because of the interaction between the gravitational field and the rotation, the space-time environment is dragged around the spinning black hole. As a consequence an observer stationed near a black hole, but outside the ergosphere, should feel carried along with the rotating black hole relative to distant stars. Outside the ergosphere he or she can apply a force opposite to the direction of the black hole's rotation in order to remain at rest relative to the pattern of distant stars (against which the black hole's rotation is measured).

FIGURE 15.18
Cross-sectional view of rotating black hole. Between the outer event horizon (static limit), surface of infinite redshift produced by rotation, and the inner event horizon produced by the collapsing mass lies the ergosphere.

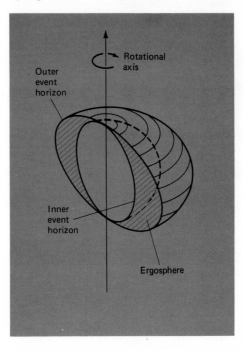

Once inside the ergosphere, however, he or she cannot prevent rotation with the black hole. The other phenomenon is that energy can be extracted from objects spiraling into rotating black holes (hence the reason for the name *ergo*, which is Greek for "work"). Since more energy is removed than is put in, the rotational energy decreases—that is, the spin slows down. When rotation stops, all energy possible has been extracted from the black hole, and the outer event horizon has shrunk to coincide with the inner event horizon.

15.7
Evolution of Binary Systems

MASS TRANSFER: ROCHE LIMIT

In Chapter 14 we saw that many stars lose matter to their surroundings on either a slow time scale, such as in stellar winds, or a rapid time scale, such as in a supernova outburst. Since at least half of all stars appear to be members of binary- or multiple-star sys-

SCHWARZSCHILD RADII FOR BLACK HOLES

The Schwarzschild radius is the radius at which the space-time geometry about a nonrotating collapsing mass folds over itself so that contact with the outside world is lost; it is the radius of the event horizon. The numerical value in centimeters of the Schwarzschild radius R can be found from the relation

$$R = \frac{2GM}{c^2} \approx 1.5 \times 10^{-28}M,$$

where G is the gravitational constant (6.67×10^{-8} cubic centimeters per gram-second squared), c is the velocity of light ($c^2 = 8.99 \times 10^{20}$ square centimeters per second squared), and M is the mass of the collapsing body in grams. If we express the mass in units of the sun's mass, then R is in kilometers and is given by

$$R(\text{km}) \approx 3.0M.$$

The density of the body when it has collapsed to the size of the Schwarzschild radius for R again in centimeters and M in grams,

$$\rho = \frac{M}{\frac{4}{3}\pi R^3} = \frac{3c^6}{32\pi G^3 M^2} \approx \frac{7.3 \times 10^{82}}{M^2},$$

or it is inversely proportional to the square of the mass. Table 15.4 lists the Schwarzschild radius for various objects even though there is no evidence that any of them (except perhaps the last) will ever form a black hole.

TABLE 15.4
Some Schwarzschild Radii

Object	Mass	Schwarzschild Radius	Density (g/cm³)
Hydrogen atom	2×10^{-24} g	3×10^{-44} Å	10^{130}
Human being	7×10^4 g	1×10^{-15} Å	10^{73}
Earth	6×10^{27} g	0.9 cm	2×10^{27}
Sun	2×10^{33} g	3 km	2×10^{16}
Galaxy	$10^{11}M_\odot$	0.03 ly	2×10^{-6}
Cluster of galaxies	$10^{14}M_\odot$	60 ly	5×10^{-13}
Closed universe	$10^{22}M_\odot$	16×10^9 ly	6×10^{-30}

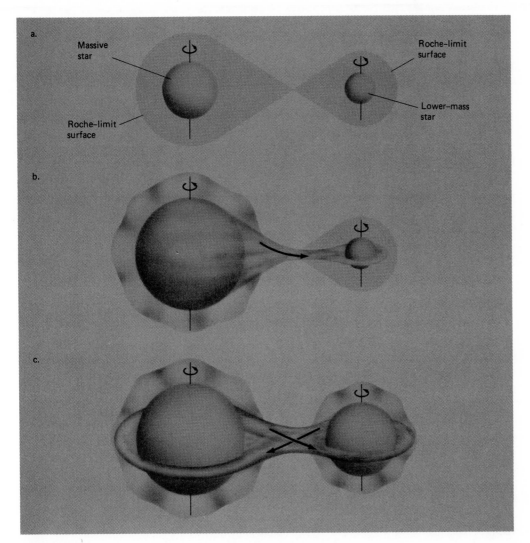

FIGURE 15.19
Roche-limit surfaces for three possible close binary stars. (a) In this system neither star fills its lobe. After the larger star expands, it may fill its lobe (b) and begin to transfer mass to the companion. (c) In this system both stars fill their lobes and may exchange mass as shown.

tems, it is important to inquire whether the various mass-loss processes can actually result in the exchange of mass between the components of a binary system. If this does occur, then how does that process affect the evolution of the stars involved? Finally, what signs should we look for to verify that mass transfer is occurring?

In a binary system gravitational theory predicts that there exists for each star a teardrop-shaped surface surrounding it known as the *Roche limit*. The significance of this encompassing surface is that, if either star expands outside the Roche limit, its outer layers, under the gravitational attraction of the companion, can flow away onto the companion star. If the two stars in the binary system are relatively close, on the order of several tens of their radii apart, the Roche limiting surface is close enough to each star (Figure 15.19) that the course of normal evolution for either star can cause it to overflow its lobe of the Roche surface.

Computations for binary systems in which the two stars are close to each other suggest that significant amounts of mass can be transferred from one star to the other several times during the evolution of the system. This mass-transfer process can in the extreme case completely change the course of evolution for both stars. Let us briefly outline the process of evolution in a close binary before discussing some examples that are of current interest.

EVOLUTION OF BINARY STARS

For a binary system in which the two components have different masses the larger-mass star will have the shorter main-sequence lifetime. For example, a $10M_\odot$ star spends about 1 percent as long as a $2M_\odot$ star. At the end of hydrogen burning the envelope of the star will expand as the star restructures itself to become a red giant. In that process the star may fill its lobe of the Roche surface and begin the process of transferring its outer layers to its companion. Matter will probably form a gaseous disk about the companion in the plane of the orbit before splashing onto the companion's surface. Theoretical studies of the evolution show that the mass transfer may include some or all of the envelope of the massive star and will occur in a relatively short period of time.

What remains of the original primary is the evolved core and possibly some of the original envelope. The original primary now appears for its new mass as an overluminous main-sequence star, while the original secondary is just a massive main-sequence star with possibly some atmospheric chemical anomalies. If the mass of the core of the original primary is less than $1.4M_\odot$, the object could also be a white dwarf, depending upon how complete the transfer process was.

After mass transfer the original secondary has become more massive than the original primary. The now-massive secondary may lose mass by intense stellar winds, a portion of which is captured by the former primary. Also it finishes its main-sequence evolution in a shorter time than it otherwise would have. When it ceases hydrogen burning, the now-massive secondary will expand to become a red giant, filling its lobe of the Roche surface. This can lead to a reverse transfer of matter, back to the original primary, which by then may be an extremely compact object such as a white dwarf.

Matter coming from the normal star is heated as it falls through the intense gravitational field of the compact companion. This process converts kinetic energy to thermal energy and is extremely efficient in producing X rays. In fact the inward-falling matter can produce more energy in the form of X rays than it could in thermonuclear processes. In some cases it is possible to envision several exchanges of matter between the two components of the binary, and in each exchange the course of evolution would be greatly altered. Another possibility is that during the exchange process matter is completely lost from the system so that two white dwarfs are left at the end.

There are obviously a number of different endpoints of the evolutionary process that can occur in binary systems involving mass transfer. Let us examine some of the more interesting possibilities.

ARE NOVAE MEMBERS OF CLOSE BINARIES?

There is strong spectroscopic evidence that many—and perhaps all—novae are the hotter members in binaries whose members are very close to each other. It appears that the two stars are an expanding, red giant star and a hot, white dwarf. During its evolution the red giant has swollen and overfills its lobe of the Roche surface. Hydrogen-rich gas streams from the cool giant onto its small, hot companion. Eventually enough matter collects on or about the white dwarf to unsettle its equilibrium or to initiate thermonuclear reactions in its outer layers. The energy necessary to heat the gas comes from the kinetic energy of the in-falling matter; the burning occurs in degenerate matter, causing a thermonuclear runaway, as in the helium flash. The end result is a violent explosion by the white dwarf as a visible nova.

After a relatively quiet period that may last months or centuries another buildup of hydrogen-rich matter on the white dwarf comes with similar consequences. In recent years several X-ray novae have been identified by orbiting spacecraft; this helps to confirm astronomers' belief that the nova outburst is due to mass transfer onto a compact star like a white dwarf.

COULD SOME SUPERNOVAE BE EXPLODING WHITE DWARFS?

It has been suggested that the Type I supernovae may be due to the explosion of a white dwarf in a binary system. If the compact white dwarf acquired enough hydrogen-rich matter where its mass exceeds $1.4M_\odot$, the result might be the detonation of the white dwarf as a supernova.

Astronomers estimate that probably half of all O- and B-type stars are members of a binary system, with the companion being close to the same mass. It does not appear that very massive stars form a binary system with lower-mass stars. Of these O and B binary systems, approximately 20 to 30 percent are old enough for the most massive member to have gone through a supernova outburst and have become a neutron star or a black hole. If 10 percent or so of these binary systems survive the supernova outburst (that is, they are still gravitationally bound to each other after the outburst), then there are binary systems transferring matter in which one of the pair is far more likely to be a neutron star or a black hole than a white dwarf.

BINARY PULSARS

The first pulsar in a binary system was discovered in 1975. The second star in the binary system unfortunately is not visible although the guess is that it also is a compact star. From the orbital parameters of the binary system, like those of a spectroscopic binary, the pulsar is a neutron star with a mass about 1.4 times that of the sun, with the companion's having about the same mass. Besides providing a mass for a neutron star, an observed shift in the position of periastron for the binary pulsar provides another test to confirm the validity of general relativity. Finally the period of the binary motion is observed to be decreasing, indicating a loss of energy in the system. This could be due to another prediction of general

relativity known as gravitational radiation (Chapter 19).

Since the discovery of the first binary containing a pulsar, two more such systems have been identified. One interesting point is that only 3 of the some 300 pulsars are part of a binary system. If half of all stars are members of binaries, then there should be many more binary pulsars. This is probably good evidence that the supernova outburst is so violent that it flings the two stars apart.

X-RAY BINARIES

Among the most fascinating stellar objects are the X-ray binary stars, whose orbital periods range from a fraction of a day to many days. The usual X-ray binary includes a visible, hot, blue star, from which matter is streaming to the neighborhood of a nearby, gravitationally collapsed body, such as a white dwarf, neutron star, or even a black hole. Let us examine the remarkable behavior of such systems (Table 15.5).

A typical eclipsing and pulsating X-ray binary is SMC X-1, which is some 190,000 light years away. It was discovered in the Small Magellanic Cloud by the *Uhuru* satellite in 1971. The rapidly spinning neutron star emits X rays from a hot spot on its surface. As the

TABLE 15.5
Some X-ray Binaries

Binary System	Period (days)	Mass of X-ray Source (sun = 1)	Mass of Companion (sun = 1)	Spectral Type of Companion
HD 77581[a]	9.0	1.9	23	B0 I
SMC X-1	3.9	1	16	B0 I
1538-52	3.7	2	20	B0 I
Cen X-3	2.1	1	17	O7 III
LMC X-4	1.4	2.6	25	O7 III–V
HD 153919[b]	3.4	1.3	27	O6
Cyg X-1	5.6.	6	18	O9 I

[a]Vela X-1
[b]4U 1700-37

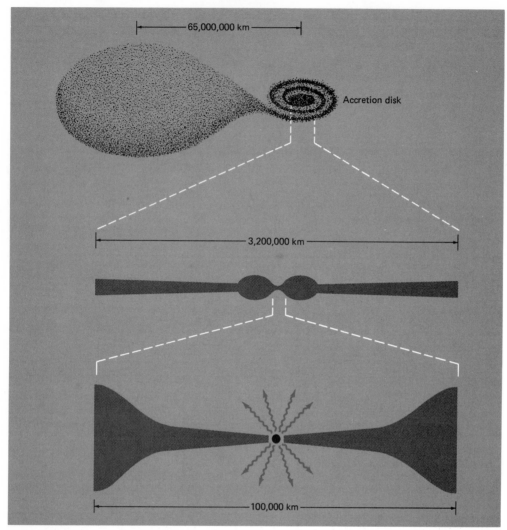

FIGURE 15.20
Proposed model for Cygnus X-1. A black hole in a binary system with a very massive, blue supergiant star pulls matter escaping the supergiant into an accretion disk, as shown in the lower figure. The X rays probably come from as close as several hundred kilometers from the black hole.

inward-falling material collides with the neutron star, it causes the hot spot on the surface from which X rays are emitted. The total amount of energy radiated is 3×10^{38} ergs per second, the highest of any known neutron star.

Another example of an eclipsing X-ray binary is Herculis X-1, in which the X-ray source is eclipsed by a blue giant companion, HZ Herculis, for about 6 hours every 1.7 days. The binary system is about 12,000 light years from us. Analyzing the orbital data

gives upper limits for the mass of $2.8 M_\odot$ for the visible component and $2.2 M_\odot$ for the X-ray component. The low mass of the X-ray source suggests that it is a neutron star.

Still other examples of X-ray binaries abound, each one with its own unique behavior. Some X-ray binaries are more or less stable in their behavior, while others suddenly turn on briefly and then turn off for weeks, months, or even years.

A BINARY CONTAINING A BLACK HOLE?

Finally there are the suspected black-hole candidates known as Cygnus X-1 (period of 5.6 days) and Circinus X-1 (period of 16 days). Both systems are eclipsing binaries containing a blue supergiant component (which is visible to us) and an X-ray source associated with a body of many solar masses that is massive enough to be considered a black hole rather than a neutron star. These systems exhibit variable X-ray emission believed to originate in a thin disk surrounding the black hole. For Cygnus X-1, whose estimated distance is 10,000 light years, a suggested model is shown in Figure 15.20. In this system the black hole mass, derived from orbital parameters, is definitely over $5M_\odot$ and probably lies in the range of 9 to $15M_\odot$, while that of the blue supergiant is about $30M_\odot$ or greater. Gas is apparently being pulled from the overextended blue supergiant by the black hole's powerful gravitational field. Most of the mass escapes from the system, except that headed toward the black

hole. As matter streams toward the black hole, a large part of it is captured and forced to swirl around the black hole to form a thin, gravitationally compressed disk. Friction within the spiraling gas heats it to very high temperatures so that it emits variable X-ray radiation. Like water draining out of a bathtub, the tightening vortex of the X-ray-emitting gas is finally sucked into the black hole; as it crosses the event horizon, it disappears from view.

The end of the lives of stars as white dwarfs, neutron stars, or black holes is fascinating in itself. But because stars are components of galaxies, their end has important meaning for the life of a galaxy. In Chapters 16 to 18 we shall discuss galaxies, beginning with our Galaxy, in terms of their structure and organization as a universe of galaxies.

SUMMARY

The end of main-sequence life. Main-sequence life of a star ends with exhaustion of hydrogen in the core. This may take from 200 billion years to as short as a few million years. When the hydrogen fuel is exhausted in the core, hydrostatic equilibrium shifts to favor gravity so that the core contracts, raising its thermal energy content. When the star's temperature reaches 120 million K, helium burning can begin and contraction ceases. Depending on the star's main-sequence mass, helium burning initiates nonexplosively (high-mass stars) or explosively (low-mass stars).

Red giant evolution for low-mass stars. As a result of internal restructuring after hydrogen burning ceases, the surface temperature decreases somewhat but the luminosity increases significantly as the star becomes a red giant. The helium-rich core has a high density, and the gas no longer obeys the perfect gas laws. It is now degenerate and acts more like a solid. Helium burning in a degenerate gas is a runaway process called a helium flash, which removes the degeneracy. The H-R diagrams for open clusters confirm the evolution after hydrogen exhaustion away from the sequence. Horizontal-branch stars are globular cluster stars of about $0.5M_\odot$ burning helium.

Death of low-mass stars like the sun. Unable to burn carbon, eventually the outer shell of the low-mass star separates from the core, expands, and becomes the shell of a planetary nebula. Remaining behind is the hot, dense, degenerate remnant of the

former star—a white dwarf. The only source of energy is the white dwarf's thermal energy so that the interior begins to cool. A white dwarf is limited to be less massive than $1.4M_\odot$. When its thermal energy is exhausted, a white dwarf becomes a black dwarf.

Death of high-mass stars. When hydrogen is exhausted from the core of a massive star, the core shrinks drastically, elevating temperature and density and exaggerating the differences between central regions and outer layers of the star. Evolution across the H-R diagram is almost horizontal toward the red giant region when helium burning begins. Stars up to about $9M_\odot$ will also burn carbon, while more massive stars undergo oxygen, neon, and silicon burning depending on the mass of the star.

Supernovae, pulsars, and neutron stars. The end of the nuclear burning can be explosive in massive stars. These explosions are supernovae. The remaining star will be either a neutron star or a black hole; the outcome depends on the mass of the star. A neutron star is composed of degenerate neutrons with a density as high as that of the nucleus and a radius of a few tens of kilometers. The rapidly spinning neutron star is thought to be observed as the pulsar.

Stellar black holes. Once nuclear fuel has been exhausted and the star undergoes a supernova outburst, if the remaining core is more than $3M_\odot$, there is no force to counteract the force of gravity. The result, when this happens to extremely massive stars, is a black hole. The geometry of space-time is so warped by the immense gravitational field that nothing can escape the event horizon of the black hole—neither light nor matter. Black holes cannot be detected directly, consequently but can be observed as part of a binary system.

Evolution of binary systems. Computations for two stars relatively close to each other show that they may exchange matter back and forth and influence each other's evolution. For example, novae may be the hotter members of close binaries for which mass exchange is occurring. Binary pulsars and X-ray binaries containing a neutron star have been detected. Finally, there may be binaries in which one member is a black hole, such as Cygnus X-1.

KEY TERMS

black dwarf
black hole
carbon burning
Cepheid instability strip
Chandrasekhar limit
ergosphere
event horizon
gravitational redshift
helium burning
helium flash

Hertzsprung gap
high-mass stars
horizontal branch
low-mass stars
neutron star
pulsar
Roche limit
Schwarzschild radius
space-time
turnoff point

CLASS DISCUSSION

1. Describe what happens to stars on the main sequence as they consume their hydrogen fuel. What changes are occurring in their internal structure and in their position in the H-R diagram? What role does their mass play?

2. What is the probable evolutionary path of the sun after the main sequence?

3. What is the sequence of nuclear fuels stars may fuse in their deep interiors? Do all stars go through the same burning stages? Why must they burn different fuels?

4. What determines the structure of a white dwarf, a neutron star, or a black hole?

5. Why is the evolution of binary star systems important in explaining many of the unusual events astronomers are observing?

READING REVIEW

1. What is the significance of the turnoff point?

2. How does a red giant differ from a main-sequence star?

3. How do open- and globular-cluster H-R diagrams support the astronomers' story of stellar evolution?

4. What will the sun be like at the end of its life?

5. How does the white dwarf resist gravity?

6. Is the evolution of a $50M_\odot$ star the same as a $10M_\odot$, a $3M_\odot$, or a $1M_\odot$ star?

7. What precipitates a supernova outburst in a massive star?

8. Why do astronomers believe the pulsar is a neutron star?

9. What is a black hole? Are rotating and non-rotating black holes different?

10. Is there any evidence that black holes really exist? What is it?

PROBLEMS

1. The $15M_\odot$ star in Figure 15.1 (near the zero-age main sequence; it is a B0 V star) remains at approximately constant luminosity after it ceases hydrogen burning. Using the relation for the luminosity of a star derived from the Stefan-Boltzmann law, show that the star's radius must be increasing as it goes from a surface temperature of 28,000 to 4000 K. What is the relative change in radius? What is the change in units of the solar radius? (Assume the sun's surface temperature to be 5800 K.)

2. From Wien's law, what is the wavelength of the peak of the blackbody energy curve for Sirius B (see Table 15.2)? What wavelength region—that is, visible, infrared, etc.—does the wavelength fall in?

SELECTED READINGS

Iben, I. "Globular Cluster Stars," *Scientific American*, July 1976.

Kaler, J. B. "Bubbles from Dying Stars," *Sky & Telescope*, February 1982.

Kaufman, W. J. *Black Holes and Warped Spacetime*. Freeman, 1979.

Margon, B. "The Bizarre Spectrum of SS433," *Scientific American*, October 1980.

Schramm, D., and W. Arnett. "Supernovae: The Origin of the Chemical Elements, Cosmic Rays, Neutron Stars, and Maybe Even Black Holes," *Mercury*, May–June 1975.

16

The Milky Way Galaxy

Although the title of this chapter may suggest a first discussion of the Milky Way Galaxy, the preceding chapters have in many ways also been about our Galaxy: Among the known objects stars are the principal components of the Galaxy, and in all their variety they, along with the interstellar gas and dust, determine the Galaxy's general appearance. In this chapter we bring together our knowledge of stars, of the interstellar medium, and of the mechanics of the Galactic system in order to understand our Galaxy as a whole. We begin with a brief survey of the way we view the Galaxy from our home deep within it.

"Lo," quoth he, "cast up thine eye,
See yonder, lo! the galaxie,
The which men clepe the Milky Way
For it is white; and some parfay
Callen it Watling streete."

Geoffrey Chaucer

16.1
Overview of Our Galaxy

THE MILKY WAY

If you pick a clear, moonless night, away from city lights, and look up, arching across the sky you will see a misty, irregular, beltlike cloud of light, the Milky Way. In temperate latitudes we view different parts of the Milky Way inclined to the horizon at different angles, depending on the time of night and season of the year. About 30 percent of the Milky Way lies below our horizon; thus to see it all we must move to the southern hemisphere. Visualizing the parts of the Milky Way Galaxy in the sky is easier if you are familiar with the constellations (see the star maps in Appendix 2).

For midnorthern latitudes in the late summer months, an hour or so after sunset, the Milky Way shines its richest. Then the Northern Cross of Cygnus is directly overhead, along with the summer triangle of bright stars: Deneb in Cygnus, Vega in Lyra, and Altair in Aquila. The bright star Arcturus is setting in the west, and the great W of the constellation of Cassiopeia is rising in the northeast (Figure 16.1). The shimmering band of the Milky Way can be traced from Cassiopeia and Cephus in the northeast through Cygnus and then down toward the southern horizon through the constellations of Aquila, Ophiuchus, Sagittarius, and Scorpius. From Cassiopeia to Cygnus it is a single silvery band of varying width, but between Cygnus and Sagittarius it divides into two bright lanes

◀ A detail from *The Origin of the Milky Way*, painted around 1578 by Tintoretto.

flanking the dark band of interstellar clouds known as the Great Rift (Section 14.2). The center of our Galaxy lies beyond the great *star cloud* in Sagittarius (an area of the sky where great numbers of stars are to be seen, giving the impression of an almost continuous cloud of brightness; Figure 16.2).

In winter we see the dimmer and more sparsely populated portion of the Milky Way. It starts with Canis Major in the southeast, passes through the winter triangle of bright stars (Sirius in Canis Major, Procyon in Canis Minor, and Betelgeuse in Orion), and continues through Orion, Taurus, Auriga with its bright star Capella, and Perseus in the northwest. The direction between Orion and Auriga points away from the Galactic center (Figure 16.1). Notice that the brightness and width of the band of the Milky Way differ greatly from one section to another.

Historically it was not immediately obvious from the sky's appearance that the sun was immersed in a disk-shaped system of stars even though the bandlike appearance of the Milky Way is an edgewise view of that disk. Not until 1917 was the sun shown not to be at the center of the disk (Figure 1.8). The historical path to understanding the structure of our Galaxy began with the study of the motions of the stars in the neighborhood of the sun. We shall take up that topic after giving an overview of the Galaxy.

GENERAL FEATURES OF THE GALAXY

From a variety of different measurements astronomers have amassed the information summarized in Table 16.1. From these figures astronomers are aware that the Galaxy is major in size and similar in many respects to our large neighbor in the constellation Andromeda, some 2 million light years away (Figure 17.17).

In general appearance the Galaxy is a large, flattened, or disk-shaped, system of approximately 400 billion stars. There is a large spherical *bulge* of

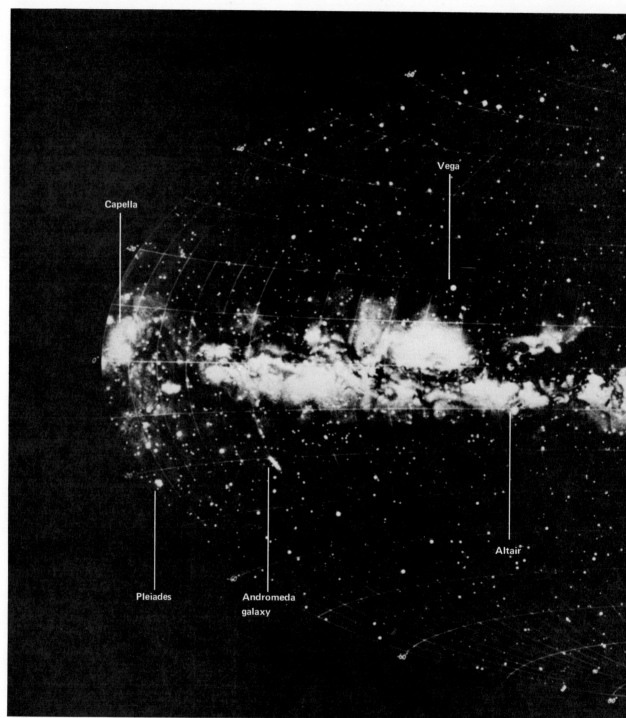

FIGURE 16.1
Panoramic view of the Milky Way Galaxy. The co-ordinates shown are Galactic longitude and lati-tude.

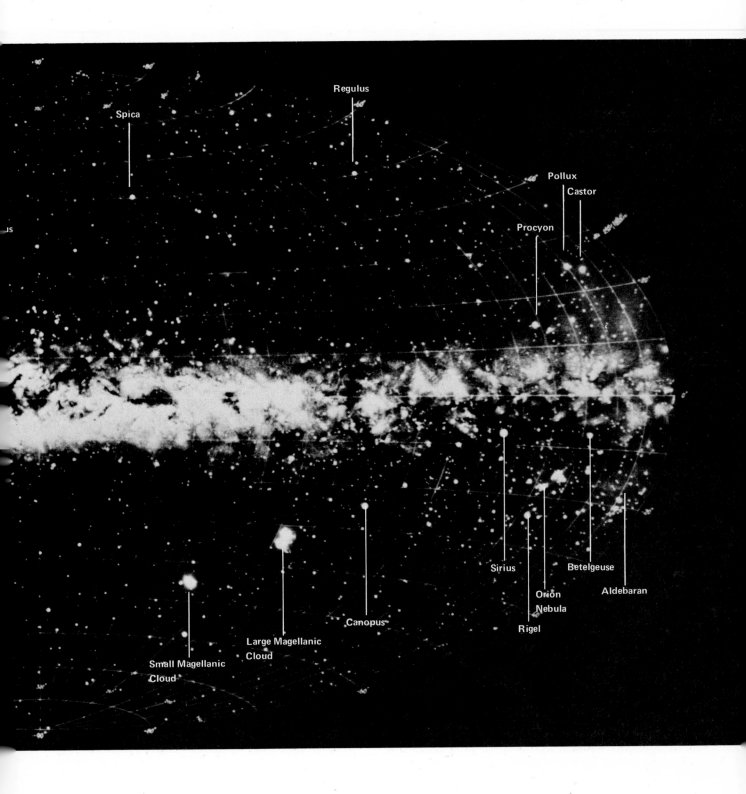

FIGURE 16.2
Star clouds in the constellation of Sagittarius, marking the direction toward an edge of the nucleus of our Galaxy. North is at the top.

TABLE 16.1
The Galaxy

Galactic Property	Numerical Value
Age of Galaxy	≈ 12–15×10^9 yr
Radius of disk	50,000 ly
Radius of halo[a]	$\approx 60,000$ ly
Radius of nucleus	13,000–15,000 ly
Thickness of nucleus	10,000–12,000 ly
Thickness of disk near sun	3000 ly
Volume of disk	$\approx 4 \times 10^{13}$ ly^3
Volume of halo	$\approx 9 \times 10^{14}$ ly^3
Luminosity of Galaxy	$\approx 20 \times 10^9 L_{\odot}$
Mass of Galaxy[b]	150–200 $\times 10^9 M_{\odot}$
Mass of halo	10–30% of total mass
Number of stars	$\approx 400 \times 10^9$
Mean density of matter in disk	$5 \times 10^{-3} M_{\odot}$/ly^3
Mean density of stars	$4.7 \times 10^{-3} M_{\odot}$/ly^3
Mean density of interstellar matter	$0.3 \times 10^{-3} M_{\odot}$/ly^3
Mean star density in disk	5×10^{-3} star/ly^3
Distance of sun from center	30,000 ly
Velocity of sun around center	250 km/s
Period of sun's revolution	230×10^6 yr

[a] Radius of possible massive corona $\approx 350,000$ ly; see page 402.
[b] Mass of possible massive corona $\approx 10^{12} M_{\odot}$; see page 402.

stars in the center, in which is located the *nucleus*, which occupies about 10 to 15 percent of the inner radius. Extending out from the nucleus is a thin grindstone-shaped *disk* of stars through which thread even thinner spiral arms. If our Galaxy is like other spiral galaxies that we see, there are two *spiral arms* (coming out of opposite sides of the nucleus) that wrap around the nucleus in such a fashion that they trail as the Galaxy rotates. Surrounding the disk and nucleus is a spherical *halo*. Stars are most numerous in the nucleus, and their numbers decline outward through the disk and even more rapidly away from the disk into the halo. Within the Galaxy stars occur singly, in multiple-star systems, or in clusters, such as the open clusters of the disk or the globular clusters in the halo. The oldest stars are in the halo, and the youngest in the spiral arms. Our sun is a middle-aged yellow

dwarf star that is part of the disk population of stars.

About 95 percent of the Galaxy's observable mass is apparently tied up in stars. The remainder is gas and cold dust grains strewn about the disk with a density greater in the spiral arms than in the regions between the arms. There are dark interstellar clouds in the arms and bright H II emission nebulae. Their bright patchy appearance forms the spiral framework of the Galaxy.

For many years, as was discussed in Chapter 13, astronomers felt that the way to visualize substructure in the Galaxy was to see it as made up first of a very thin spiral pattern of the youngest objects, called Population I, which was embedded in a succession of thicker lens-shaped groups of older and older stars. The final group of the oldest stars of Population II was believed to be spherical in shape, as shown in Figure 16.3. This structure was thought to reflect the actual

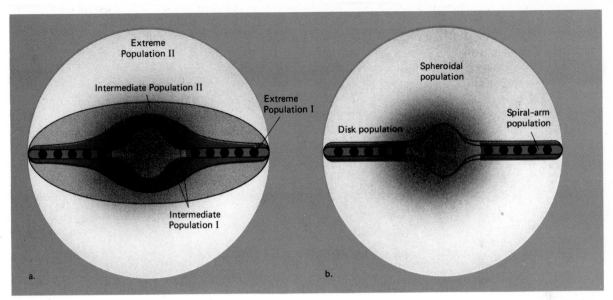

FIGURE 16.3
Locations of the various stellar populations in the Galaxy. In (a) the locations are identified for the older concept of Population I and Population II stars. Each population group is divided into two subgroups, with extreme Population II the oldest stars and extreme Population I the youngest. On the other hand (b) shows the current view of a spheroidal population group distributed about the Galactic center and a disk population group spanning the equatorial plane.

MASS OF OUR GALAXY

Another application of Kepler's modified third law is using it to obtain an approximate mass for our Galaxy. Assuming that most of the mass lies toward the center in roughly a spherical shape, we can write

$$P^2_{Gal} = \frac{4\pi^2 R^3_{Gal}}{G(M_{Gal} + M_\odot)},$$

where P_{Gal} is the period of orbital revolution at the sun's distance, R_{Gal}; M_{Gal} is the Galaxy's mass; $M_\odot$ is the sun's mass; and G is the gravitational constant. This example is similar to the one on planetary motion on page 530. Because the sun's circular orbital velocity is $V_\odot = 2\pi R_{Gal}/P_{Gal}$, we can restate the preceding equation after some transformation; we also can neglect $M_\odot$, which is insignificant compared to M_{Gal}, so that

$$M_{Gal} = \frac{V_\odot^2 R_{Gal}}{G}.$$

Substituting in the equation $V_\odot = 2.46 \times 10^7$ centimeters per second, $R_{Gal} = 2.84 \times 10^{22}$ centimeters, and $G = 6.67 \times 10^{-8}$ cubic centimeter per gram per second squared, we find $M_{Gal} = 2.58 \times 10^{44}$ grams $= 1.30 \times 10^{11} M_\odot$. This represents the mass inside the sun's Galactic orbit, which is somewhat less than, but reasonably close to, the value given in Table 16.1.

evolution of the Galaxy from roughly spherical shape through the succession of lens-shaped groups to the present thin spiral, always leaving stars to occupy the volume defined by the Galaxy at the time of their birth. Views on population groups have changed and continue to change. Today astronomers visualize the Galaxy as divided into a *spheroidal population* group of old stars, whose characteristics vary roughly in the radial direction out from the Galactic center, and a *disk population* group of younger objects, whose characteristics vary roughly perpendicular to the mid-plane of the Galaxy (Figure 16.3). Table 16.2 provides

a summary of the characteristics of the two population groups.

The revolution period for the sun and the solar neighborhood of 230 million years is a reasonable standard for measuring changes in our Galaxy. This time scale is called the *Galactic year,* and in its units the sun and the solar system are about 20 Galactic years old. The young O and B stars used to trace the spiral structure are less than 0.5 Galactic year in age, while the oldest stars in the halo are 40 to 70 Galactic years in age. From the dynamics of the Galaxy it appears that relatively little change can occur in the

TABLE 16.2
Stellar Population Groups in the Galaxy[a]

Property	Disk Population Group				Spheroidal Population Group
	Spiral arm	Young	Intermediate	Oldest	
Representative objects	Interstellar gas and dust Open clusters and stellar associations O and B stars Supergiants Classical Cepheids T Tauri stars Some A stars	A stars F stars A–K giants Some G, K, and M dwarfs and white dwarfs	Sun Most G dwarfs Some K and M dwarfs and white dwarfs Some subgiants and red giants Planetary nebulae	Some K and M dwarfs and white dwarfs Some subgiants and red giants Moderately heavy-element-poor stars Long-period variables RR Lyrae with periods less than 0.5 days	Globular clusters RR Lyrae variables with periods greater than 0.5 days Some subdwarfs Extreme heavy-element-poor stars Some red giants
Age in 10^9 years	<0.1	≈1	≈5	≤10 (?)	≈10 − 15
Age in Galactic years[b]	<0.4	≈4	≈20	≤40 (?)	≈40 − 65
Location	Spiral arms	Thin disk	Disk	Nucleus and thick disk	Nucleus and halo
Concentration toward Galactic center	Very weak	Weak	Moderate	Considerable	Strong
Concentration toward Galactic plane[c]	Very strong (≤1000 ly)	Strong (≤ 1500 ly)	Moderate (≤ 3000 ly)	Weak (≤ 5000 ly)	Very weak
Galactic orbits	Circular	Low eccentricity	Moderate eccentricity	Considerable eccentricity	Very eccentric
Heavy-element content $(Z/Z_\odot)$[d]	1 to 2	1 to 2	0.5 to 1	0.2 to 0.5	0.05 (halo)–1 (nucleus)
H-R diagram	h and χ Persei	Hyades	M67	NGC 188	NGC 188 (nucleus)–M92 (halo)

[a] Adapted from D. M. Mihalas and J. J. Binney, *Galactic Astronomy*, 2nd ed. (San Francisco: Freeman, 1981).

[b] A Galactic year is the orbital period of the sun, which is about 230 million years.

[c] Approximate width of band centered on Galactic plane in which majority are found.

[d] Heavy-element content by mass, with $z_\odot \approx 0.02$.

Galaxy's structure in periods of less than a few Galactic years. And to an external observer probably very significant changes have occurred in the appearance of our Galaxy in periods of tens of Galactic years.

16.2 Stellar Motions in the Solar Neighborhood

Most of the changes we see in the night sky result from the earth's daily rotation and annual revolution, but the stars too are moving. Do they move chaotically, or is there some pattern to their motion? There is, as we shall see, a slight individual random motion relative to their neighbors superimposed on a larger, systematic motion, in which all stars participate as they revolve around the Galaxy's center.

MEASURING STELLAR MOTIONS

Astronomers cannot directly observe the actual motions of the stars relative to the sun, but they can detect their motions by comparing photographs of the same star field taken years apart. If all the stars of the solar neighborhood moved at about the same rate, then the nearer ones should sweep across your line of sight faster than a more distant one. From photographs astronomers can measure the minute change in position; it is expressed as the angular change in position in arc seconds per year, called the star's *proper motion*.

While stars close to the sun have proper motions large enough to measure, those for more distant stars are too small to measure. Barnard's star has the largest known proper motion, 10.3 arc seconds per year. Proper motions have now been measured for more than a quarter of a million stars.

The constellations appear to us very much the same as they did in historical times, but eventually these familiar patterns of stars will change because of proper motion. This is illustrated in Figure 16.4 for the Big Dipper, whose shape has changed drastically over the last 100,000 years and will change equally dramatically over the next 100,000 years.

A star's proper motion coupled with its distance is sufficient to tell us its speed at right angles to our line of sight. This is called its *tangential velocity*. We can

also find its line-of-sight motion, or *radial velocity*, from the Doppler shift of the absorption lines in the star's spectrum. The radial velocity of stars can be measured to an accuracy of a few kilometers per second or less, and there exist measurements for some 20,000 of them. Together, a star's tangential velocity and its radial velocity define its speed and direction, its *space velocity*, relative to the sun (Figure 16.5). Typical values for the space velocities of stars in the solar neighborhood are around 30 kilometers per second, with most under 60 kilometers per second.

WHERE ARE THE STARS GOING?

Even without knowing how far away they are, we can make some generalizations about the movements of

FIGURE 16.4
Changes in the appearance of the Big Dipper due to proper motions. The five inner stars are parts of an open star cluster whose members are moving on parallel tracks in space.

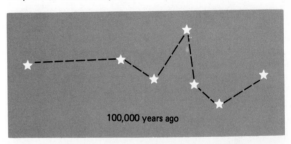

100,000 years ago

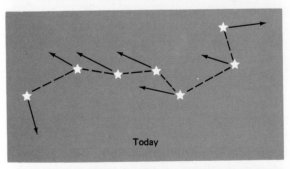

Today

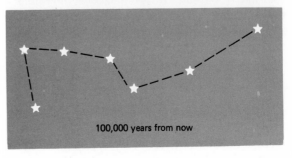

100,000 years from now

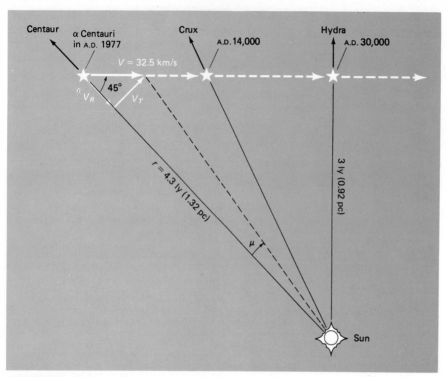

FIGURE 16.5
Future space motion of Rigel Kent (α Centauri).
The components of its space velocity are
r = distance in light years, μ = proper motion,
V_R = radial velocity, V_T = tangential velocity = 1.45
μr km/s, and V = space velocity = $\sqrt{V_R{}^2 + V_T{}^2}$.

the stars in the solar neighborhood from their proper motions and radial velocities. By statistically analyzing proper motions, we find that nearby stars generally appear to diverge outward from a point on the sky that is the point toward which the sun is headed. Simultaneously, on the opposite side of the sky, nearby stars appear to be converging toward a point on the sky that is the point from which the sun is receding. This is the result of the *solar motion* (Figure 16.6), which is taking the sun toward a point on the celestial sphere that lies in Hercules within 10° of the bright star Vega. Radial velocities of the sun's closest neighbors indicate that the sun's motion relative to them is below the average space velocity, being about 16 kilometers per second (3.3 AU per year).

Overall, the motion of the stars of the solar neighborhood appears to be almost random in direction as seen from the solar system; that is, as many stars are going in one direction as in any other direction. As we have stated, the typical speed for this random motion is about 30 kilometers per second.

Is there an organized motion of stars when we consider a scale larger than the several hundred light years of the solar neighborhood? There is; from data on stars at great distances from the sun we know that the sun and its neighbors are moving at about 250 kilometers per second in nearly circular orbits around the center of the Galaxy (about one orbit in 230 million years). The solar neighborhood is also approximately 30,000 light years (about 9000 parsecs) from the Galactic center, or about two-thirds of the radius of the visible disk of the Galaxy. Hence for the nearby stars slight variations in their individual Galactic orbits create the appearance that they and the sun move randomly within a local frame of reference. With respect to the sun's stellar neighbors we see it moving toward Hercules, but relative to the whole Galaxy the sun is a participant in the Galaxy's rotation in the direction of Cepheus.

GALACTIC COORDINATES

To help in the understanding of the rotation of the Galaxy and its organization, astronomers have devised a coordinate system (see Figure 16.1) from the perspective of the Galaxy (see Appendix 3). Figure 16.7 shows an idealized sketch of the Galaxy, both in a

> Geographers . . . crowd into the edges of their maps parts of the world which they do not know about, adding notes in the margin to the effect that beyond this lies nothing but sandy deserts full of wild beasts, and unapproachable bogs.
>
> Plutarch

face-on view to the disk and in a cross-sectional view through the disk. On the right-hand side of the figure are shown the Galactic coordinates, which are known as Galactic longitude and latitude. *Galactic longitude* is the angular distance measured in the central plane of the disk, starting from the Galactic center and measuring along the Milky Way through the constellations shown. The direction 90° to the Galactic center in Sagittarius is toward the constellations of Cepheus and Cygnus; 180° is toward Taurus, Auriga, and Perseus, not far from the Pleiades open cluster; 270° is toward Canis Major and Puppis. *Galactic latitude* is the angular distance above and below the central plane of the Milky Way. The North Galactic Pole, at 90° N Galactic latitude, lies in the constellation of Coma Berenices and in the same hemisphere as the North Celestial Pole. The South Galactic Pole, 90°S, lies in the constellation of Sculptor. The Galactic coordinates will help you to visualize the nature of our Galaxy.

16.3
The Disk Component
of the Galaxy

ROTATION OF THE SOLAR NEIGHBORHOOD
One means of determining the Galaxy's rotation is to use a frame of reference outside the Galaxy, such as distant galaxies. From systematic study of the Doppler shifts of distant galaxies the solar neighborhood is found to be moving toward the galaxies in the direction beyond the stars of Cygnus and away from those beyond the stars of Canis Major. As stated, the velocity of the solar neighborhood, which is about 30,000

light years from the center of the Galaxy, is about 250 kilometers per second (about 0.0008 light year per year), giving a revolution period of about 230 million years.

At present astronomers estimate that within about 3000 light years of the Galactic center the central regions rotate somewhat like a solid wheel so that from the center outward the orbital velocity increases. From about 4000 light years through the position of the sun at 30,000 light years and on to about 60,000 light years, the velocity ranges from 200 to 300 kilometers per second. Beyond about 60,000 light years, if indeed the Galaxy should extend beyond that distance, the velocity is probably nearer to 200 kilometers per second.

DO ALL PARTS OF THE GALAXY ROTATE?
Astronomers have compared the rotation of the halo of our Galaxy, whose most visible occupants are the hundred or so globular clusters, with that of the disk. From this it seems that the rotation of the halo, if any, cannot be much more than about 50 kilometers per second. In summary the spheroidal population group in the Galactic halo rotates the slowest, if at all; the

FIGURE 16.6
Effect of the sun's motion on nearby stars. The stars appear to scatter outward from the point of the sky (apex) toward which the sun is headed in Hercules. The nearby stars appear to converge toward the opposite point (antapex) in Columba.

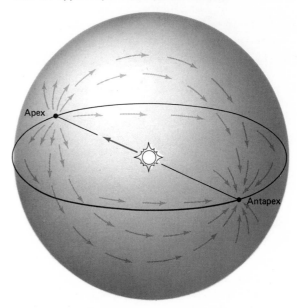

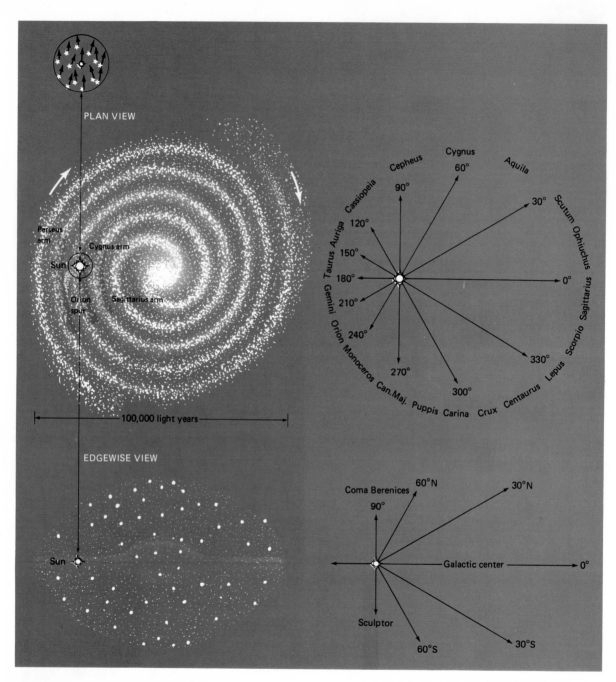

FIGURE 16.7
Our Galaxy and its structure. Shown here are a plan view of an idealized model of the Galaxy, with a corresponding diagram showing Galactic longitudes, and an edgewise view and corresponding diagram of Galactic latitudes. Our Galaxy is a huge system of stars having the shape of a disk with a central bulge, out of opposite sides of which wind two spiral arms made up of bright stars interspersed with clouds of gas and dust. Surrounding it is a spheroidal halo of stars and scattered globular clusters. The entire Galaxy is rotating, and the stars are moving in orbit around the Galaxy's center. The sun is located on the inner edge of one arm. It is in slow random motion with respect to its neighbors. The sun and its neighbors as a whole, however, are orbiting around the Galactic center at 250 kilometers per second toward Cygnus-Cepheus on a line 90° from the direction toward the Galaxy's center in Sagittarius. The sun completes its journey around the center in approximately 230 million years.

older stars of the disk population group rotate a little faster; and the youngest stars of the disk population group appear to rotate the fastest. Thus not only different rates of rotation are involved, but these different rates can be identified with distinctive parts of the Galaxy, as defined by the ages and characteristics of its component stars.

GALACTIC ORBITS FOR VARIOUS POPULATION GROUPS

Galactic rotation is the composite of the orbits of individual stars like the sun. The sun's orbit about the Galactic center is very nearly a circle or an ellipse of small eccentricity. In addition two or three times per orbit the sun oscillates up and down perpendicular to the Galactic plane, sometimes above the midplane (0° Galactic latitude) and sometimes below. The variation is probably not more than 1000 light years, or one-third of the thickness of the disk of the Galaxy, as shown in Figure 16.8.

The youngest stars, such as the O and B stars, H II emission nebulae, some open clusters, and the interstellar medium—all of which are disk population objects—move in nearly circular orbits about the Galactic center. These objects are also strongly confined to the midplane of the Galactic disk. Somewhat older stars, say up to about 8 billion years in age, are also orbiting in small-eccentricity elliptical orbits like that of the sun. Additionally they oscillate farther above and below the midplane of the Galactic disk than do younger stars. It is the stars and interstellar matter of the disk population group that truly define the rotation of our Galaxy.

On the other hand the stars of the spheroidal population move in strongly elliptical orbits that carry them to higher Galactic latitudes than is the case for the disk population stars as shown in Figure 16.8.

OBSCURING EFFECT OF INTERSTELLAR MATTER

As was discussed in Chapter 14, interstellar dust concentrated in interstellar clouds in the plane of our Galaxy obscures from our view distant portions of the disk (as one can see in Figure 16.1). The direction toward the center of the Galaxy in Sagittarius is completely hidden in visible wavelengths. Light from distant galaxies that lie in a direction vertical to the Galactic plane is dimmed by about 15 percent. The dimming becomes even greater in directions farther from the

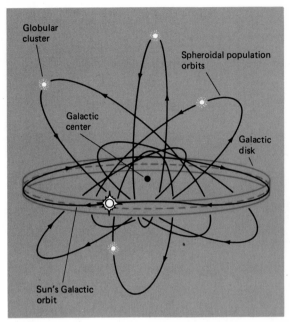

FIGURE 16.8
Galactic orbits for the sun and spheroidal population objects. The sun moves in the Galactic disk in an elliptical orbit with a very small eccentricity, its period being 230 million years (a Galactic year); the sun is 30,000 light years from the Galactic center and oscillates up and down from the midplane of the Galaxy in a band about 1000 light years wide. The other stars of the disk follow pretty much the same kind of orbit as the sun does. However, the spheroidal population stars, such as the globular clusters, move in eccentric orbits inclined at all angles to the Galactic plane.

vertical until it is nearly total along the plane of the Galaxy, especially toward the Galactic center. Since interstellar dust occurs in clouds, the dimming of light by it is not uniform but depends on our line of sight through scattered interstellar clouds (Figure 16.9). As well as dimming the light passing through it, interstellar matter also reddens the light by scattering blue-wavelength photons more proficiently than it does red-wavelength photons.

TRACING THE SPIRAL ARMS OF OUR GALAXY

In spite of obscuring dust, optical astronomers in the early 1950s identified segments of spiral arms that weave through the Galactic disk by tracing the locations where the O and B associations and their H II emission nebulosities are most prominent. One of them, the Orion arm, as shown in Figure 16.10, con-

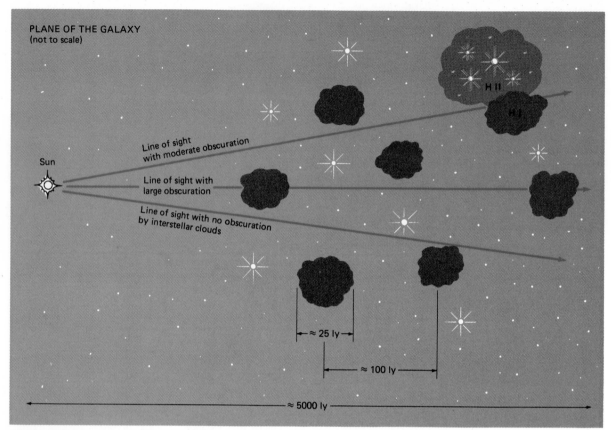

FIGURE 16.9
Lines of sight through dark interstellar clouds. Much of the obscuration in the plane of
the Galaxy is produced by dark interstellar clouds within about 5000 light years. The
clouds contain higher concentrations of interstellar gas and dust than do the regions between clouds. Thus a line of sight that passes through about five to ten clouds every
1000 light years of distance is heavily obscured. Lines of sight that do not pass through
interstellar clouds are still obscured somewhat by the interstellar medium lying between
the clouds. The farther into space we try to observe, the greater the obscuration. In addition to the obscuration the light from distant stars is reddened since blue light is more
readily absorbed by the interstellar medium than is red light.

tains the sun on the inside portion that lies toward the
Galactic center. The segment we see arches over an
angle covering some 10,000 light years. Near the sun
the arm has a short extension, called the "Orion
spur," that extends outward. A segment of a second
arm, known as the Perseus arm, arcs in the same general direction as the Orion arm and lies nearly 6000
light years beyond the Orion arm, closer to the edge
of the Galaxy. A segment of a third arm, the
Sagittarius-Carina arm, lying inside the Orion arm, has
been observed toward the Galactic center about 6000
light years from the sun. There is also some evidence
for a segment of an arm even farther out, beyond the
Perseus arm, and another segment of an arm (maybe

two) inside the Sagittarius-Carina arm, closer to the
Galactic center.

One of the obstacles to the use of O and B stars as
tracers of spiral arms is that they are all too far away to
determine their distances by trigonometric parallax.
Hence the less reliable inverse-square law of light
must be used, and consequently their distances are
known only to about plus or minus 10 percent; that is,
about 1000 light years out of every 10,000 light years.
Since the spiral arm is only a few thousand light years
across, spiral structure becomes impossible to distinguish from the locations of O and B stars that are
more distant than 15,000 to 25,000 light years from the
sun.

Are there other tracers of spiral structure? Yes, there are several possible ones. The Cepheid variable stars are very luminous supergiant stars with readily recognizable brightness variations (Section 12.5). They are also stars that have evolved away from the main sequence and are consequently somewhat older than the O and B stars but nevertheless relatively young and still near their birthplaces. Thus the Cepheids ought to be good candidates for spiral-arm tracers. But the results for Cepheids in our Galaxy are not convincing. Other spiral-arm tracers on which astronomers are working are the dark interstellar clouds, red supergiants, supernovae remnants, pulsars, and a number of X-ray sources. For example it appears that the inner edges of most spiral arms are marked by the presence of dark interstellar clouds. However, it is probably the radio-wavelength view of the Galaxy that provides astronomers with the second-best evidence for spiral structure after the O and B associations and their H II emission nebulae.

RADIO MAPPING OF SPIRAL STRUCTURE

Radio astronomers analyzing Doppler shifts in 21-centimeter radiation emitted by neutral hydrogen

FIGURE 16.10

Spiral-arm features in the vicinity of the sun. Looking down on the plane of the Galaxy, we see that the features outlining its spiral structure are the O associations, H II regions, and open clusters. The sun's position is shown on the inside of an arm, with portions of an arm or arms visible inside and outside the sun's arm.

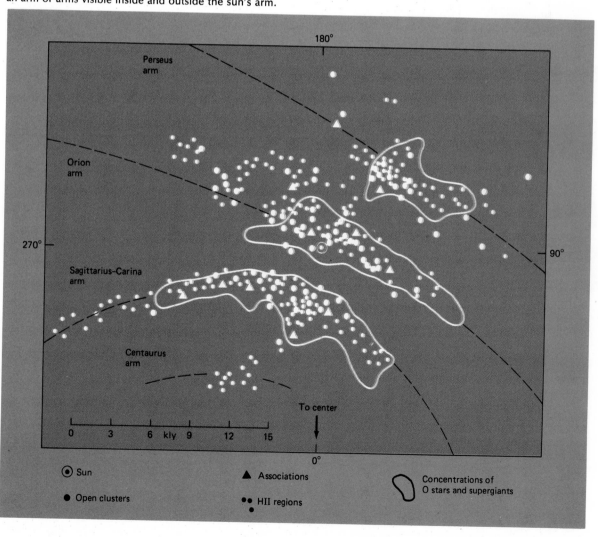

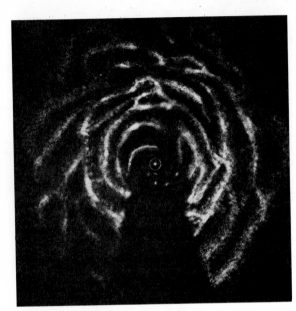

FIGURE 16.11
Radio picture of the spiral structure of our Galaxy derived from the 21-centimeter observations made by Dutch and Australian radio astronomers. The sun is represented by the small white dot and enclosed circle above the Galaxy's center.

in many directions through the Milky Way have mapped the distribution of atomic hydrogen in detail (Figure 16.11); and more segments of arms have been discovered in 21-centimeter surveys of the Galaxy than can be recognized in visible light.

By way of illustrating radio-mapping techniques, let us consider a small segment of the arm structure as revealed in Figure 16.12 by the 21-centimeter-line profiles along Galactic longitude 85°. What are the distances from the sun to the three observed maxima (labeled *A*, *B*, and *C*) in Figure 16.12b, where the intensity of the Doppler-shifted 21-centimeter radiation is greatest? To determine this, we can use the Galactic rotation data derived by optical astronomers in Figure 16.12c to establish a model of the arm structure with an approximate scale of distances. From this model we can calculate what the radial velocities should be where the spiral arms cross our line of sight at Galactic longitude 85°, assuming that the hydrogen gas moves in nearly circular orbits around the center of the Galaxy as do the O and B stars.

The arm region at *A* in Figure 16.12a is orbiting at approximately the same distance from the Galactic center as the sun is. The difference in the motions

between the sun and region *A* projected along our line of sight produces a slight negative Doppler shift of the 21-centimeter line, corresponding to a distance of about 1600 light years. Since the arm region at *B* is farther from the Galactic center than is the sun, its orbital motion projected along our line of sight produces a large negative Doppler shift, corresponding to a distance of about 12,000 light years. Finally in region *C*, which is farthest from the Galactic center, the orbital motion is the slowest. Its projection along our line of sight produces the largest negative Doppler shift, corresponding to a distance of about 25,000 light years.

As discussed in Chapter 14, emission from carbon monoxide at 2.6 millimeters is also available to radio astronomers as a probe of the Galaxy's spiral structure. This is because regions suitable for the formation and maintenance of the carbon monoxide molecule are also reasonable locations for the existence of molecular hydrogen, which is a very poor emitter in the radio wavelengths. Approximately one-half of all the hydrogen in the interstellar medium is in molecular form, and it is not heavily concentrated where there are heavy concentrations of atomic hydrogen so that it must be inferred from its association with carbon monoxide. The molecular clouds containing carbon monoxide identify the location of the three spiral arms evident in visible light in Figure 16.10. However, just as one sees, when examining Figures 16.10 and 16.11, that the spiral features evident from atomic hydrogen are not identical with the visible-light spiral arms, the carbon monoxide spiral features are not precisely those seen in either atomic-hydrogen radiation or visible light.

Regardless of how well young stars, atomic hydrogen, or molecular hydrogen outline the same spiral features, astronomers have no doubt that the Galaxy has a spiral structure. In summary the spiral structure begins at 10,000 to 15,000 light years out from the Galactic center and extends out through the disk to distances of at least 50,000 light years. The arms are a few thousand light years across and are separated by distances of about 10,000 light years. If the Milky Way Galaxy is similar to nearby spiral galaxies, like M31 (Figure 17.18), M33 (Figure 17.19), M51 (Chapter 17 opener), and M81 (Figure 1.4), then it probably has two arms alternately weaving their way out through the disk. Astronomers are, however, still a long way from fitting the segments of arms we observe into a coherent pattern.

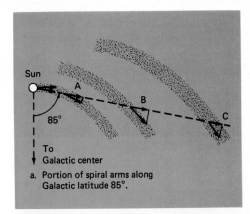

a. Portion of spiral arms along Galactic latitude 85°.

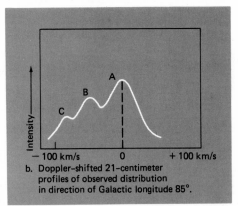

b. Doppler-shifted 21-centimeter profiles of observed distribution in direction of Galactic longitude 85°.

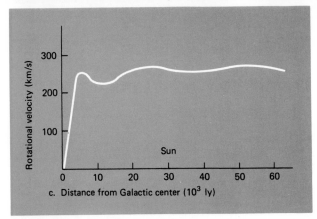

c. Distance from Galactic center (10³ ly)

FIGURE 16.12

Mapping spiral-arm structure with 21-centimeter radio emission. In (b) the observed Doppler-shift peaks labeled A, B, and C correspond to the positions in (a) of the spiral arms labeled A, B, and C. Their distances are determined in (c) by using the Galactic rotation curve from optical astronomy. They are at about 1,600, 12,000, and 25,000 light years, respectively.

WHAT MAINTAINS SPIRAL STRUCTURE?

Gravity is probably the dominant force in shaping and preserving the structure of spiral arms in normal spiral galaxies. Since the arms trail behind in rotation, the more rapid motion of the spiral arms close to the nucleus should wind them so tightly after several Galactic years that the spiral pattern of the Galaxy should soon disappear. From the large number of spiral galaxies that exist we can see that spiral structure is fairly stable and is probably not a transient feature of our Galaxy. There are two promising theories to explain why the spiral arms persist.

One theory says that *spiral density waves,* or waves of compression, move through the gaseous and stellar matter of the disk as a result of gravitational variations in the disk. These waves spiral outward from the center at a constant rotational rate somewhat slower than the rotation of the stars and interstellar matter. The waves do *not* consist of moving matter but are overtaken by the faster moving interstellar matter. As a result, when interstellar clouds move through the compressional waves, gas and dust pile up in a spiral pattern that is dense enough to initiate star formation (Figure 16.13). On the other hand between the spiral arms interstellar matter is not dense enough to contract under its own gravitation to form stars. To help visualize a spiral density wave, think of the following analogy: Imagine a slowly moving road crew painting white divider strips on a freeway where traffic moves in one direction. The traffic piles up wherever the crew is working before it can proceed normally. From an airplane the congestion would seem to be moving slowly forward as the crew plods along.

The extensive dust lanes frequently observed along inner edges of the spiral arms seem to confirm that these compressional waves are real. The newly formed blue supergiants and H II regions, like brilliant

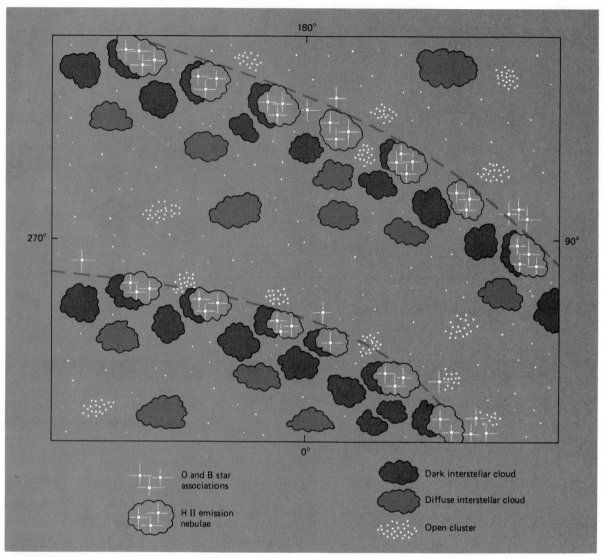

FIGURE 16.13
Spiral density waves forming spiral arms. The dashed lines represent the positions of the
compressional waves. As the interstellar clouds overtake the density waves, they are
compressed and initiate star formation. Thus the young O and B stars and their association H II emission nebulae form on the outer edge of the arm, while the dark interstellar clouds mark the inner edge of the arm.

beacons, illuminate and indicate the spiral arms'
present locations. In time the wave lags behind the
star-formation region, and the H II regions and the
short-lived massive blue stars soon disappear. Long-
lived stars of small mass, like the sun, are left to mix
with, and become part of, the disk population of stars.
Thus there is a continual re-forming of the arms.

A second theory for the persistent spiral-arm structure is that of a repeating process in which clusters of
stars are created from shock waves generated by supernova outbursts within the differentially rotating
Galactic disk. Star formation is supposedly triggered
by the rapidly expanding shock-wave fronts from the
supernova explosions that compress interstellar
clouds (Section 14.4). In the stars formed there are
very massive ones that soon live out their lives and
explode as supernovae, generating shock waves that
trigger the formation of the next generation of stars.

Thus the process is perpetuated from one generation to the next, while the differential rotation of the Galaxy stretches new star groups into the recognized spiral features.

It may well be that within our Galaxy both theories are operating in part. One possible combination is that the density waves initiate the positions of the spiral arms and supernovae outbursts intensify arm formation. Astronomers are probably many years away from a definitive theory for the maintenance of spiral structure.

16.4
The Spheroidal Component of the Galaxy

As stated earlier, the spheroidal component of the Galaxy varies in its characteristic properties in a radial direction from the Galactic center. It contains those regions of the Galaxy identified as the halo, the bulge, and the nucleus, through which the density of stars decreases outward very significantly. Globular clusters and RR Lyrae variables are typical of the spheroidal population group and can be found throughout its volume. In the solar neighborhood subdwarfs and high-velocity stars are part of this population group. Although they are only detected near the sun, those stars presumably exist everywhere within the spheroidal population. All its stars are old, from 10 to 15 billion years in age, and have typical masses of $0.8 M_\odot$ since all its more massive stars have already evolved to become white dwarfs. The spheroidal component contains little, if any, gas and dust except in the nucleus.

NUCLEUS OF THE GALAXY

The nucleus of our Galaxy lies far beyond the stars of Sagittarius, where it is obscured from our view by immense quantities of interstellar dust that lie within 10,000 to 15,000 light years of the sun. Thus the great star cloud in Sagittarius (Figure 16.2) probably represents only the outer edge of the true nucleus. What would the nucleus look like if we could see it unobscured? Estimates are that it would be about the brightness and size of the full moon, fading away rapidly into the halo and merging gradually with the disk stars along the Galactic plane.

Nature is a network of happenings that do not unroll like a red carpet into time, but are intertwined between every part of the world; and we are among those parts. In this nexus, we cannot reach certainty because it is not there to be reached; it goes with the wrong model, and the certain answers ironically are the wrong answers. Certainty is a demand that is made by philosophers who contemplate the world from outside; and scientific knowledge is knowledge for action, not contemplation. There is no God's eye view of nature, in relativity or in any science: only a man's eye view.

J. Bronowski

Little is known about the stellar composition of the nucleus since stars emit most of their radiant energy in the visible part of the spectrum, which is prevented from reaching us by interstellar dust. Since old spheroidal population stars, like globular clusters and RR Lyrae variables, increase in number toward the Galactic center, it is probable that the nucleus is a dense concentration of old stars. Its brightest stars should be red giants of relatively low mass. To support this belief, astronomers have a reasonably unobscured view of the nucleus in some of the nearby spiral galaxies. An H-R diagram for the nucleus should resemble that of the old open cluster NGC 188 in Figure 15.4 and not that of the globular clusters, such as M92 in Figure 15.5. The density of nucleus stars is likely to be up to a few million times that of the solar-neighborhood stars; that is to say, the nucleus stars are spread a few thousand astronomical units apart, while the spread is a few hundred thousand astronomical units for stars in the solar neighborhood. The overall shape of the nucleus is that of a slightly flattened sphere.

In the infrared region of the spectrum a much larger fraction of photons than in the visible can complete the journey from the nucleus to the sun. And in

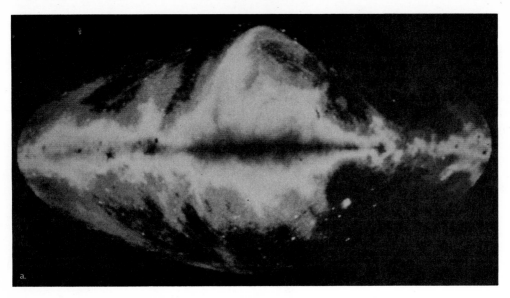

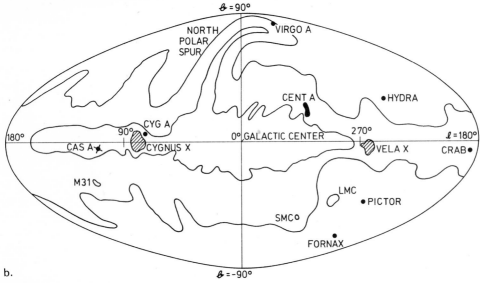

FIGURE 16.14
Radio survey of the Galaxy. In (a) various intensities of radio radiation at a wavelength of 73 centimeters emitted by the Galaxy are shown in Galactic longitude and latitude. (b) A key to the survey shows both Galactic and extragalactic radio sources. Compare this with Figure 16.1 of the Galaxy in visible light.

the radio, X-ray, and gamma-ray regions very few photons are lost during the 30,000 year trip through the plane of the Galaxy. However, the X-ray and gamma-ray photons have trouble penetrating the earth's atmosphere; so most of what astronomers know about the central regions of our Galaxy has been obtained by radio and infrared observations.

NUCLEUS IN VARIOUS WAVELENGTH REGIONS

As discussed in Chapter 5, data received by a radio telescope are processed by a computer to produce contour maps that show the location of the most intense regions of radio emission (Figure 16.14). A similar technique is used for infrared radiation. Results of such studies indicate that on the outer boundary of

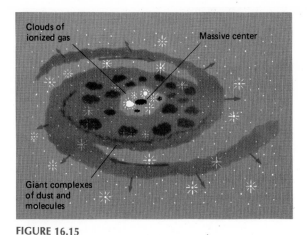

FIGURE 16.15
Central regions of our Galaxy shown schematically (not to scale). The innermost light year or so is dominated by a massive object, possibly a black hole, surrounded by a region of ionized gas clouds and dust clouds with bright red giants and normal stars. This disklike region is less than 100 light years across and is permeated by a sparse distribution of ionized gas. Lying outside this is a ring of neutral gas that contains giant complexes of dust and molecules. In some of the complexes young stars may have formed. This ring is probably several thousand light years across. About 10,000 light years from the center is what appears to be an expanding and rotating spiral arm or ring of hydrogen (shown as spiral arm) with a similar feature on the other side of the center. Sprinkled throughout the region shown are billions of old stars forming the central bulge of our Galaxy.

(Labels on figure: Clouds of ionized gas; Massive center; Giant complexes of dust and molecules)

the nucleus, 15,000 light years from the Galactic center, a ring of giant H II emission nebulae exists, like the Orion Nebula, and a number of giant molecular clouds. Inside this ring, about 10,000 light years on this side and 8000 light years on the other side of the center, is a rotating and expanding arm of neutral hydrogen. Figure 16.15 attempts to picture these arms and the region into the center as one might see it from above the Galactic disk north of the sun.

Farther in toward the Galactic center lies a disk of gas extending about 4000 to 5000 light years from the center. Another disk of gas having a higher temperature lies farther inside the first one; it extends out from the center about 1000 light years. Embedded within this second disk are giant complexes of dust and molecules, in which are apparently located very hot young stars and their associated H II emission nebulae. Estimates of the mass inside a radius of about 1000 light years indicate that it amounts to perhaps 1

to 2 billion solar masses, with a couple of percent in gaseous form. That is, it is 1 percent or so of the observable Galaxy's total mass confined to 0.01 percent of the volume of the disk.

Even closer to the Galactic center, radio contour maps show an elongated distribution of radio emission centered on the Galactic equator. These sources of intense radio emission are clouds of hot interstellar gas like the H II emission nebulae in the spiral arms. Somewhat different from the others, the strongest source is known to radio astronomers as Sagittarius A. It is a small, very bright source of radio emission that is not visible optically. Sagittarius A is the most powerful source of nonthermal radiation in the Galaxy, and it can be seen on all contour maps of radio emission covering the Galactic center. Most astronomers now think that Sagittarius A is the actual center of our Galaxy.

What else do we know that would confirm this supposition? Approximately coinciding with the location of Sagittarius A is a source of infrared, ultraviolet, and X-ray radiation. The infrared observations suggest that the source is at most a few light years in size with an estimated mass of several million solar masses. To some astronomers the central region of our Galaxy is the seat of violent events, then one interpretation of the compact massive object at the center is that it contains one or more black holes. Considerable amounts of matter in the form of stars and interstellar matter should then be swallowed by the black holes from surrounding regions.

HALO OF THE GALAXY
Surrounding the disk of the Galaxy is the halo, whose shape is that of a slightly flattened sphere. It contains some individual stars and 100 or so globular clusters that are situated far above the disk, with possibly another 100 or so globular clusters in or near the disk and nucleus.

The halo is apparently a remnant of the Galaxy's very early history. The stars and globular clusters in it are some of the oldest stars in the Galaxy (Chapter 15) and show how conditions in the Galaxy were very different at the time the globular clusters formed. The halo stars are found to be very deficient in elements heavier than helium compared with the youngest stars in the spiral arms of the Galactic plane. The stars of the spheroidal component appear to have less of the heavy elements the farther they are from the Galactic center. Stars in the nucleus once also thought to be deficient in heavy elements apparently are not, even though most are quite old. Stars elsewhere, then, formed under conditions that were different from those in the distant parts of the halo.

It was thought for many years that the halo's mass was only a small percentage of that in the disk. But there is evidence suggesting that the mass of the halo must be at least comparable with the disk's in order for the Galaxy's disk to maintain stability over its life span. If that were true, then the halo might add about 100 billion solar masses to the total mass of the Galaxy. At one extreme this mass could be tied up in stars of very low luminosity and small mass or at the other extreme it could be hot gas. This is not the end of the surprises about the halo that appear to be unfolding.

EVIDENCE FOR A MASSIVE CORONA

As early as 1974, some in the international community of astronomers were suggesting that the observable Galaxy—disk, nucleus, and halo—is embedded in a still larger and more massive, but unseen, *Galactic corona*. Estimates of the corona's size include values for its radius of up to 350,000 light years (almost 6 times that of the halo) and a mass value up to 1 to 2 $\times 10^{12} M_\odot$ (or about 7 to 10 times the mass of the observable Galaxy). If the observable Galaxy is only 10 to 15 percent of the total, what are the arguments astronomers have for the existence of all that unseen matter in a giant corona? And in what form is the matter that it cannot be seen directly?

The basic argument for the existence of a massive corona engulfing our Galaxy is one of stability and permanency. First, the Galaxy should have a great deal of mass to stabilize gravitationally for the billions of years of the Galaxy's existence the relatively thin and delicate disk with its spiral structure. Second, if the observable Galaxy is all the gravitational attraction in the system, then the orbital velocity of stars in the disk should decrease beyond 20,000 light years in analogy with the Keplerian orbits of the distant planets in our solar system. However, as pointed out, the orbital velocity does not decline but is approximately constant or slightly increasing well beyond the distance of the sun from the Galactic center. Third, the motions of the nearest galaxies to our Galaxy, which are small galaxies, suggest that it is much more massive than what we observe in visible light. And finally, the rotational velocity of other spiral galaxies leads astronomers to believe that there is invisible matter far from those visible galaxies. If these arguments are not invalidated by further study and observation, then our Galaxy is a much larger and more massive system than we once thought.

What is the form of this unseen matter (some refer to it as the "dark population group")? It does not appear to be cool gaseous matter containing neutral hydrogen and carbon monoxide since it does not emit 21-centimeter radio radiation. If some of it was hot gaseous matter, then it should emit short-wavelength radiation in the ultraviolet or X-ray regions. Indeed ultraviolet observations with the *Ultraviolet Explorer Satellite* suggest that the Galactic corona contains hot gas with a temperature of about 100,000 K, that extends 25,000 to 35,000 light years on either side of the Galactic disk and cycles in and out of the disk like giant fountains. But such gas is not nearly the amount needed. The best guess at this point is that, if the corona exists, most of it is in the form of old stars of very low luminosity, such as red dwarfs, white dwarfs, and black dwarfs. Whatever form the unseen matter is in, it has presented astronomers with a fascinating mystery story.

16.5 Stellar Evolution in Our Galaxy

A BRIEF HISTORY OF CHEMICAL EVOLUTION

All classes of objects moving around the Galactic center continue to occupy to some extent the volume within the Galaxy that they inhabited at the time of their formation. For that reason astronomers believe that in this they can see the progress of the Galaxy's evolution over its inferred life. From this we can also track the progress of the evolution of the chemical elements since stars are the furnaces in which heavy elements are synthesized from lighter ones.

Early in the Galaxy's life gaseous matter was much denser than now, which favored star formation. The existence of a small amount of heavy elements in the oldest stars suggests that either the Galaxy formed from matter already possessing some heavy elements or an early population group of short-lived supermassive stars formed and has vanished, leaving only their nucleosynthesis efforts for us to see.

Most of the matter in the stars' central parts is converted in stages from hydrogen to helium to carbon and oxygen and even to iron in some instances. During supernova outburst by massive stars all the elements not already synthesized are produced very rapidly and this chemically evolved matter is thrown back into space to mix with interstellar matter.

Out of this gaseous mixture enriched with heavy elements the next generation of stars is born. The new stars in turn synthesize in their interiors more heavy elements, some of which are expelled into interstellar space by loss of mass due to stellar winds or supernova outbursts. And finally the spiral-arm-population stars appear inside the gas and dust clouds located in the thin disk portions of the Galaxy; these stars are the youngest and contain the largest abundance of heavy elements of all the stars.

With the various nucleosynthesis processes astronomers can account reasonably well for the relative abundances of the chemical elements observed in successive generations of stars. The compositions they obtain from spectroscopic analysis of the sun, stars, and nebulae suggest that in the solar neighborhood the abundance of elements heavier than hydrogen and helium has increased with time. There was an increase by a factor of five or so between 5 and 12 billion years ago, and the heavy-element abundance has been roughly constant since then. Also, the heavy-element abundance appears to decline in disk stars the more distant they are from the Galactic center. Extensive spectroscopic studies of globular clusters suggest that the clusters near the Galactic center have larger heavy-element abundances than do those far out on the boundary of the Galactic halo.

This pattern of element abundance indicates that the Galaxy started with hydrogen and some helium in a roughly spherical form and in about 5 billion years of star formation produced the spheroidal component of the Galaxy. Star formation later became pronounced in a disk-shaped region and continues today in the spiral arms of the disk. The cyclic production and re-distribution of elements diagrammed in Figure 16.16 indicates the way in which chemical evolution has proceeded in the Galaxy. The cycle will stop when most of the matter in the Galaxy is tied up in dead stars so that new stars can no longer be formed.

Nearly the whole of that part of our everyday world made up of elements heavier than hydrogen and helium was synthesized in the cores of stars billions of years ago. These elements were spewed back into space during supernova outbursts and stellar winds, mixed with the interstellar matter, and became part of the solar system when it formed. The matter in our bodies is but one link in our Galaxy's cycle of chemical evolution.

Just as the advent of radio astronomy revealed not a Galaxy of stars but one of nebulous patches of gas and dust dotted with discrete sources, ultraviolet, X-ray, and gamma-ray astronomy promises to reveal to us a third or fourth aspect to our Galaxy. Radiation at very short wavelengths is coming from stars in primarily exotic phases of their life cycle (such as birth, during or after violent outbursts, mass exchange with a companion, and so on) or hot gaseous matter that is part of some violent or energetic event. Studies using ultraviolet, X-ray, or gamma-ray observations are greatly expanding astronomers' knowledge of stellar evolution and its message of Galactic evolution.

THE GALAXY IN ULTRAVIOLET LIGHT

Although yet in its infancy, astronomical studies in the ultraviolet portion of the electromagnetic spectrum promise to be an extremely valuable tool in the study of our Galaxy.

Most of the events astronomers have detected with satellites involve hot stars. Many of these stars are old, compact stars, such as white dwarfs, subdwarfs, and neutron stars, with some as components in binary-star systems that had not even been recognized as such. Others are young, high-mass stars that are very hot and very luminous, such as shown in Figure 16.17. Not all are hot stars, a number are highly evolved red giants. And for many of these sources of ultraviolet radiation not only are stars involved but also included is circumstellar matter possibly expelled from stars and large nebulosities surrounding the stars, as in Figure 16.17. For example, the best evidence for mass loss by very luminous O-type stars comes from observations of their stellar winds in the ultraviolet.

In general what we are learning about the Galaxy is

> There is nothing more incomprehensible than a wrangle among astronomers.
>
> H. L. Mencken

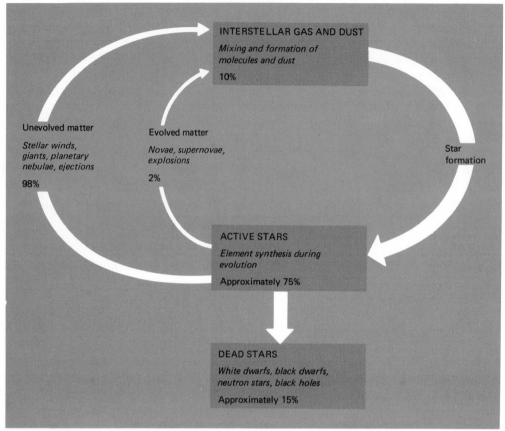

INTERSTELLAR GAS AND DUST

Mixing and formation of molecules and dust

10%

Unevolved matter

Stellar winds, giants, planetary nebulae, ejections

98%

Evolved matter

Novae, supernovae, explosions

2%

Star formation

ACTIVE STARS

Element synthesis during evolution

Approximately 75%

DEAD STARS

White dwarfs, black dwarfs, neutron stars, black holes

Approximately 15%

FIGURE 16.16
Chemical-evolution cycle. Matter cycles in the Galaxy between the interstellar medium and stars and back to the interstellar medium. Some matter is taken out of the cycle by dying stars. Percentages in the boxes give the present division of matter among the interstellar medium, active stars, and dead stars; percentages on the left of the drawing give the division between chemically evolved matter and unevolved matter that has been expelled by various processes and is returned to the interstellar medium.

that the birth and evolutionary processes for stars, particularly the massive ones, are much more exotic than once thought, and they involve large volumes of the space surrounding stars. As a result of knowing more about the evolution of stars, we can, like working on a large jigsaw puzzle, identify large-scale changes in the Galaxy. Thus we can make an educated guess as to how the Galaxy might have appeared in the past and what it will look like in the future.

OUR GALAXY FROM X-RAY STUDIES

The great penetrating power of X rays makes possible the observation of X-ray objects through the interstellar haze. But the earth's atmosphere is not transparent so that X-ray observations are made from satellites like the *Einstein Orbiting Observatory* (page 113). The brighter X-ray sources trace out the position of the plane of the Galaxy (Figure 16.18). Unfortunately, reliable distances for X-ray sources are not easy to estimate unless the sources can also be identified optically or are companions in binary systems whose distances can be derived by one or more well-established methods. These brighter X-ray objects are spread on either side of the Galactic equator by no more than about 5°, with the exception of Scorpius X-1 (the first X-ray source found in the constellation Scorpius) and Hercules X-1. If the bright sources are within 1000 light years of the Galactic plane, then 5° means that they are within about 15,000 light years of the sun. Scorpius X-1 and Hercules X-1 are thought to be about 3000 to 4000 light years from us.

The second feature one notes on an X-ray map of the sky (Figure 16.18) is that the X-ray sources tend to cluster around the direction toward the Galactic center and toward the constellation Cygnus (the direction toward which Galactic rotation carries the solar system). Clustering toward Cygnus suggests that many of the X-ray sources lie along the spiral arm containing the sun (the Orion arm) and are relatively nearby (less than 10,000 light years). Generally objects at the high Galactic latitudes are extragalactic. Some faint objects (smallest dots) as well as some bright objects (large dots) at low Galactic latitudes may also be extra-

galactic. Identifications of these sources with an optical object can settle the issue. Compared to other galaxies our Galaxy is not conspicuously bright in X rays in spite of all its X-ray sources.

The rich variety of known X-ray sources, numbering over 3000, was unanticipated, and more sources are constantly being discovered. To our human eyes the visible Galaxy is a galaxy of stars, while to our radio eyes it is a gaseous and diffuse-appearing Galaxy. The discrete radio sources are exotic objects or energetic events, certainly not ordinary stars. The production of X-ray photons requires hundreds to thousands of times more energy per photon than does that of visible photons. Therefore, like discrete radio sources, it was thought that the vast majority of X-ray sources should be exotic objects, like black holes, or energetic events, such as interstellar shock waves plowing into gigantic clouds of gas and dust. Indeed those objects are seen, but the vast majority of X-ray sources observed are stars. They are stars of all kinds, all spectral classes, all luminosity classes, and all ages; that is, they come from all parts of the H-R diagram. Most of these stars are not shown on the X-ray map of the

FIGURE 16.17
Comparison of hot stars in the vicinity of the H II emission nebula NGC 7000 (North American nebula) in visible and ultraviolet light. The nebula is located about 3° E of the bright star Deneb in Cygnus. The photograph on the left is in the blue emission line of hydrogen, while that on the right is in the far ultraviolet. The ultraviolet emission of the nebula is about three times that of the star labeled HD 199579, which is an O6-type star. The ultraviolet spectrum of the nebula is continuous, suggesting that it is due to scattering of starlight by dust grains.

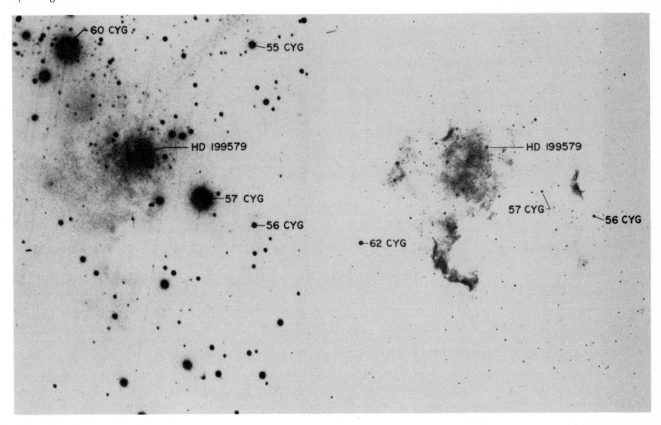

a.

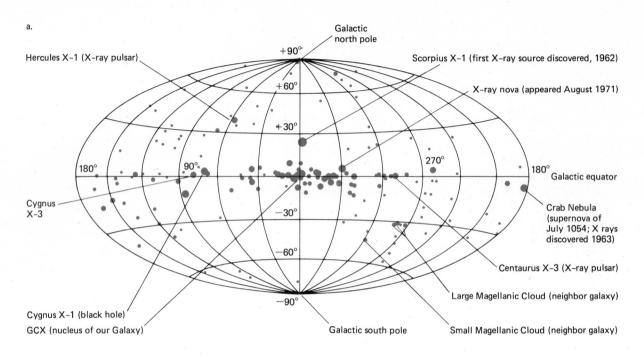

b.

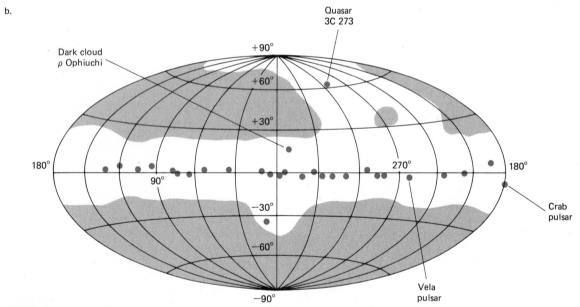

FIGURE 16.18

Sky distribution of X-ray and gamma-ray sources. In general for both (a) and (b), those objects near the Galactic equator (in the plane of the Milky Way) belong to our Galaxy; those well removed from it are beyond our Galaxy. The larger the dot, the more intense is the X radiation. Most of the gamma-ray sources in (b) are in our Galaxy.

Galaxy in Figure 16.18, which exhibits primarily the 200 or so most exotic sources.

Let us now summarize the various classes of X-ray sources found so far in our Galaxy, but because of the newness of X-ray discoveries, these groupings may well not be exclusive categories.

1. *Stars.* The stars that are significant emitters of X-ray radiation come from all regions of the H-R diagram. They not only occur along the main sequence but include much of the giant branch and many of the blue and red supergiants. The stronger the X-ray emission is, apparently the younger the star.

2. *Binary-star systems.* Many of the exotic X-ray sources in our Galaxy are binary-star systems. Most of these systems are thought to be composed of a compact star, such as a white dwarf, a neutron star, or a black hole, in orbit about a massive blue supergiant (Section 15.7). The mechanism proposed for the continuous X-ray emission is that gas from the normal star flows into an accretion disk around the compact star, converting its kinetic energy into thermal energy that results in the X-ray emission.

3. *X-ray transients.* These X-ray objects are ones for which there is a sudden flaring of X-ray emission followed by a gradual decline. They usually fade to invisibility within a few weeks or months. Many, if not all, of these X-ray stars are probably binary systems also. There must be some differences between the transients and the other X-ray binaries as the observational characteristics clearly suggest.

4. *X-ray bursters.* In 1975 a class of X-ray stars was discovered that emits intense and repeated bursts of X-rays of only a few seconds duration each. The peak energy in the burst is up to 1 million times the sun's luminosity, and it rises very rapidly to peak emission with a slower fading. While the bursts are basically similar, an individual burster often has its own characteristic pattern with bursts recurring at intervals of hours to days or seconds to minutes. The short duration and rapidity of the bursts imply that the source is very small, such as a neutron star. The proposed burst mechanisms suggest that gravitational potential energy is being converted to thermal energy when matter from a massive normal star crashes spasmodically onto the surface of a compact star in a binary system or a thermonuclear outburst results from the accumulation of in-falling matter.

5. *Globular-cluster X-ray sources.* Until 1976 globular clusters were thought to be made up of old, dying stars without much activity going on inside the cluster. This view changed when X-ray bursts were observed coming from the center of the globular cluster NGC 6624 in the constellation Sagittarius. In addition to NGC 6624 there are six or so globular clusters that are emitting X-ray bursts from sources near their centers. It has been proposed that the X rays are produced when accreted gas is fed into a $100M_\odot$ black hole at the center of the cluster. The gas comes either from aging stars in the cluster or from the interstellar medium.

6. *Supernova remnants.* One of the *Einstein Orbiting Observatory*'s priorities was to obtain X-ray images of supernova remnants. Figure 16.19 shows X-ray emission from three supernova remnants; the Crab Nebula, Cassiopeia A, and Tycho's 1572 supernova. Part of the X-ray emission is synchrotron radiation, and part is the result of the expanding nebulae plowing into interstellar clouds. The pulsar in the Crab Nebula is clearly visible in Figure 16.19 (the bright circle near the center).

7. *Nebulae and superbubbles.* The Carina and other nebulae are suffused with an X-ray glow resulting from many supernova outbursts. In the constellation of Cygnus lying about 7000 light years from the sun (beyond the bright star Deneb) and partially hidden behind the dark interstellar cloud complex known as the Great Rift is a rarefaction in the interstellar medium known as a superbubble. It is about 1000 light years in diameter and contains gas at temperatures of about 2 million K. It has been hypothesized that the superbubble was created by a chain of supernova explosions occurring within the last 3 million years. Such bubbles may occupy as much as 10 percent of the entire Galactic disk and thus are important to the dynamics and evolution of the Galaxy.

8. *Galactic-bulge X-ray sources.* These are X-ray sources that populate the inner reaches of the Galaxy and are concentrated around its center. X rays from the bulge sources are coming to us in a rather steady fashion; that is, they are not in the form of bursts, not on and off

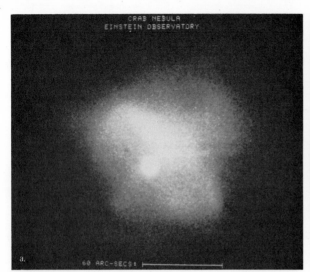

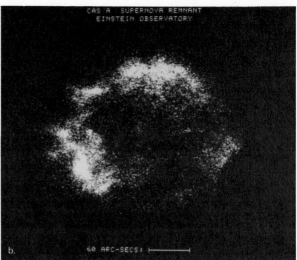

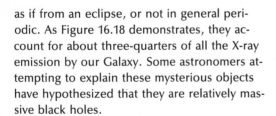

FIGURE 16.19
Supernova remnants in X rays.

THE GALAXY REVEALED BY GAMMA-RAY EMISSION

Gamma rays are the most energetic photons of any in the electromagnetic spectrum. Because they are so energetic, they readily penetrate interstellar matter as do X-ray photons, and as is true of X rays, the surface of the earth is shielded from them by our atmosphere. Therefore astronomers are forced to fly gamma-ray detectors in satellites.

The result of several satellite studies is a map of the sky as shown in Figure 16.18 in which the area searched (unshaded area) reveals 29 discrete sources of gamma rays. Some sources could be multiple and some could actually be rather extended in size (not star size).

Clearly the gamma-ray sources define the Galactic plane and in fact spread only about 5° on each side of the Galactic equator. Most lie in the direction of the Galactic center, four having been identified with known objects. They are the quasar 3C 273 (extragalactic, page 450), the Rho Ophiuchi dark-interstellar-cloud complex, and the Crab and Vela pulsars. Estimates of how many gamma-ray sources the Galaxy could contain lie between 100 and 1000. The typical source probably lies between about 6000 and 20,000 light years from us.

Astronomers have not been able to identify most of the gamma-ray sources with known optical objects, radio sources, or X-ray sources; therefore a gamma-ray source is considerably brighter in gamma rays than

as if from an eclipse, or not in general periodic. As Figure 16.18 demonstrates, they account for about three-quarters of all the X-ray emission by our Galaxy. Some astronomers attempting to explain these mysterious objects have hypothesized that they are relatively massive black holes.

As this catalog of different X-ray sources indicates, there is a great deal going on in our Galaxy for which we have only sketchy explanations. The full significance of these events for the structure and evolution of our Galaxy still eludes astronomers. We can speculate that X-ray studies of other spiral galaxies may shed a great deal of understanding on our own system.

it is in any other wavelength region. This suggests that the gamma-ray emission is not the direct result of the temperature of the object but probably results from the collision with interstellar matter of such subatomic particles as protons.

Gamma-ray astronomy is still in its infancy—about where X-ray astronomy was 10 years or so ago. New satellites in the planning stages should enhance our knowledge of the Galaxy as revealed by gamma rays.

GALACTIC COSMIC RAYS

Closely related to gamma-ray production in our Galaxy is the constant bombardment of the earth by subatomic particles from all directions in space. First detected in our upper atmosphere early in this century, these energetic charged particles were misidentified as high-energy photons and named *cosmic rays* and are still called that. The earth receives as much energy in the form of cosmic rays as from starlight. Some of the low-energy cosmic rays are coming from the sun. There is reason to believe that the highest-energy cosmic rays are extragalactic, but most are of apparently Galactic origin. The very energetic cosmic rays coming from distant parts of the Galaxy carry information about their origin and are the only source of matter from outside the solar system.

With some exceptions the abundance of elements in cosmic-ray particles closely matches that of matter within the solar system. Since all the nuclei in the periodic table are present in cosmic rays, the chemical composition of cosmic rays suggests a thermonuclear origin in some explosive event. Supernova outbursts have long been thought to be a prime source of cosmic-ray particles from our Galaxy. There are at least two or three supernova events in our Galaxy per century. Thus they are sufficiently frequent for the principal source of gamma-ray emission to be the interaction of cosmic-ray particles with the interstellar medium. We suppose that the cosmic-ray particles are ejected at high speeds in supernova outbursts and are accelerated by interaction with interstellar magnetic fields. A succession of accelerations by magnetic fields tends to randomize the directions of the cosmic-ray particles and to hide the direction of their origin, so that they do not come from the Galactic center as the gamma rays do. Thus our Galaxy's magnetic fields confine all but the most energetic particles within it. That is why astronomers believe that most cosmic-ray particles originate in our Galaxy.

We close this chapter by noting again that astronomical research with ultraviolet, X-ray, and gamma-ray photons and cosmic-ray particles is dramatically broadening the astronomer's knowledge of the Galaxy. But we are still far from understanding the full significance of this work.

SUMMARY

Overview of the Galaxy. From the perspective of the earth, our Galaxy is visible as the Milky Way. In shape, the Galaxy is a disk with a spherical bulge of stars in the center and spiral arms weaving out from the bulge through a disk of stars, wrapping around and trailing as the Galaxy rotates. Surrounding the disk and nucleus is a spherical halo and, perhaps, an immense corona of unseen matter. Stars of the Galaxy are divided into old stars of the spheroidal population and the younger stars of the disk population.

Stellar motions in the solar neighborhood. The revolution period of the sun and the solar neighborhood is 230 million years, known as a Galactic year. From proper motions, distances, and rotational velocities of stars the space velocities relative to the sun can be measured. Although the local motions of stars in the solar neighborhood are small and random, the entire neighborhood of the sun moves with a large velocity in a circular orbit around the center of the Galaxy.

The disk component of the galaxy. Close to the Galactic center the Galaxy rotates almost like a solid body. However, out through the outer 90 percent of the radius, the velocity of rotation does not vary dramatically and does not decrease as in Keplerian motion. Stars in the spheroidal group in the halo rotate slowest, if at all. Older stars in the disk population rotate faster. The youngest stars of the disk population rotate fastest. Orbits in the disk population are nearly circular. Orbits in the spheroidal population are more elliptical. Study of 21-centimeter radio emission has allowed astronomers to map the radio spiral arms and compare them with the visible arms marked by O associations and H II regions. There is not a strong similarity. Two theories to explain the formation of spiral arms are spiral density waves and a succession of supernova outbursts. Both processes may be at work.

The spheroidal component of the galaxy. The spheroidal component includes the halo, bulge, and nucleus. The nucleus of the Galaxy is obscured by interstellar dust but seems to be the location of some very energetic events having great importance for the Galaxy. The halo is apparently a remnant of the Galaxy's very early history. Evidence for a massive corona is not yet conclusive but is suggestive.

Stellar evolution in the galaxy. Stellar evolution has resulted in the chemical evolution of the Galaxy with the consequence that the heavy element composition has continued to increase. Ultraviolet, X-ray, gamma-ray, and cosmic-ray studies have revealed much about the evolution of stars and hence the Galaxy. Such studies reveal a far more active and structured Galaxy than once thought possible.

KEY TERMS

cosmic rays
disk population
Galactic center
Galactic disk
Galactic halo
nucleus
proper motion

radial velocity
space velocity
spheroidal population
spiral arms
spiral density waves
tangential velocity
X-ray source

CLASS DISCUSSION

1. What are the differences between the objects of the spheroidal population group and those of the disk population group? How do these differences indicate the evolution of the Galaxy?

2. Describe the motion of the various population groups in the Galaxy. What defines the rotation of the Galaxy? Contrast the motions of the spiral arms and the halo stars.

3. Why do some astronomers believe there is a black hole (or holes) at the center of the Galaxy? Is it possible that the belief is resorting to the exotic when a much more ordinary explanation would suffice?

4. What is the so-called evidence for the existence of a massive corona in which the visible Galaxy is embedded? How would you rate such evidence—is it convincing, plausible, doubtful, fantasy? Why?

5. Trace the evolution of the chemical elements in our Galaxy over its lifetime. Are you physically a totally new entity in the world? What relation exists between you and a supernova?

READING REVIEW

1. How was the mass of our Galaxy determined?

2. What is the relative percentage of stars by mass and interstellar matter in the Galaxy?

3. How did we discover the rotation of the Galaxy?

4. What is the range in age for the objects of the various subgroups of the disk population group?

5. What are the primary tracers of spiral arms?

Which are the best? Do they reveal the same spiral arms?

6. What role do supernovae play in maintaining spiral structure?

7. Are all the stars of the nucleus old? Is there a range of age?

8. Are the orbits of globular clusters low- or high-eccentricity orbits?

9. Are there any representatives of an early population group known?

10. Does the Galaxy present the same image in visible light and ultraviolet, infrared, radio, X-ray, and gamma-ray radiation?

PROBLEMS

1. If the Galaxy can be idealized as a spherical halo of radius 60,000 light years, a spherical nucleus of 12,000-light-year radius, and a disk of 50,000-light-year radius and 5000-light-year thickness, what is the volume of each of the three components? What percentage of the Galaxy's total volume does each component occupy?

2. If the entire luminosity of the Galaxy in Table 16.1 were provided by O stars, how many would it take? If it were all provided by red giants or white dwarfs, how many would be required? How do these figures compare with the estimated numbers of each type in Table 13.3 (express your answer as the percentage of the estimated to the needed number)?

SELECTED READINGS

Blitz, L. "Giant Molecular-cloud Complexes in the Galaxy," *Scientific American*, April 1982.

Bok, B. J. "The Milky Way Galaxy," *Scientific American*, March 1981.

deBoer, K. S., and B. D. Savage. "The Coronas of Galaxies," *Scientific American*, August 1982.

Geballe, T. R. "The Central Parsec of the Galaxy," *Scientific American*, July 1979.

Jaki, S. L. *The Milky Way, an Elusive Road for Science.* Science History Publications, 1972.

17
Galaxies

In Chapter 16 we considered our Milky Way Galaxy and its contents. Although it is home to us, our Galaxy is an extremely small entity in the vastness of the universe. It is only one of billions of galaxies, which are the building blocks of the universe. Like the *stars* of our Galaxy, these immense *collections of stars* can and do exert sufficient gravitational attraction to be bound together permanently. Consequently galaxies occur in pairs, small groups of a few tens, great clusters of hundreds to thousands, and superclusters, which are composed of a number of clusters, small groups, and individual galaxies. Galaxies constitute the bulk of the matter that is emitting visible light. And as far as we know, the visible galaxies in all their diversity contain the bulk of the matter in the universe.

17.1 Island Universes

NEBULAE: INTERNAL OR EXTERNAL?

More than two centuries ago, Thomas Wright and Emanuel Swedenborg imaginatively suggested that the thousands of small, nebulous patches of light that with the telescopes of their time could be seen interspersed among stars might be other stellar systems like our own. In 1755 Immanuel Kant extended this concept when he proposed that these objects might be distant star systems like the Milky Way (island universes), distributed at random angles of inclination to the line of sight. The nearby spiral galaxy M101 in Ursa Major is shown in Figure 17.1).

The first definitive observational evidence that the spiral nebulae are stellar systems came in 1912 and the years following, when Vesto Slipher (1875–1969) of the Lowell Observatory recorded spectra for about 40 spiral nebulae. The spectra consisted of absorption lines on a continuous background, the kind that would be expected from the composite light of a vast number of stars. The peculiar aspect of these spectra was that the majority had absorption lines whose wavelengths

FIGURE 17.1
The spiral galaxy M101 in the constellation of Ursa Major.

were shifted toward the red (or redshifted), with recessional velocities up to 1800 kilometers per second. These large radial velocities were much greater than those observed for other objects, but there was no ready explanation for the redshifts until years later.

However, controversy continued over whether the spiral nebulae were well outside our own Galaxy or were really associated with it. By 1924 the answer came when Edwin Hubble of the Mount Wilson Observatory succeeded in resolving individual stars in NGC 6822 in Capricornus and in the peripheral portions of the two large spirals M31 in Andromeda (Figure 17.17) and M33 in Triangulum (Figure 17.18). He was able to identify some of these stars as Cepheid variables, whose distances could be derived from the Cepheid period-luminosity relation. As a result, astronomers could accept that the small nebulous patches of light were in reality great star systems far outside our Galaxy.

Although western culture has been speculating for over 200 years on the basic structure of the universe, it is only within the lifetime of your grandparents that astronomers have known that we live in a universe of

◀ The Whirlpool galaxy, M51, and its close companion in Canes Venatici, about 32 million light years away. They appear to show the effects of a close encounter in which matter is exchanged between them. The inset is a negative reproduction in the infrared.

EDWIN P. HUBBLE   (1889–1953)

Hubble attended high school in Chicago, where he excelled as a student and as an athlete. In 1906 he received a B.S. degree in mathematics and astronomy from the University of Chicago. At the university he earned such an excellent reputation as a

boxer that a sports promoter wanted him to train for a fight with Jack Johnson, then the world's heavyweight champion. He was awarded a Rhodes scholarship for Oxford, England, where he studied jurisprudence.

Hubble returned to the United States in 1913 to practice law in Louisville. A year later he went back to his alma mater to begin studies for an astronomical career at the Yerkes Observatory. No one, he stated, should go into astronomy without a genuine call, and the only way to test a call is by having another calling to be called away from.

In 1919 Hubble joined the staff of the Mount Wilson Observatory. Toward the end of 1924 Hubble made his first great discovery: With the new 100-inch reflector he was able to sort out a number of bright Cepheids in some large spirals, and employing the Cepheid period-luminosity relation, he demonstrated that these spirals were other galactic systems. This put an end to the raging controversy about whether these objects be-

longed to our Milky Way system or were beyond it.

In 1925 Hubble established a classification system of the galaxies into four groups: the spiral, the lenticular, the elliptical, and the irregular galaxies (Figure 17.2). Four years later he made his greatest discovery: After the tedious process of determining the distances of a number of galaxies and observing their redshifts, Hubble and others used them to show that there existed a proportional relationship between distance and velocity, now known as the Hubble velocity-distance law of recession. Although at first Hubble doubted the notion that this relationship was evidence for an expanding universe, he later accepted this view after cosmologists had pointed out that this was the only logical explanation.

A person of extraordinary talents—scholar, athlete, lawyer, and astronomer—Hubble is best remembered as the founder of observational cosmology and explorer of the distant cosmos.

FIGURE 17.2
Hubble sequence of galaxies. The four major classes are ellipticals (E), lenticulars (S0), spirals (S and SB), and irregulars (Irr). The arm structures of the spirals are graded a to c, tight to open; nuclei, a to c, large to small. The ellipticals are graded from 0 to 7, spherical to most flattened.

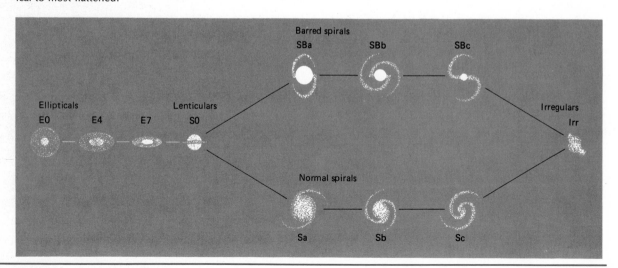

galaxies. This is as important an advance in cosmological thought as the Copernican revolution, and in fact it may be called a milestone in all intellectual history.

NAMING THE GALAXIES

Astronomers had observed, described, and cataloged small nebulous patches and bright knots of light even before the advent of photography. The French comet hunter Charles Messier (1730–1817), in order to avoid confusion with the comets for which he was searching, carefully described these objects and assembled them in a catalog located in the front inside cover of this book. Completed in 1781, the catalog, known as the *Messier catalog*, listed 103 objects that we know today as galaxies, star clusters, and gaseous nebulae. Between 1888 and 1908 Danish astronomer John Dreyer (1852–1926) compiled the *New General Catalog* (NGC), the most comprehensive of the older catalogs still used, and two supplemental *Index Catalogs* (IC).[1] These three catalogs list about 13,000 star clusters, planetary nebulae, diffuse nebulae, and galaxies. A more recent catalog includes about 200,000 extended optical sources, which are galaxies, star clusters, nebulae of all descriptions, or quasistellar objects. Even this number is dwarfed by the total number, running into the hundreds of millions or billions, that can be photographed with modern telescopes.

Besides their catalog designations a number of galaxies are identified by a proper name, such as the Andromeda galaxy or the Large and Small Magellanic Clouds. Somewhat less confusion exists in identifying galaxies than with star names.

17.2
Types of Galaxies

CLASSIFICATION OF GALAXIES:
THE HUBBLE SEQUENCE

From his extensive collection of photographs Hubble chose about 600 well-defined bright galaxies on which to base a classification scheme for galaxies. He arranged them in an orderly progression, which we now

[1]The catalog numbering system is simple: M13 is Messier's thirteenth catalog entry, for the globular cluster in Hercules, which is NGC 6205, or entry 6205 in Dreyer's *New General Catalog*.

The fires that arch this dusky dot —
  Yon myriad-worlded way —
The vast sun-clusters gather'd blaze,
  World-isles in lonely skies,
Whole heavens within themselves, amaze
  Our brief humanities.

Tennyson

call the *Hubble sequence*. Shaped like a tuning fork, his sequence ran from essentially spherical configurations through lens-shaped systems, to very flat spiral systems to irregular ones, as shown in Figure 17.2.

Today astronomers, having elaborated on Hubble's original scheme, classify galaxies according to several criteria: the degree of flattening, the relative size of the nucleus compared to the disk, how tightly the spiral arms are wound, the type of stars, and the relative amount of interstellar matter (gas and dust). With these characteristics in mind, we now look at the four major classes of galaxies: elliptical (E), lenticular (S0), spiral (S), and irregular (Irr).

ELLIPTICAL GALAXIES: GIANTS AND DWARFS

The *elliptical galaxies* are distinguished from the other three classes in that they are characterized by a very smooth and symmetrical texture (with no evidence for internal structure) and are tightly bound, like NGC 4486 in Figure 17.3. That is, they possess a bright center from which the brightness fades in a reasonably uniform fashion; in short photographic exposures they can appear quite small, whereas in longer exposures they can be enormously larger. They do not show evidence for a disk, or plane, containing spiral structure as in our own Milky Way Galaxy, which is characteristic of the spiral class. Elliptical galaxies are designated by an E followed by a number that increases from 0 to 7 as the shape appears more elongated, that is, spheroidal to ellipsoidal.

The ellipticals progress in size from dwarf systems of tens of millions of stars up to giant ellipticals containing trillions of stars covering a diameter that is thousands of times that of the dwarf ellipticals. In the elliptical galaxies the constituent stars are analogous to the spheroidal population stars (old stars) in our

NGC 4486 Type E0

NGC 147 Type E6

Dwarf elliptical galaxy in Sextans

FIGURE 17.3
Elliptical galaxies. They are made up primarily of older stars. The number beside the E designation indicates the degree of flattening, from 0, the minimum (spherical), to 7, the maximum. Globular clusters are faintly visible in the outer fringes of NGC 4486. Seen edgewise, as in NGC 147, most elliptical galaxies show no dark dust lane across the middle.

ally ceased in the typical elliptical galaxy; that is, there are no stars analogous to the disk population stars of our Galaxy.

The closest ellipticals are reddish in color and can be resolved into individual stars. The brightest of these are older spheroidal population stars, such as globular-cluster stars, which can be readily photographed in the outskirts of nearby giant ellipticals; in Figure 17.3 globular star clusters are faintly visible in the outer fringes of NGC 4486. The dwarf ellipticals have a stellar content similar to that of the giant elliptical galaxies, but it is one in which there is less of the heavy elements in general. Dwarf ellipticals are also a much less compact grouping of stars compared to their larger counterparts. It is interesting to note that, if one considers only the spheroidal stars in our Galaxy, they form a collection of stars that is very much like an elliptical galaxy in the nature of its stars and its size, except that an elliptical galaxy is much more luminous.

Rotation rates have been measured for some elliptical galaxies, and it appears that as a class they are slow rotators compared with the spiral galaxies. For some time most astronomers had thought that the more nearly spherical ellipticals were probably the slowest rotators, while the more pancakelike ones rotated faster: It was the rate of rotation that caused

FIGURE 17.4
Elliptical galaxy NGC 5266, showing what might be a dust ring around its middle.

Galaxy, and most such galaxies do not have the pronounced dust streak across their middle that is so prevalent in spirals seen edge on (an exception is NGC 5266 in Figure 17.4). Astronomers' interpretation of these two observations is that star formation has virtu-

NORMAL SPIRAL GALAXIES

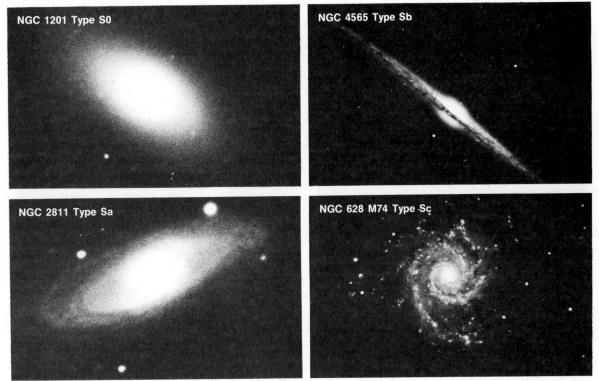

FIGURE 17.5
Representative lenticular galaxy and normal spiral galaxies. They range from S0, with only a hint of a disk, through the tight spirals, Sa and Sb, to the loose, small-nucleus Sc. The Milky Way is an intermediate Sb type. Notice that the more loosely coiled the spiral arms, the smaller is the nuclear bulge. Viewed in profile, the Sa, Sb, and Sc spirals should all show the characteristic central dust streak.

flattening in an elliptical galaxy. However, recent data indicate that there is no significant difference between spherical and ellipsoidal galaxies in rate of rotation. Thus their shape may not be the result of rotation, so this still remains a perplexing question for astronomers.

If one were to consider the 1000 or so brightest galaxies, one would find that most are spiral galaxies (to be discussed below). Since the dwarf elliptical galaxies are so faint and difficult to see at great distances, they might be expected to be the most numerous. Yet a study of all galaxies within 30 million light years, a distance 600 times the radius of our Galaxy, found that only about 20 percent are elliptical, including dwarf ellipticals. The percentage of each class of galaxy among all the galaxies in the universe is an important question yet to be answered.

LENTICULAR GALAXIES: A TRANSITION GROUP

A type of galaxy that seems to represent a transition between elliptical and spiral galaxies is the S0, or *lenticular, galaxies*. Such galaxies, like NGC 1201 in Figure 17.5 and NGC 2859 in Figure 17.6, possess a smooth, abbreviated, disklike extension out from a nuclear bulge that shows no sign of spiral structure. From the edge the S0 and SB0 galaxies appear reasonably flat and are generally without the dark streak indicating the presence of interstellar dust. In SB0 galaxies like NGC 2859 there is a stump, or faint suggestion of a disk without spiral arms, in place of a clearly defined bar, and such galaxies are often surrounded by a faint halo or ring structure. Unlike the elliptical galaxies, which as a class are slow rotators, the S0 and SB0 galaxies rotate more rapidly, as do the spiral galaxies. The S0 galaxies contain a disk population group of

BARRED SPIRAL GALAXIES

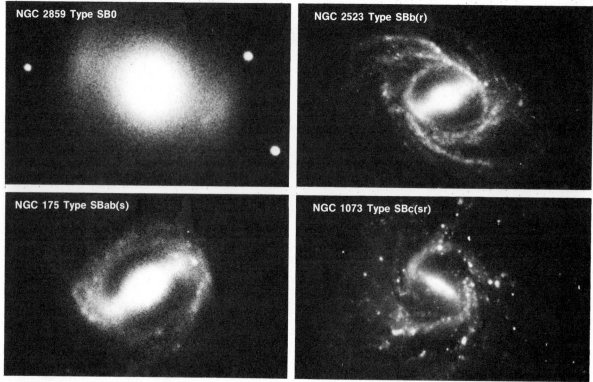

FIGURE 17.6
Representative lenticular and barred spiral galaxies. The arms of barred spirals trail from spindle-shaped spherical hubs. The arms may be thick or, sometimes, fine-drawn, encircling the hub. As is true in the normal spiral galaxies, the less tightly wound the arms of the galaxy are, the smaller the nuclear bulge.

stars like those of our Galaxy in addition to the spheroidal population in the nuclear bulge, which is like the elliptical galaxies.

SPIRAL GALAXIES: COSMIC PINWHEELS

Among the conspicuous galaxies in the neighborhood of the Milky Way most are spirals; two distinct kinds are recognizable: One is the *normal spiral* (S) *galaxies*, as shown in Figure 17.5, and the other, which is somewhat less numerous, is the *barred spiral* (SB) *galaxies*, in Figure 17.6. Normal spirals generally have two (sometimes more) arms emerging from opposite sides of a nucleus and winding through a disk in much the same fashion as that observed for our Galaxy). Extending out of opposite sides of the nucleus in barred spirals is a barlike group of stars, from whose ends spiral arms emerge, usually one from each end. Like the elliptical galaxies, the spirals are brightest at the center, with a gradual fading in brightness radially out

through the disk and a very rapid fading away from the disk.

The normal and barred spirals vary in size, as do the elliptical galaxies, but they do not exhibit the enormous differences in size and mass that is observed for the ellipticals. The largest spiral recognized has a visible diameter of almost 1 million light years, about 10 times that of the Milky Way and the Andromeda galaxy, M31 (Figure 17.17). There are recognizable categories of both giant and dwarf spirals.

Both normal and barred spirals in Figures 17.5 and 17.6 are graded by the lower-case letters a, b, c, and occasionally d for the tightness of the arms (a, tight, to c, open) and the size of the nucleus (a, large, to c, small). That is, the more tightly wound arms generally go with the largest nucleus, and the loosely wound arms with the smallest nucleus, as depicted in Figure 17.2. The Milky Way Galaxy appears to be a normal Sb spiral.

In the nuclei of spiral galaxies, both normal and barred, the brightest stars are red giants or supergiants, while in the arms the brightest stars are the highly luminous spiral-arm population stars that are blue in color. The brightest objects in the spiral arms are the H II emission nebulae that are interspersed with the O associations, and both are strung along the arms like brilliant beads in subclasses b and c. Obscuring dust lanes usually lie on the inner edges of spiral arms, as can be seen in M51 in the frontispiece to this chapter. In spirals seen edge on, such as NGC 4565 in Figure 17.5, this opaque interstellar dust is clearly evident as a long dark streak threading across the galaxy's middle. Figure 17.7 shows the spiral galaxy NGC 4622 in a face-on attitude so that the long, well-wrapped arms are clearly evident. Subclass a has a somewhat smooth and unbroken texture made of older disk population stars and sparse occurrences of H II emission nebulae or O associations. The halo is more conspicuous in subclass a than it is in c.

Just as spirals in visible light can be seen to possess varying amounts of interstellar dust, radio observations reveal that they also contain varying amounts of interstellar gas, primarily atomic and molecular hydrogen, and some interstellar molecules. As in our Galaxy, we have been able to detect atomic hydrogen in some nearby spirals by its emission of 21-centimeter radiation and at the appropriate wavelengths emission from such molecules as carbon monoxide (CO), water vapor (H_2O), and others. Whereas visible light from stars in spiral galaxies is centrally concentrated, the 21-centimeter radiation from atomic hydrogen (H I) is not centrally concentrated but instead shows a pronounced minimum, or hole, at the center of the galaxy (somewhat larger than the size of the nucleus). In our own Galaxy the location of atomic hydrogen is correlated with the location of the spiral arms seen in visible light. This same situation more or less occurs in other spirals. We can summarize the presence of interstellar matter by saying that the amount increases from subclass a to subclass c.

Just as globular clusters have been observed in elliptical galaxies, they have also been discovered in the halos of nearby spiral galaxies. This raises the question whether or not other spirals besides our Galaxy show evidence for a massive corona. The strongest argument for such coronae is that the rotational velocity measured for several spirals does not decline in the outer reaches of the visible disks as it should if most of the mass lies toward the nucleus of the galaxies, but instead is approximately constant out to the edge of the visible disks. This is interpreted as meaning that there are significant amounts of mass at large distances from the nuclei. It is assumed that this mass is very faint matter contained in roughly spherical coronae that engulf the visible galaxies and whose radii are several times that of the visible galaxies.

No less a significant question for spiral galaxies is why the spiral arms, which appear to trail as the galaxy rotates, do not wind up (this question was discussed in relation to our Galaxy on page 397). Of the two mechanisms proposed for maintaining spiral structure—the spiral-density-wave theory and the succession of supernova outbursts—most of the evidence supports the spiral-density-wave theory. It is more likely to produce symmetrical two-armed spirals such as M51 (chapter opener) or NGC 4622 (Figure 17.7), while the supernova-outburst theory should explain a more chaotic or ragged arm structure, such as that for M33 in Figure 17.18.

As a means of summarizing the structure of spirals, Figure 17.8 shows schematically how a normal spiral galaxy, like our Milky Way, is composed of different subunits. Although the figure is somewhat idealized, it is a reasonable representation of the composition of spiral galaxies.

IRREGULAR GALAXIES

About 3 percent of the known galaxies are classified as *irregular galaxies*. They exhibit little symmetry, as can be seen in Figure 17.9, and can be divided into two kinds: The group Irr I contains highly luminous blue stars, star clusters, and some interstellar gas with very little dust (a prime example is the nearby Magellanic Clouds, described in Section 17.5); the second group, Irr II, is marked by more deformity in its structure, fairly conspicuous dust lanes, and a composite spectrum of white unresolved stars.

WHAT DOES THE CLASSIFICATION OF GALAXIES MEAN?

Hubble's classification arranged the galaxies in a progressive morphological sequence, but the galaxies do not necessarily evolve from one form into another. We think it likely that the different forms in the sequence reflect differences in their initial situations at the time of formation and how galaxies have evolved under a variety of conditions and environments—not different stages in evolution.

FIGURE 17.7
Spiral galaxy NGC 4622 in the constellation Centaurus. This galaxy is a member of the Centaurus cluster of galaxies. Note its remarkably smooth and thin spiral arms, including either a coalescence of the arms on the right side or a forking of the arms. Such symmetry and finely wrapped arms are more likely due to spiral density-wave formation than to supernova outbursts. The distance of the galaxy is about 200 million light years. This photograph was taken with a 4-meter telescope at Cerro Tolola, Chile.

What then is the Hubble sequence? It is, first, a *dynamic arrangement*, ordering the galaxies according to the degree of rotation and structure; and it is, second, a *population sequence*, characterized by the progress of stellar evolution in each galaxy.

PECULIAR GALAXIES: SOME DON'T SEEM TO FIT

Not all galaxies fit into Hubble's classification scheme. In recent years astronomers have discovered over 10,000 *peculiar galaxies*, which do not fit the scheme because of either their optical appearance or their excessive radio emission as compared to a Hubble-type galaxy. It is extremely difficult at this juncture to estimate what percentages of distant galaxies (and thus of all galaxies) are peculiar galaxies.

A subgroup of the peculiar galaxies are the *active galaxies* with their compact appearance and bright nuclei. Nearly all the active galaxies are strong radio emitters; some are strong X-ray and infrared emitters of the type associated with violent or energetic events. Figures 17.10 and 17.11 show some typical examples of peculiar galaxies.

Other peculiar galaxies have tidally distorted forms, many of them connected to each other by luminous bridges of stars and interstellar matter. From computer-generated models for gravitational interactions between grazing galaxies, such encounters apparently can produce streams of stars and interstellar matter curving in opposite directions from the galaxies. Many of these simulated appearances closely resemble actual forms, as shown in Figure 17.12. In Sec-

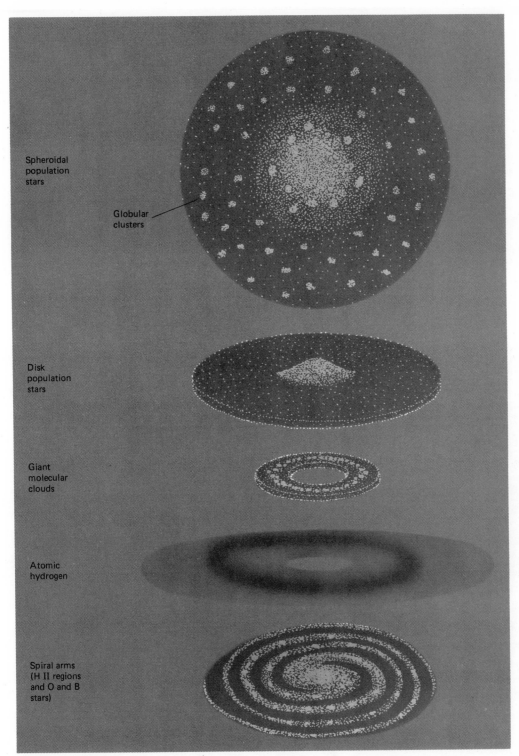

FIGURE 17.8
Composition of a normal spiral galaxy. Idealized illustration separating spheroidal population of old stars from disk population of intermediate-age stars and spiral arms of
young stars. Composing the interstellar medium are dust clouds, neutral hydrogen (H I
regions), and ionized hydrogen (H II regions).

IRREGULAR GALAXY

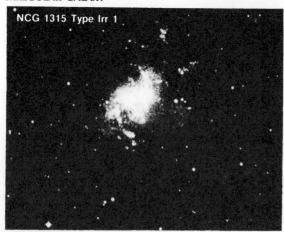

FIGURE 17.9
Irregular galaxy. The two types of irregular galaxies are shapeless; most contain turbulent gas clouds and brilliant, blue stars, but some are poor in gas and have a few old, red stars.

FIGURE 17.10
Ring and theta galaxies.

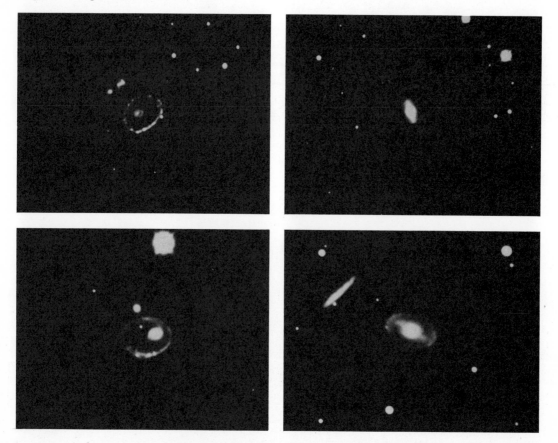

tion 18.2 we shall discuss the abnormalities associated with the active galaxies.

17.3
Properties of Normal Galaxies

DISTANCE INDICATORS:
INVERSE-SQUARE LAW OF LIGHT

It is hardly surprising, in view of the enormous expanses involved that distances to remote galaxies are at best crude estimates. Distances to neighboring galaxies, however, are fairly well established. Accuracy, of course, diminishes as the distance grows. Just as precise determination of the astronomical unit (the earth's mean distance from the sun) sets the scale for our solar system, so we must have an accurate yardstick for measuring distances of the galaxies in the universe. Essential cosmic data depend on how re-

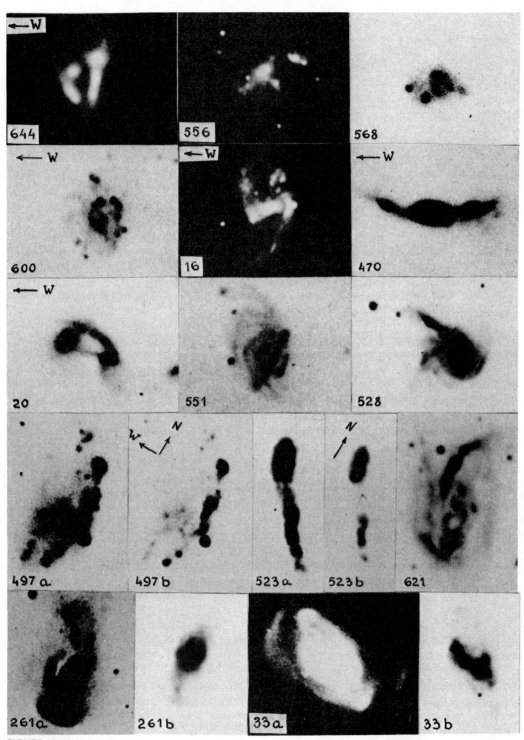

FIGURE 17.11
Peculiar galaxies. Photographs of some interacting galaxies taken at the prime focus of
the 6-meter telescope in the Soviet Union. Some pictures are negatives (a dark galaxy on
a bright background); others are positives (a bright galaxy on a dark background). The
interacting galaxies of the top two rows are referred to as nests, while the next two rows
are chains of interacting galaxies. The last row is enigmatic as to structure and origin.

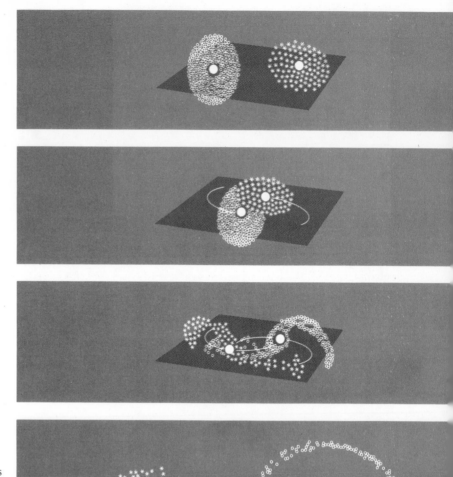

FIGURE 17.12
Computer simulation of two interacting galaxies. The near collision was
made to simulate the appearance of
NGC 4038/9 shown at the bottom.
Circles are material from one galaxy,
while the stars are material in the
other. The interval between successive pictures is about 200 million
years. Colliding galaxies can interact
for as long as 1 billion years.

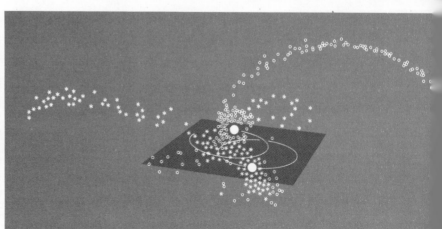

liable this yardstick is. The linear dimensions, spatial distribution, intrinsic luminosities, and masses of the galaxies; the physical and evolutionary differences among them; the average density of matter in the universe; the rate of expansion of the universe; and the type of cosmological world model—all depend on the correct scale of distance.

To find the distances of galaxies, we assume that similar objects in our Galaxy and in other galaxies have the same physical characteristics. From investigations in our own Galaxy and nearby galaxies we know the absolute luminosity of these objects (or *distance indicators*, as they are called). To determine the distance to a galaxy, we take the absolute luminosity of one of the distance indicators within the galaxy and its observed apparent brightness. Then, using the inverse-square law of brightness, we find the distance by way of the distance modulus, $m - M$, as described in Section 12.2.

Table 17.1 lists some distance indicators and the maximum distances to which they theoretically may be applied in estimating distances to galaxies with the 5.1-meter Hale reflector. Except perhaps for supernovae, in practice astronomers cannot measure distances beyond roughly 50 million light years by using individual objects in a galaxy. This is because the objects are so hard to resolve in the more remote galaxies. If they cannot be resolved in the galaxy, then they cannot be used. There are a variety of things that affect the resolution, such as orientation of the galaxy. Cepheid variables are still our most dependable criteria in finding distances. We can use them today for only the closer galaxies (less than about 15 million light years), where they can be identified readily.

OTHER DISTANCE METHODS

Even when we cannot distinguish isolated objects in a galaxy, astronomers can often estimate the distance from the galaxy's total luminosity, which they obtain from its surface brightness, apparent size, or mass-luminosity ratio. They must be careful to recognize exactly the class of galaxy involved because galaxies differ greatly in brightness and size. Lumping all galaxies into one luminosity class would introduce large errors in the estimated distance. We can distinguish between the image of a galaxy at the threshold of visibility and a star's pointlike image by the galaxy's fuzziness or nonspherical shape. However, this still may not allow the astronomer to classify the galaxy, as the images in Figures 1.5, 17.10 and 17.11 clearly show.

Another method for determining distance is to take advantage of a correlation between the luminosity of a galaxy and the wavelength width of the 21-centimeter emission line produced by atomic hydrogen in the interstellar medium of the galaxy. The correlation was established for nearby galaxies of known distance

TABLE 17.1
Distance Indicators

Brightest of Its Type (distance indicator)	Absolute Magnitude (M)	Theoretical Maximum Distance (10^6 ly)[a]
RR Lyrae variables	0	1
Spheroidal population red giants	−3	5
Cepheids	−6	20
Red supergiants	−8	50
Blue supergiants	−9	80
Novae	−9	80
Globular clusters	−10	130
H II emission nebulae[b]	−12	320
Brightest supernovae	−19	8000
Brightest galaxy in cluster	−21	20,000
Supergiant elliptical galaxy in cluster	−26	200,000

[a] Assuming that the faintest object that the 5.1-meter Hale telescope can photograph has apparent magnitude +23.
[b] Value depends on the apparent size of emitting H II region.

and can now be used as a means of determining distances of more remote galaxies.

At large distances the distance to a large cluster of galaxies can be determined much more accurately than that to its individual galaxies. We can select, say, the observed average apparent magnitude of the 10 brightest galaxies as a criterion for luminosity instead of depending on one galaxy. If we assume that their mean absolute magnitude is a particular value, like $M = -21$, based on knowledge gained from similarly constructed nearer clusters, the distance of the cluster can be found from the inverse-square law.

COSMIC DISTANCE SCALE

Astronomers have set up a cosmic distance scale by building a chain of overlapping distances, proceeding from the nearest objects to the farthest. Figure 17.13 depicts this overlapping system of distance indicators and shows practical distances that are less than the

FIGURE 17.13
Cosmic distance scale (not drawn to scale) in millions of light years. An overlapping scale of distances has been established by using the various distance indicators in Table 17.1. In practice the maximum distances for these distance indicators are not achievable. The scale of distances begins with objects in our Galaxy and is expressed in millions of light years. The next objects on the distance scale are the galaxies of the Local Group. The following points are nearby groups of galaxies, the Virgo cluster of galaxies, and rich clusters. The final link is the distant clusters, whose distances are primarily known from the Hubble law of recession.

theoretical limits in Table 17.1. The cosmic distance scale begins with the distances of relatively close stars set by trigonometric parallax and by the distance of the nearby Hyades open star cluster. These distances serve as a basis for the next step, which comes from variable-star data, chiefly from the Cepheids, and from the spectroscopic and intrinsic brightness of stars in our Galaxy. This sequence in turn serves as a basis for distances of the neighboring galaxies of the Local Group, which are determined from characteristics of their brightest stars, Cepheid variables, and other stellar data.

The next sequence uses the distances of the more remote galaxies (in what astronomers refer to as "nearby groups"), taking as criteria their brightest stars, surface brightness of the galaxy, and the apparent size of their bright gaseous nebulae. The next link in the chain is a cluster of galaxies such as the Virgo cluster, taking the brightest galaxy of the cluster or the cluster's luminosity type as a standard of comparison. Finally we connect this distance scale to distances of the most remote clusters of galaxies by means of the Hubble constant derived from expansion of the universe, which will be discussed in Section 17.4.

GALACTIC LUMINOSITIES AND MASSES

Galaxies have a very large range in size, brightness, and mass, as shown in Table 17.2. As noted earlier, the largest and the smallest galaxies are elliptical. The true dimensions of galaxies are derived from their apparent diameters and known distances. By measuring ga-

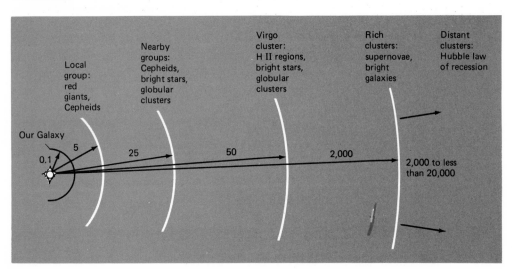

TABLE 17.2
Properties of Classes of Galaxies

	Elliptical (E)	Lenticular (S0 and SB0)	Spiral (S and SB)	Irregular (Irr)
Mass $(M_{Gal})^a$	10^{-6}–50	10^{-3}–2	10^{-3}–2	10^{-4}–0.2
Radius $(R_{Gal})^a$	10^{-2}–10	0.2–2	0.2–2	0.05–0.3
Luminosity $(L_{Gal})^a$	10^{-5}–7	10^{-3}–2	10^{-3}–2	10^{-5}–0.1
Mass/luminosity (solar units)^b	10–100^c (1–10 dwarfs)	10–50	10–30^d	1–10
Typical rotation rates (km/s)	$\leqslant$100	100–200	100–300	50–150
Color	Red	Red	Red (nucleus), blue (spiral arms)	Blue
Spectral class of central regions	K	G	F–K	A–F (G–K Irr II)
Stellar population	Spheroidal	Spheroidal, disk	Spheroidal (halo, nucleus), disk	Disk, some spheroidal
H I in total mass (%)	$\leqslant$0.1	$\leqslant$1	1–10^e	15–20
Dust content	Very little or none	Little to none	Some	Some

^{a} Milky Way Galaxy: $M_{Gal} = 200 \times 10^9 M_\odot$, $R_{Gal} = 5 \times 10^4$ ly, $L_{Gal} = 20 \times 10^9 L_\odot$.
^{b} Solar units: $(M/L)_\odot = 1$.
^{c} Highest in giant ellipticals.
^{d} Highest in center.
^{e} Decreases from Sc to Sa.

lactic apparent magnitude and combining that with known distances, we obtain luminosity.

The simplest way for astronomers to estimate the masses of galaxies, particularly for spirals that are seen more or less edge on, is to find the Doppler shifts of their spectral lines. From this we can measure the difference in radial velocity at the opposite disk ends (the difference between the end approaching us and the one receding along our line of sight). With the velocity difference and the known size of a galaxy we can calculate the mass from Newton's modified form of Kepler's third law.

Binary galaxies, in which two galaxies orbit each other, provide another means of measuring masses of galaxies. The pair's difference in radial velocity, com-bined with their known separation, leads to an esti-mate of their individual masses through the use of Kepler's third law. This method is similar to the one we use in finding the masses of binary stars (Section 12.6).

A third method makes use of the correlation be-tween mass and luminosity known as the mass-luminosity ratio (representative values are given in Ta-ble 17.2): The mass of a galaxy can be estimated from its observed luminosity. The mass-luminosity ratio is an important indicator of the kinds of stellar popu-lations in different galaxies. The variation in value of the mass-luminosity ratio reflects differences in spec-tral types of the stars that make up various classes of galaxies. It ranges in solar units (for the sun's mass divided by luminosity is unity) up to 100 for the giant

If a man will begin with certainties, he shall end in doubts; but if he will be content to begin with doubts; he shall end in certain-ties.

Francis Bacon

ellipticals, from about 10 to 50 for S0 galaxies, from about 10 to 30 for the spirals, and from 1 to 10 for the irregular galaxies. A higher value means that a larger proportion of the stars are faint dwarf stars whose mass contribution to a galaxy is much greater than their luminosity contribution. We must remember from the discussion of stars in our Galaxy that the luminosity of a galaxy is provided by the intrinsically bright stars. Thus in pictures of galaxies the distribution of light does not necessarily reflect the distribution of mass in those galaxies.

17.4
Expansion of the Universe

REDSHIFTS OF GALAXIES: COSMOLOGICAL REDSHIFTS

In 1912 Slipher found that the absorption lines in the composite spectrum of all the stars in the Andromeda galaxy were Doppler shifted toward the blue, indicating a velocity of approach between that galaxy and ours. But by 1928 he had found large redshifts in the absorption lines of all but 5 of 41 galaxies he had been studying (the 5 having blueshifted spectra). Even larger redshifts have since been found for the fainter galaxies as one can see in Figure 17.14.

After Edwin Hubble succeeded in estimating distances for a number of galaxies whose redshifts had been measured, he found that a straight-line relationship existed between their redshifts interrupted as recessional velocities and their distances, such that the farther away the galaxy is, the faster it is moving away from us (Figure 17.15). The only exceptions were several nearby galaxies that exhibited velocities of approach. Hubble was able to show that for the intrinsically more luminous galaxies, the recessional velocity is also correlated with its apparent brightness. This is in the general sense that the greater the recessional velocity, the fainter the galaxy appears and the more distant it is. Hence, unlike the stars of our Galaxy, where those that appear brighter are not necessarily closer to us but are intrinsically bright, the faint galaxies in general are truly the distant ones.

HUBBLE VELOCITY-DISTANCE LAW

If we interpret it literally, the relationship in which velocity is proportional to distance, or the *Hubble*

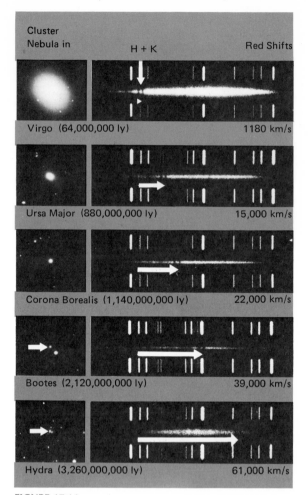

FIGURE 17.14
Relation between the redshifted spectral lines of the galaxies and their distances. The length of the arrow indicates the amount of the recessional velocity. The arrow tip marks the position of the shifted H and K lines of ionized calcium.

velocity-distance law, indicates that the universe is expanding (we shall discuss cosmological implications in Chapter 19). Hubble's original results have been extended by other investigators and now include hundreds of more distant galaxies and several dozen distant clusters of galaxies.

HUBBLE VELOCITY-DISTANCE LAW: *The farther away a galaxy is from our Galaxy, the faster that galaxy is receding from us; that is, recessional velocity equals a constant times distance.*

Are the redshifts of the distant galaxies caused by the Doppler effect? The answer is no; these redshifts are the result of the expansion of the universe: an expansion of space. Thus the redshifts should be called *cosmological redshifts* to distinguish them from Doppler or gravitational redshifts. In addition to a recessional velocity, galaxies also exhibit a small peculiar velocity superimposed on the expansion velocity. For nearby galaxies the peculiar velocity is larger than the recessional velocity, and we observe some blueshifted galaxies. However, for the very distant galaxies, the recessional velocity is much larger than the peculiar velocity, and the peculiar velocity may be neglected in comparison. The Hubble law deals with the recessional velocity caused by the expansion of the universe and not a local Doppler effect producing the peculiar velocity.

In the mathematical expression of Hubble's law ($v = H \times r$), in which the recessional velocity v is proportional to the distance r, the constant of proportionality H is called the *Hubble constant*. The value of Hubble's constant is the slope of the line in Figure 17.15. In a straight-line relationship between v and r, we see that a galaxy 2 billion light years away is receding twice as fast as one that is 1 billion light years away. As we shall discuss in Chapter 19, the numerical value of the Hubble constant H and any departure from the straight-line relationship at great distances have a lot to do with the type of cosmological model that applies to the universe.

Recent estimates of the Hubble constant place its value between 15 and 30 kilometers per second per million light years (50 to 100 kilometers per second per million parsecs) with a likely value of about 17 kilometers per second for each 1 million light years of distance, or about 55 kilometers per second for each million parsecs. When the distance indicators of Table 17.1 are unresolvable, astronomers can use the velocity-distance relation for a specified value of H to estimate the distance of a remote galaxy from its redshifted spectral lines. For example, the largest redshifts yet found for supposedly normal galaxies have values ($\Delta\lambda/\lambda$) equal to about 1.2, which corresponds to a recessional velocity of about 200 thousand kilometers per second. Assuming a Hubble constant of 17 kilometers per second per million light years, these galaxies are about 11 billion light years away and we are observing them as they were 11 billion years ago. That is, half the estimated age of the universe has passed since the light that we observe today left those galaxies. Some of the quasars to be discussed in Section 18.2 have redshifts over 3, corresponding to distances of almost 15 billion light years.

17.5
The Local Group of Galaxies

WHAT IS A GROUP OF GALAXIES?

As was mentioned at the beginning of this chapter, gravity binds galaxies together in pairs, in groups containing several tens of galaxies, and in clusters containing hundreds to thousands of galaxies. Our Galaxy, along with about 20 nearby galaxies, belongs to a physical group of galaxies known to astronomers as the Local Group (Table 17.3). Most of the galaxies in the Local Group are dwarf elliptical systems along with three spirals and four irregulars. The Local Group may have still more galaxies, hidden by obscuring material in the disk of our Galaxy because they are faint, dwarf elliptical systems.

The largest galaxies in the Local Group are the three spiral galaxies: the Andromeda galaxy (M31; Figure 17.17), our Galaxy, and the Triangulum galaxy (M33; Figure 17.18), in that order of size. Smallest are

FIGURE 17.15
Hubble velocity-distance relation derived from clusters of galaxies. The constant of proportionality in this relation is known as the Hubble constant; its current value is about 17 kilometers per second per million light years, or 55 kilometers per second per million parsecs.

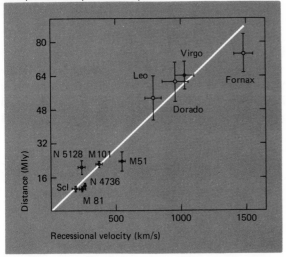

TABLE 17.3
The Local Group of Galaxies

Galaxy	Type	Distance (10³ ly)	Diameter (10³ ly)	Approximate Luminosity ($10^6 L_\odot$)[a]	Approximate Mass ($10^6 M_\odot$)[a]
Milky Way	Sb	—	100	20,000	200,000
Large Magellanic Cloud[b]	Irr I	170	30	2,900	25,000
Small Magellanic Cloud[b]	Irr I	190	25	630	6,000
Ursa Minor system[b]	E4 dwarf	220	8	0.4	0.1
Draco system[b]	E2 dwarf	220	3	0.8	0.1
Sculptor system[b]	E3 dwarf	270	8	2	3
Carina system[b]	E3 dwarf	500	4	?	?
Fornax system[b]	E3 dwarf	650	20	23	20
Leo II system[b]	E0 dwarf	700	4	0.7	1
Leo I system[b]	E4 dwarf	700	6	3	4
NGC 6822	Irr I dwarf	1,630	10	100	1,000
NGC 147[c]	E6	2,000	10	70	350
NGC 185[c]	E2	2,000	8	90	450
IC 1613	Irr I dwarf	2,150	15	65	250
Andromeda galaxy (M31, NGC 224)	Sb	2,250	150	30,000	300,000
NGC 205[c]	E5	2,250	16	330	3,000
NGC 221 (M32)[c]	E3	2,250	8	250	2,100
Andromeda I[c]	E0 dwarf	2,250	≈2	2	2
Andromeda II[c]	E0 dwarf	2,250	≈2	2	2
Andromeda III[c]	E0 dwarf	2,250	≈2	2	2
Triangulum galaxy (M33, NGC 598)	Sc	2,350	60	4,000	39,000

[a] $L_\odot = 4 \times 10^{33}$ erg/s; $M_\odot = 2 \times 10^{33}$ g.
[b] Satellite of our Galaxy.
[c] Satellite of the Andromeda galaxy.

the dwarf elliptical galaxies. The Local Group spans a region of space about 3 million light years across. And in three dimensions most of the galaxies are located around either our Galaxy or the Andromeda galaxy, as shown in Figure 17.16. This is presumably a result of the large mass and the consequent gravitational pull of our Galaxy and the Andromeda galaxy. All the galaxies in the Local Group are moving relative to each other, and it is probable that the small ones orbit the two large spirals, which in turn orbit each other with very long periods.

In addition to the galaxies listed in Table 17.3 as being part of the Local Group there are at least eight additional galaxies lying within about 5 million light years of the Milky Way. Some astronomers have ar-

gued that these galaxies are also gravitationally attached to the Local Group, but there is still uncertainty about that point. There is a growing sense among astronomers that maybe all galaxies are part of small groups (such as the Local Group), large clusters, or superclusters. In reality there may be few, if any, isolated galaxies.

Although it is not possible to present all that astronomers know about the galaxies of the Local Group, the following brief discussions will give some sense of the state of our knowledge. As is true for stars, most galaxies may be fitted into one of several classes, but upon closer inspection they all have their individual distinguishing characteristics, such that no two are precisely alike.

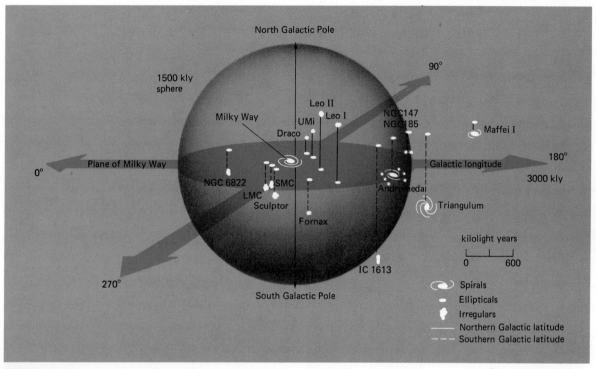

FIGURE 17.16
A three-dimensional plot of the locations of the members of the Local Group relative to the plane of our Galaxy.

ANDROMEDA GALAXY, M31

Visible with the unaided eye, as a faint, hazy patch of light, the Andromeda galaxy (Figure 17.17 and Plate 25) is the largest member of the Local Group (Figure 17.16) and is an Sb-type normal spiral like our own Galaxy. Its longest angular dimension covers nearly 5° on the sky, the same as the distance between the two end stars in the bowl of the Big Dipper. The central plane of the galaxy is inclined at an angle of about 12° to the line of sight.

On short-exposure photographs Andromeda's small, brilliant nucleus looks quite similar to the one detected by infrared and microwave observations of our Galaxy's nucleus. Two spiral arms appear to wind out of the nuclear bulge for several turns, trailing the direction of rotation. In recent years astronomers have identified hundreds of H II emission nebulae, associations of young, blue stars, and some 400 open clusters—all marking the position of the arms in the disk.

The galaxy's dust lanes exhibit chaotic patterns and do not appear to fit any spiral pattern well. Instead

they are most noticeable as nearly a ring around the nucleus. Atomic hydrogen gas extends well out beyond the optical image of the galaxy; it does not extend into the center but stops about 12,000 light years from it. On the other hand the hot, ionized-hydrogen gas clouds (H II emission nebulae) taper off at about 50,000 light years, and also have a hole in their center about 12,000 light years in radius. This doughnut-shaped distribution of cold dust clouds, atomic hydrogen, and hot, ionized-hydrogen clouds is now known to be characteristic of spiral galaxies (Figure 17.8). The absence of interstellar matter in the nucleus, the hole in the doughnut, indicates that star formation has ceased there. Although there is still great uncertainty as to whether the abundances of the chemical elements are similar to those in our own Galaxy, they do not appear to be dramatically different.

Spheroidal population red giants and a number of planetary nebulae are the brightest objects in the galaxy's central regions; the outlying portions are dominated by bright, blue, spiral-arm population supergiants (Figure 17.17). In the intermediate areas be-

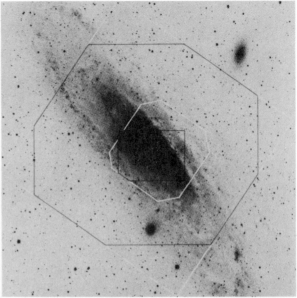

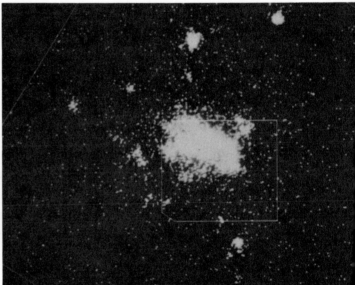

FIGURE 17.17
The Andromeda galaxy as seen in visible, X-ray, ultraviolet, and radio wavelength radiation.

This page:
In a visible photograph (*above*), we see primarily starlight, its obscuration by interstellar dust, and H II emission nebulae.

This negative print (*top right*) of a visible photograph provides a somewhat different impression of the galaxy. It also contains the outline of X-ray pictures, shown in white, of which a portion, shown in black, are reproduced in the photographs below and on the facing page, top left.

A 60-minutes-of-arc field is shown (*middle right*) that reveals at least 20 discrete X-ray sources outside the overexposed nucleus. Many are probably X-ray binaries as in our Galaxy.

Facing page:
A narrow-angle view in X-rays (*top left*) shows that the nucleus contains some 18 discrete sources or somewhat more than discovered in our Galaxy.

In 21-centimeter radio radiation (*top right*) we see the location of atomic hydrogen in the interstellar medium. Note the hole in the center where the galactic bulge is located.

A radio contour map (*middle*) at 11 centimeters overlying a visible picture in which the strongest source is the galactic center. Within about 300,000 light years of the center, stars are exploding and being born.

An ultraviolet picture at 2000 Å (*bottom*) shows the locations of H II regions and their associated young 0 and B stars.

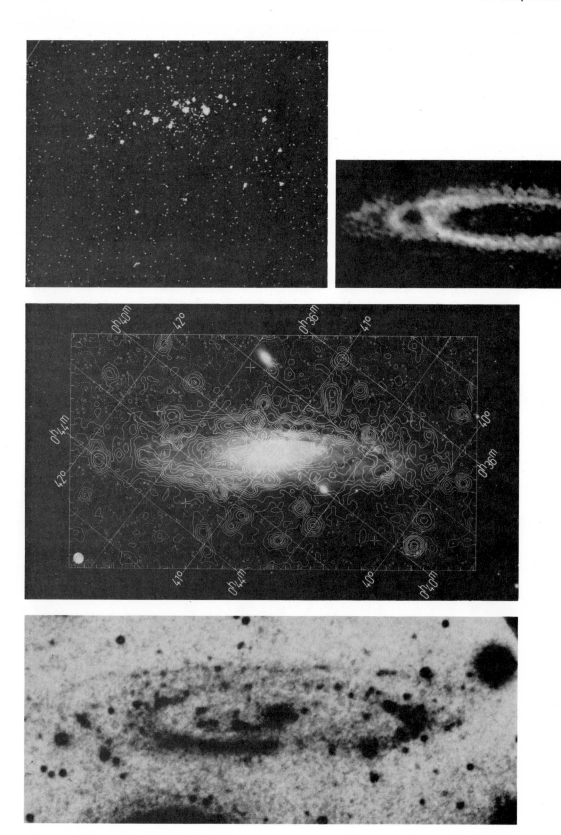

tween the arms, which are relatively free from obscuring material, there is a mixture of different-age disk population stars. These interarm regions are sufficiently transparent for us to see remote galaxies through them. Completely surrounding the galaxy are about 400 globular clusters.

From the work of the *Einstein Orbiting Observatory* about 80 sources of X rays have been found in the Andromeda galaxy. They appear to be comparable to the brightest X-ray sources in our Galaxy. Figure 17.17 shows both a wide- and a narrow-angle view of the nuclear bulge in Andromeda. As can be seen, some of the X-ray sources are in the spiral arms and some in the nucleus. Although in both places X-ray sources signal the occurrence of violent events, possibly dying stars, they are not exactly the same kind of event. Finally about 20 of the X-ray sources are in globular clusters.

As in the Milky Way, the central bulge of the Andromeda galaxy rotates much like a solid body. The outer disk shows an almost constant rotational velocity out to great distances from the center. This uniformity, along with the extension of atomic hydrogen well beyond the optical image, suggests to astronomers that Andromeda possesses a very large and massive corona of unseen matter encircling the halo like that suspected for our Galaxy. In almost everything our stellar system and the Andromeda galaxy are alike: major features, physical characteristics, and composition of celestial objects.

TRIANGULUM GALAXY, M33

The two spiral galaxies besides our Galaxy in the Local Group, Andromeda and Triangulum, do not possess the most elegant or dramatic spiral patterns, but they

RELATIVISTIC DOPPLER REDSHIFTS AND COSMOLOGICAL REDSHIFTS

The Doppler effect must be modified when the relative speed of recession or approach is an appreciable fraction of the speed of light. The respective equations for nonrelativistic velocities and for relativistic velocities are

$$z = \frac{\Delta\lambda}{\lambda} = \frac{v}{c} \quad \text{(nonrelativistic)},$$

$$= \sqrt{\frac{1 + (v/c)}{1 - (v/c)}} - 1 \quad \text{(relativistic)},$$

where $\Delta\lambda$ is the measured wavelength shift at λ, v is the radial velocity of the object, and c is the velocity of light. If a Doppler redshift from 3000 to 9000 angstroms is observed in the spectrum of the object, then $\Delta\lambda/\lambda = (9000 - 3000)/3000 = 2$, the object's correct velocity is $0.8c$ from the relativistic Doppler equation. If we insert a Doppler redshift of 2 in the nonrelativistic Doppler equation, the velocity is twice the speed of light ($v/c = 2$), which is not physically possible for local velocities producing the Doppler redshift.

When considering cosmological redshifts, it is convenient to discuss the redshift as only the proportional shift in wavelength, z ($=\Delta\lambda/\lambda$), rather than in terms of velocity. For various values of the redshift z we can calculate a quantity called a universal *scaling factor*, denoted by R. The scaling factor can be thought of as measuring a characteristic scale size for the universe such as a typical separation between galaxies or clusters of galaxies. For an expanding universe, which the Hubble law indicates is what is actually happening, the scaling factor R increases with time and at any instant has the same value everywhere in space. The value of the scaling factor at the present can be denoted by R_0 so that just R means some earlier time. The cosmological redshift is related to the scaling factor by the relation

$$z = \frac{R_0 - R}{R}.$$

Since the wavelength of light is a distance λ, the length increases by $\Delta\lambda$ while the wave moves through space because space itself is expanding. Thus the cosmological redshift must be related to a basic measure of the scale size of space that is the scaling factor R.

FIGURE 17.18
The nearby spiral galaxy M33 in Triangulum. It is slightly more than 2 million light years away and is one of three spirals in the Local Group.

have the advantage of being nearby. The Triangulum galaxy (M33) is a Hubble-type Sc galaxy (Figure 17.18), and we see it almost face on. About a tenth the size and mass of the Andromeda galaxy, it contains only a few tens of billions of stars.

Although smaller than M31, the Triangulum galaxy emits almost as much energy in the form of radio waves as does the Andromeda galaxy. This may not be surprising: With more interstellar matter in an Sc galaxy like M33 than in an Sb like M31, more stars should be forming as is evidenced by the larger H II regions in Triangulum. Supporting the hypothesis of an enhanced star-formation rate is the fact that red giants, stars having evolved off the main sequence, are about

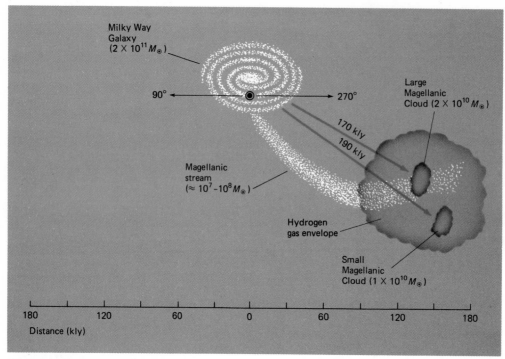

Milky Way
Galaxy
$(2 \times 10^{11} M_\odot)$

90°   270°

Large
Magellanic
Cloud $(2 \times 10^{10} M_\odot)$

170 kly
190 kly

Magellanic
stream
$(\approx 10^7 - 10^8 M_\odot)$

Hydrogen
gas envelope

Small
Magellanic
Cloud $(1 \times 10^{10} M_\odot)$

180   120   60   0   60   120   180

Distance (kly)

FIGURE 17.19
Relationship of the Magellanic clouds to our Galaxy and the Magellanic Stream. An immense stream of hydrogen gas has been detected lying between the Large Magellanic Cloud and our Galaxy. It is possible that gas is being pulled out of the Large Cloud by the Milky Way because of the latter's greater gravitational attraction.

100 times more prevalent in Triangulum in a region the size of the solar neighborhood than they are in Andromeda or our Galaxy.

Emission of 21-centimeter photons by hydrogen atoms comes from all parts of the disk, but it is somewhat heavier from what appears to be a multiarm—though broken and disordered—spiral pattern. Studies of the ionized-hydrogen emission nebulae show a multitude of filaments, loops, and arcs as part of the structure of individual H II nebulae. This supports the 21-centimeter results of a rather chaotic spiral structure.

LARGE AND SMALL MAGELLANIC CLOUDS

The closest extragalactic neighbors (both are visible to the naked eye) are two objects in the southern skies called the Magellanic Clouds. They were named in honor of the explorer Ferdinand Magellan, whose observations of them in 1522 introduced these galaxies to the western world. The Large Magellanic Cloud and the Small Magellanic Cloud are a physically related double system immersed in a common envelope of neutral hydrogen gas. They are considered satellites of our Galaxy and move in known orbits about the Milky Way.

The orbital plane for the Large Magellanic Cloud appears close to perpendicular to our Galactic plane; its closest approach is about a quarter of a million light years. Figure 17.19 shows the Magellanic Clouds relative to the Milky Way. Also shown is an arc of atomic hydrogen, containing some tens of millions of solar masses of gas, that appears to follow a great circle through the Large Magellanic Cloud. The structure is known as the Magellanic Stream and may represent gas pulled out of the Large Magellanic Cloud by our Galaxy. This is certainly feasible if our Galaxy is as much as 10 times more massive than we presently suppose.

The central plane of the Large Magellanic Cloud is tilted nearly 90° to our line of sight; we also see the Small Magellanic Cloud at an oblique angle. Both galaxies have a ragged, disklike structure somewhat flattened by rotation. In them there are stars of all descriptions and ages, including thousands of Cepheid variables as well as gaseous nebulae, star clusters, and several supernova remnants. Very prominent in

The world, the race, the soul—
    Space and time, the universes
All bound as in befitting each—all
    Surely going somewhere.

Walt Whitman

the Large Magellanic Cloud are numerous blue and red supergiants and obscuring interstellar dust. Thus some star formation is still going on.

In fact, young blue stars are seen in some globular clusters in addition to the usual red giants. This means that these globular clusters formed within the last billion years or so. The Magellanic Clouds seem to be in a different, perhaps earlier, stage of evolution than our Galaxy is.

DWARF ELLIPTICAL COMPANIONS OF OUR GALAXY

In 1937 Harvard astronomer Harlow Shapley discovered a swarm of extremely faint stars in the constellation of Sculptor. He was able to show that it resembled a diffuse globular cluster with the stars at its center distributed 1000 times more sparsely than the density of stars in the solar neighborhood. But the swarm was some 250,000 light years away and therefore not part of our Galaxy. In succeeding years six more of these dwarf elliptical galaxies were found. They are basically featureless, roughly symmetrical, and not highly concentrated toward their centers.

The largest of the dwarf ellipticals is the one in Fornax, which has a mass of about 20 million solar massed compared to the 200 billion for the Milky Way and is about a fifth of the diameter of our Galaxy. The smallest of the dwarf ellipticals is comparable in mass to that of a globular cluster, but larger in diameter than a globular cluster. The stellar content of these dwarf systems is dominated by spheroidal population objects. These results coupled with the absence of interstellar gas and dust in the dwarf systems suggest that star formation has long since ceased.

SUMMARY

Island universes. The idea is an old one that stellar systems like our own might exist elsewhere in the universe. In the nineteenth century, however, controversy still existed over whether galaxies existed or not. The final answer came in 1924 when Cepheid variables were identified in the two large spirals in Andromeda and Triangulum, allowing Hubble to calculate distance from the period-luminosity relation.

Types of galaxies. The Hubble sequence is an arrangement of galaxies from spherical to lens-shaped to flat spiral to irregular systems. Elliptical (E) galaxies have no apparent internal structure and vary in shape from spherical to ellipsoidal. Ellipticals contain old spheroidal population stars and rotate relatively slowly. Lenticular (S0) galaxies are a transitional group between elliptical and spiral galaxies. S0 galaxies contain a disk population and a spheroidal population, and they rotate faster than E galaxies. Spiral galaxies are of two types: normal spiral (S) and barred spiral (SB). They vary in size of their nucleus and in the tightness of the spiral arms. Spiral galaxies are a mixture of very old spheroidal population and young disk population stars. Irregular (Irr) galaxies constitute a small (3 percent) class with little symmetry. There are peculiar galaxies that for one or several reasons do not fit into the Hubble sequence. The Hubble sequence probably reflects differences in the initial conditions in which galaxies formed and how they have evolved.

Properties of normal galaxies. Measuring the properties of galaxies depends on estimating their distance. Several methods have been developed for approximating the enormous distances involved for galaxies. Distance indicators have been identified in galaxies and can be used in the inverse-square law of light to determine distance. Galaxies possess a very large range in size, brightness, and mass. The mass-luminosity ratio gives important information about the population of stars in a galaxy. In general, the ellipticals are less luminous than the lenticulars, which are less than the spirals and the irregular galaxies per unit of mass.

Expansion of the universe. Cosmological redshifts of galaxy absorption lines are interpreted as the result of the expansion of the universe. They are not Doppler or gravitational effects. The farther away the galaxy, the greater the redshift and, hence, the greater the recessional velocity. In the Hubble velocity-distance law (velocity = distance × the Hubble constant), the value of the Hubble constant is a fundamental parameter in understanding the nature of the universe, such as its age.

The local group of galaxies. Gravity binds galaxies together in pairs, groups, and clusters. Our Galaxy belongs to the Local Group. Close study of the Local Group shows that most of its galaxies are dwarf elliptical ones. Galaxies of the Local Group tend to cluster about the Andromeda galaxy (M31) and our Galaxy. In most respects Andromeda and our Galaxy are very much alike.

KEY TERMS

barred spiral galaxies
cluster of galaxies
cosmological redshift
elliptical galaxy
group of galaxies
Hubble constant
Hubble law

Hubble sequence
irregular galaxy
lenticular galaxy
Local Group
normal spiral galaxy
peculiar galaxy
supercluster of galaxies

CLASS DISCUSSION

1. Describe the distinguishing characteristics of the four classes of galaxies. What do they look like? How much mass is present? How bright are they?

2. What, if anything, does the Hubble sequence of galaxies represent?

3. Discuss the distribution and locations of the various stellar populations and interstellar matter in the four classes of galaxies in the Hubble sequence.

4. What is the Hubble velocity-distance law of recession for galaxies? What, if any, is its significance? What is the Hubble constant?

5. What is the Local Group of galaxies? Is it unique in the realm of the galaxies, or is it commonplace? Describe the motions of the galaxies in the Local Group.

READING REVIEW

1. How many classes in the Hubble sequence? Name them.

2. Which are the largest galaxies in the Hubble sequence? Which are the smallest?

3. What is the significance of the numbers following the E designation for elliptical galaxies, the a, b, and c following the S, and the B following the S in the Hubble sequence?

4. Do all galaxies fit into the Hubble sequence?

5. What is a distance indicator? Upon what does its use depend?

6. What, if any, is the significance of the mass-luminosity ratio for different classes of galaxies in the Hubble sequence?

7. What is the Hubble law of recession for galaxies? What is the most likely value of the Hubble constant and its units?

8. Which Hubble class is most represented in the galaxies of the Local Group?

9. How does the Andromeda galaxy compare in size with our Galaxy?

10. Is the spiral galaxy in Triangulum the same Hubble type as our Galaxy?

SELECTED READINGS

Berendzen, R., R. Hart, and D. Seeley. *Man Discovers the Galaxies.* Neale Watson Academic Publications, 1976.

Ferris, T. *Galaxies.* Sierra Club Books, 1980.

Hodge, P. W. "The Andromeda Galaxy," *Scientific American,* January 1981.

Talbot, R. J., E. B. Jensen, and R. J. Dufour. "Anatomy of a Spiral Galaxy," *Sky & Telescope,* July 1980.

18
A Universe of Galaxies

> My suspicion is that the universe is not only queerer than we suppose, but queerer than we can suppose.
>
> John Haldane

Galaxies seemingly stretch away into space without end. On photographs taken with large telescopes astronomers have found literally billions of galaxies; in fact in some directions they outnumber the foreground stars of our Galaxy. After galaxies had been recognized as islands of stars in the immense, black sea of space and time, astronomers began to wonder about their arrangement in the universe. Certainly the galaxies of the Local Group are not uniformly spread out in space. But on the largest scale of the universe are galaxies uniformly or nonuniformly distributed, and how does that distribution change with time if at all? For astronomers the simplest assumption is that on the largest scale the galaxies are uniformly distributed in both space and time. Any number of arguments can be advanced to support such a contention, and that is more or less what astronomers have assumed to be true about the large-scale organization of the universe.

With the advent of radio astronomy as a major tool for astronomical research it became evident that in the universe some tremendously energetic events were occurring and that some galaxies were extremely active objects. The existence of these active galaxies bears heavily on the question of the large-scale distribution of matter in the universe, as we shall try to demonstrate. A second facet of this question comes from studies of clusters. *Clusters of galaxies* are gravitationally bound collections of hundreds to thousands of galaxies. Astronomers find that within a cluster of galaxies the thousand or so brightest galaxies are moving at thousands of kilometers per second. Such high velocities for objects so densely packed in clusters imply that the member galaxies of the cluster are bound together by much larger gravitational forces than can be accounted for by the total mass of visible galaxies. So a major question in astronomy to-

◀ Edwin Hubble (around 1925) sitting at the observer's station of the Mount Wilson telescope, with which he discovered the Cepheids in M31 and the expansion of the universe.

day is in what form and where are the colossal amounts of matter necessary to keep the clusters from flying apart? And what effect does this "missing mass" have on the large-scale structure of the universe?

18.1
Active Galaxies:
The Mysterious Ones

VIOLENT EVENTS IN THE COSMOS

When radio astronomers started probing the heavens in the late 1940s, our picture of the universe began to change abruptly from a quiet, orderly cosmos to one punctuated by extraordinarily violent events.

There are galaxies with extremely active nuclei, exploding galaxies, galaxies with jets, and quasars. Because the radiation from these active galaxies may be very intense throughout, or in particular portions of, the electromagnetic spectrum, they stand out from normal galaxies. One of their most prevalent features is a continuous spectrum produced by nonthermal synchrotron radiation.

Astronomers distinguish two kinds of radiation source: thermal and nonthermal. The intensity of thermal radiation grows weaker with increasing wavelength in accordance with the blackbody distribution of energy (Figures 4.12 and 11.4). Most of the thermal radiation from a galaxy comes from its stars and is located in the visible and ultraviolet portions of the electromagnetic spectrum.

Radiant energy from isolated Galactic sources, like the Crab Nebula and other supernova remnants, or from the Galactic nucleus is primarily nonthermal. Unlike the intensity of thermal radiation, the intensity of nonthermal radiation grows stronger—or at least does not drop so rapidly—with increasing wavelength. Most often nonthermal radiation is what scientists refer to as *synchrotron radiation,* which is radiation emitted by rapidly moving free electrons spiraling around magnetic lines of force (Figure 18.1).

Astronomers find abundant nonthermal emission coming from many galaxies whose optical appearance and spectra are unusual, the so-called *active galaxies.* Even though these active galaxies possess a variety of different structural forms, in general they exhibit large redshifts in their spectra, some shifts being among the largest found for any type of galaxy. Because active

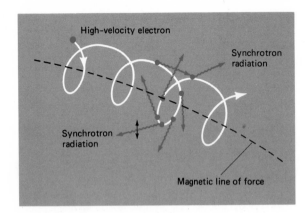

FIGURE 18.1
Synchrotron radiation, depicted by the wiggly lines, emitted continuously along the spiral path of the electron. (It is shown here for only one loop of the electron's path.) The double arrow indicates the plane of polarization for the radiation, which is perpendicular to the magnetic line of force and to the direction of the radiation.

galaxies tend to be very far from our Galaxy, they must be emitting immense amounts of energy to have their observed apparent brightnesses. Astronomers feel reasonably confident that the same energy source, whatever it is, is present in varying degrees in all of them. The most luminous of the active galaxies are the quasars, which emit up to 100 times more light than does a normal galaxy like the Milky Way. Indeed, as pointed out in Chapter 16, the same energy source as in the quasars may operate at the center of our Galaxy but on a greatly reduced scale. In the following sections we shall discuss the various types of active galaxies, whose general properties are summarized in Table 18.1. The Seyfert galaxies are our first example.

SEYFERT GALAXIES:
SPIRALS WITH COMPACT NUCLEI

More than 120 *Seyfert galaxies* have been identified since Carl Seyfert first called attention to this subclass of spiral galaxies. These galaxies have unusually small and bright nuclei, yet with a reasonably normal-looking disk (NGC 4151 is shown in Figure 18.2). Their redshifts range up to a few tenths, the largest corresponding to a velocity of a fifth the speed of light and a distance of about 1 billion light years.

The visible spectra of Seyfert galaxies have Doppler-broadened emission lines of such elements as hydrogen and narrow emission lines of highly ionized elements like iron that indicate that very hot matter is being expelled at high velocities (several thousand kilometers per second) from extremely small central regions. Supporting this is the fact that the nuclei are emitters of intense and variable X-ray, infrared, and radio radiation.

The luminosity of Seyfert galaxies is compared with that of other active galaxies in Table 18.2. Although in the visible region of the electromagnetic spectrum their absolute brightness is comparable to that for normal spirals, the Seyfert galaxies are a hundred or so times brighter than normal spirals when compared in the infrared and radio regions. Seyfert galaxies are

TABLE 18.1
Some Properties an Active Galaxy May Possess

1. High luminosity

2. Nonthermal emission; possible excessive emission in X-ray, ultraviolet, infrared, and/or radio spectral regions in comparison to normal galaxies

3. Variable emission; rapid variability in time, suggesting a small size for emitting region, but possibly some slow variability

4. Compact appearance: peculiar appearance with high contrast in brightness between nucleus and rest of galaxy

5. Explosive features: jetlike extensions out of galaxy's center or sometimes broad emission lines in spectrum of nucleus, indicating an outflow of matter, and in general nonstellar spectra

6. Gravitational disturbance; large internal velocities and/or disturbed visual appearance at one or several points across the galaxy

7. Large redshifts; average redshifts of active galaxies in following range for particular groups:

   Seyfert galaxies, $0.0 \lesssim z \lesssim 0.2$;
   N galaxies, $0.0 \lesssim z \lesssim 0.3$;
   BL Lacertae objects, $0.1 \lesssim z \lesssim 0.5$;
   radio galaxies, $0.0 \lesssim z \lesssim 0.8$;
   quasars, $0.1 \lesssim z \lesssim 3.5$

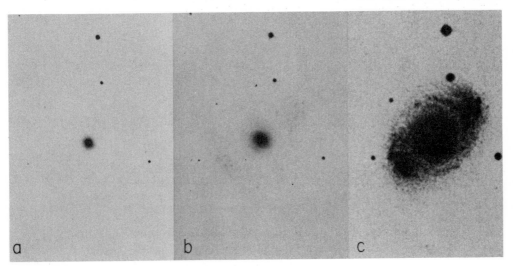

FIGURE 18.2
Seyfert galaxy NGC 4151 as shown in a negative print. The succession of longer exposures — a, b, c — illustrates the unusually bright, starlike nucleus of these spiral galaxies. As shown in (c), the disk of the galaxy is reasonably normal in appearance. In (b), the right H II emission nebulae in the galaxy spiral arms are the first features of the disk to be seen.

also sources of X-ray emission, with about 25 reliable identifications among them. Figure 18.3 shows the X-ray sky (the same as Figure 16.18) with several of the extragalactic sources identified, including several Seyfert galaxies. Because much of the X-ray emission varies in periods of hours or days, the X-ray-emitting region must be very small. For example the emitting region in NGC 4151 is estimated to be about twice the distance between the earth and the sun. Since this Seyfert galaxy emits thousands of times more energy in the form of X rays than does our Galaxy, the nature of the emitting region is extremely mysterious. Adding to the mystery is the fact that NGC 4151 is also a source of gamma rays.

There is some thought that Seyfert galaxies may be in an early, disruptive stage through which all spirals, including our own, must pass. Thus, rather than a few percent of all spirals' being Seyfert galaxies, all spirals spend a few percent of their lives being Seyfert galaxies. If so, then there might well be an evolutionary sequence for active spiral galaxies that leads to normal spiral galaxies.

N GALAXIES: STARLIKE APPEARANCE

N galaxies have some properties in common with Seyfert galaxies and quasars although they are somewhat less luminous. One of them was originally identified in 1958 by W. W. Morgan (1906–   ) at the Yerkes Observatory as an optical object responsible for powerful radio emission. They have a bright, sharply defined nucleus surrounded by a small nebulous envelope, and of the couple of dozen or so that have since been identified all are at great cosmological distances according to their large redshifts, which range up to distances of about 4 billion light years. Some astronomers have pointed out that a very distant Seyfert galaxy might resemble an N galaxy. N galaxies fluctuate rapidly in brightness and color, at times within a few days, suggesting that the variable source must be less than a light week (the distance light travels in a week) in diameter. Other similarities to the quasars and Seyfert galaxies are their large output of energy in the X-ray and radio regions of the electromagnetic spectrum.

Not chaos-like, together crushed
    and bruised,
But as the world harmoniously confused;
Where order in variety we see,
And where all things differ, all agree.

Alexander Pope

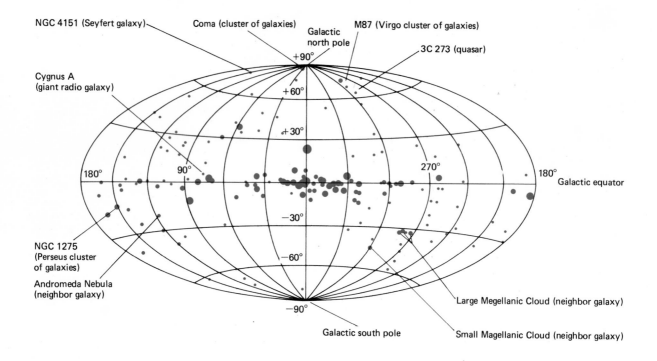

FIGURE 18.3
X-ray sources beyond our Galaxy. This figure is the same as that of Figure 16.18 except that the extragalactic objects are identified rather than those in our Galaxy.

BL LACERTAE OBJECTS: COSMIC FLARES

A small number of extremely compact galaxies, which closely resemble the Seyfert and N galaxies, are the *BL Lacertae objects*. They are named after their proto-type, BL Lacertae, which in 1929 was incorrectly identified as a variable star. Its true nature was not revealed until 1969, when radio astronomers identified it as a very active radio source. About 40 such objects are presently known. They are characterized by a sharply defined and brilliant nucleus that is emitting nonthermal radiation and whose continuous visible spectrum has no emission or absorption lines. Surrounding the bright nucleus is a faint halo whose starlike spectrum resembles that of a typical elliptical

galaxy — at least this is true in the case of BL Lacertae. The spectrum of this fuzzy halo has absorption lines from which redshifts can be measured. Such measurements place the BL Lacertae objects with halos at great distances, comparable to the nearer quasars.

Although the emission from BL Lacertae objects is strong in all wavelengths, they radiate most of their energy in the optical and infrared wavelengths. They undergo rapid changes in brightness in visible light, the infrared, and X rays. At peak brightness their luminosity rivals that of the brightest quasars. Radio measurements by very-long-baseline interferometry point to a central source, which is at most only a few light years in diameter. This is also inferred from rapid fluctuations in brightness. For example the X-ray brightness variation in periods of several hours for a particular BL Lacertae object seen in the southern skies suggests that the emitting region is less than 6 light hours in extent. For comparison the Pluto-sun distance is about 5.5 light hours.

RADIO GALAXIES: BEACONS OF THE UNIVERSE

Catalogs now list more than 10,000 discrete radio sources that are emitting significant amounts of energy by nonthermal radiation processes. Those that are extragalactic objects are called *radio galaxies,* and they emit from 10 to 100 million times more radio radiation than any normal galaxy. Radio galaxies can be divided into two groups according to their physical appearance: compact sources and extended sources. Compact sources are very small, and most coincide with galactic nuclei or quasars (see Section 18.2); most of the extended sources consist of two (sometimes more) immense radio-emitting regions located on opposite sides of a normal-looking elliptical galaxy. In this section we shall concentrate on the extended sources. At least four kinds of extended radio galaxy are recognized (Figure 18.4):

1. Giant ellipticals with extensions in the form of jets, such as in M87

2. Double radio-source galaxies whose radio-emitting regions are displaced from the galaxy, such as Cygnus A and Centaurus A

3. Bent double radio sources, in which the radio-emitting regions are almost V-shaped with the visible galaxy at the vertex, such as NGC 7720

4. Radio-trail-source galaxies, in which the emitting regions appear to have been swept back from the optical galaxy, such as in NGC 1265

The positions of radio-emitting regions, referred to as radio lobes, vary with the different radio galaxies. In most radio galaxies the radio lobes appear to be completely separate from the optical galaxy; but for several radio galaxies optical emission has been detected within the radio lobes, or bright, threadlike features connect the optical galaxy to the radio lobes. Possibly high-energy subatomic particles are ejected along these visible features by the central source. For those radio galaxies in which one can see a sequence of jets or blobs along the lobes it suggests that a number of explosions have occurred at different times. The axis along which the matter flows out of the galaxy following an explosion is probably the axis of rotation of the visible galaxy although there is no certainty that it is the rotation axis. As matter is ejected, it traps and carries magnetic fields along. These magnetic fields help to confine the particles and to produce the synchrotron radiation that is the characteristic signature of radio galaxies.

Desist from thrusting out reasoning from your mind because of its disconcerting novelty. Weigh it, rather, with a discerning judgment. Then, if it seems to you true, give in. If it is false, gird yourself to oppose it. For the mind wants to discover by reasoning what exists in the infinity of space that lies out there, beyond the ramparts of this world. . . . Here, then, is my first point. In all dimensions alike, on this side or that, upward or downward through the universe, there is no end.

Lucretius

FIGURE 18.4
Radio galaxies. The illustrations show the four kinds of radio galaxies discussed in the text.

On the negative print of the visible galaxy NGC 7720 is shown its associated radio source known as 3C 465 (*top left*) with the radio emission contour lines forming almost a right angle with the visible galaxy at its vertex.

The optical image of radio galaxy Cygnus A (*top right*) shows a fuzzy double object; the radio map reveals two large external sources within each of which there is an additional structure that suggests multiple explosions. Cygnus A, about 500 million light years away, was the first discrete radio source outside our Galaxy to be discovered (1945).

NGC 7720

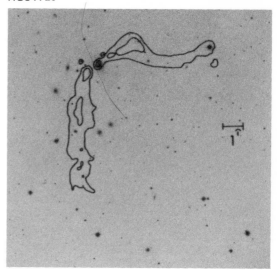

Giant elliptical galaxy in Virgo, M87; also known as Virgo A

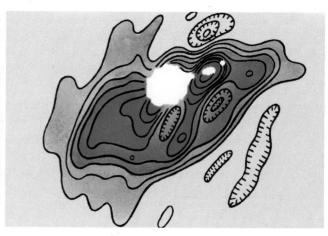

Radio galaxy Centaurus A

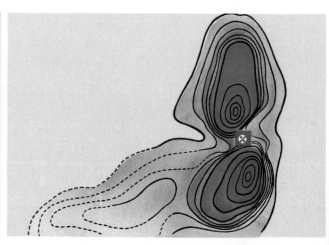

Radio galaxy Cygnus A

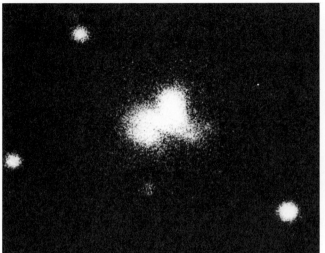

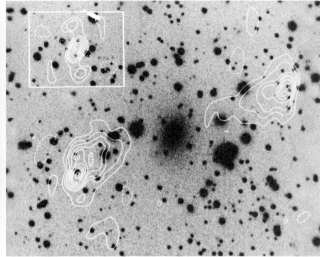

NGC 1265

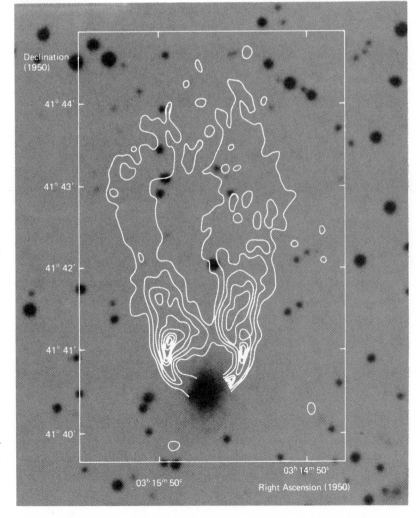

Virgo A is a giant elliptical galaxy about 64 million light years away. In the optical picture (*middle left*), taken with a short exposure, we see a small intense core and a luminous jet that emits synchrotron radiation; the new, computer-enhanced photograph in the inset shows that not a single jet but a series of objects is being ejected. The radio contour map (*middle right*) reveals a small bright core at the center of the superimposed optical image surrounded by a weaker halo that extends well beyond the optical source. Notice that the visible jet is a region of strong radio emission.

Radio galaxy Centaurus A (*bottom left*), one of the strongest radio sources, is about 16 million light years away. The optical picture shows a strong obscuring lane cutting across the middle of the galaxy. A small double radio source appears on either side of the obscuring dust lane of the galaxy, drawn in the rectangle at the center of the radio map. The central compact source fluctuates in brightness over a wide range of wavelengths. Farther out, on a line perpendicular to the dust band, appear two intense, oval-shaped radio lobes surrounded by a fainter, large elliptical region.

Radio trail source for the radio galaxy NGC 1265 in the Perseus cluster (*bottom right*). The region responsible for the radio emission attributed to the visible galaxy trails behind the galaxy somewhat like the galaxy plowing through unseen matter.

18.2
Quasars: A Cosmic Enigma

DISCOVERY OF QUASARS

By 1960 positions of extragalactic radio sources could be determined with reasonable accuracy, through radio interferometry, so that astronomers were then able to identify some compact radio sources with visible starlike objects, which they named *quasistellar radio sources*, popularly called *quasars*. The quasars, which resemble ordinary stars on a photograph, had been photographed many times before attention was drawn to them, but no one had suspected how peculiar they were. The first recognition of their unusual nature came in 1960, with an object labeled 3C 48 (the forty-eighth entry in the third *Cambridge Radio Catalog;* Figure 18.5). It looked like an ordinary star of the sixteenth apparent magnitude, but its spectrum, undecipherable at that time, consisted of emission lines superimposed on a continuous spectrum.

In 1962 Maarten Schmidt (1929–   ) of the Hale Observatories made a spectrogram of the small thirteenth-magnitude quasar 3C 273 (it is also shown in Figure 18.5) from positions accurately pinpointed by radio astronomers with the 64-meter radio telescope at Parkes, Australia. Within weeks he had untangled the puzzle of the optical spectrum of 3C 273, when he recognized in three emission lines the characteristic spacing of the second, third, and fourth lines of the Balmer series of hydrogen (Figure 18.6). To his amazement they were displaced toward the red end of the spectrum by a large redshift, corresponding to a velocity of 15 percent of the speed of light. With this key to the optical spectrum in hand astronomers soon decoded spectrograms of other quasars, revealing even larger redshifts.

APPEARANCE AND DISTRIBUTION

More than 1500 quasars have now been identified and the number continues to grow. They are still very enigmatic with their starlike optical appearance and large redshifts. Work over the last 20 years has shown that fewer than 10 percent of the quasistellar radio sources exhibit strong radio emission. The majority of them are not intense radio sources, but they are nevertheless still referred to as quasars. The radio-quiet quasars can be identified by their unusually bright ultraviolet emission and large redshifts. All the quasars possess emission that varies in periods of days or weeks in one or several regions of the electromagnetic spectrum.

From the Hubble velocity-distance law the redshifted spectral lines indicate that the quasars are very distant cosmologically and thus must emit extraordinarily large amounts of radiant energy to be as bright as they appear in our sky. In the radio region of the electromagnetic spectrum quasars emit as much energy as the brightest radio galaxy, while in visible light they are more luminous than the brightest giant elliptical galaxies. Quasars outshine all other types of galaxies in gamma rays, X rays, and infrared radiation. As shown in Table 18.2, the luminosity of the quasar can be the equivalent of 100,000 billion suns, or a thousand times that of normal galaxies. However, it is

FIGURE 18.5
Quasars — four quasistellar radio sources photographed with the 5.1-meter Hale reflector. From left to right, 3C 48, 3C 147, 3C 273, and 3C 196 (3C stands for the third *Cambridge Radio Catalog*). Compare 3C 273 with the photograph in Figure 18.7.

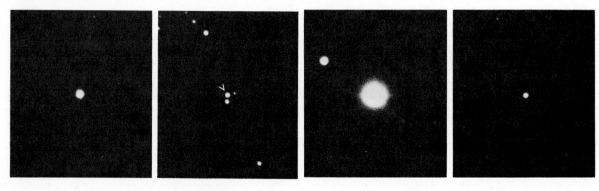

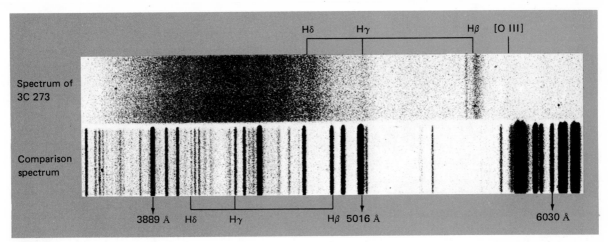

FIGURE 18.6
Spectrogram (negative) of the quasar 3C 273. Notice the pronounced redshift of the
Balmer hydrogen lines (Hβ, Hγ, and Hδ) in relation to the comparison spectrum.

their brightness variations that make them the striking cosmic objects that they are since it indicates that these immense radiation sources are located in volumes as small as a light day across.

Since quasars are in general quite distant objects (if their redshifts are cosmological), there is speculation that, like the Seyfert galaxies, which possess smaller redshifts, a quasar may be the brilliant nucleus of a faint galaxy and that it is because of its great distance that we are unable to resolve the surrounding galaxy. In a few of the nearer quasars some nebulous extensions have been photographed, whose faint starlike spectra have the same redshifts as the quasars them-

selves. This suggests that visible quasars may be overly luminous centers of galaxies. Figure 18.7 shows in negative photographic reproduction the quasar 3C 273, whose redshift of 0.16 corresponds to about 3 billion light years. The quasar, marked Q, seems to have not only a faint nebulous halo but also a jet to the lower right. It is hoped that *Space Telescope* (page 112), which will be able to resolve objects 50 times smaller in angular size than the best ground-based telescope can, will permit astronomers to decide whether or not quasars are the nuclei of very distant galaxies.

Despite their starlike appearance in visible light,

TABLE 18.2
Luminosities of Active Galaxies

Object	Spectral Region and Rate of Radiation (ergs/s)					Luminosity $(L_\odot)$[a]
	Radio	Infrared	Optical	X-Ray	Gamma-Ray	
Supernova	10^{34}	10^{35}	10^{36}	10^{37}	6×10^{33}	10^{4}
Normal galaxy	10^{38}	10^{40}–10^{42}	10^{42}–10^{44}	10^{38}–10^{39}	10^{38}–10^{39}	10^{10}
Seyfert galaxy	10^{40}–10^{42}	10^{43}–10^{46}	10^{42}–10^{45}	10^{41}–10^{43}	10^{45}	10^{13}
N galaxy	10^{43}–10^{45}	10^{42}–10^{46}	10^{42}–10^{45}	10^{42}–10^{45}	?	10^{13}
BL Lacertae objects	10^{42}–10^{45}	10^{44}–10^{46}	10^{43}–10^{45}	10^{43}–10^{46}	?	10^{13}
Radio galaxy	10^{42}–10^{45}	$<10^{43}$	10^{44}	10^{41}–10^{45}	10^{44}	10^{12}
Quasar	10^{44}–10^{46}	10^{46}–10^{47}	10^{45}–10^{47}	10^{44}–10^{47}	10^{46}	10^{14}

[a]Because these figures are so approximate, the solar luminosity has been assumed to be 10^{33} ergs/s rather than the more precise value in Table 11.1.

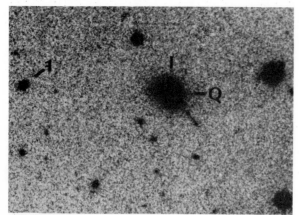

FIGURE 18.7
Quasar 3C 273 in a negative photograph. The quasar (Q) appears to possess a faint nebulous halo and a jet extending to the lower right. The galaxy marked 1 is thought to be a companion to 3C 273. The rough appearance of the picture is due to excessive enlargement, which reveals the grains that compose the photographic emulsion.

radio images of quasars are frequently large, structured, and noncircular. Quite a few quasars have a radio structure composed of minute discrete components, while others have extended radio lobes on either side of the optical image, somewhat like radio galaxies. Measurements by very-long-baseline radio interferometry disclose that in some quasars there is apparently a rapid separation of close, minute sources that—if interpreted literally—is occurring at velocities greater than the velocity of light. Figure 18.8 shows a radio image of 3C 273, in which over a 3-year period first one and then a second knot of radio emission are seen separating from the main lobe of radio emission; the separation velocity appears to be about eight times the speed of light. However, as we discussed in Chapter 3, relativity theory is predicted on the assumption that neither mass nor energy can be transported at speeds in excess of the speed of light, and the number of observational tests of relativity the-

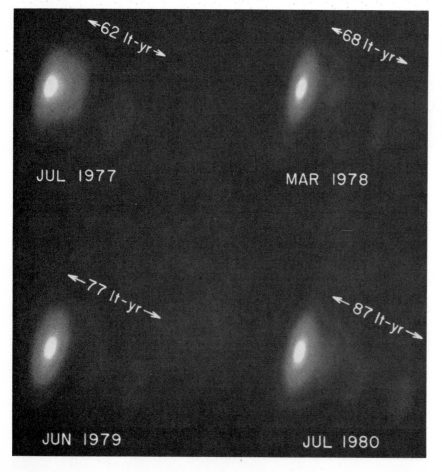

FIGURE 18.8
Quasar superluminal velocities. The radio image was made with the very-long-baseline interferometer over a period of about 3 years with the intervals between being about 1 year. The rate at which the blob of radio emission is separating from the main source is 8 times the velocity of light if 3C 273 is at the cosmological distance given by its redshift.

ory is sufficient to enable us to conclude that its foundations are secure. Therefore the problem lies not with relativity theory but with the interpretations of the quasar observations. There are a number of alternative interpretations that have been advanced, but so far none is without some difficulty in explaining the observations.

The only way of estimating the distances of remote quasars is to use the Hubble velocity-distance law. Sample counts of quasars show that their numbers increase much faster with distance than would be the case if their distribution in space were uniform. A plausible interpretation is that the quasistellar objects were formed in large numbers within a couple of billion years after the universe began to expand. The universe may then have had a thousand times more quasars than it does now. Of those quasars within the visible universe most would by now have evolved into normal galaxies. Therefore only very remote quasars, which for us represent an earlier epoch in the universe, would still be observable.

MYSTERY OF QUASAR SIZE AND DISTANCE

Quasars are striking in that an incredible flood of energy gushes out of sources no bigger than a fraction of a light year in diameter. Spectra of the visible portion of the radiation from quasars exhibit broad emission lines of familiar ionized elements (such as carbon, magnesium, oxygen, neon, silicon, and helium) and of hydrogen—all overlying a continuous spectrum. The bright-line spectrum apparently originates in hot gases surrounding the source of energy; and as noted, the lines are highly redshifted. We can identify the lines from their relative spacings, which are the same as those in the spectra of planetary nebulae, novae, stellar coronae, and other hot sources of radiation. Measured redshifts for quasars range from a few percent up to 91 percent ($z = 3.53$) of the velocity of light. Why no quasars have so far been discovered beyond this limit is puzzling since the technique for measuring their redshifts, at least up to 95 percent of the velocity of light, is available.

About 10 percent of the quasistellar objects also have narrow absorption lines in their spectra, usually with redshifts less than, or equal to, those of the emission lines. An obvious interpretation is that the absorbing is done by cool shells or clouds moving outward from the quasar at different speeds up to appreciable fractions of the speed of light. Since the

Twinkle, twinkle, quasi-star
Biggest puzzle from afar
How unlike the other ones
Brighter than a billion suns
Twinkle, twinkle, quasi-star
How I wonder what you are.

George Gamow

shells are coming toward us, the absorption redshifts are the result of subtracting a Doppler shift (a blueshift) from the true cosmological redshift indicated by the emission redshifts. On the other hand, if the absorption is not physically related to quasars, it presumably must be the result of intergalactic absorption in the extended halos of galaxies between us and the quasar. Since the absorption lines are close to the wavelength of the emission lines in many cases but not all, the first of these hypotheses seems more acceptable for those cases; but the question is far from settled.

ARE THE QUASAR REDSHIFTS COSMOLOGICAL?

Most astronomers believe that the redshifts of the lines in the spectra of quasistellar objects correspond to true velocities resulting from the expansion of the universe, a view known as the *cosmological interpretation*. However, a few astronomers think that these objects may not be so distant after all. If they were closer to us, their energy output could be smaller, bringing them more in line with other cosmic objects. One noncosmological interpretation, the *local Doppler hypothesis*, proposes that these bodies were violently expelled at high speeds from nearby normal galaxies or from the center of our Galaxy. Thus quasar redshifts are Doppler redshifts and not cosmological. If the quasars were ejected from nearby galaxies, though, some should have velocities of approach. Yet no blueshifted quasars have been observed.

If we consider what appear to be strings of galaxies stretching across part of the sky that include normal galaxies, active galaxies, and quasars, we have an alternative proposition that quasars have been ejected from the nuclei of large abnormal galaxies. In these apparent strings very different redshifts have been measured for the quasar and the allied galaxy or galaxies; the redshifts vary by as much as many thousands of kilometers per second (Figure 19.7). In each of these cases the quasar is always redshifted more than is the allied galaxy or galaxies. (However, several quasars that appear to belong to a cluster of galaxies are known to exhibit the same redshift as the apparently normal galaxies in the cluster.) If enough discrepancies in redshift between physically related normal galaxies and quasars could be substantiated (for reasons other than chance alignments of bodies at different distances), the cosmological interpretation that redshifts are caused entirely by the expansion of the universe would be questionable. This would indicate that some unrecognized effect produces part of the redshift, possibly depending on the type of allied galaxy involved.

ENERGY SOURCE FOR QUASARS AND OTHER ACTIVE GALAXIES

If the cosmological explanation for the quasar redshifts is correct, how can we account for the vast amounts of energy pouring out of such distant but incredibly small regions? The energy is immense and is being produced more or less continuously over many millions of years. Something like a trillion solar masses of hydrogen, equivalent to 100 average-sized galaxies, would have to be converted into energy by thermonuclear fusion to equal just the radio energy being emitted by some quasars. It seems safe to hypothesize that the source of energy in quasars and other active galaxies involves, separately or in combination, intense gravitational fields, powerful magnetic fields, immense explosions, and rapid rotation.

One of the most fascinating speculations on the nature of the energy source for quasars is one proposing that the central source is a rotating black hole, with a mass of up to a billion solar masses. Matter, such as gas or even stars from its crowded surroundings, falls toward the surface of the rotating black hole. In the process rotational energy from the black hole is extracted, and the black hole slows its rotation.

> Life is the art of drawing conclusions from insufficient premises.
>
> Samuel Butler

On the other hand most astronomers believe that one or several less exotic explanations are sufficient to provide the immense amount of energy radiated away into space by the quasars and other active galaxies. These energy-source mechanisms include:

1. A random succession of supernova outbursts by massive stars in the center of a quasar

2. Collisions of stars in a dense clustering of stars at the center of a quasar

3. Gravitational collapse followed by a supernova outburst by one or several very massive (many thousands of solar masses) objects

4. A rapidly rotating, magnetized, and very massive object's transferring its rotational energy to its surroundings

5. Gas flowing onto the accretion disk of a very massive black hole

Each of these energy-producing mechanisms has a certain merit in explaining some of the observed characteristics of quasars and other active galaxies in that they could produce a lot of energy from a very small volume, could produce nonthermal radiation, could be variable in time, and would not last forever. In their present form, however, none seems to satisfy all the requirements.

18.3 Clusters of Galaxies

SMALL GROUPS OF GALAXIES

The clumping of matter in the universe is certainly evident on all the levels of size we have discussed so far. Starting with the nucleus of the atom, we find dense clumps of matter such as people, the earth, stars, clusters of stars, galaxies, and clusters of galaxies.

Galaxies occur in a wide variety of gravitationally bound systems, ranging from binary pairs through small groups containing a few tens of galaxies to rich clusters composed of hundreds to thousands of galaxies. These clouds of galaxies in turn vary widely in the number and spacing of member galaxies. From small groups to rich clusters astronomers find that the typical separations of bright galaxies vary from a few tens

TABLE 18.3
Properties of Some Nearby Small Groups of Galaxies

Brightest Spiral Galaxy	Constellation	Distance (10^6 ly)	Approximate Diameter (10^6 ly)	Number of Galaxies	Approximate Mass ($10^9 M_\odot$)	Approximate Luminosity ($10^9 L_\odot$)
Local Group	—	—	4	21	500	53
M81	Ursa Major	11	4	15	300	44
M101	Ursa Major	23	3.2	7	300	53

of thousands of light years to a few million light years (a few tenths of our Galaxy's diameter to a few tens of times our Galaxy's size). We must talk about typical separations among bright galaxies since those are the ones we see; the number of faint dwarf galaxies lying between the bright ones is simply not known.

In the category of *small groups of galaxies* probably the most common are the rather loosely packed groups containing fewer than 10 bright galaxies with typical separations of a few million light years (or a few tens of diameters for a large spiral). Examples of small groups of galaxies are those centered on the bright spiral galaxies M81 (Figure 1.4) and M101 (Figure 17.1). Table 18.3 lists some properties of these two small groups of galaxies as compared with our Local Group. Within about 30 million light years there are a number of small groups similar to those listed in Table 18.3, and fewer than 25 percent of all the galaxies within that distance appear not to belong to a small group. Thus it is tempting to ask whether all galaxies are, or were at one time, members of some kind of grouping. We believe that it is only an insufficient amount of time available on large telescopes that prevents astronomers from answering this question. With the ability of *Space Telescope* to observe galaxies 10 times fainter than the best earth-based telescopes can, the astronomical community may have an answer to this question in the not too distant future.

RICH CLUSTERS OF GALAXIES
Considerably larger and more complex than the small groups of galaxies are the *rich clusters*, containing hundreds to thousands of galaxies. Even in the early part of this century these rich clusters were clearly evident to the pioneers of extragalactic astronomy. Today astronomers have identified several thousand rich clusters—out to about 4 billion light years of

distance—which contain at least 10 percent or more of all galaxies. Whether more galaxies are members of small groups than of rich clusters is still unsettled. Fourteen representative rich clusters are listed in Table 18.4. (The actual number of galaxies in a cluster may be much larger than noted if the dwarf galaxies could be seen.)

To help organize clusters of galaxies as a field of study, astronomers have devised a classification scheme for them, analogous in principle to stellar spectral classification (Section 12.3) and Hubble's galaxy-classification scheme (Section 17.2). Many cluster properties have been taken into consideration, such as richness (number) of galaxies; shape; dominance of bright galaxies in a cluster population; variation of galactic type; variation of the density of galaxies outward from a cluster's center, and radio and/or X-ray emission. These various criteria can be summed up in three basic classes: regular, intermediate, and irregular, where the name describes their general appearance as elliptical and the like do for galaxies (Table 18.5).

Regular clusters are all rich, high-density clusters, having galactic populations numbering on the order of a thousand galaxies or so. Regular clusters are roughly spherical in shape with a high concentration of galaxies toward the center. These clusters are composed largely of dust-free galaxies—the E and S0 Hubble types—with fewer than 20 percent of their membership as spirals. Regular clusters are dominated at their centers by giant elliptical galaxies, often referred to as *supergiant galaxies*, with extended halos and few, if any, have spiral galaxies in the cluster central region.

An example of the class of regular clusters is the rich cluster known as A2199,[1] whose dominant central

---

[1]Number 2199 in the catalog of rich clusters published in 1958 by George O. Abell.

TABLE 18.4
Representative Rich Clusters of Galaxies

Cluster	Approximate Distance (10^6 ly)	Approximate Diameter (10^6 ly)	Recessional Velocity (km/s)	Estimated Number of Bright Galaxies	Density of Galaxies (no./$(10^6$ ly$)^3$)
Virgo	65	15	1,180	100[a]	2
Pegasus I	210	5	3,700	100	20
Pisces	220	35	5,000	100	5
Cancer	260	15	4,800	150	15
Perseus	320	20	5,400	500	10
Coma	400	25	6,700	800	10
Hercules	570	10	10,300	300	5
Ursa Major I	880	10	15,400	300	5
Leo	1,000	10	19,500	300	5
Corona Borealis	1,100	10	21,600	400	10
Gemini	1,100	10	23,300	200	5
Boötes	2,100	10	39,400	150	5
Ursa Major II	2,200	10	41,000	200	10
Hydra	3,300	?	60,600	?	?

[a] For the same limiting luminosity as used for the other clusters.

TABLE 18.5
Classification Scheme for Clusters of Galaxies[a]

Cluster Property	Regular	Intermediate	Irregular
Richness (number)	High to moderate	High to low	High to low
Shape	Spherical	Flattened to irregular	Irregular
Galactic content	Elliptical-rich	Spiral-poor	Spiral-rich
Spirals, %	20	30	50
Dominance of brightest galaxies over faint galaxies	High contrast[b]	Intermediate	Less contrast
Central concentration	Strong	Moderate	Very little
Overall galactic density	High	Intermediate	Low
Brightness variation from center	Rapid decline	Intermediate decline	Constant decline
Radio emission	High	Low	Low
X-ray emission	High	Intermediate	Low
Presence of subclustering	Not evident	Some evident	Often present
Examples	Coma, Perseus, Corona Borealis	A194, A539[c]	Virgo, Hercules

[a] Adapted from N. A. Bahcall, *Annual Review of Astronomy and Astrophysics, 1977*, p. 505.
[b] Most often dominated by giant elliptical galaxy with bright core surrounded by extensive halo, known as cD galaxies.
[c] The number in the catalog of rich clusters by Abell.

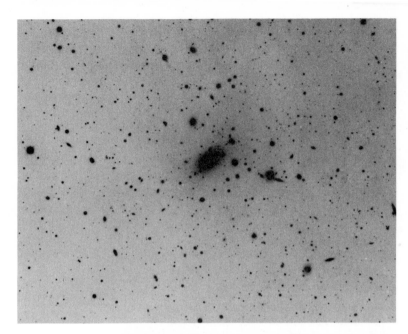

FIGURE 18.9
Rich cluster of galaxies, A2199, in the constellation Hercules. This cluster is an example of the class of regular clusters, being rich and dense with galaxies strongly concentrated toward its center. Its center is dominated by the supergiant elliptical galaxy NCG 6166, whose diameter is approximately 2 million light years, making it one of the largest galaxies known. A2199 is roughly 600 million light years away. The photograph is a negative print.

FIGURE 18.10
Central region of the cluster of galaxies in Coma Berenices, whose distance is estimated to be 400 million light years. It is classified as a regular cluster and contains thousands of galaxies, most of which are faint, making it one of the richest clusters known.

galaxy is NGC 6166 (the largest galaxy in Figure 18.9). This cluster is about 600 million light years from our Galaxy, a distance that implies that NGC 6166, a supergiant elliptical galaxy, is almost 2 million light years in diameter, or one of the largest galaxies known. Surrounding the galaxy is a large cloud of gaseous matter that is emitting radio and X-ray photons. The X-ray emission is apparently the result of thermal processes, indicating that the gas is extremely hot (20 to 200 $\times$ 10^6 K), rather than due to nonthermal. Surprisingly, the gas contains an excess quantity of iron that has lost 24 of its 26 electrons. This suggests that the gas is not primordial gas, which would be predominantly hydrogen and helium. But it is gas that has been enriched with heavy elements produced in supernova outbursts and thrown out of the galaxies in the central regions of the cluster and has collected around NGC 6166. The cluster A2199 is not the only regular cluster in which hot gas fills the center of the cluster. Other examples are the Coma cluster (Figure 18.10), the Perseus cluster (Figure 18.11), and the Corona Borealis cluster (Figure 1.5).

FIGURE 18.11
Perseus cluster. Located some 320 million light years from us, Perseus is classified as a regular cluster. The center of the cluster is dominated by the supergiant elliptical galaxy NGC 1275 (inset), which possesses strong emission in both X-ray- and radio-wavelength radiation.

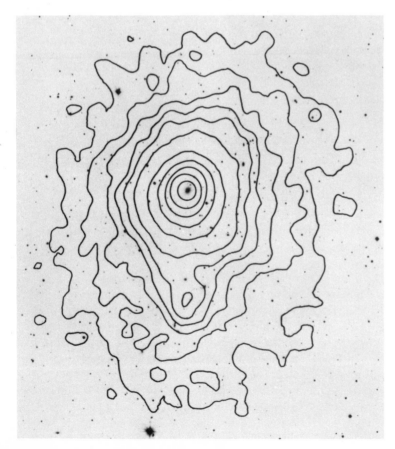

FIGURE 18.12
X-ray emission from the regular cluster A85. Located in the constellation of Cetus, A85 is a regular cluster, hot gas filling its center from which the X-ray emission arises. Such a situation suggests that regular clusters are more evolved than are irregular clusters. The visible photograph is a negative print.

Figure 18.12 reveals the X-ray-emitting gas in the cluster A85 in the constellation Cetus by contour lines superimposed on a negative reproduction of the cluster as seen in visible light. The presence of the hot gas in regular clusters provides a natural explanation for the many trail radio galaxies, such as NGC 1265 (Figure 18.4g) in the rich Perseus cluster. Apparently these radio galaxies are really plowing through a gas that sweeps the radio-emitting region back behind the visible galaxy.

The class of *intermediate clusters,* whose general features are outlined in Table 18.5, are (as their name implies) intermediate between the regular clusters and the irregular clusters. In terms of the types of galaxy found in the cluster, over half are the lenticular S0 Hubble type with about 15 percent being elliptical galaxies and 30 percent spirals. They, like the regular clusters, tend to contain mostly ellipticals and S0s in the central regions with spirals more toward the periphery of the cluster. The brightest galaxy or galaxies are usually normal-looking giant ellipticals.

Irregular clusters are the third class in the classification scheme for clusters. (Don't confuse the irregular cluster with the irregular galaxy in Hubble's galaxy-classification scheme.) They are marked by an irregular shape with an overall lower density of galaxies that show little tendency to concentrate toward the center of the cluster. Spiral galaxies are more frequently found in irregular clusters, and they amount to over half of all galaxies in an irregular cluster. In fact the brightest galaxies are generally spirals, and they are fairly uniformly spread from the center to the outer edge of a cluster. Radio and X-ray radiation have been detected coming from irregular clusters, but it is weaker than from regular clusters. Also it seems to be more nearly associated with the individual galaxies of a cluster than with gas lying between them.

THE RICH CLUSTER IN VIRGO

The nearest of the rich clusters, an irregular cluster, is that in Virgo, containing 100 or so bright galaxies, spread over a few tens of millions of light years, and lying some 64 million light years from us (Figure 18.13). Its recessional velocity is about 0.3 percent of the velocity of light. About $(100°)^2$ of sky are covered by the Virgo galaxies—roughly the area covered by a reasonable-size book held at arm length.

Most of the galaxies in the Virgo cluster are large normal spirals and dwarf elliptical systems. Its bright-

FIGURE 18.13
Cluster of galaxies in the constellation Virgo. The two large elliptical galaxies are M84 and M86. Thousands of galaxies, beyond the edges of this photograph, form a rich, loose, irregular cluster, which appears to have no central concentration. It possesses a number of small subgroups. All types of galaxy are represented in the cluster.

est members are giant ellipticals, such as M87 (also in Figure 18.4), M84, and M86 (Figure 18.13). The X-ray emission from the Virgo cluster is concentrated around individual galaxies, particularly M84 and M86. The strong radio galaxy M87 in the Virgo cluster is also a strong source of X rays. Another irregular cluster of galaxies, about 570 million light years from us, is the Hercules cluster (Figure 18.14).

WHY THE DIFFERENCES AMONG RICH CLUSTERS?

The masses of the rich clusters of galaxies are on the order of $10^{15} M_\odot$, being somewhat smaller for irregular than for regular clusters. Cluster luminosities range between 10^{12} and 10^{13} times that of the sun, with as much as 1 percent in the X-ray region for the strongly X-ray-emitting regular clusters. As a result, the aver-

FIGURE 18.14
Irregular cluster of galaxies in the constellation of Hercules. It is about 570 million light years away and is composed of many spiral galaxies.

age mass-luminosity ratio is about 200, which indicates that clusters of galaxies have larger masses compared to light output than is typically true for individual galaxies (Table 17.2).

There is a strong concentration of elliptical galaxies toward the center of regular clusters, with a dominant supergiant elliptical galaxy located there. Several lines of evidence, including X-ray emission, point to the existence of a large mass of hot gas in the centers of the regular clusters. This suggests that over the life of the cluster the massive central galaxy has grown to colossal size by capturing stars and gas from other galaxies in the cluster. The hot gas filling the center is a by-product of this cannibalizing process. Thus the regular clusters are evolving more rapidly, while the evidence points to the opposite situation for the irregular clusters, where change is much slower.

MISSING MASS IN CLUSTERS?
In 1933 Fritz Zwicky (1898–   ) noted that there did not appear to be enough mass in the form of galaxies to bind them gravitationally into a cluster. Zwicky's note introduced the problem of "missing mass" into the study of clusters of galaxies.

Attempts to estimate the total mass of an entire cluster of galaxies have led to conflicting figures. One common method of estimating the mass is to average the Doppler line shifts arising from internal motions of the individual galaxies in the cluster. The spread of values about the average can be shown to be a measure of the cluster's entire mass. Another method of estimating the mass uses the observed luminosities of the individual galaxies with their mass-luminosity ratios in Table 17.2. Knowing the number of particular galaxy types in a cluster, we can convert the amount of light that they emit to mass and add up the individual masses to find the cluster's mass.

The "dynamical mass" found by analyzing the observed radial-velocity differences is many times greater than the "luminous mass" obtained from the light emitted. For the latter the mass derived includes only luminous matter within a cluster, but the former method includes matter that may be extraneous to individual galaxies in a cluster. The luminous mass of all galaxies in a cluster amounts to only 3 to 5 percent of the mass needed to provide gravitational stability for a cluster. In other words, if the gravitational binding of a cluster, was truly as weak as the luminosities of the galaxies in it suggest, then the cluster galaxies would not be concentrated into a relatively tight unit that is a few million light years in diameter. Instead the galaxies should disperse, spreading out over tens of millions of light years. This result does not depend upon the classification of the cluster but is true for all three classes of clusters.

In what form might this unseen matter be? Perhaps a large amount of undetected matter exists in a cluster as intergalactic material lying between galaxies, or as subluminous galaxies, or as undetected extended coronae for the observed galaxies. As for intergalactic matter, if it were cool atomic hydrogen, then we should be able to detect it by its emission of 21-centimeter photons. If it were molecular hydrogen, there would be the possibility of detecting its ultraviolet spectrum with the orbiting ultraviolet observatories. No evidence exists for extensive quantities of either atomic or molecular hydrogen in clusters that could account for this puzzling discrepancy in the mass of clusters.

A third possibility for intergalactic matter is that the missing mass is in the form of a very hot gas. A hot gas should emit X-ray photons but no 21-centimeter radiation. As noted in the preceding subsection, dozens of clusters of galaxies are powerful X-ray sources. The richer the cluster, the greater is the X-ray emission. The source of the X-ray emission is apparently hot, intergalactic gas, but estimates are that the mass of hot gas is about equal to the mass of all the galaxies. Therefore, it is not gravitationally sufficient to prevent the cluster from eventually coming apart.

If the so-called missing mass in clusters were in the form of subluminous galaxies, there would have to be millions to billions of them for each bright galaxy; and they should collectively be visible as a faint glow spread over the cluster. Such an effect is not observed; so subluminous galaxies are unlikely to be the answer to the missing-mass problem.

Finally, if the missing mass were to be found in immense coronae surrounding visible galaxies, such as is suspected for our Galaxy, the large central galaxies of the cluster should have rapidly cannibalized the other galaxies earlier in the life of the cluster. In clusters where several large central galaxies exist these large galaxies should have merged long ago since they would be 10 times more massive than they are thought to be. Hence clusters ought to appear very different from the way they do now. Although the evidence grows that massive coronae encircle at least some galaxies, the coronae raise many perplexing questions if they are the location of the missing mass. The problem of the missing mass in clusters of galaxies is thus as important to our understanding of the universe as any in astronomy today.

ACTIVE GALAXIES IN CLUSTERS

The active galaxies discussed in the earlier parts of this chapter play an important role in our understanding of the organization of galaxies into clusters and the evolution of cluster structure. For example, if active galaxies were found only among the 25 percent or so of galaxies that are not obviously members of clusters of galaxies, then astronomers would draw a conclusion about the evolution of galaxies and the universe very different from that if active galaxies were found in clusters of galaxies. Evidence to date shows that active galaxies do occur in clusters; but whether it is in all, most, some, or just a few is not known for sure. When dealing with the most distant quasars, astronomers

We know our immediate neighborhood rather intimately. With increasing distance, our knowledge fades, and fades rapidly. Eventually, we reach the dim boundary — the utmost limits of our telescopes. There we measure shadows, and we search among ghostly errors of measurement for landmarks that are scarcely more substantial.

Edwin Hubble

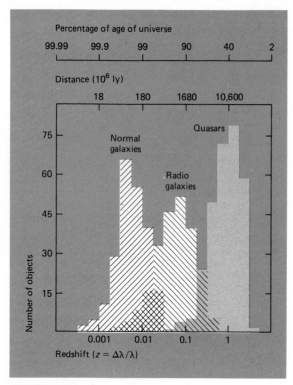

FIGURE 18.15
Histogram for quasars, radio galaxies, and normal galaxies with redshift, a study of 663 normal galaxies, 230 radio galaxies, and 265 quasars showing the range of redshift for each group. The vertical axis for the normal galaxies has been contracted by a factor of three to allow easier comparison with the radio galaxies and quasars. Note that, although there is a range of redshift, the typical redshift for normal galaxies is about 0.003 (1000 km/s), corresponding to a distance of about 50 million light years; that for the radio galaxies, 0.08 (25,000 km/s), or about 1.5 billion light years; and that for the quasars, 1.0 (180,000 km/s), or about 10 billion light years.

are seeing objects so much brighter than normal galaxies that a cluster of galaxies about the quasar would just not be detectable.

Among the nearer active galaxies we find that about 20 percent of all the strong radio galaxies are in rich clusters, many of these being the supergiant elliptical galaxies in the hearts of the clusters. Seyfert galaxies, N galaxies, BL Lacertae objects, and quasars, however, are relatively rare in rich clusters, but some of each type are located in small groups containing presumably normal galaxies. To give a sense of where active galaxies are located in distance, Figure 18.15 shows a histogram of the redshifts for typical samples

of normal galaxies, radio galaxies, and quasars. Characteristically the normal galaxies are nearby, radio galaxies are farther away, and quasars are very distant. Because of the look-back effect, the important point is that each type of object seems to be prevalent at a different time in the history of the universe.

18.4
Superclusters:
Clusters of Clusters

LOCAL SUPERCLUSTER IN VIRGO
As we have just seen, clustering of galaxies seems to be the rule rather than the exception, about three-quarters of all galaxies being in clusters. Thus it is logical to ask whether there is any evidence for clusters of clusters. Within 20 million light years of us there are many small groups much like the Local Group. And about 64 million light years away is the giant Virgo cluster, shown in Figure 18.13. The proximity to the Virgo cluster of some 50 nearby small groups, including the Local Group, and apparently isolated galaxies suggests that they all may form an enormous flattened cluster of clusters of galaxies, known as the *Local Supercluster*. The Supercluster's center appears to coincide with that of the Virgo cluster (Figure 18.16), and its equatorial plane is almost perpendicular to our Galactic plane. Its diameter is about 130 million light years, and its collective mass is estimated to be about $10^{15} M_\odot$. The Local Group, which is near the edge of the Local Supercluster, appears to be revolving around its center at about 400 kilometers per second.

FINDING SUPERCLUSTERS
Although the details of the Local Supercluster are still uncertain, the acceptance of its existence is secure. One means found to be useful for searching for other superclusters and an even larger hierarchy of clumping is to plot the locations on the sky of galaxies that are brighter than a chosen limiting apparent magnitude. Presumably what one should see projected on the plane of the sky are all the galaxies in a certain volume of space centered on our Galaxy.

The most extensive such survey was conducted at Lick Observatory over a 12-year period and includes

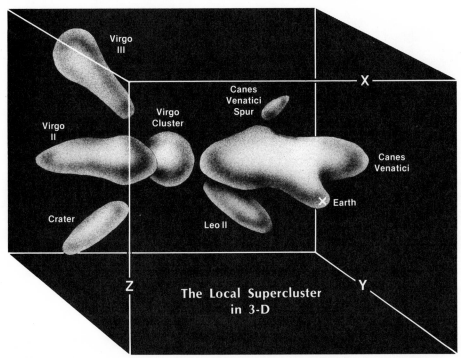

FIGURE 18.16
Local Supercluster. Pictured schematically is the clustering of groups of galaxies and isolated galaxies about the Virgo cluster, the *xy* plane being the equatorial plane of the Local Supercluster. About 80 percent of the galaxies are contained in the Virgo cluster and the five clouds of small groups and isolated galaxies, which tend to point radially toward the Virgo cluster. The Local Supercluster is highly flattened, its diameter being about 130 million light years.

more than 1 million galaxies brighter than nineteenth magnitude. Figure 18.17 is based on that survey and shows a volume of space extending to about 500 million light years, or about five times the distance of the Virgo cluster. Even without distinguishing distances for the different galaxies, the map clearly reveals a clumping of galaxies in the form of knots, clouds, and long filaments. On scales of hundreds of millions of light years galaxies are found in interlocking chains, but even more surprising are the great cosmic voids, up to tens of millions of light years in size, from which galaxies appear to be absent.

To advance beyond the results of Figure 18.17, astronomers must know the distances of galaxies in order to find their true arrangement in space. From the Hubble velocity-distance law of recession, if we know the redshift of a galaxy, we can determine the distance. Before converting redshift to distance, one must know the value of Hubble's constant, a number not so well known. The redshift alone suffices as a measure of distance so that, much as we use Kepler's third law to construct a relative scale for the solar system, we can construct a relative scale for the universe with just the redshift. And by measuring redshifts for all galaxies brighter than a chosen limiting apparent magnitude, we can in principle determine the relative space distribution of galaxies for a volume of space set by the limiting apparent magnitude.

COMA, PERSEUS, AND HERCULES SUPERCLUSTERS
Such studies have been done and reveal the probable existence of several other superclusters. One such supercluster contains the Coma cluster of galaxies. The Coma cluster (about 30 million light years across; see Figure 18.10) is apparently part of a vast complex of galaxies spanning some 200 million light years of space. The method of demonstrating this result is shown in Figure 18.18, in which redshift velocities are plotted against angular position on the sky to more

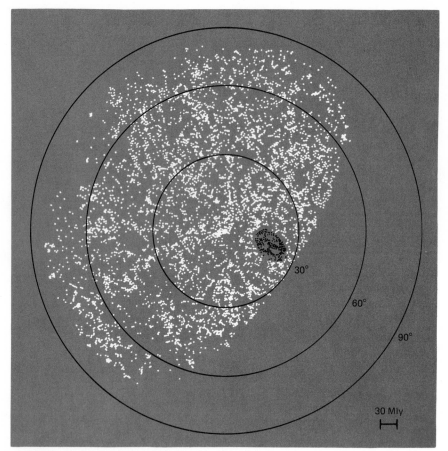

FIGURE 18.17
The distribution of galaxies with apparent magnitudes brighter than 19. These galaxies are typically nearer to us than 300 million light years (cosmological redshifts less than 0.02). The plot is an equal-area polar projection, with the center the north Galactic pole and the outer circle the Galactic equator. The line at the bottom is a distance of 30 million light years, seen from 300 million light years. The region outlined in black is the central region of the Virgo cluster, which is about 64 million light years from us. From the plot it is apparent that on scales of less than 1 billion light years the galaxies are not uniformly distributed but are definitely arranged in clumps.

than 20° from the Coma cluster and all the way to the rich cluster A1367 in the constellation of Leo. It seems apparent that the two rich clusters, Coma and A1367, are members of the same supercluster and that a large *cosmic void* lies in front of it. In a similar fashion such diagrams have been done for regions around the rich clusters Perseus (Figure 18.11) and Hercules (Figure 18.14). Figure 18.19 shows that the Hercules supercluster is about 330 million light years in diameter with the center of the supercluster some 700 million light years distant. The astonishing feature is the immense cosmic void, easily 300 million light years deep, in front of the supercluster. The void is as large as the Hercules supercluster itself!

The largest of the suspected cosmic voids is one lying in the direction of the north Galactic pole at a distance between 520 and 780 million light years. The void, if real, has a volume approaching 30×10^{24} cubic light years. If the evidence can be substantiated, then the existence of such immense realms of space that are apparently empty has the most profound significance for the nature of the universe.

HOW LARGE ARE SUPERCLUSTERS?
The redshift velocities of the Perseus, Coma, and Hercules superclusters tend to group around 5000, 7000, and 11,000 kilometers per second. If, as some astronomers have done, one assumes that galaxies, small groups, or clusters having redshift velocities in the neighborhood of these three values are physically related, then these three superclusters become only part of even larger structures, which approach a substantial fraction of a billion light years in length. And rather than being spherical in shape, they are more like long filaments, with the Coma and Perseus superclusters as dense knots along interlocking chains with vast cosmic voids separating the various strands of the cosmic tapestry (Figure 18.20).

Since the superclusters in Perseus, Coma, and Her-

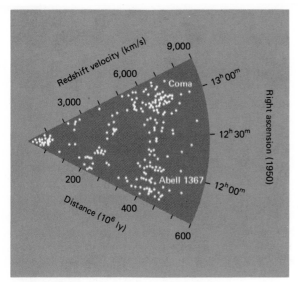

FIGURE 18.18
Coma supercluster. Redshifts are plotted for Coma cluster galaxies with angular position on the sky (right ascension). The two rich clusters, Coma and A1367, appear to be part of the same supercluster, which has a large empty space, or cosmic void, in front of it. The galaxies with redshift velocities of 1000 kilometers per second (in the apex of the triangle) are part of the Local Supercluster.

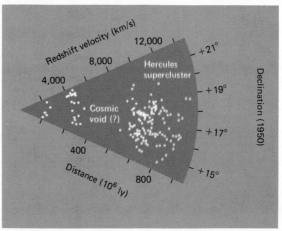

FIGURE 18.19
Hercules supercluster. Redshifts for galaxies in the direction of the Hercules cluster are plotted against their position on the sky in declination. Note the immense cosmic void lying between us (apex of the triangle) and the Hercules supercluster.

FIGURE 18.20
The cosmic tapestry. A schematic presentation of the clumping of clusters of galaxies to form superclusters on scale sizes less than 1 billion light years. Clusters form interweaving filaments of galaxies, the heart of the superclusters being the knots of several intersecting filaments. The spaces between are the great cosmic voids, which do not apparently contain any galaxies. The reality of such spatial distribution of clusters and superclusters is not generally accepted, but the evidence grows.

cules are within 1 billion light years of our Galaxy (cosmological redshifts less than 0.05), they could be a relatively recent feature in the universe. What evidence exists that superclusters were part of the early universe? It has recently been pointed out that several pairs of quasars with nearly equal redshifts lie very close to each other on the sky. This suggests that they are physically associated, and for their observed redshifts they are within 15 to 100 million light years of each other in space. Such separations suggest that the physical association is provided by the fact that they are both members of a supercluster. Thus all quasars could be members of superclusters. Those pairs with redshifts of about 3 are seen as they existed some 15 billion years ago, or from the early history of the universe. Although this is a very indirect argument that superclustering has been a common feature throughout most of the life of the universe, it is still an important consideration in understanding the nature of the universe.

Until more redshifts have been measured for galaxies, it will be difficult to resolve many of the questions surrounding the cosmic design.

SUMMARY

Active galaxies. Active galaxies are ones that may have extremely active nuclei, or are exploding, or have jets of ejected matter, or they may be quasars. They are distinguished by having a continuous spectrum produced by nonthermal synchrotron radiation. They also exhibit some very large cosmological redshifts. Although active galaxies are widely distributed in space and have various structures, they probably all have the same energy source. Seyfert galaxies, N galaxies, and BL Lacertae objects are compact galaxies with unusually bright nuclei, while for radio galaxies the emitting regions are usually large and lay well outside the visible galaxy.

Quasars. Astronomers identified quasistellar radio sources (quasars) when they could precisely locate extragalactic radio sources. On photographic plates, quasars look like ordinary stars, but their large redshifts and variable brightness make them mysterious objects. Because of their great redshift, they must be very distant objects and emit extraordinarily large amounts of radiant energy. They may be active nuclei of distant and faint galaxies, but their apparent small size and variable brightness are still a puzzle. Massive black holes, rotating magnetic bodies, or other such exotic proposals have been suggested as their source of energy. It is possible, however, that the redshift of quasars is not cosmological but rather is a local Doppler effect, and thus they are not so energetic or so distant.

Clusters of galaxies. Galaxies occur in a variety of gravitationally bound systems. Small groups of galaxies (10 or fewer) form loosely packed structures. Rich clusters of galaxies (10^2 to 10^3) have been discovered with more complex structure. They are classified as regular, intermediate, or irregular. Differences in the structure of the cluster may be due to the rate at which the cluster evolves, the irregular clusters changing the slowest. There appears to be not enough mass in clusters to bind them gravitationally for any length of time as evidenced by the difference in their dynamical mass and luminous mass. The "missing" mass may be contained in intergalactic material, subluminous galaxies, or coronae about the visible galaxies. Although not common to rich clusters, active galaxies, even quasars, do occur in clusters of galaxies.

Superclusters. Clustering of galaxies is the rule. There are also clusters of clusters, or superclusters, like the Local Supercluster in Virgo of which our Galaxy and the Local Group are part. The superclusters are like brilliant knots in long filaments of galaxies separated by immense cosmic voids that snake across space intertwining like the pattern of a knitted garment. How far in space or back in time this cosmic design extends is one of the great cosmic mysteries.

KEY TERMS

active galaxy
BL Lacertae object
cosmic void
cosmological redshift
irregular cluster
Local Supercluster
massive black hole
nonthermal source

quasar
radio galaxy
regular cluster
rich cluster of galaxies
Seyfert galaxy
supercluster
synchrotron radiation

CLASS DISCUSSION

1. What is the evidence that quasars are very distant objects? If distant objects, what is the most difficult aspect of quasars to understand? What do quasars imply about the evolution of the universe?

2. Are the active galaxies significantly different from normal galaxies? What are the significant differences? What pecularities do the active galaxies have in common?

3. How common is the phenomenon of the clustering of galaxies? What aspects of clusters are used to classify them? Can small groups of galaxies be classified in the same scheme with rich clusters?

4. Do astronomers find that some members of a cluster of galaxies are active galaxies? In those clusters that have one or more dominant central galaxies is that central galaxy ever an active or peculiar galaxy?

5. What are superclusters? Is our Galaxy part of a supercluster? Describe the structure of superclusters. Are they a common phenomenon throughout the universe?

READING REVIEW

1. What are the distinguishing characteristics of active galaxies?

2. In active galaxies where does the center of activity appear to be?

3. Are all the quasars (quasistellar radio sources) strong radio sources, as their name implies?

4. What is the largest redshift measured for a quasar? Do most quasars have large redshifts?

5. Do all astronomers believe quasar redshifts to be entirely cosmological? What are some of the alternative explanations?

6. Do all clusters of galaxies contain hundreds to thousands of member galaxies?

7. What is the problem of the missing mass in clusters of galaxies?

8. Are all the active galaxies located in clusters of galaxies? If not, are any to be found in clusters?

9. What is the center of the Local Supercluster? How far is it from our Galaxy?

10. How many other superclusters have been identified?

SELECTED READINGS

Blandford, R. D., M.C. Begelman, and M. J. Rees. "Cosmic Jets," *Scientific American*, May 1982.

Chincarini, G., and H. J. Rood. "The Cosmic Tapestry," *Sky & Telescope*, May 1980.

Gregory, S. A., and L. A. Thompson. "Superclusters and Voids in the Distribution of Galaxies," *Scientific American*, March 1982.

Groth, E., P. J. Peebles, M. Seldner, and R. Soneira. "The Clustering of Galaxies," *Scientific American*, May 1977.

Osmer, P. S. "Quasars As Probes of the Distant and Early Universe," *Scientific American*, February 1982

...ught.

...ms to me that if the matter of o[r] Sun
...Universe was eavenly scattered though
...e had an innate gravity towards all the
...wch this matter was scattered was but
...f this space would by its gravity tend
...ide & by consequence fall down to y[e]
...ere compose one great sphærical mass
...diffused through an infinite space, it
...mass but some of it convene into one
...to make an infinite number of great
...s from one another throughout all y[e]
...e Sun & fixt stars be formed supposin
...ure. But how the matter should divide it
...it wch is fit to compose a shining body
...e make a Sun & the rest wch is fit
...graps, like a

> Man is not born to solve the problems of the universe, but to find out where the problems begin, and then to take his stand within the limits of the intelligible.
>
> J. W. von Goethe

With the study of cosmology we have arrived at the largest-scale sizes in the universe. Here we are far beyond the level of stars in galaxies or, for that matter, the galaxies themselves. Even the largest galaxies are at most a few million light years in diameter and can be considered as the first level of cosmic clumping; a second level is the clusters of galaxies, in which the richest are a few tens of millions of light years across. At the end of Chapter 18 we discussed a third level of cosmic clumping, the superclusters, yet another factor of ten larger in the cosmological scheme, on the order of a few hundred million light years across. Galaxies are thus at their largest only a small percentage of the size of superclusters. As we tried to illustrate in Figure 18.20, the superclusters may be part of an even larger organization of interlocking chains that begins to approach a billion light years in size. The cosmological scale is one of billions to tens of billions of light years, where the clusters and superclusters participate in a general cosmic flow known as the *expansion of the universe*. And in this view the superclusters are only a few percent, the clusters a few tenths of a percent, and the largest galaxies a few hundredths of a percent of the cosmological scale. Another way of saying this is that, if the size of the visible universe were to shrink to that of this page, a supercluster would then be about the size of a capital letter and any galaxy would be much smaller than a period.

Are there even larger scales of cosmic clumping, something on the order of billions of light years in size? This is the first part of the cosmological question on the structure of the universe, and the second part of the question is whether the universe has undergone any changes in its billions of years of existence. Cosmology is an attempt to reveal the underlying unity among seemingly diverse phenomena, from subatomic particles to superclusters of galaxies. In this quest astronomers are confronted by an immense

◀ A 1692 letter to Richard Bentley, who later became Master of Trinity College, Cambridge, from Issac Newton.

problem: If light had infinite velocity, then it would bring us information about all parts of the universe instantaneously, but since light has a finite velocity, we observe the universe at different stages in its evolution, depending upon the distance that various parts are from us (that is, the more distant regions are observed earlier in cosmic time than the nearer regions are). The effort in cosmology is basically to develop from a theory for gravity, such as general relativity, and some simplifying principles a unifying theory of the universe, which contains its various observed aspects at different ages.

Before discussing cosmological theories, we ought to examine two observed phenomena that have proved to be of immense importance to our understanding of the large-scale distribution of matter and to our devising a cosmological theory. These phenomena are the cosmic background radiation and the X-ray background radiation.

19.1
Design of the Cosmic Tapestry

COSMIC BACKGROUND RADIATION

We cannot help but speculate that at some time in the past, as a consequence of the observed expansion of the universe, it was a lot smaller and much denser than it is now. In the beginning the universe should have been packed into a hot, superdense state from which the expansion initiated, that process being referred to as the *big bang*. One should not conceive of this big bang as a state in which all of the universe was confined to a small point in space from which it began to expand to fill space. Space does not exist independent of, nor does it precede, the universe. Instead the universe began by filling all space, and it continues to fill all space, except that after the big bang space is expanding. Following the big bang, the universe was filled with an overwhelming abundance of high-

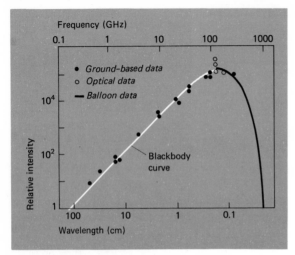

FIGURE 19.1
Cosmic background radiation. The observed distribution of the microwave background radiation closely matches a blackbody energy curve with a temperature of about 3 K. The dots represent ground-based measurements, while the open circles are from interstellar cyanogen (CN) absorption lines observed in the visible spectrum of O-type stars. The most recent data suggest that the cosmic background radiation may be about 10 percent brighter near the peak and 20 percent fainter on the short-wavelength side than a 3-K blackbody-radiation curve is.

energy photons that we now call the *primeval fireball.* Subsequent expansion of the universe cooled this radiation so that today most of its energy lies in the microwave region of the electromagnetic spectrum and in a background sea of neutrinos. In 1934 George Gamow (1904–1968) used such a scenario to predict the existence of a low-energy *cosmic background radiation* that permeates all space.

In 1965 the low-temperature cosmic background radiation was discovered accidentally by physicists Arno Penzias (1933–   ) and Robert Wilson (1936–   ) at the Bell Telephone Laboratory.[1] The cosmic background radiation appears to be coming from all directions in space and to have about the same intensity from all directions. The possible existence of the cosmic background radiation was also, as it happened, being theoretically reinvestigated by Robert Dicke (1916–   ) and his Princeton co-workers. They soon learned that Penzias and Wilson had detected the background radiation. The discovery

[1]For their discovery Penzias and Wilson shared the Nobel Prize in physics in 1978.

of the cosmic background radiation is the most significant cosmological discovery since Hubble showed that the universe is expanding.

Following 15 to 20 billion years of expansion by the universe, the initial high-temperature radiation of the primeval fireball has been cooled so that it now corresponds to blackbody radiation at a temperature only a few degrees above absolute zero. And if the expansion was the same in all directions, the cosmic background radiation should be isotropically distributed— the same coming from all directions. It appears that to a reasonably high degree the cosmic background radiation is indeed observed to be isotropic. The observed spectral distribution of the cosmic background radiation (its photons have a typical wavelength of 1 millimeter) is compared to a 3-K blackbody radiation curve in Figure 19.1. The part of the curve at the peak and toward short wavelengths in the range of 0.13 to 0.03 centimeter was first derived in 1975 from measurements made in a balloon 30 kilometers above the earth. (The short-wavelength side of the peak must be observed from outside the earth's atmosphere because water vapor and oxygen molecules absorb microwave radiation.) Later balloon measurements suggested that the observed spectral energy-distribution curve may not be entirely smooth, as a blackbody curve should be, for reasons as yet unknown.

NEW ETHER-DRIFT EXPERIMENT

The cosmic background radiation can be used as a backdrop for determining the absolute motion of the earth or the solar system or the Galaxy. Unlike Michelson and Morley in their celebrated experiment to detect the motion of the earth relative to ether, the material medium thought at that time to fill all space (see pages 58–59), contemporary astronomers have apparently succeeded in measuring the earth's drift in the sea of cosmic-background-radiation photons.

In the direction of the earth's motion the cosmic background radiation should be slightly hotter, or of shorter wavelength, as a result of the Doppler effect; in the opposite direction, slightly cooler, or of longer wavelength. Consequently there should be a slight departure from isotropy (anisotropy) in the cosmic background radiation due to the earth's motion. What scientists have observed is a minute anisotropy recorded as a temperature difference of approximately 0.0035 K from the average value. The maximum (hottest) and minimum (coolest) values are in the directions of the constellations of Leo and Aquarius, respectively. The data in this experiment reveal that our Milky Way Galaxy and the Local Group of galaxies are traveling at a velocity between 300 and 600 kilometers per second toward a point in the constellation of Hydra some 45° from the center of the Virgo cluster, the center of the Local Supercluster. These are remarkable results in that the isotropic sea of cosmic background photons can provide a backdrop against which motion in the Local Supercluster can apparently be measured.

X-RAY BACKGROUND RADIATION

Early rocket-borne X-ray experiments revealed a diffuse background of X rays coming from every direction in space. This radiation was in addition to the many discrete sources of X rays in our Galaxy, other galaxies, and clusters of galaxies. The uniformity of this radiation over the sky was a strong argument that it could not be from our Galaxy but must be coming from far beyond. Some astronomers have suggested that this background radiation is telling us about an all-pervading gas that contains more mass than all the galaxies put together. The thermal spectrum of the background X rays corresponds to temperatures of several hundred million kelvins for that pervading gas.

There is an opposing view that insists that the X-ray background radiation is just the sum of very many faint unresolved discrete sources, such as quasars. The deep-sky survey by the *Einstein Orbiting Observatory* seems to be finding large numbers of discrete X-ray sources. As a result, it now seems possible to account for at least a third of the diffuse X-ray background by a large number of quasars at cosmological distances. A number of astronomers now believe that with new technology on the horizon it is possible that we shall find that the universe is not pervaded by hot gas but that all the X-ray background is the product of a vast array of quasars spread through distant space.

LARGE-SCALE DISTRIBUTION OF MATTER

The chapter opening photo is a statement by Newton on what can be expected of gravity in shaping the universe. Although there are now additional considerations to the cosmological question of which Newton was not aware, his statement is an argument for an infinite, centerless, and edgeless universe, which can be unstable on a local scale but must be stable and homogeneous on the cosmic scale. Gravity tends to pull matter into clumps, making any irregularities in the universe more irregular. Thus like a box within a box within a box, gravity works on the smallest scales, clumping up matter, and then advances to the next largest scale and the next and so on. Finally, if given long enough and nothing opposes it, gravity could produce the superclusters and the large-scale patterns of the cosmic tapestry (Figure 18.20).

As we learned in Chapter 18, the distribution of matter on sizes of several hundred million light years resembles a highly irregular cosmic net. Galaxies are embedded in superclusters, which appear to be connected with each other and snake their way across space in long filaments. Within these filaments the dynamic units are individual galaxies and small groups of galaxies, up to rich clusters of galaxies. Separating the superclusters are immense cosmic voids apparently as large as the superclusters themselves. This is apparently how matter is distributed within 1 billion or so light years of us, and it is apparently the same regardless of the direction in which we look. Various astronomers have argued that the universe possesses a cellular structure in which the galaxies and clusters of galaxies line the cell walls, the center of the cell being relatively empty space. Overall there could be a rather inhomogeneous distribution for matter, possibly all the way to the largest scales visible in the universe.

Three arguments have been advanced to claim that the universe was homogeneous at least in its earlier days if not now: The first argument is that the radio

There is a single general space, a single vast immensity which we may freely call Void: in it are innumerable globes like this on which we live and grow; this space we declare to be infinite, since neither reason, convenience, sense-perception nor nature assign to it a limit.

Giordano Bruno

galaxies, which are a fair fraction of all galaxies, are spread remarkably uniformly across the sky; they are objects that possess from modest to large redshifts. Secondly the diffuse X-ray background radiation is also uniform across the sky, regardless of whether it comes from an all-pervading hot gas between the galaxies or from discrete objects, like quasars. If this radiation derives from high-redshift quasars, then at an early phase in the life of the universe matter was apparently distributed in a fairly homogeneous fashion. Finally the cosmic background radiation is highly isotropic. Although the cosmic background is the result of the big bang and is not emitted by matter, it confirms that the universe in its earliest moments was homogeneous and has remained so on the largest scale as it expanded.

In 1933 E. A. Milne (1896–1950) noted that assuming that the large-scale distribution of matter was homogeneous had considerable philosphical merit and was consistent with Hubbles's law of recession. Although Milne was not the first to propose a homogeneous universe, he gave the assumption the name of *cosmological principle,* and it has become a basic postulate of cosmology; that is, an article of faith:

COSMOLOGICAL PRINCIPLE: *Since we observe the universe to be isotropic here, all observers anywhere in space should see the universe in its essential features in the the same way in all directions; that is, the universe is isotropic, and hence the universe must be homogeneous so that all places in the universe are alike.*

This assumption says that our sample of the universe, except for local or small-scale variations, is no different at present from another sample selected at random at a different place in the universe. Since most of the cosmological models we shall consider in the next sections utilize general relativity as their theory of gravity, we should restate the meaning of general relativity:

PRINCIPLE OF GENERAL RELATIVITY: *Curved space-time tells matter how to move, and matter tells space-time how to curve.*

A homogeneous and isotropic universe means that any curvature to large-scale space must be uniform; that is, it may not vary from place to place throughout the universe at one time. Also the Hubble velocity-distance law is the same at any time anywhere in the universe. Let us now consider some cosmological models, scenarios for the evolution of the universe.

19.2
Static Cosmological Models

In 1917 Einstein solved the mathematical equations from which he derived the notion of a static, or non-expanding, universe. Einstein's solution was a reasonable approach before cosmological redshifts had been discovered. He assumed that the random motions of the galaxies cancel out, leaving the universe in a static condition, and that the average density of matter spread out over the universe remained constant.

If the universe is static, the scaling factor is in essence the radius of the universe and it does not change with time. What would such a universe look like? The model at which Einstein arrived was a spherically closed universe, with matter thinly spread out, rather than an infinite universe of Euclidean geometry. He pictured the cosmos as finite in extent but centerless and edgeless, just as in two dimensions a sphere's surface is limited in size but does not possess either an edge or a center.

Einstein found that he had to add a slight repulsive force between material particles, which acts over the distances separating galaxies, to keep the universe from collapsing by its own self-gravitation; so he introduced a "cosmological constant" into his equations. He found that the radius of his model universe is inversely proportional to the square root of the mean (spread-out) density of matter. For reasonable estimates of the mean density, which are between 10^{-29} and 10^{-31} grams per cubic centimeter, the radius of Einstein's model universe is between 10 billion and 100 billion light years.

Einstein's model universe is truly static and predicts no cosmological redshifts for galaxies. However, Eddington showed in 1930 that it is also unstable and should either expand or contract if perturbed. Soon after Einstein's work another apparently static model universe was worked out by the Dutch astronomer Willem de Sitter (1872–1934). His model universe, which contained no matter in a Euclidean space, predicted a redshift proportional to distance, and it drew a great deal of attention as redshift data were obtained for galaxies. Retrospectively we know that de Sitter's model was not genuinely static (for reasons that are too complicated to discuss here); it was in reality a forerunner of the nonstatic models.

19.3
Expanding Cosmological Models

Around 1930 Eddington in England and Georges Lemaître (1894–1966), a Belgian, both proposed nonstatic cosmological models. Eddington's universe was basically obtained by perturbing Einstein's static cosmological model so that it would awake from its static sleep and begin an expansion lasting forever, as in de Sitter's universe. On the other hand Lemaître's universe began with a big bang (for which he is often referred to as "the father of the big bang"); it expands for a while, hesitates in a state resembling Einstein's static universe, and then expands a second time that lasts forever. In pursuit of his model universe Lemaître rediscovered the cosmological equations of the Russian cosmologist Alexander Friedmann (1888–1925), whose nonstatic models predicting cosmological redshifts were constructed in the early 1920s but seem to have gone unnoticed.

Nonstatic, or time-dependent, cosmological models can be such that the universe is infinite—either curved or flat—and open or is finite, curved, and closed. Which of a very large number of cosmological models is the correct one? The answer depends on several observable factors, such as the mean density of matter in the universe, the Hubble constant in the velocity-distance law of recession, and cosmological constant.

EXPANDING UNIVERSES

Redshifts had been measured for a number of galaxies even before Hubble discovered in 1929 the velocity-distance law for their recession. Consequently cos-

mologists became more interested in nonstatic, or expanding, models of the universe. The nonstatic-model universes do not require that the cosmological constant be other than zero, but a nonzero value of the constant greatly expands the number of cosmological models possible. The mere proliferation of models has been used to argue that the cosmological constant must be equal to zero; and Einstein is said to have felt that introducing the constant was a great mistake that misled him.

Let us reexamine the Hubble law of recession. We showed in Figure 17.15 that a galaxy's velocity of recession is proportional to its distance; the farther away the galaxy is, the faster it is receding from us because of the expansion of the universe. To help visualize this law on a cosmological scale, consider a two-dimensional analogy: tiny, flat microbes living on an ink spot that serves as their galaxy. Along with other more or less uniformly distributed ink spots (other galaxies) the ink-spot galaxy of the microbes is located on the surface of a balloon that is being inflated (their expanding universe). Suppose that at one time four galaxies, denoted by G_1, G_2, G_3, and G_4, are observed by the microbe astronomers from their home galaxy G. They occupy the positions shown in Figure 19.2a, and their initial distances are those given in the second column of Table 19.1. We assume that the four galaxies are originally confined to a region of space small enough to be considered by the microbe astronomers as essentially flat Euclidean space. Let us assume that after 1 second the balloon has doubled in size (Figure 19.2b) such that the distances between galaxies have doubled, as shown in Table 19.1. In the last column of the table we see that the speed of recession of the four galaxies as observed from the home galaxy is proportional to their distance. This is the Hubble velocity-distance law of recession, and it is the same from whichever galaxy we consider it. A proportionality between relative velocity and distance is to be expected in any theory that satisfies the cosmological principle.

If, as the balloon continues to expand, the microbe astronomers measure the velocities of the very remote galaxies, they might discover that the straight-line relation between velocity and distance no longer holds, indicating that the Hubble constant was different in the past from its present value. Suppose the little astronomers find that for the most distant galaxies velocity increases faster than does distance; thus

TABLE 19.1
Recessional Motion of Two-dimensional Galaxies

Galaxy	Initial Distance from Home Galaxy G (cm)	Distance from Home Galaxy 1 Second Later (cm)	Speed of Recession (cm/s)
G_1	1	2	1
G_2	2	4	2
G_3	3	6	3
G_4	4	8	4

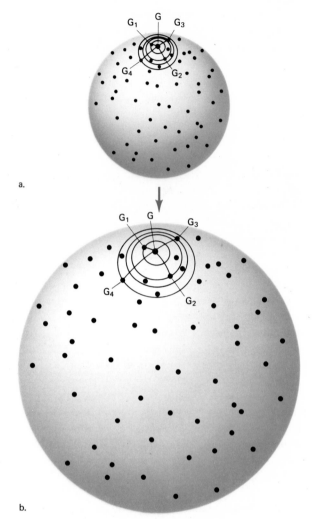

a.

b.

FIGURE 19.2
Hubble's velocity-distance law for the galaxies on a two-dimensional spherical surface. (a) A partially inflated balloon has a random distribution of spots (galaxies) on its surface at a certain time. (b) The same balloon after inflation shows the apparent distribution of spots (galaxies) 1 second later. The analogy between the surface of the balloon and actual receding galaxies is inexact in that, although the space between the galaxies stretches, the galaxies themselves do not expand.

the Hubble constant was larger in the past. This would be an indication that the flat, two-dimensional Euclidean universe they see locally is part of a larger curved space.

CURVED SPACES

In our three-dimensional space we are familiar with two-dimensional surfaces, such as the imaginary flat world on the surface of a balloon. Such two-dimensional surfaces can be used to illustrate the types of space that are uniform everywhere—that is, homogeneous and isotropic—and that the universe might exhibit on the largest scale. Flat *Euclidean space* can be illustrated in two dimensions by a plane extending to infinity in all directions (Figure 19.3a); it exhibits *no curvature,* throughout all the universe. It is the type of space of our everyday experience. A second type of space is known as *spherical space,* for which the surface of a sphere is a two-dimensional illustration (Figure 19.3b). Spherical space exhibits a global curvature that is referred to as *positive,* while locally it may strongly resemble flat Euclidean space. The last type of uniform space is called *hyperbolic space,* which possesses a global curvature known as *negative* but which locally may also be Euclidean. The two-dimensional illustration of hyperbolic space is a saddle-shaped surface (Figure 19.3c) that is uniform locally but not in a global sense. In that regard it is not a good representation for a hyperbolic universe. As pointed out in Chapter 3, space-time is four-dimensional and its geometry may be Euclidean, spherical, or hyperbolic on a global scale as well as locally.

The three choices for the large-scale structure of the universe are:

1. The universe may possess zero curvature globally; thus it is described by a flat Euclidean geometry, the space geometry most familiar to us, and it is infinite in extent.

2. The universe may possess positive curvature globally; thus it is described by a spherical ge-

ometry, like the two-dimensional surface of a sphere, and it is finite in extent but unbounded.

3. The universe may possess negative curvature globally; thus it is described by a hyperbolic geometry, like the two-dimentional geometry of a saddle-shaped surface, and it is infinite in extent.

FRIEDMANN UNIVERSES

Beginning in 1922 with the work of the Russian cosmologist Alexander Friedmann, most modern cosmologies are based on nonstatic solutions of Einstein's field equations of general relativity and the mathematical formulation developed by Friedmann, which was rediscovered later by Lemaître. When the cosmological constant is set equal to zero, three types of cosmological model, known as *Friedmann universes*, are possible. The three types of Friedmann model are based on the effect of the average density of matter in the universe on the universe's expansion. This effect arises from the mutual gravitational attraction of matter, which tends to slow, or decelerate, the expansion of the universe.

If the average density is *less than* a certain critical value, then there will be an insufficient amount of matter to stop the expansion, and the universe will expand forever. Thus the universe is an open universe of negative curvature and therefore is infinite in extent (curve *A* in Figure 19.4); the geometry of space is hyperbolic. The meaning of the critical density is that, if the average density is *equal to* the critical value, there is enough mass in the universe to halt the expansion but not before it has become infinite in extent. Such a universe is a marginally open one, for which the geometry is flat Euclidean space of zero curvature (curve *B* in Figure 19.4). Finally, if the average density is *greater than* the critical value, then the gravitational attraction produced by matter in the universe is sufficient to halt the expansion at a finite size and to cause the universe to collapse. It is a closed universe of spherical geometry with positive curvature (curve *C* in Figure 19.4).

In Friedmann universes the value of the Hubble constant decreases with cosmic time at different rates, depending on the model, as the expansion is slowed by self-gravitation of the universe. The critical density is proportional to the square of the Hubble constant, and for a present value of about 15 kilometers per second per million light years the critical density is about 5×10^{-30} gram per cubic centimeter. This corresponds to about three atoms per cubic meter. In Figure 19.4 all the models have the same typical separation between galaxies and rate of expansion at the present epoch, as shown by the point at which the

GEOMETRY OF THE UNIVERSE

The geometries of interest in cosmology are those that are uniform everywhere and hence satisfy the cosmological principle. Locally they can resemble Euclidean geometry but on a global scale be quite different.

Euclidean geometry postulates that through any given point one and only one line can be drawn that will never intersect a given line (that is, a parallel line) as in Figure 19.3a, which defines a *flat space*. Other assumptions are possible that also produce a uniform space: For example no line can be drawn through the external point that will not intersect the given line; this situation defines a *spherical space*. Or through any point not on a given line any number of lines can be drawn that will never intersect the given line, which is the definiton for *hyperbolic space*.

Spherical space possesses what is referred to as *positive curvature*, as in the two-dimensional example of the surface of a sphere, where "straight" lines are arcs of great circles. Spherical space is centerless and edgeless even though it is finite in extent. In spherical space a beam of light sent off in one direction will eventually return to you from the opposite direction. Of an opposite nature, hyperbolic space possesses *negative curvature*, as on a two-dimensional saddle-shaped surface, where "straight" lines are arcs of hyperbolae. Hyperbolic space is of infinite extent and a beam of light sent off in some direction will never return. Both spherical and hyperbolic space contrast globally with Euclidean space, in which straight lines possess *no*, or *zero*, curvature. Like hyperbolic space, Euclidean space is infinite in extent so that a light beam will never return if sent off in some direction.

a. TWO-DIMENSIONAL FLAT SPACE, NO CURVATURE

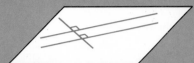

Lines that start
off parallel
remain parallel

Sum of angles of a triangle $= 180°$
Circumference of a circle $= 2\pi r$
Area of a circle $= \pi r^2$

To make a flat map of
the surface of a plane,
we need make no
alternations since there
are no distortions of space.

b. TWO-DIMENSIONAL SPHERICAL SPACE, POSITIVE CURVATURE

Lines that start
off parallel converge
and ultimately intersect

Sum of angles of a triangle $> 180°$
Circumference of a circle $< 2\pi r$
Area of a circle $< \pi r^2$

To make a flat map of
the surface of the sphere,
we must stretch space
away from us since
distances increase less
rapidly.

c. TWO-DIMENSIONAL HYPERBOLIC SPACE, NEGATIVE CURVATURE

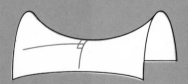

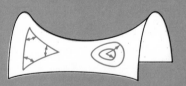

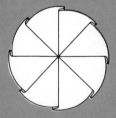

Lines that start
off parallel diverge

Sum of angles of a triangle $< 180°$
Circumference of a circle $> 2\pi r$
Area of a circle $> \pi r^2$

To make a flat map
of the surface of
hyperbolic space, we
must shrink space
away from us since
distances increase
more rapidly.

FIGURE 19.3

Two-dimensional world geometries demonstrate some of the properties of curved uniform space. Unfortunately, the two-dimensional saddle-shaped surface as a representation of hyperbolic space is not uniform everywhere as are the infinite plane and the surface of a sphere. Therefore, the saddle-shaped surface is homogeneous and isotropic only locally, but with that restriction it serves as a good two-dimensional illustration.

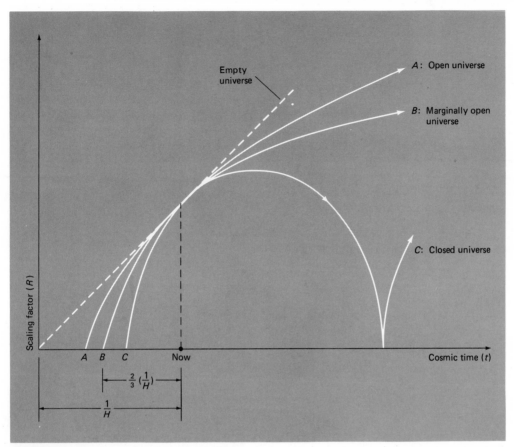

FIGURE 19.4
Big-bang cosmological models, showing how the typical separation between galaxies varies with cosmic time. In the case of the open universe (A), the mean density of matter is less than the critical velocity so that the universe expands forever. Thus the universe is infinite with negative curvature. For the closed universe (C), the mean density of matter is greater than the critical so that the universe stops expanding at some finite size, the space-time curvature is positive, and expansion is followed by contraction. For the marginally open universe (B), the mean density is equal to the critical value for the expansion to slow such that it comes to rest in the infinite future. The geometry for this model is a flat Euclidean one. For comparison an empty universe is shown; for example, one in which there is no matter and the expansion is constant. All the models are made to agree for the present epoch, marked "Now." If the universe had always expanded at the same rate, then the empty-universe model shows that the expansion began at $1/H$ in the past. If H equals 15 kilometers per second per million light years, then the age of the universe is 20 billion years.

curves coincide, labeled "Now." If the universe had a rate of expansion that was the same for all cosmic time, it would have begun about 20 billion years ago for H equal to 15 kilometers per second per million light years. However, if the mean density of the universe is just the critical value necessary to stop the expansion at infinity, then the age of the universe is more like 13 billion years. An open universe would of course be older than this and a closed one younger.

OPEN FRIEDMANN UNIVERSE:
NOT ENOUGH MATTER

In an *open* universe of negatively curved space-time expansion settles down to increase proportionately to time ($R \propto t$). There is but one expansion, beginning supposedly with a big bang in a superdense core of superhot matter. The amount of matter in this Friedmann universe is not sufficient to halt the expansion by self-gravitation; the expansion proceeds inexorably toward an infinite value for the scaling factor. While space stretches out with the expansion, the average density of matter decreases as the distances between the galaxies increase. The open Friedmann universe is summarized in Table 19.2.

Light arriving from the most distant objects will just manage to reach us and will be redshifted almost beyond perception. This outermost limit of observability is our *horizon*. It is not an edge to space; the universe no more has an edge than it has a center. Light is the fastest means of transporting information in the universe; but we can be affected only by events that are so close to us that light has had enough time to reach us since the beginning of the universe. If objects are at a distance that requires a greater length of time than the age of the universe for their light to reach us, then they are beyond the horizon. Figure 19.5 uses a space-time diagram to illustrate this concept. The distance of the horizon is of the order of the velocity of light multiplied by the age of the universe, but the size of the universe can be shown from Friedmann's cosmological models to be proportional to the one-half or two-thirds power of the age. If we go back in time, the distance to the horizon shrinks. Consequently the

AGE OF THE UNIVERSE

The reciprocal of the Hubble constant can be used to estimate an age for the universe, assuming a uniform expansion over its lifetime. Since the relation between distance, velocity, and time is

$$r = v \cdot t, \quad \text{or} \quad t = \frac{r}{v},$$

for nonaccelerated motion, then by analogy from the Hubble law of recession (see page 428)

$$v = H \cdot r, \quad \text{or} \quad \frac{1}{H} = \frac{r}{v}.$$

Thus the reciprocal of the Hubble constant has units of time, and it is known as the *Hubble time*. For a value of H equal to 15 kilometers per second per million light year (or 50 kilometers per second per million parsec) the Hubble time is

$$\frac{1}{H} = 20 \times 10^9 \text{ yr},$$

and for H equal to 18 kilometers per second per million light year, or 60 kilometers per second per million parsec, the Hubble time is

$$\frac{1}{H} = 16 \times 10^9 \text{ yr}.$$

These values are compatible with the ages of the oldest globular clusters in our Galaxy (see Table 12.7) of 10 to 20 billion years. If the expansion has slowed with time, then the universe must be younger than the above values. For the marginally open model, which has a flat Euclidean geometry, the age of the universe is

$$t_{\text{Euclid}} = \tfrac{2}{3} \frac{1}{H},$$

which for H equal to 15 kilometers per second per million light year yields

$$t_{\text{Euclid}} \approx 13 \times 10^9 \text{ yrs},$$

and for H equal to 18 kilometers per second per million light year

$$t_{\text{Euclid}} \approx 11 \times 10^9 \text{ yrs}.$$

The age of an open universe lies between the Hubble time and the age of the marginally open universe, while the age of the closed universe is less than that of the marginally open universe.

> It is fairly certain that our space is finite though unbounded. Infinite space is simply a scandal to human thought.
>
> Bishop Barnes

horizon encloses a smaller and smaller portion of the universe as we approach its beginning. This means that for objects at large redshifts the horizon is proportionally less of the universe, or there are objects whose separation from our position at the beginning is such that they are not yet inside the horizon.

MARGINALLY OPEN FRIEDMANN UNIVERSE: JUST ENOUGH MATTER

The scenario for the Friedmann type of universe that is *marginally open*, or a Euclidean universe of zero curvature (also known as the Einstein–de Sitter universe), starts with a big bang. The expansion is proportional to the two-thirds power of time ($R \propto t^{2/3}$). This Friedmann universe is therefore expanding more slowly than is an open universe. Self-gravitation of the universe causes the expansion to cease at some time in the infinite future, when the universe has an infinite extent, as the average density falls from an infinite density when time began to zero at the end. The marginally open Friedmann universe is also summarized in Table 19.2.

CLOSED FRIEDMANN UNIVERSE: MORE THAN ENOUGH MATTER

In a *closed* universe of positively curved space there exists a finite amount of matter, a finite limit to space, and a finite time for expansion and collapse. As the universe's expansion is slowed by self-gravitation, it eventually reaches a maximum size, depending on how much larger the average density of matter is than the critical value at any time. Contraction takes over, slowly at first and then accelerating toward a spectacular climax, called the *big crunch,* as all matter collapses toward a superheated, superdense state. The greater the average density of matter, the less time it should take to reach this state. Some cosmologists have suggested that the universe would then rebound into a new cycle and that the recycled universe need not have the same physical details as the previous one. The universe could have different physical and chemical properties, in which only the fundamental constants of nature—the velocity of light, the gravitational constant, and the Planck constant of radiation—might remain unchanged; or it might begin anew with no remembrance of its past cycle of

TABLE 19.2
Friedmann Universes

Friedmann Universe[a]	Beginning	Curvature of Space	Geometry of Space	Present Mean Density[b] (g/cm^3)	Age of Universe[b] (yr)	Fate	Spatial Extent
Open	Big bang	Negative	Hyperbolic	$<5 \times 10^{-30}$	$>13 \times 10^9$	Expands forever	Infinite
Marginally open	Big bang	Zero	Flat Euclidean	5×10^{-30}	13×10^9	Expands forever	Infinite
Closed	Big bang	Positive	Spherical	$>5 \times 10^{-30}$	$<13 \times 10^9$	Expands, stops, and contracts	Finite

[a] Friedmann model with cosmological constant equal to zero.
[b] For H_0 = 15 km/s/Mly or 50 km/s/Mpc.

expansion and contraction in such things as fundamental constants and physical laws (see summary in Table 19.2).

19.4 Alternatives to a Big-Bang Cosmology

STEADY-STATE COSMOLOGY

In big-bang cosmologies, if the cosmological constant is zero, the age of the universe cannot exceed $1/H$, which at the time Hubble first measured it corresponded to an age of 1.8 billion years. When it became apparent that the earth was at least 4.5 billion years old and star clusters were on the order of billions of years old, it was clear that something was wrong in cosmology. In 1948 Herman Bondi (1919–   ), Thomas Gold (1920–   ), and Fred Hoyle (1915–   ) sought an alternative to the big-bang cosmology that was not based on Einstein's general relativity. They devised a nonstatic model of the universe, whose general appearance remains unaltered forever; that is, there is no beginning and no ending. This is the *steady-state universe,* which is not static but allows for expansion as indicated by the cosmological redshifts (Figure 19.6). To guarantee the unchanging laws of physics for all time, Bondi and Gold extended the cosmological principle to even more stringent conditions:

PERFECT COSMOLOGICAL PRINCIPLE: *Not only does the universe appear the same to all observers everywhere, but it looks the same to them at all times.*

The steady-state model has no singularity, no beginning, and no end. Space expands exponentially with time toward infinity. The Hubble constant does not vary with time as in the evolving models, where it decreases with time. Galaxies form, evolve, and disappear, while the average density of matter in space remains constant. To keep the population of galaxies, or the mean density of matter, constant, we have to assume that new matter—hydrogen—is continuously being created. This creation compensates for the expansion that thins out matter. The average rate of creation in a large classroom would be about one hydro-

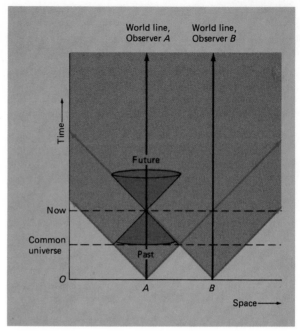

FIGURE 19.5
Space-time diagram for observer *A* and observer *B.* Time is plotted vertically and space horizontally on scales such that a light ray travels along a 45° line. For velocities less than the velocity of light more time elapses (farther up the diagram) than distance traversed (across the diagram) as compared to light. Thus velocities less than light are confined to the shaded cones about the world lines for *A* and *B,* or these are the horizons of *A* and *B.* For *A* information about a *B* event's having occurred at the origin (*O*) enters his horizon at "Now" since the separation of *A* and *B* is such that it took the time interval zero to now for light to bring him the information. At the point marked "Common universe" *A* and *B* begin to observe common parts of the universe, and only after now do *A* and *B* enter each other's horizon. Finally, since the universe is expanding, the world lines for *A* and *B* are diverging as time advances (up in the diagram).

gen atom in 50 million years (2.8×10^{-46} gram per cubic centimeter per second), a rate hopelessly beyond detection. Because galaxies are being formed at a steady rate to keep pace with the expansion, the average separation between them remains unchanged with the passage of time.

The steady-state model in its original form now seems inconsistent with the observed properties of the universe. The counts of several thousand radio sources and quasars at great distances reveal that these remote sources seem more numerous in the past (that is, their numbers increase with distance) than they are today. This result contradicts the steady-

the universe. The cosmic background radiation is a logical consequence of the big bang, and it cannot be reconciled with the steady-state theory because of the steady-state precept that the universe was never in a superdense condition, as required in big-bang cosmologies, but has always been pretty much the same as it is today.

NONEXPANSION INTERPRETATIONS OF THE REDSHIFT

Several cosmologies have been proposed that offer an interpretation of galactic redshifts other than that involving expansion of the universe. These are known as *nonexpansion cosmologies*. As we pointed out earlier, the redshift is so fundamental to cosmological thought that there is always the nagging, but remote, possibility that it means something different from expansion of the universe. One of the first alternatives proposed to the expansion interpretation was the *tired-light hypothesis*. In this cosmology photons lose energy, while traveling from distant galaxies, by an amount that is proportionate to the length of the path they have traversed through the universe. Thus their wavelengths are lengthened proportionate to the distance of the source, giving the impression of a cosmological velocity of expansion.

Although astronomers are reasonably certain that no known physical processes could produce the tired-light effect across the entire electromagnetic spectrum, there is always the argument that this is a "new law of physics" operating on a cosmological scale. Historically an argument on a new law of physics has been difficult to address since objections raised to it have often been dismissed by proponents' adding new wrinkles to it to meet the objections and making more bizarre a contention that is already contrary to collective scientific experience. Ultimately the only convincing counterargument has been to raise enough legitimate objections that the so-called new law of physics becomes self-contradictory and dies from lack of adherents. The tired-light hypothesis has been with us for some time, but there are very few astronomers who believe that it offers real hope of enhancing cosmological thought.

Although we could go on for many pages to review alternative interpretations of one sort or another, we shall stop with one more, which is based on observation and thus able to command a great deal of attention. It is the apparent association of high-redshift

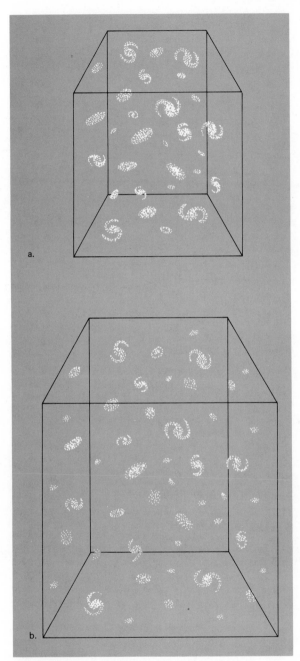

FIGURE 19.6
The steady-state cosmological model in which new matter is created in (b) to keep the mean density constant as the universe expands from (a).

state tenet, which requires that the counts remain the same in all regions of space.

The most telling argument in favor of the evolutionary models is the presence of the low-temperature, isotropic, cosmic background radiation filling

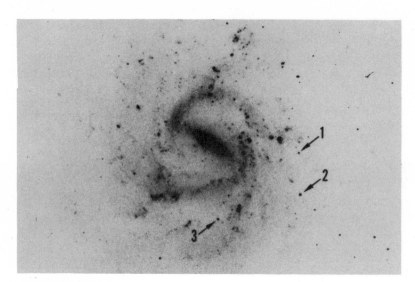

FIGURE 19.7
Are these objects of different red-shifts associated? This negative photograph of the spiral galaxy NGC 1073 shows three quasars that are less than 2 minutes of arc from the galaxy's center.

quasars with low-redshift galaxies mentioned in Chapter 18 (see pages 451–452). Figure 19.7 shows an example of what appear to be related extragalactic objects with different redshifts from the work of Halton Arp (1927–   ), of the Hale Observatories. The compelling point of this argument is not just the close proximity of the quasar to low-redshift galaxies but that in most of the examples the associated galaxy or galaxies are peculiar in some regard. A number of cases exist in which it almost appears as if several quasars and small peculiar galaxies have been ejected from a giant peculiar galaxy, with all of them having different redshifts. Thus a new proposal in cosmology is that only a portion of the observed redshift is due to cosmological expansion. Most of the redshift for active and peculiar galaxies is an intrinsic property of the galaxy, possibly associated with its compact structure. The majority of astronomers have not accepted this contention, but there is a reasonable body of observational material that is suggestive if not convincing.

19.5
Tests of Cosmological Models

MEAN DENSITY OF MATTER IN THE UNIVERSE

Faced with at least three different cosmological models, the Friedmann universes, that begin with a big bang and a number that do not, how do astronomers set about deciding which world model is most nearly correct? Although there are a number of tests, there are four widely recognized, or classic, tests of the various cosmological models: mean density of matter, luminosity distances versus redshifts, number counts of distant radio sources, and angular size. We begin our discussion with the most obvious, the mean density of matter in the universe.

We pointed out in the discussion of the Friedmann cosmological models that the primary discriminator between the open and closed models was the total mass of matter in the universe. If there is sufficient mass, its self-gravitation will eventually stop the expansion of the universe at a finite size and initiate a contraction. On the other hand, if there is insufficient mass, the universe's expansion will continue into the infinite future. Another way of discussing total mass is its mean density.

As stated earlier, one means of determining the critical value of the mean density is through the Hubble constant since in the Friedmann cosmological models the critical value is proportional to the square of the Hubble constant. Astronomers believe that the Hubble constant lies between 15 and 30 kilometers per second per million light years so that the critical mean density lies between 5×10^{-30} and 14×10^{-30} gram per cubic centimeter, the more widely accepted value being about 5×10^{-30} gram per cubic centimeter, or about three atoms per cubic meter.

To measure the mean density directly, astronomers must select volumes of space comparable to clusters of galaxies or larger. The values obtained by various studies of the mean density of matter are about 5 to 10 percent of the critical value or about six-hundredths to two-tenths of an atom per cubic meter (that is, 2×10^{-31} to 5×10^{-31} grams per cubic centimeter). Thus our universe should be an open, or hyperbolic, universe that will expand forever.

However, as was discussed in the problem of the missing mass in clusters of galaxies, these estimates are based primarily on luminous matter and do not account for nonluminous matter. The apparent stability of the clusters of galaxies suggests that there is 10 times more matter present than is visible. This is tantalizingly close to the amount needed to close the universe. The most perplexing problem is the form in which this additional matter resides since it represents about 90 percent of all the matter in the universe. X-ray observations of clusters reveal hot gas lying outside the component galaxies that may amount to as much matter as is visible in the galaxies. Although this is a step in the right direction, it is far short of the required amount.

One of the more thought-provoking suggestions for the missing mass is that it is supplied by neutrinos, which have masses 100,000 times or so smaller than electrons but fill the universe in incredible numbers. Although when we discussed nuclear burning in stars we stated (see page 321) that neutrinos are massless and travel at the speed of light, let us suppose that they do have mass, one too small to measure previously. A consequence of neutrinos' having mass is that they might decay into some other kind of elementary particle, possibly accompanied by emission of high-energy photons, such as X rays or gamma rays. Numerous laboratory experiments as well as astronomical observations are in progress to look for the decay and the time scale over which it might occur. The prize to be gained in this search may be as much as 10 to 100 times more mass in the universe than we now estimate, or enough to close it. If the additional mass simply does not exist or is only a few times the amount of visible matter, then astronomers can only conclude that the universe is open.

MEASURING THE GEOMETRY OF SPACE-TIME: LUMINOSITY DISTANCES

A distinctly different approach to testing which world model is more nearly correct is efforts to measure the large-scale geometry of space. Over distances comparable to the sizes of clusters of galaxies the curvature of space-time is negligible, except for local deformations caused by black holes. It is not until we reach the cosmological scale of distances that the effect of the curvature of space-time becomes appreciable. Thus as astronomers observe more and more distant galaxies, space-time curvature has a greater effect on distance measurements.

An equation that permits us to measure space-time curvature by observation can be derived from Einstein's field equations, which connect the bolometric apparent magnitudes (see Section 12.2) of galaxies with their redshift z, or $\Delta\lambda/\lambda$. The connection assumes that galaxies maintain an approximately constant bolometric absolute magnitude throughout their lifespan—possibly a questionable assumption.

What we are is in part only of our making: The greater part of ourselves has come down to us from the past. What we know and what we think is not a new fountain gushing fresh from the barren rock of the unknown at the stroke of the rod of our own intellect: It is a stream which flows by us and through us, fed by the far off rivulets of long ago. As what we think and say today will mingle with and shape the thoughts of Man in the years to come, so in the opinions and views we are proud to hold today we may be looking back, trace the influence of the thoughts of those who have gone before.

Sir Michael Foster

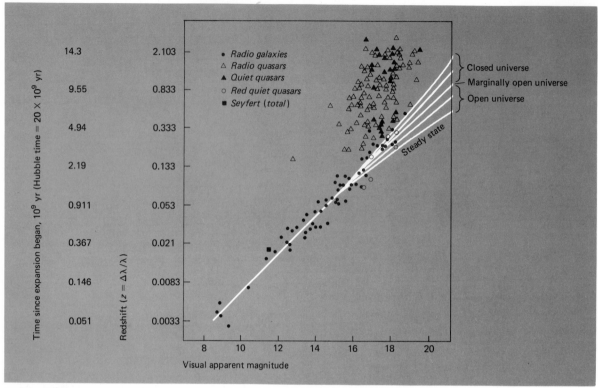

FIGURE 19.8
Hubble diagram for galaxies and quasars. The curves show the values predicted by the different cosmological models. The steady-state model is shown for comparison with the three big-bang models, the closed universe, the marginally open universe, and the open universe. The radio quasars and radio-quiet quasars are too bright compared with the radio galaxies of the same redshift.

Graphically the equation is a plot of redshift ($\Delta\lambda/\lambda$) against visual apparent magnitude (see Figure 19.8) instead of a velocity-against-distance plot, as in the original Hubble diagram for the law of recession (Figure 17.15). If we assume that remote galaxies with similar characteristics have about the same luminosity, then apparent magnitude is a measure of distance, and it is relatively easy to measure.

For several values of the mean density of matter the equation is plotted in Figure 19.8. For a positively curved space (a closed universe), Figure 19.3 shows that distances increase less rapidly as we move farther into space. As a result, galaxies at a given redshift, or recessional velocity, would have smaller apparent distances than they would in a flat space. This can be seen in the case of the surface of a sphere, which is a two-dimensional illustration. To make a flat map of the surface of a sphere, one must draw points together or reduce the distance. For a negatively curved space (an

open universe), the opposite effect occurs, as Figure 19.3 shows. As a result, galaxies at great distances, or a given redshift, should have larger apparent distances than they do in a flat space. Consequently in Figure 19.8 the theoretical curve for the closed universe bends up, showing positive-curvature geometry, while the open universe bends down at large apparent visual magnitudes, revealing negative-curvature geometry.

Once astronomers can fit sufficiently reliable observational data to one of the theoretical curves in the plot of Figure 19.8, the type of world model follows. Currently the observational data agree with the theoretically calculated slope of the curve along the beginning portions. Farther along the various curves, however, it is not possible to differentiate unambiguously between the models by departures from linearity. Although the quasars have about the same slope as the galaxies, they do not conform to the plot of the galaxies. They are much too bright for their measured redshifts. If consideration is restricted to just the brightest galaxies in clusters, the results are more promising, as shown in Figure 19.9. The scatter in the points is still large although somewhat suggestive of a closed universe.

The possibility has been raised that galaxies may become intrinsically fainter or brighter as they age. They should become fainter as they use up their interstellar matter and form fewer massive blue stars. Galaxies should become brighter with age if they are massive enough to cannibalize matter in smaller galaxies. Because of the look-back effect, galaxies in the diagram are younger as one goes to larger apparent magnitudes (that is, more distant galaxies). Thus it may be galactic evolution, not the curvature of space, that astronomers are measuring.

NUMBER COUNTS OF RADIO SOURCES

Another approach to measuring the curvature of space is illustrated in Figure 19.10. If galaxies or clusters of galaxies or radio galaxies are distributed at random, then by making a flat map of space we should find that the numbers of galaxies are evenly distributed over the map if space is flat. For a positively curved universe, there will appear to be more nearby galaxies than distant ones, while there will appear to be more distant than nearby galaxies if the universe is negatively curved. The number count for radio galaxies, using radio apparent brightness along with theoretical predictions for open and closed cosmological models suggests that there are more of them at great distances than even the open-universe model predicts. Yet it is hard to infer from the number counts that our universe is anything but an open one.

FIGURE 19.9
Hubble diagram for the brightest galaxies in 103 clusters. These galaxies suggest, somewhat weakly, that space is positively curved, or the universe is closed.

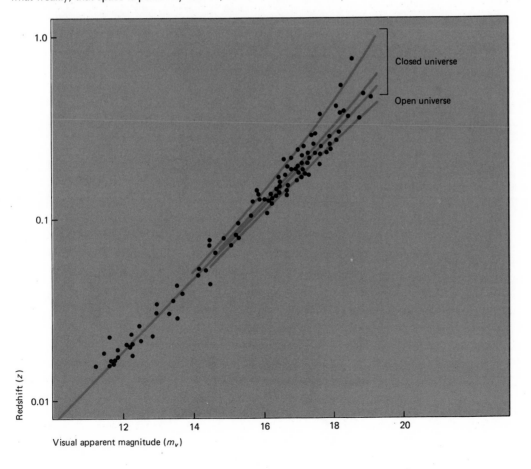

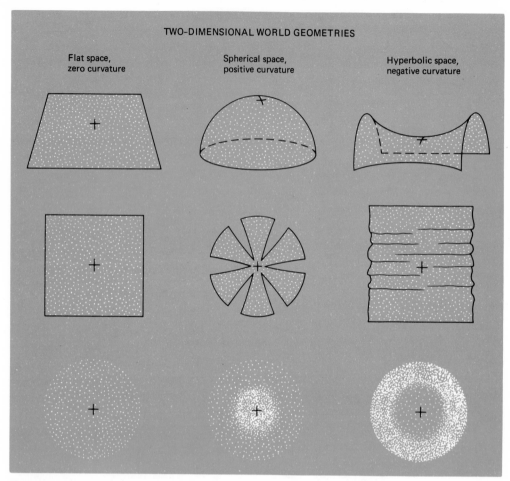

FIGURE 19.10
Galaxy counts as a measure of the curvature of space. If galaxies (or clusters of galaxies) are distributed at random throughout space and flat maps are made of their distribution, then for a flat space we find that the points are evenly distributed over the paper. If space is positively curved, then the plotted positions seem to show more galaxies nearby than far away. Finally, if space is negatively curved, there appear to be more galaxies at great distances than nearby.

ANGULAR-SIZE DISTANCES

Finally another measure of the curvature of space is the variation of angular size with redshift. Going back to our two-dimensional illustrations of spatial geometry (Figure 19.11), we are aware that in everyday experience in flat Euclidean space the angular diameter declines as distances of objects, such as clusters of galaxies having the same linear diameter, grow larger. In mathematical terminology angular diameter is inversely proportional to distance. On the surface of a sphere angular diameter declines with distance up to the equator for a microbe astronomer at the pole, but

beyond the equator it increases. For positive curvature of a spherical universe the angular diameter is larger at large redshift than it would be for a Euclidean space with zero curvature. The opposite is the case for a hyperbolic universe with negative curvature, where the angular diameter is smaller than that for Euclidean geometry. Angular diameters have been estimated for the core regions of rich clusters of galaxies possessing different redshifts. Since nearby rich clusters appear to have the same-size core region, if we assume that this is the case for all rich clusters, even those at great distance, Figure 19.12 suggests that the universe is

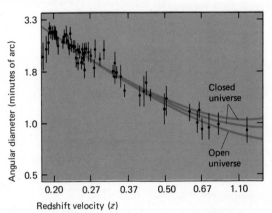

FIGURE 19.12
Diagram of angular diameter versus redshift for clusters of galaxies.

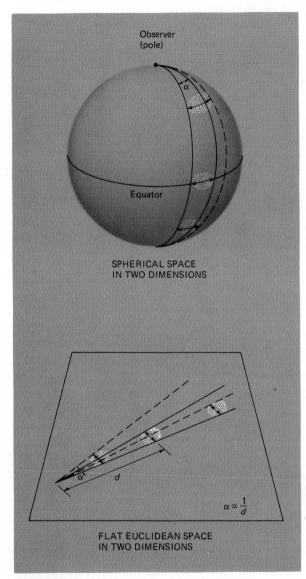

FIGURE 19.11
Angular diameter as measured on the surface of a two-dimensional spherical universe. Assume that the observer is located at the pole of the sphere and observes at the equator a cluster of galaxies subtending the angle defined by the solid lines (meridians of longitude). If that cluster were closer to the observer, the actual diameter of the cluster would lie between the solid line on the left and the dashed line on the right; that is, the angular diameter would be large.

open. But we must be cautious in assuming no evolutionary effects for clusters of galaxies when there are many reasons to believe that clusters evolve.

19.6
The Early Universe

From what astronomers now know about the universe it certainly appears that cosmological models based on the expansion of the universe from a big bang are correct. There is yet no definitive answer to the question of whether the universe is open or closed. If we believe that the big bang took place, what was it like?

BIRTH OF THE UNIVERSE

The initial state of the universe almost defies comprehension and analysis. Supposedly it began as unimaginably chaotic and superdense, with a high temperature to match (greater than 1500 billion K), at a time about 15 to 20 billion years ago, when the fireball erupted. In the first one-hundredth of a second the constituents may have been the quarks, which are thought to be the building blocks of subatomic particles like protons and neutrons. Thereafter the superdense, hot cosmic fluid was a mixture of protons, neutrons, photons, electrons, neutrinos, and others. Matter and antimatter were almost evenly divided, with matter somewhat more plentiful than antimatter by one part in a billion. If the quantities of matter and antimatter had been equal at the start, some cosmologists suggest matter would have been almost

> The evolution of the world can be compared to a display of fireworks that has just ended; some few red wisps, ashes, and smoke. Standing on a cooled cinder, we see the slow fading of the suns, and we try to recall the vanished brilliance of the origin of the worlds.
>
> Georges Lemaître

completely annihilated and the universe would have consisted of radiation and very little matter, far less than is still around. As far as we know now, nothing says that the universe is not fragmented into islands of matter and antimatter, but there are no compelling arguments for that arrangement either.

At nearly the instant when time began, the fireball erupted from the annihilation and conversion of elementary particles into powerful gamma-ray photons. Events occurring then (the *hadron era* in Table 19.3) are unclear mainly because they are dominated by incompletely understood interactions among strongly interacting heavy nuclear particles, like protons and neutrons.

The *lepton era* began after the hadrons of lowest mass annihilated each other. It continued as the lighter-mass particles, like electrons and positrons, were destroyed, and it ended when neutrinos produced in these reactions broke away to form a ghost world of their own, moving about eternally and independently of the other constituents of the universe. At their present temperature of about 2 K, these relic neutrinos number about 1200 per cubic centimeter, but they cannot be detected with present technology. As mentioned earlier, if neutrinos were to have mass, they would dominate all else in determining the mass of the universe.

About 1 second after the beginning of the universe, the *radiation era* began, and expanding space was filling mostly with photons and neutrinos with a small amount of matter. In the words of cosmologist Edward Harrison (1919–   ) "Matter was like a faint precipitate suspended in a world of dense light." During the early radiation era the relatively small amount of matter consisted mostly of protons and neutrons; the neutrons combined with protons to produce deuterium, and then helium was synthesized when the universe was some 200 seconds old. Much later, when the universe was between 100,000 and 1 million years

old and the temperature had dropped to about 3000 K, the electrons were captured by the protons to form atoms of hydrogen. Electrons could then no longer scatter photons, and the universe became transparent to radiation. The fireball radiation that had flooded the expanding universe for the first million years now appears as a relic, known as the cosmic background radiation.

HYDROGEN, DEUTERIUM, AND HELIUM

Most of the hydrogen, deuterium, and helium nuclei were synthesized during the first 200 seconds of the radiation era, at a fireball temperature of about 1 billion K, when the rapidly expanding universe had cooled enough to allow deuterons to survive and form helium. The calculated percentage of deuterium and helium created depends critically on the values we adopt for the early matter density and present temperature of the cosmic background radiation, but the results are reasonably consistent with present observed cosmic ratios of deuterium and helium to hydrogen.

The measured density of deuterium leads to estimates of the present average matter density in the universe that are less than the critical density needed to close the universe. Thus we again find evidence suggesting an open universe that will expand forever, always assuming that neutrinos do not have mass.

The theoretically calculated abundance ratio of helium to hydrogen is about 25 percent by weight, or 1 helium atom to 12 hydrogen atoms—about the same as the observed ratio in most stars, including the sun, and in planetary nebulae and H II regions. The predicted abundances for the heavier elements, however, are very low in comparison with the observed values. It seems, then, that the heavier elements were primarily created long after the big bang by thermonuclear reactions inside stars and in supernova outbursts.

TABLE 19.3
Evolution of the Universe Since Birth

Epoch	Time	Density (g/cm^3)	Temperature (K)	Event
Big bang	Begins at zero	Almost infinitely high initially	Extremely high	Birth of the universe
Hadron era	$<10^{-4}$ s	$>10^{14}$	$>10^{12}$	Annihilation of matter and antimatter
Lepton era	10^{-4} s–1 s	10^{14}–10^5	10^{12}–10^{10}	Rapid expansion and cooling; thermal equilibrium of electrons, positrons, neutrinos, and photons
Radiation era	1 s–10^6 yr	10^{5a}–10^{-22}	10^{10}–3000	Formation of helium and deuterium; radiation uncouples from matter at end of era
Matter era	$>10^6$ yr	$<10^{-22}$	$<3000^b$	Quasars and clusters of galaxies condense
Present era	15–20 × 10^9 yr	5×10^{-30}–5×10^{-31}	3^b	Galaxies and stars have formed; stars still forming

^{a}At the beginning of the radiation era, when the universe was 1 second old and $T = 10^{10}$ K, the radiation density was equivalent to about 10^5 g/cm^3, while the matter density was only about 1 g/cm^3.
^{b}Temperature of cosmic background radiation, which is no longer coupled to matter and its temperature.

MATTER ERA: FORMATION OF CLUMPS OF MATTER

The era of galaxies began when matter was more plentiful than radiation, when the universe was about 1 million years old and about 1 million times more dense than it is at present. How was this primordial matter (with a density of less than 10^{-22} gram per cubic centimeter, approximately 100 atoms per cubic centimeter, and a relatively low temperature) distributed in space? How was it formed into clusters of galaxies and individual galaxies?

Abbé Georges Lemaître noted that, if a homogeneous and isotropic world model expanding from a big bang is a reasonable place to begin the universe, then the critical step may be to account for departures from homogeneity in order to form galaxies and clusters of galaxies. One approach to the formation of galaxies is to have matter accumulate out of churning primordial eddies. The most favorable period when very large turbulent eddies could have formed occurred at the beginning of the *matter era*, when the temperature had dropped to about 3000 K. We can summarize the extremes of a spectrum of ideas concerning the formation of galaxies in a big-bang cosmology. In the first, superclusters formed and then fragmented into individual galaxies; in the second, small clumps formed first and accreted into galaxies, which in turn accreted into the clusters in overdense regions. In both, the universe grew more irregular under the dominion of gravity.

EVOLUTION OF CLUSTERS

Each cluster is either a regularly shaped, dense cluster of galaxies resembling the Coma cluster, or it is an irregularly shaped, semistable collection resembling the Virgo cluster. How are clusters evolving? With more matter concentrated toward their centers, regular clusters are probably enhancing that difference by capturing more matter into the center at the expense of the outer portions of the cluster. Such does not seem to be the case in irregular clusters.

Galactic evolution in clusters may be affected by forces between galaxies, by merging of colliding galaxies, by cannibalization of one galaxy by another, and by movement through an intergalactic medium. If hydrogen gas is swept out of a galaxy, then star formation will cease, and the galaxy will have prematurely ended that phase of its existence. This will certainly influence subsequent evolution of galaxies and hence the cluster.

High gas density in a forming galaxy favors rapid star formation before pronounced gravitational contraction of the galaxy can occur to reduce the rate, and the end result is possibly an elliptical galaxy. We observe that in elliptical galaxies star formation has virtually ceased. Star formation in a forming galaxy that begins life with a lower gas density is slower, leaving time for gas to settle toward an equatorial plane about the axis of rotation. This gas forms the disk portion of a spiral galaxy, which is surrounded by a spherical halo of stars created before the collapse. The degree of flattening in an elliptical galaxy or in the disk of a spiral galaxy depends presumably on the amount of spin the material had before the contraction that formed the galaxy started.

MINI BLACK HOLES

Combining gravity with quantum mechanics has enabled us to probe much deeper into the bizarre complexities of the black hole (see page 370). Stephen Hawking (1942–   ) in England has developed a quantum theory to analyze the properties of a black hole. His calculations have led him to the concept of *mini black holes* capable of emitting particles and photons in a thermal fashion, the same as from other hot bodies. The temperature of a black hole is not very high but is larger the smaller the black hole's mass. It turns out that particles from inside the black hole can tunnel their way out into open space as if there were no event horizon.

Possibly the enormous pressures present during the early stages of the expanding universe could have created mini black holes. A typical mini black hole could be the size of a proton ($\approx 10^{-13}$ centimeter) and have the mass of an asteroid (≈ 1 billion tons) and a temperature of millions of degrees. If mini black holes were formed in large numbers after the big bang, they would be quite bright—not black—and their gamma-ray emission might be detectable with today's instruments. Particle and antiparticle pairs—such as electron-positron pairs, neutrino-antineutrino pairs, and photon-antiphoton pairs—are created in the powerful gravitational field of a black hole. The black hole would thus appear to be emitting particles and antiparticles. The resulting emission decreases the

GRAVITATIONAL RADIATION: TESTING RELATIVITY THEORY

One of Einstein's predictions lay idle for nearly half a century as too difficult to verify: Extremely weak gravitational waves are radiated into space with the velocity of light by rapidly accelerated or orbiting bodies. These waves might be detectable with sensitive apparatus, and large astronomical objects undergoing violent activity, such as supernova outbursts or the nucleus of an active galaxy, may be the best places from which to detect them. Any gravitational wave passing through an object momentarily deforms its space and causes the object to vibrate slightly.

More than a decade ago experiments tried to pick up, inside large, suspended, aluminum cylinders, infinitesimal oscillations that would be produced by gravitational waves

striking the cylinders. At first it seemed that they had succeeded in detecting gravity waves simultaneously in a Maryland laboratory and at the Argonne National Laboratory near Chicago. Most of these were reported to be coming from the Galaxy's center in Sagittarius. It now appears that the observed oscillations were much too large to be consistent with current physical theory. So far, other and far more sensitive gravity-wave detectors, which can detect deformations as small as 10^{-17} centimeter, have failed to find any evidence of gravitational radiation coming from the center of our Galaxy or anywhere else. However, indirect evidence of gravitational radiation has been found in the radio observations of a binary pulsar (see page 377). The

gravitational interaction between the pulsar and its close companion, perhaps a neutron star or white dwarf, results in part of the orbital kinetic energy's being radiated away in the form of gravity waves. The loss in energy decreases the orbital separation between the components. Radio monitoring during the period 1974 to 1979, covering some 1000 orbital revolutions, shows a decrease in the orbital period of about 101 microseconds per year. Allowing for the uncertainties in the mass of each component and the inclination of their orbital plane, the result is in reasonable agreement with general relativity's prediction of 76 microseconds per year.

> Of the real universe we know nothing, except that there exist as many versions of it as there are perceptive minds.
>
> Gerald Bullitt

mass of the mini black hole, while its temperature rises. Eventually a runaway effect develops, which results in a rapid and enormous release of energy followed by the complete evaporation of the black hole down to zero mass within a finite time.

For a black hole of one solar mass the process described above is extremely slow and unobservable; the temperature is a tiny fraction above absolute zero, and its calculated lifetime is 10^{66} years, trillions of trillions of times the accepted age of the universe (15 to 20 billion years). A mini black hole of about 1 billion tons could develop a temperature of 100 billion K and have a lifetime expectancy roughly equal to the age of the universe. The flash of gamma radiation arising at the time of its rapid demise down to zero mass might now be observable. It is conceivable that we can find, as another means of verifying the big-bang cosmology, powerful gamma-ray radiation pouring out of primordial mini black holes that were created during the early universe.

A FINAL NOTE

We end with a word of caution. According to the principle of Occam's razor, we have introduced certain simplifying assumptions: Local physical laws are valid everywhere; the universe is isotropic and homogeneous; and the observed redshift of the galaxies is due to the expansion of the universe. But our understanding is confined to the collective limit of human experience. It is within that framework that we search for answers. Although to us human perception and reasoning ability seem almost awesome at times, we are nevertheless finite creatures who have only the smallest direct experience with the universe. One of the great hopes of astronomers is that contact with other intelligent life forms, as we discuss in Chapter 20, will increase the boundaries of our experience by many orders of magnitude. It is interesting to speculate whether they—if they exist—see the universe as we do.

SUMMARY

Design of the cosmos. The effort in cosmology is to develop a unifying theory for the structure and evolution of the universe from simplifying principles and a theory for gravity, such as the general theory of relativity. Observations of the expanding universe suggest cosmological models in which the universe was once much smaller and denser than it is today. According to the big-bang theory, a smaller space was filled by primordial matter from which the expansion began billions of years ago, and it is still expanding. Cosmic background radiation remaining from that big bang has been detected. The cosmic background radiation is apparently the same in all directions (isotropic). Isotropy implies that the universe is also homogeneous, and both are basic tenets of the cosmological principle. A homogeneous and isotropic universe means that any large-scale curvature must be uniform throughout space, and the Hubble velocity-distance law is the same anywhere in the universe.

Static cosmological models. Before redshifts were discovered, Einstein proposed a static cosmological model of a closed, spherical universe with matter thinly distributed. Although finite in extent, it is centerless and edgeless. Discovery of redshifts, however, required an expanding model of the universe.

Expanding cosmological models. Eddington proposed an expanding model by perturbing Einstein's static model from which expansion would continue forever. Lemaître, working with the Friedmann equations, proposed an expanding model beginning with a big bang. Deciding whether the universe is infinite (curved or flat) or finite (curved and closed) requires data about the mean density of the universe, the Hubble constant, and the cosmological constant. Most modern cosmologies are based on Einstein's field equations of general relativity and Friedmann's mathematical descriptions of space.

With a zero cosmological constant, three types of model, known as Friedmann universes, are possible as determined by the mean density of matter's effect on the universe's expansion. Since mutual gravitational attraction of matter slows expansion, the universe may possess an open hyperbolic or a flat Euclidean geometry if too little matter exists, and a closed spherical geometry if enough matter exists to stop the expansion. Depending on the value of the Hubble constant, the age of an open universe is greater than 13 billion years and is less than that for a closed universe.

Alternatives to the big-bang cosmology. Bondi, Gold, and Hoyle proposed an expanding universe, whose general appearance remains unaltered forever, the steady-state universe. New matter (hydrogen) is continuously created, eventually creating new galaxies. The cosmic background radiation contradicts the original steady-state hypothesis, and it, along with other objections, has made the steady-state model less probable.

Tests of cosmological models. Four classic tests of cosmological models are (1) mean density of matter in the universe, (2) luminosity distances versus redshifts, (3) number counts of radio sources with distance, and (4) angular size variation with distance. The primary discriminator between open and closed Friedmann universes is the total mass of the universe (that is, mean density). The critical value is proportional to the Hubble constant. However, it appears that only one-tenth of the mass necessary to close the universe has been observed. The other three tests are attempts to measure the curvature of space directly. So far they are unconvincing in that evolutionary effects cannot be accounted for. Whether the universe is open or closed is still undecided. The strongest lines of evidence point to an open universe.

The early universe. Mathematical models of the birth of the universe have been proposed, starting with the first one-hundredth of a second after the start of expansion in the big bang and describing its evolution as matter and radiation in the universe cooled and uncoupled. Physical events in that early universe made our existence possible and determined the present nature of the universe. Evolution of the universe is divided into the hadron era, lepton era, radiation era, matter era, with its formation of inhomogeneities of matter—formation of galaxies and clusters of galaxies.

KEY TERMS

big bang
closed universe
cosmic background radiation
cosmological constant
cosmological principle
cosmological redshift
Euclidean space
horizon of the universe
Hubble time
hyperbolic space

marginally open universe
matter era
negatively curved space
mini black hole
open universe
positively curved space
radiation era
spherical space
steady-state universe
X-ray background radiation

CLASS DISCUSSION

1. What is the large-scale distribution of matter like in the universe? What is the significance of the cosmic background radiation in relation to the cosmological principle?

2. What are the differences among the various Friedmann universes, and how do they differ

from Einstein's static universe? Is the geometry of space and the age of the universe the same for all Friedmann universes?

3. Describe the cosmologies alternative to a big-bang cosmology. How does the perfect cosmological principle differ from the cosmological principle?

4. Describe some of the tests to distinguish among cosmological models. Is there a favored cosmological model on the basis of the various tests?

5. What was the early universe like? Did it appear much as it does today, or has it changed?

READING REVIEW

1. Relative to superclusters, what are the sizes of individual galaxies? Relative to the visible universe, what size are they?

2. What presumably is the reason for the existence of the cosmic background radiation?

3. Can one infer from the cosmological principle that the universe has a center and an edge?

4. If the mean density of matter in the universe at present is less than the critical value, what is the geometry of space? Is the universe finite or infinite?

5. Do all the Friedmann universes begin with a big bang? Do they all end with a big crunch?

6. Does the observed mean density of the universe suggest the universe to be open or closed? If the "missing mass in clusters" were true, would the universe be open or closed?

7. Do astronomers have ways of measuring the large-scale curvature of space? If so, what kind of universe do they suggest?

8. In the early universe during the radiation era was there more matter than photons present?

9. The hydrogen present in all living things first appeared in the universe when it was approximately how old?

10. If the universe begins as homogeneous and isotropic, is there any difficulty in producing galaxies and clusters of galaxies in it?

SELECTED READINGS

Gingerich, O., ed. *Cosmology +1.* Freeman, 1977.

Harrison, E. R. *Cosmology: The Science of the Universe.* Cambridge University Press, 1981.

Kaufmann, W. J. *Relativity and Cosmology*, 2nd ed. Harper & Row, 1977.

Schramm, D. N. "The Age of the Elements," *Scientific American,* January 1974.

Shipman, H. *Black Holes, Quasars, and the Universe,* 2nd ed. Houghton Mifflin, 1980.

Weinberg, S. *The First Three Minutes.* Basic Books, 1977.

20
Exobiology: Life on Other Worlds

The belief that other worlds may have life has been with us since ancient times. Throughout this chapter are reprinted quotations from the near and distant past that speak of a belief in extraterrestrial life. Take a few moments to read them. In Chapter 10 we explored the possibility of other life forms *in* the solar system, and we shall continue this theme by considering the possibilities of life *beyond* the solar system. The first question we may wish to ask is about the environment, or platform, on which the slow processes of chemical and biological evolution can take place. An initial thought would be to assume that the life process might begin in a planetary system about a star and on a solid surface of one or more of its planets—as did our own development. Although this kind of beginning appears to have the highest probability, it certainly does not exhaust all possibilities; so if we were to consider all the possibilities on each topic in this chapter, it could quickly become a whole book in itself. Our failure to mention a wide variety of possibilities is not, then, a narrowness in thinking or a silent judgment on the merits of the speculation; it is, rather, a recognition that space constraints force us to limit ourselves to the primary considerations in this field and leave to you the task of researching details and variety.

20.1
Beyond the Solar System

WHICH STARS MIGHT HAVE VIABLE PLANETARY SYSTEMS?

If we decided to select the stars in the solar neighborhood near which life might be found, how should we proceed? Some criteria have been proposed that could be used when looking for stars with planetary systems on which life could be possible. The criteria do not argue against the existence of planets as such but rather address the viability of the planet for supporting life.

◄ "To consider the earth as the only populated world in infinite space is as absurd as to assert that on a vast plain only one stalk of grain will grow," said the Greek philosopher Metrodoros of Chios in the fourth century B.C.

> If there are places in the universe where people make genuine sense of their existence as a species and where they comprehend the delicate connections between the individual and collective existence, all the treasures on earth would be a small price to pay for the clues.
>
> Norman Cousins

First, we should probably assign a very low probability to the vast majority of binary and multiple stars—about half the stellar population—because the orbits of planets around them might not be stable enough to maintain a planet in a thermally habitable zone. A spectroscopic analysis of 123 sunlike stars out to 85 light years, discussed in Section 12.6, has led to the speculation that 20 percent of these stars may have planetary systems rather than stellar companions. Thus a planetary system may well be an alternative to a companion star or stars, with one in five sunlike stars perhaps possessing a planetary system.

A second criterion for assigning a low probability to a star's having a viable planetary system is the length of time the star is on the main sequence, which is the longest phase in the star's life. Thus stars of spectral classes O, B, and A would be rejected for life-bearing planetary systems because their time on the main sequence is much too short (less than 1 billion years) to permit chemical and biological evolution (which would probably take several billion years). Even if the time scale for biological evolution was assumed to be 1 billion years or less, the quantity of high-energy photons (X rays and ultraviolet) emitted by hot stars would probably prevent the formation of the complex organic molecules.

Even though the cool stars on the lower end of the stellar temperature scale, such as the M stars, have a lengthy life span on the main sequence, they are much too cool to support the development of life except in orbits that would be very near the star and probably unstable. For this reason they are not assigned a very high probability for possessing planets that support life.

A third criterion would be how rapidly the host star changes so that all stars after the main-sequence stages of evolution are probably not good candidates. If planetary systems had developed and sustained life, then the expansion of stars during red giant evolution should destroy the thermal environment in which they had existed. And subsequent phases in a host star's existence would be too short for life to develop a second time somewhere else in that system.

HABITABLE ZONES AROUND STARS

We are left with main-sequence stars between spectral classes F2 and K5 as the best possibilities for supporting advanced forms of life. Possible *ecoshells* are illustrated in Figure 20.1 for late main-sequence stars. These habitable zones or shells of life, are essentially defined by where water would remain in a liquid state. The inner boundary is where water would boil, and the outer boundary is where water would freeze.

To define a habitable zone for life, why do we seek a temperature range in which water is in liquid form? Why not some other organic solvent, such as alcohol or ammonia? First, water is a simple molecule, consisting of just three atoms, of which two, hydrogen, are the most abundant element in the universe, and the third atom, oxygen, is the next most reactive element. Second, liquid water can store a great deal of thermal energy before it vaporizes. Thus it acts as a buffer to day-night temperature changes that occur when a planet rotates. Finally, water has a high surface tension, which can help concentrate solids at its boundaries. In a similar vein carbon chemistry is expected to be a more widespread basis for life than is silicon chemistry or germanium chemistry (both silicon and germanium behave chemically somewhat like carbon). This is because carbon is far more abundant cosmically than either silicon or germanium.

LIFE AMONG NEARBY STARS

Only one technique is presently available to astronomers searching for planetary bodies, and that is picking up minute wobbles in a star's proper motion. These are the deviations caused by orbital motion of its supposed planets. This periodic motion is in reality the movement of the star around the center of mass of the star-planet system and is detectable only where the ratio of planetary mass to that of the star is not insignificant (see page 294).

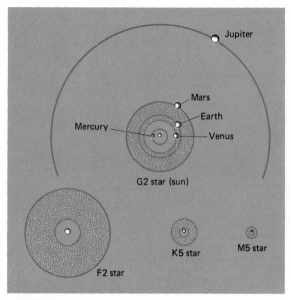

FIGURE 20.1
Possible ecoshells around stars of different spectral types are shown in solid-color circles.

A few nearby red dwarfs of spectral class M have wavy proper motion, which suggests the presence of unseen companions. These are astrometric binaries (see pages 294–295). It is possible to derive an estimate for the mass of an invisible companion from the amplitude of the wiggle of a visible star. Best known of this class is Barnard's star, a red dwarf with the largest observed proper motion. After half a century of studying its proper motion, one study reports a minute wiggle caused by two invisible, planet-size bodies. Both are well outside the ecoshell of the parent star. Another analysis of the proper motion of Barnard's star, using less extensive but independent photographic data, shows no evidence for planetary motion. The disagreement tells us of the great difficulty in making such delicate measurements.

Of the 31 stellar systems within 15 light years of the sun only 3 seem to meet the criteria needed for an ecoshell. These are the main-sequence stars Epsilon Eridani (K2), Epsilon Indi (K5), and Tau Ceti (G8). If we take the solar neighborhood as a representative sample (in which 3 stars out of the 31 stellar systems within 15 light years of the sun have potentially habitable planets), then the average distance between biologically suitable stars is about 17 light years. Within a radius of 1000 light years, then, we should expect to find about 1 million stars having suitable planets harboring some kind of life. Even if only 1 in 1000 of these planetary systems has an intelligent species, that still leaves 1000 sites of intelligent life within 1000 light years. If we conservatively extend this argument, estimating that only 1 million civilizations with a technology at least equal to ours are distributed throughout the Galaxy, the average separation between them would require 600 years either to send or to receive a message—hardly a hurried conversation!

20.2
Possible Forms of Life

CONSIDERATIONS FOR BIOLOGICAL EVOLUTION

During the course of the expanding universe, life became possible only after the galaxies had formed and their stars had existed long enough to synthesize the heavy atoms found in organisms. The six most important elements for earth life and their percentages by numbers of atoms in human beings are the following: hydrogen, 61 percent; oxygen, 26 percent; carbon, 11 percent; nitrogen, 2 percent; phosphorus, less than 1 percent; and sulfur, also less than 1 percent. Had the primeval condensate that immediately followed the big bang not cooled rapidly in the first few minutes, life could not have occurred in the universe because hydrogen would mostly have been converted into helium, leaving little available for stellar nuclear synthesis; heavy elements would not have been subsequently formed in stellar interiors. Thus the first consideration is whether enough time has elapsed for stellar evolution to build a significant abundance of the biologically significant atoms from which chemical evolution will flow if given the opportunity. Clearly it has at our position in the Galaxy, or we could not exist.

If there are approximately 400 billion stars in our Galaxy and 85 percent are main-sequence stars, then there are about 340 billion main-sequence stars. Astronomers estimate that about 90 percent of the main-sequence stars are spectral classes F, G, K, and M, or what amounts to about 300 billion stars. One-third of these, or 100 billion, are apparently F-, G-, and K-type stars; and one-half of these, or 50 billion, are probably not members of a binary system. Thus there is potentially a large number of life-supporting planetary systems. And if our cosmic environment is duplicated in many regions of our Galaxy, chemical evolution could have developed much as it did in our case.

The second consideration is whether chemical evolution up to the macromolecule stage will be followed by biological evolution, given a suitable environment and sufficient time. Although biological evolution leading to simple life forms is obviously less probable than chemical evolution is, it seems reasonable that biological evolution has occurred many times just from the huge number of opportunities. In our own case biological evolution is accomplished by mutations, which introduce new factors into the genetic message of DNA, by recombination, which is the rearrangement or new association of message units by sexlike processes (inheritance from different parents), and by selection, which is the weeding out of inferior traits in a population through successive generations (not in individuals in their lifetimes). Indeed biologists appear to find that mutation, recombination, and selection are necessary to go even from macromolecules to simple cellular systems. No one knows exactly why mutations occur; but DNA molecules are remarkably stable: mutations occur about once per gene for every 100,000 cell divisions in most mammals.

From the earth's fossil record two important points are evident: Species do not repeat although essential parts of organisms, such as the eye, may have several independent entries into biological populations; and some species have remained essentially unchanged since their appearance, while others have shown significant change. Thus the third consideration is whether or not there is a definite direction to biological evolution. That is, is it obvious that the occurrence of simple cellular life signals the later inevitable appearance of an intelligent humanoid? It is not obvious; and in fact it seems that evolution must be pushed to do anything, and, when doing something, can work only with what is already there. If so, the result would be that evolution cannot repeat itself except under absolutely identical situations, which never occur.

But if we assume that chemical evolution is followed by biological life that evolves toward greater

There is no easy way to the stars from the earth.

Seneca

complexity, what can we say about the development of intelligence? Since intelligence presumably means a capacity to communicate, it is the intelligent extraterrestrial life forms that are of interest to us.

THE RISE OF INTELLIGENCE

The great leap forward for man came only in the last few million years, when our humanoid ancestors learned to walk upright, freeing hands for the manipulation of tools. As their brains evolved and their mental capacities expanded under pressure for survival, a collective culture and civilization set human beings apart from other living creatures. Yet human beings, although the most advanced, are not the only intelligent creatures on the earth: One of the most important characteristics of intelligence is the ability to collect and transfer information, and there is a whole spectrum of this ability among earth's creatures, signifying that they possess varying intellectual capacities. We find what is apparently a "language" capacity in gorillas, chimpanzees, and dolphins; chimpanzees demonstrate a crude form of tool making and utilization.

How much time is necessary for intelligence to emerge? The time from the first mammals to ourselves is roughly 100 million years, or about 2 percent of the age of the earth. Thus if human beings were eliminated from the earth but left some life forms intact, we estimate that a species with intelligence comparable to that of man should appear within 100 million years. Would it again be man?

It is unlikely that another intelligent creature would closely resemble us. As we have pointed out, the fossil record does not indicate that evolution repeats itself. The late Loren Eiseley (1907–1977) noted in a more general sense what we might expect along such lines for the universe at large: "Life, even cellular life, may exist out yonder in the dark. But high or low in nature, it will not wear the shape of man. That shape is the evolutionary product of strange, long wandering through the attics of the forest roof, and so great are the chances of failure, that nothing precisely and identically human is likely ever to come that way again."

SPECULATION ON PHYSICAL FORM

Evolution is nonrepeatable, and so any extraterrestrial beings need not have humanoid characteristics. For us sophistication in an organism increases in proportion to its ability to react to changes in its environ-

> For a moment of night we have a glimpse of ourselves and of our world islanded in its stream of stars—pilgrims of mortality, voyaging between the horizons across the eternal seas of space and time.
>
> Henry Beston

ment, an ability partly due to its storing and recalling experiences. Thus the nature of the environment should be crucial in determining the form of extraterrestrial life.

It is quite possible that highly evolved species on other planets would differ in many ways from terrestrial life forms (even if the key molecule DNA were there) because of the enormous number of combinations of nucleotides possible in the structure of DNA and because of differences in environment. On earth, however, different life forms have developed many similar structural features; so we might suspect that an evolved, land-based, extraterrestrial species, subject like ourselves to gravity and electromagnetic radiation, might have some of our biological characteristics: such as side-to-side or front-to-back symmetry or compactness rather than extension.

Evolved intelligent creatures have to have a somewhat symmetrical body structure due to gravity. In addition they should have some kind of protected internal communication system, such as a central nervous system, a brain, and sensory organs up front. Regardless of whether such creatures are bipedal, quadripedal, or multipedal, it is difficult to believe that their electromagnetic sensors would be equally distributed over their bodies. As a result, they probably would not be capable of moving in any direction without turning unless they possess multiple brain centers. The type of sensory organ would be dictated by the environment and by the electromagnetic radiation to which the species was exposed. They would enjoy the advantages of mobility and manipulative ability, but they need not have the same symmetrical appendages as earth creatures. There should be a biological upper limit to size; for too bulky a creature demands a large source of energy for locomotion and a massive skeletal support of some sort to resist gravity.

Figure 20.2 illustrates some speculations on what extraterrestrial life might look like.

FIGURE 20.2
Speculations on the physical forms of extraterrestrial life. These pictures are not the full spectrum of speculation and are not based on scientific principles, but they are representative and amusing.

Top row:
From the 1950s movie "This Island Earth" comes a humanoid form with an obviously large brain cavity.

Another movie extraterrestrial from "The War of the Worlds" — a supposed martian.

Middle:
The extraterrestrial from "Earth vs. the Flying Saucers" is enclosed in a space suit of sorts. This implies that fluid pressure in its body is not equal to the atmospheric pressure of our atmosphere, or that the humanoid needs to control its environment for survival.

Bottom row:
An artist's renderings of a description of UFO occupants. These figures are surprisingly humanlike in form and facial appearance — maybe more so than we would have reason to expect.

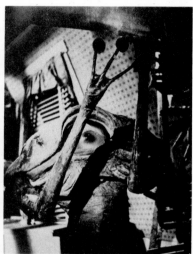

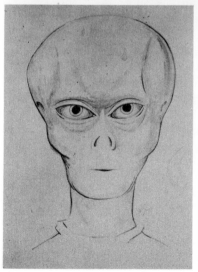
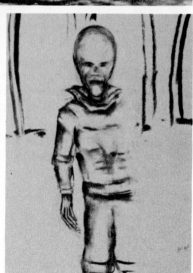

20.3
How Many Civilizations
in Our Galaxy Now?

In order to estimate the number of extraterrestrial civilizations in our Galaxy now, we need a definition of the term. We shall define "civilization" as one technologically capable of, and inclined by curiosity to, communication with other Galactic civilizations. We can start by making an inquiry into the possible number of communicative civilizations now existing in our Galaxy. An equation due to Frank Drake (1930–   ), known as the *Drake equation*, is a general formula expressing the number of such Galactic communities in terms of several factors:

number of communicative societies in the Galaxy now

$$= \begin{pmatrix} \text{astronomical} \\ \text{factors} \end{pmatrix} \cdot \begin{pmatrix} \text{biological} \\ \text{factors} \end{pmatrix} \cdot \begin{pmatrix} \text{sociological} \\ \text{factors} \end{pmatrix}$$

$$N = (R_s \cdot f_p \cdot n_p) \cdot (f_l \cdot f_i) \cdot (f_c \cdot L).$$

The first of the astronomical factors, R_s, is the number of stars in our Galaxy divided by the lifespan of the Galaxy. This factor amounts to about 400 billion stars divided by 10 billion years, or about 40 stars per year. It is a rough measure of the rate at which stars form in the Galaxy.

The second astronomical factor, f_p, is the fraction of stars that live long enough for life to develop and to have a planetary system in which it can develop. From our earlier arguments F, G, and K main-sequence stars (about 100 billion stars) live long enough for life to develop, but only half of these probably have planetary systems (about 50 billion stars). Therefore f_p is about 0.125, or 50 billion divided by 400 billion.

The third astronomical factor, n_p, is the number of planets in each planetary system suitable for life; it is the product of the average number of planets per planetary system and the fraction that are suitable for life. For the solar system, that product equals 1, so that we conservatively estimate one planet for each system.

The first of the biological factors, f_l, is the fraction of planetary systems in which life actually appears. The factor $f_l = 0.5$ is arrived at on the assumption that under the proper conditions sooner or later life will take hold, flourish, and evolve into a myriad of thriving forms in every other system.

The second of the biological factors is f_i, which is the fraction of evolving systems that evolve at least one intelligent species. We guess that the probability that nature, with say 4 billion years of effort, will create at least one intelligent species on a planet is 50 percent. Therefore we set the factor f_i equal to 0.5.

Now for the sociological factors: The factor f_c is the fraction of Galactic societies technologically able and willing to take part in interstellar communications; this factor we also guess to be 0.5, or there is a 50 percent chance that the species will develop technological capability and will want to try communicating with other Galactic civilizations.

The factor L is the length of time the civilization continues in its communicative phase. Our own interest in interstellar communication dates back only a few decades in a period of more than 6000 years of civilization.

For the number of intelligent communicative societies, then recognizing a great deal of uncertainty exists in each term, we obtain

$$N = (40 \times 0.125 \times 1 \times 0.5 \times 0.5 \times 0.5)L$$

$$= 0.625L \approx 1.0L.$$

In other words the number of communicative civilizations in our Galaxy now approximates the average number of years spent in the communicative phase, and the factor L is probably the most uncertain of all to evaluate. Although one may wish to question the assumptions leading to the particular values quoted above, most astronomers believe that they are reasonable. We need to remember also that this result applies only to our Galaxy and does not include the billions upon billions of other galaxies in the universe.

When we think about the possibilities of our own destruction by nuclear holocaust, by biological disasters from new mutant strains, by changes in the planet's ecology and climatology due to human stupidity and blunders, by terrestrial and extraterrestrial catastrophes, and by other calamities that could befall a civilized society, it is tempting to predict that the moment of civilized glory may indeed be brief in the span of an intelligent species.

Assuming that we have a fair grasp of the values for the product $R_s f_p n_p$, we can calculate the average separation between communicative societies for various values of the product $f_l f_i f_c$ and L. Table 20.1 lists val-

TABLE 20.1
Approximate Distance in Light Years to Nearest Communicative Civilizations

L (yr)	$f_l f_i f_c$ [a] ← High Density of Civilizations in the Galaxy				
	10^{-1}	10^{-2}	10^{-3}	10^{-4}	10^{-5}
10^3	3,000	10,000	30,000	100,000	300,000
$3 \cdot 10^4$	1,000	3,000	10,000	30,000	100,000
10^5	300	1,000	3,000	10,000	30,000
10^6	100	300	1,000	3,000	10,000

(Long-lived civilizations →)

[a] Assuming $R_s f_p n_p = 40 \cdot \frac{1}{8} \cdot 1 = 5$ so that $N = 5(f_l \cdot f_i \cdot f_c)L$.

ues that cover a range of reasonable values for the two variables—the product $f_l f_i f_c$ and L. Note that, in the first row, if a species survives in a communicative phase for only 1000 years, then the length of time for messages to travel the distance between civilizations exceeds the lifetime of the communicative phase. Thus only the combinations below and to the left of the line in the table present any reasonable chance for an exchange of messages.

20.4
Extraterrestrial Communication

EFFORTS AT RECEPTION OF SIGNALS

Radio techniques had been so spectacularly improved after World War II that a few astronomers and physicists privately considered the feasibility of detecting extraterrestrial signals from intelligent life. The subject finally surfaced in the British scientific journal *Nature* in September 1959, when the physicists Giuseppe Cocconi (1914–   ) and Philip Morrison (1915–   ) presented logical reasons why effort should be made to search for interstellar signals generated by intelligent life.

For now it seems more practical for us to listen for signals than to transmit them. Perhaps messages that older, more advanced extraterrestrial civilizations have been transmitting for centuries have by now reached the solar system. The most advanced celestial communities could avail themselves of energy sources far more sophisticated and powerful than any we can realize today, perhaps even using the energy output of their parent stars by modulating their light as signals. We may be no more aware of them than New Guinea aborigines, who use drums for communication, are aware of the international radio traffic constantly passing overhead.

The first modern attempt in the United States to detect artificial signals from space was conducted by Frank Drake at the National Radio Astronomy Observatory at Green Bank, West Virginia. This undertaking was called *Project Ozma*, after the legendary princess in the imaginary land of Oz. The 26-meter radio dish was aimed at Tau Ceti and Epsilon Eridani for 150 hours of observation from May through July 1960. The observation proved to be unrewarding. Since then, other attempts to locate intelligent signals have been carried out. None has succeeded, but the several hundred stars examined are a very tiny sample of the possible sources. The problem in locating signals is to pick not only the right star but also the right frequency and the right time to observe. Even if a communication were received from another world, it would take great amounts of time to exchange messages; so perhaps the first step in acknowledging contact with an extraterrestrial society would be to transmit a duplicate of the received message back to its source to inform the sending society that its inquiry had been received and recognized as originating from an intelligent source.

CHOOSING THE FREQUENCY

In terms of radio signals the earth is a very chatty planet. It pours out a constant flood of signals into space from FM, television, radar, and commercial stations. Such radiation might be intercepted by nearby civilizations up to about 50 light years away, since that is the distance that the radio waves have traveled since the early days of very-high-frequency broadcasting. We, in turn, should be looking for chance or unintended radio signals from nearby stars that might have advanced civilizations on habitable planets.

Deciding which microwave region to use for the intentional or preplanned interstellar communication is not the impossible task that one might at first imagine—although it is not simple either. Radio astronomers suggest that the wavelengths between 3 centimeters and 30 centimeters (1000 to 10,000 megahertz) should be suitable. In that interval the atmospheric absorption and Galactic noise from synchrotron radiation, the 3-K cosmic-background radiation, and other unwanted sources are at a minimum. A logical frequency with universal significance should be 1420 megahertz (21 centimeters), the emitting frequency of the dark, neutral hydrogen clouds of our Galaxy.

The quietest part of the microwave spectrum is very nearly between the lines of hydrogen (H) at 1420 megahertz and of hydroxyl (OH) at 1667 megahertz. The region between the two frequencies is often referred to as the *water hole* because H and OH are the products into which water (H_2O) dissociates. If water-based life is present elsewhere, it might be reasonable to expect it to use a frequency at the water molecule's center of mass, which is 1652.42 megahertz.

A MESSAGE FROM THE EARTH

The 305-meter radio telescope (Figure 5.15) at Arecibo, Puerto Rico, is capable of transmitting a message that could be received by an identical radio telescope many thousands of light years away. After the great dish was resurfaced and its efficiency improved in late 1974, a powerful signal was beamed toward the globular cluster M13 in Hercules (see Figure 12.20). The message that it contained was transmitted in binary-code form, a long string of zeros and ones (Figure 20.3). Because the cluster is quite old and contains several hundred thousand stars, it was thought that there might be several civilizations present to receive

the astrogram. There is no expectation that we shall ever receive an answer. The message was sent as a token of our recognition of the existence of intelligent life somewhere in the Galaxy. Even if a response were to be forthcoming, we cannot expect it until approximately A.D. 52,000. Partners in interstellar communication will need a lot of patience.

PROJECT SETI

The groundwork has been laid for developing the search strategy that could ultimately bring us into communication with advanced societies. The program that has been proposed to do this is known as *Project SETI* (*S*earch for *E*xtraterrestrial *I*ntelligence). Working on the premise that a large and expensive radio-receiving system is not needed to begin with, SETI would equip existing radio telescopes with low-cost state-of-the-art receiving, data-handling, and data-processing equipment. With this apparatus it should be possible to explore the vicinity of the sun out to several hundred light years for radio leakage from an extraterrestrial civilization or for signals intentionally beamed toward us.

SETI is far more than a single effort. Like the voyages of exploration that discovered the New World or the present missions of planetary exploration, the search would involve many distinct projects with definite goals in mind. These would initially be carried

Therefore it must be acknowledged that in the same way the heaven, the earth, the sea, the sun, and the moon and all other things which exist are not unique, but rather are part of an incredible number. For the perimeters on their life are firmly fixed, and they too are made of mortal clay, just like any race of creatures which here on earth abounds in specimens differing according to their kind.

Lucretius, *De Rerum Naturae*

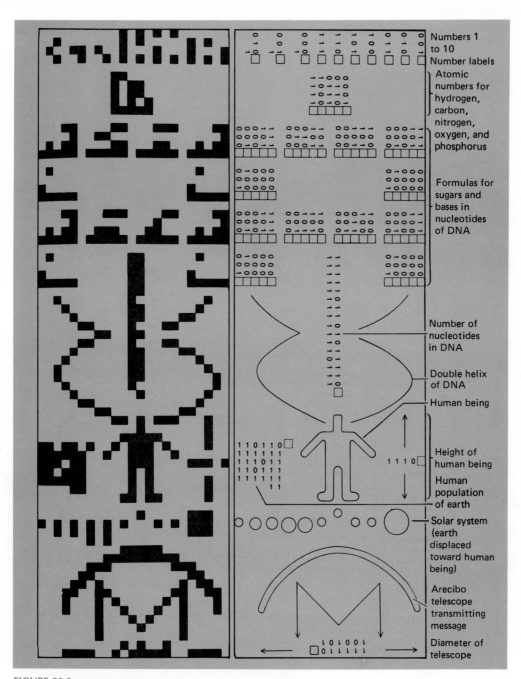

FIGURE 20.3

Arecibo message in pictures and a translation showing the binary version of the message decoded. Each number is marked with a label that indicates its start. When all the digits of a number cannot be fitted on one line, the digits for which there is no room are written under the least significant digit. (The message must be oriented in three different ways for all the numbers shown to be read.) The chemical formulas are those for the components of the DNA molecule: the phosphate group, the deoxyribose sugar, and the organic bases thymine, adenine, guanine, and cytosine. Both the height of the human being and the diameter of the telescope are given in units of the wavelength that is used to transmit the message: 12.6 centimeters.

FIGURE 20.4
A large dish built in a lunar crater on the far side of the moon could provide a collection area with complete shielding from radio emission from the earth.

out along with other astronomical investigations, but a time would come when dedicated facilities would be needed. It is estimated that a facility consisting of the collecting area equivalent to a few 100-meter radio dishes and associated data-processing equipment could do the job. In the event of a positive result or strong prospects for a positive result, a more ambitious program for SETI may be needed.

There are, of course, many ways we could attempt to communicate with other worlds. One possibility is to orbit a large antenna dish. Still another idea is to construct one or more large dishes inside lunar craters on the back side of the moon, as shown in Figure 20.4. This would provide complete shielding from radio interference from the earth. These are only two of the many proposals.

While we wait for better tools, serendipitous contact is also a possibility, although a slim one. We might stumble onto artificial signals randomly directed to other worlds; we might intercept signals being exchanged between two communicating groups; perhaps we could locate signals between a cruising spaceship and its parent planet; it is even possible that

we could intercept signals being sent back to a home station from an automated probe monitoring our solar system; or we might eavesdrop on a Galactic network exchanging information among its member societies. Once interstellar communication has begun, it will put us into a Galactic club of intelligent societies, from which we can never resign to return to our innocence. Social scientists have only begun to think about the sociological ramifications of membership in such a prestigious group.

20.5
Interstellar Travel

The next stage after communication—or it may be carried on simultaneously—would be manned interplanetary flight (that is, travel within the solar system). On the route to interplanetary flight it is a reasonable guess that we will orbit space stations about the earth or sun and perhaps establish space colonies in them.

Beyond the exploration of our solar system is the challenge of interstellar flight on the assumption that it is feasible even at some distant date in the future.

SPACE STATIONS

A design study in 1977 participated in by a group of scientists in various disciplines concluded that there are no insurmountable obstacles (except funding, amounting perhaps to $100 billion) to human beings' living in space; that space provides great reserves of matter and energy that can be used for the benefit of inhabitants on the earth and in space; and that we have both the knowledge and ability to colonize space.

As an example of a human space colony, it is technologically feasible to construct a number of cylindrical terraria in coupled pairs in orbit around the earth. Each space structure could house a complete ecological system, imitating the earth's environment with land and water areas, animals, birds, and even an artificial blue sky. Imitating the acceleration due to gravity at the earth's surface (1g) by spinning, the station could rotate around its cylindrical axis, which could constantly point to the sun. This means that with the proper positioning of arrays of windows and

mirrors the sun's light could be angled to provide both daily and seasonal cycles. A large solar-collector mirror at each end of the cylinder could furnish sufficient heat to run a steam-turbogenerator electric-power plant.

All but approximately 2 percent of the construction material could be taken from the moon and possibly the asteroids. The first proposed model would house 10,000 residents in a cylinder 1 kilometer long and 200 meters in radius, with a rotation period of 21 seconds (see Figures 20.5 and 20.6). The total mass would be 500,000 tons, equivalent to that of a present-day tanker. Later, larger units could be constructed with cylinders up to 32 kilometers long and 6.4 kilometers in diameter, capable of accommodating from 200,000 to 20 million inhabitants. However, such an effort is far short of interplanetary and eventually interstellar travel.

IS TRAVEL TO OUTER SPACE POSSIBLE?

The stars are so far from us and the technological and biological difficulties in traveling to them are so great that a trip even to the closest star seems hopeless now. Just consider the time it would take, beginning with a modest round trip to α Centauri, our nearest

FIGURE 20.5
Exterior of a proposed space habitat for about 10,000 persons. The large sphere in the center is nearly 1 mile in circumference and rotates to provide a gravity comparable with that of the earth.

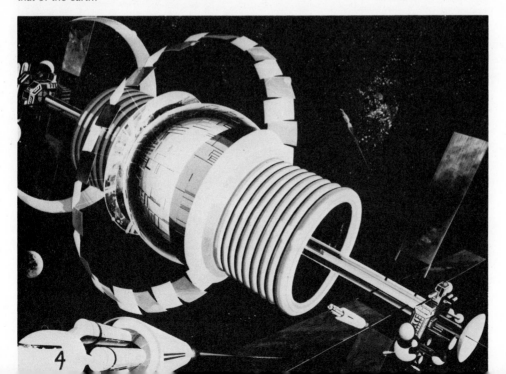

FIGURE 20.6
Space-wheel environment.

neighbor: The distance is 4.3 light years; our ship's speed is constant at 50 kilometers per second, slightly more than we need to escape from the solar system (at the earth's distance from the sun). Our round trip will take about 52,000 years.

How then do we get around time? The only alternative is to use a ship that can move at something approaching the speed of light, say, 95 percent of the speed of light. That will certainly shorten the trip and also let us take advantage of relativistic time dilation (see Section 3.5). The implications of time dilation for space travel can be illustrated by the following example of what is known as the *clock paradox*.

Astronaut A leaves the earth on a round-trip flight to star *S*, 12 light years away, at a speed of 0.6 the velocity of light (Figure 20.7). At the same time, astronaut B is to travel in the opposite direction on a round-trip voyage to star *T*, also 12 light years away, at a speed of 0.8 the velocity of light. The third person, C, will remain on the earth to monitor their flights. Before takeoff the three individuals synchronize their clocks. To avoid complications in recording the traveler's clock times, we assume that the periods of acceleration at the beginning of the outward and return voyage and the periods of deceleration in approaching each star or the earth are extremely brief

FIGURE 20.7
Hypothetical spaceflights to illustrate the clock paradox.

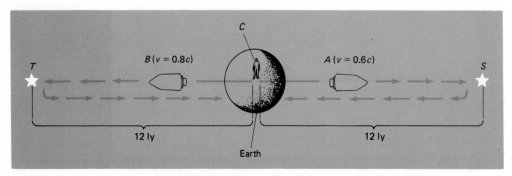

compared with the time spent in moving at constant velocity. These brief accelerations may therefore be neglected.

It takes a light ray 24 years to make the round trip between the earth and star S. A, traveling at 0.6c, will accomplish the feat in 24 ÷ 0.6 = 40 years, judged by C's clock. But A's clock runs slow by the contraction factor

$$\sqrt{1 - (0.6)^2} = 0.8,$$

so that A's round-trip time will be 0.8 · 40 = 32 years, according to A's clock.

Suppose all three astronauts are 20 years old at the start. When B returns, he finds that C is 50 and he is 38. When A returns, C is 60 and A is 52. When A and B meet again on the earth, A will be only 4 years older than B because B came back sooner than A did by 10 earth years. Because biological aging includes a measure of time in molecular cell growth, we presume the ages here are biologically correct.

The recipe for living forever is not simply to move in one direction at a speed close to the velocity of light. An observer who leaves an inertial frame of reference must return to it in order to collect the benefits of the fact that his time lags behind that of the observer who remains in the inertial frame. And while in the new frame of reference the biological clock of the body may run slower, so do all bodily activities so that the pace of living is perceived to be the same.

From the results of the above example we can refigure the trip to α Centauri discussed earlier. If we could accelerate the spaceship continuously at a constant rate of 1g (equivalent to the acceleration due to gravity that we experience on the surface of the earth),

we should feel no great discomfort. The best technique then would be to reverse the acceleration halfway out and come to rest in the vicinity of α Centauri and then on the return trip accelerate at 1g to the halfway point and decelerate at 1g the rest of the way to the earth. Taking acceleration and deceleration into account, the round trip by earth clocks would be 12 years; but the contracted time aboard the spacecraft, with a maximum speed of 95 percent of the velocity of light, would be 7.2 years.

Some representative travel times, with constant acceleration and deceleration of 1g up to a maximum velocity of 98 percent of the velocity of light, are shown in Table 20.2. From the table it appears that a journey to the Andromeda galaxy (2.2 million light years) could theoretically be made during an astronaut's life span. Unfortunately, the power a spacecraft would need for an even less ambitious interstellar flight is overwhelmingly large.

FLIGHT TO TAU CETI: HOW MUCH FUEL?

Tau Ceti, one of the biologically suitable stars, is approximately 12 light years from us. How much power would a spaceship need for a round-trip flight to it, moving at a maximum speed of 99 percent of the velocity of light? The most efficient propulsion system conceivable (it may never be achieved in practice) is a photon rocket engine powered by controlled annihilation of equal amounts of matter and antimatter. We now figure how much propellant a rocket ship might need for a payload weight of 100 tons (about twice as heavy as the *Apollo* ships) to make a round-trip flight to Tau Ceti. The fuel requirements call for a total initial

TABLE 20.2
Round-Trip Interstellar Spaceflight Times and Corresponding Distances

Astronaut's Dilation Time (yr)	Earth Time (yr)	Distance Reached (ly)	Most Distant Object Reached
5	6.5	1.7	Nothing
10	24	9.8	Sirius
20	270	137	Hyades open cluster
30	3,100	1,565	Orion Nebula
40	36,000	17,600	Omega Centauri globular cluster
50	420,000	208,460	Large Magellanic Cloud
60	5,000,000	2,479,000	Andromeda galaxy

spaceship weight of nearly 4 million tons, all of which (except the 100-ton payload) is expended by the time the crew returns home 28 earth years later. The astronauts, however, are only 10 years older.

The large initial quantity of fuel and the weight of the life-support system for the crew are not the only physical difficulties. At its maximum speed of 0.99c the exposed front section of the spaceship will be blasted by several million billion atoms each second on each square meter, even though the density of interstellar space (mostly hydrogen) is about one atom per cubic centimeter. This count is high mainly because of the relative velocities of encounter between the moving ship and the atoms and partly from the Lorentz contraction factor, which decreases the spacing between the atoms. Only massive concrete shielding could protect the astronauts inside. A two-stage chemical-nuclear propulsion system would also be necessary to move the spaceship far enough from the earth so that it could be launched without causing radiation damage to the earth (the gamma rays produced by the annihilation of matter).

If we tried to make the trip with today's most efficient chemical propulsion system, the fuel requirements would be beyond belief. To achieve the final velocity of 0.99c, a 1-ton payload would require a propellant weight of about $10^{100,000}$ tons. With the estimated total mass of the universe at about 10^{50} tons, the entire idea crumbles into futility. We can only conclude from these findings that physical contact with extraterrestrial beings may not ever be feasible within the time frame of one individual's life.

20.6
Has the Earth Been Visited?

UFO PHENOMENON

On June 24, 1947, near Mount Rainier, Washington, a salesman named Kenneth Arnold, while flying his own plane, claimed to have seen nine crescent-shaped disks flying near the mountain. His report opened the modern era of "flying-saucer," or *unidentified-flying-object* (*UFO*), reports. Arnold's report was neither the first nor the last in that year, but it is the one that caught the attention of the news services. Since that news event in 1947, there have been thousands more in the world press and something in excess of 100,000 UFO reports that never became major news events.

In reality the UFO "phenomenon" consists of reports of sightings and not the sightings themselves since for most sightings there is no way of knowing what was observed—if anything: Many sightings, even some with "photographic evidence," have later been admitted to be hoaxes. Although the majority of the sightings are by honest and sincere persons, where large volumes of reports have been studied, about 95 percent have been easily attributed to misidentification of "natural" phenomena, such as airplanes, weather balloons, artifical satellites, planets (Venus has prompted more reports than any other cause), bright stars, meteors, ball lightning, flocks of birds, clouds, reflected lights, or luminous insects.

As for the reports that cannot be readily attributed to natural phenomena, the degree of strangeness in a report seems to be inversely correlated with its credibility. That is, where the strangeness is high (such as supposed extraterrestrials waving at the observer), the credibility is low (for example only one witness saw the event, or lighting conditions were extremely bad). There are virtually no high-strangeness–high-credibility reports.

> Heaven and earth are large, yet in the whole of space they are but as a small grain of rice. . . . It is as if the whole of empty space were a tree, and heaven and earth were one of its fruits. Empty space is like a kingdom, and heaven and earth no more than a single individual person in that kingdom. Upon one tree there are many fruits, and in one kingdom many people. How unreasonable it would be to suppose that besides the heaven and earth which we can see there are no other heavens and no other earths?
>
> Teng Mu (thirteenth century Chinese philosopher)

FIGURE 20.8
UFO sighting near Santa Ana, California in August 1965. A California highway investigator sighted, photographed, and reported his sighting to his superiors. His report stated that he sighted a saucer-shaped object, estimated at 30 feet in diameter, about 8 feet thick, about 150 feet in the air. He took four photographs of this object.

Strangeness in a report is a subjective judgment, but a rough classification has been made by J. Allen Hynek (1910–   ), an astronomer who has reviewed a tremendous number of UFO reports since the early 1950s. He grades them into six categories of successively higher strangeness from nocturnal lights, daylight disks (Figure 20.8), radar and visual sightings of presumably the same object, close encounters of the first, second, and third kind. Close encounters of the second and third kind involve "physical effects" and "sighting of occupants" respectively. It is the last two categories that provide for believers the grist that the earth has been and is being visited by extraterrestrials, and for ardent disbelievers that some awfully nice people have gone off the deep end.

What is the explanation of the UFO reports? If the authors of this book could answer that question to everyone's satisfaction, their fame would be assured; but they cannot. However, we can make some simple yet reasonable arguments why the UFOs are more terrestrial (misidentifications, tricks of the mind, psychological aberrations, intentional deceptions) than extraterrestrial.

First, the reported activities and motions in a great many of the sighting reports violate the known laws of physics.

Second, as exemplifying a pattern of visitation or observation of the earth by extraterrestrials, the UFO reports are grossly lacking in sophistication. Human beings have already developed imaginative scenarios for the observation and study of extraterrestrial life, scenarios that involve remote sensing, genetically engineered viruses, and the like, which are intended to mimimize revelation of our presence—and we have not even gone exploring yet! Surely intelligent beings capable of surmounting the intellectual and physical rigors of interstellar travel must be more imaginative than to buzz around garbage dumps and play hide-and-seek with police cars.

Finally, even if we grant that visiting extraterrestrials wish to reveal themselves to us, then why the irrational pattern of the UFO reports? We find it hard to believe that visiting extraterrestrials could not soon decipher the function of an astronomical observatory, deciding that in them primitive earthlings are studying the universe and thus might be receptive to a knock on the front door for a short Galactic "town-meeting." Yet no UFO reports come from observatories, and no UFO has been found on any of the vast number of patrol photographs that cover great areas of the sky.

CHARIOTEERS FROM OUTER SPACE

If the earth is not now being visited by extraterrestrials, is it possible that extraterrestrial explorers happened this way in the past and left some sign of their visit? The sign could be some inanimate object or marking, interaction with primitive peoples if they existed, or even we ourselves if the visitors decided to seed the earth with intelligent life. Yes, this is a possibility. Then are any of the purported signs reasonable evidence for a visit in the past by extraterrestrial intelligence? They are probably not.

First are the signs in the category "engineering feats impossible to man." All the stone monuments, pyramids, buildings, and other engineering marvels frequently cited as of extraterrestrial origin are well within the capabilities of our ancestors even if it is not obvious why they would erect them.

Another group of purported signs is the "imparted knowledge" category of special building orientations, built-in numbers, and unknowable knowledge passed on through oral tradition. But none of them has been able to withstand real scrutiny.

The last category is that of "pictures of space gods," with murals, rock paintings, and pottery figures purported to show the extraterrestrial visitors (Figure 20.9). We readily admit that some of the figures are strange by today's standards. But, the anthropological picture of the culture in question can and does provide a cultural context for the putative space god so that it is neither necessary nor reasonable to resort to an extraterrestrial visitor as an explanation.

SPECULATIONS ON THE FIRST VISIT

Although it may seem that we have been unduly harsh on the possibility of a visit by extraterrestrial beings, that is not truly the case. We find nothing that says that a visit is impossible; but what we find woefully unconvincing is the supposed evidence that it is occurring now or has occurred in the past. If it has not already happened, is contact between us and other intelligent beings possible in the future? If they exist, we think it possible at some point. However, there are some very large ifs between now and that first meeting.

For the sake of discussion assume that it will happen. Then will it be a chance meeting or prearranged? We should find it highly unlikely that intelligent beings will simply happen to drop by. Instead it is more probable that the encounter will arrive through a long sequence of message exchanges and that we shall have been learning from those beings for a very long time before the face-to-face meeting.

FIGURE 20.9
Is this a picture of an extraterrestrial visitor? The picture is of the tomb lid of the Mayan king Pacal in Palenque, Mexico. Some argue that it is of an ancient astronaut in a rocket, with an oxygen mask, his hands on controls, and his feet on pedals. Each of the supposed pieces of equipment has alternative, recognized anthropological interpretations and can be found in other Mayan monuments.

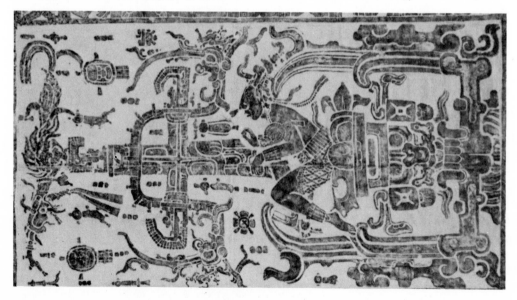

FIGURE 20.10
"We have found a strange foot-print on the shores of the unknown. We have devised profound theories, one after another, to account for its origin. At last, we have succeeded in reconstructing the creature that made the foot-print. Lo! it is our own" (Sir Arthur Eddington).

SUMMARY

Beyond the solar system. Conclusions about which stars might have viable planetary systems are based on several criteria: (1) Binary and multiple star systems would have unstable planetary orbits; (2) stars on the main sequence for less than several billion years would not provide a stable environment for the development of complex organic molecules; and (3) stars in the post-main sequence phase evolve too rapidly for life. Most likely candidates are main sequence stars between spectral classes F2 and K5. Of the 31 stellar systems within 15 light years of the sun, only 3 seem to meet criteria for having an ecoshell. On average, estimated distance between possible viable planetary systems is 17 light years.

Possible forms of life. The prime consideration in the development of life is whether enough time has elapsed for stellar evolution to synthesize sufficient quantities of the biologically significant atoms for chemical and biological evolution. The development of intelligence, likewise, is apparently determined by the course of evolution. It is therefore logically possible that intelligent life, although not possessing a humanoid form, has evolved elsewhere in the universe.

Extraterrestrial communication in our Galaxy. Exobiologists can estimate the number of extraterrestrial civilizations in our Galaxy from the Drake equation, which depends on astronomical, biological and sociological factors. Several projects have been carried out to detect signals from extraterrestrial civilizations but without success. Project SETI has been proposed as an ongoing, multifaceted effort to receive signals from extraterrestrials until contact is made.

Interstellar travel. A study in 1977 concluded that there are no insurmountable obstacles to human colonization of space. At conventional speeds, however, travel to even the closest star will take impractically long. At speeds approaching the speed of light, on the other hand, travelers would get the benefit of relativistic time dilatation. The fuel requirements for such a trip, along with other difficulties, are still overwhelming.

Has earth been visited? Reports of unidentified flying objects (UFOs) and ancient visitations abound. At this time, little exists beyond speculation to support the contentions of visits from outer space.

KEY TERMS

clock paradox
ecoshell
intelligence
life

organic molecules
Project SETI
UFO
water-hole frequency

CLASS DISCUSSION

1. What are the important considerations that might make a planetary system suitable for life? Could life develop on or about something other than a planet?

2. If chemical evolution begins, does it always lead to biological life and intelligent beings? If so, why; and if not, why not?

3. In the Drake equation give your opinion of the realism of the factors chosen and their values.

4. Should we devote time and resources to trying to communicate with extraterrestrial beings, given all the world's problems and shortages?

5. How do you react to the claims for and against UFOs and visitation by alien beings?

SELECTED READINGS

Breyer, R. A. *Contact with the Stars.* Freeman, 1982.
Edelson, E. *Who Goes There?* McGraw-Hill, 1980.
Gale, W. A., ed. *Life in the Universe.* Westview Press, 1979.
Goldsmith, D., ed. *The Quest for Extraterrestrial Life.* University Science Book, 1980.

Goldsmith, D., and T. Owen. *The Search for Life in the Universe.* Benjamin/Cummings, 1980.
Hynek, J. A. *The Hynek UFO Report.* Dell, 1977.
Klass, P. J. *UFOs Explained.* Random House, 1975.

Appendix 1
Measurement and Computation

UNITS OF MEASUREMENT: METRIC AND BRITISH

The *metric* and *British* systems of measurement are the two principal ones in use in the world today. In each there are units for length, mass, and time, which are often called the *fundamental units* of measurement since in principle all other quantities of measurement can be expressed in terms of them.

In the British system, used primarily by England, the United States, and the other English-speaking nations for engineering and commercial purposes, the fundamental units are the *foot* as the unit of length, the *pound* as the unit of weight or force, and the *second* as the unit of time. Because of the need for a common system of measurement to support worldwide science and technology, the British system is being phased out in favor of an international system of units based on the metric system.

The metric system, developed in France in the late eighteenth century, is based on the decimal system and thus is better suited to computation. Consequently the metric system has been the primary system of measurement for scientists all over the world, including the English-speaking nations. There have been two widely used versions of the metric system: In the MKS version the fundamental unit of length is the *meter,* the unit of mass is the *kilogram,* and the unit of time is the *second;* for the CGS version the unit of length is the *centimeter,* the unit of mass is the *gram,* and the unit of time is the *second.* The international system of units employs the MKS version of the metric system. These systems are summarized in Table A1.1, including the units for force and energy with their symbols in parentheses.

Standards for the meter and the kilogram are housed in the International Bureau of Weights and Measures at Sèvres, near Paris, while exact copies are maintained in various national depositories throughout the world.

The unit of time, the second, was originally defined as $\frac{1}{86,400}$ of the period for one complete rotation of earth relative to the sun. There are 86,400 seconds in one mean solar day. Because earth's rotation actually varies slightly, a new definition of the second was introduced in 1964 as the time it takes the cesium 133 atom to make 9,192,631,770 vibrations.

In the CGS version of the metric system the unit of force is the dyne, and it is the force necessary to accelerate 1 gram by 1 centimeter per second squared; that is, dyne = $g \cdot cm/s^2$. Force in the MKS system is the newton, named after Sir Isaac Newton, and it is the force necessary to accelerate 1 kilogram by 1 meter per second squared, or N = $kg \cdot m/s^2$. Since the kilogram equals 1000 grams and the meter is equal to 100 centimeters, the newton is 100,000 times the dyne in magnitude.

For units of energy the CGS system uses the erg, which is the amount of work done by a force of 1 dyne acting through a distance of 1 centimeter, while in the MKS system the unit is the joule. Since a joule is the work done by a force of 1 newton acting through a distance of 1 meter, the joule is equivalent to 10 million ergs.

Although the units in this book are derived with reference to the CGS system, it may be of interest to relate them to the MKS units and to the British units shown in Tables A1.2 to A1.3.

A number of important physical constants that are used throughout the text are given in CGS units in Table A1.4.

ASTRONOMICAL UNITS

Although the metric and British systems of measurement are useful on the scale of our everyday experience, there are many other units of measurement used by astronomers that are more appropriate to the scale of the phenomenon. Tables A1.5 and A1.6 give the conversions from the CGS system to these special units (abbreviation in parentheses).

TABLE A1.1
Measurement Systems

Quantity	British Unit	Metric Unit	
		CGS	MKS
length	foot (ft)	centimeter (cm)	meter (m)
mass	slug (sl)	gram (g)	kilogram (kg)
time	second (s)	second (s)	second (s)
force	pound (lb)	dyne (dyne)	newton (N)
energy	foot-pound (fp)	erg (erg)	joule (J)

TABLE A1.2
Conversion Factors: British to Metric

Multiply	By	To Obtain
inches	2.5400	centimeters
feet	0.3048	meters
miles	1.6093	kilometers
slugs	1.459×10^{-2}	grams
pounds	2.248×10^{-6}	dynes
foot-pounds	1.356×10^{-7}	ergs

TABLE A1.3
Conversion Factors: MKS to CGS

Multiply	By	To Obtain
meters	10^2	centimeters
kilometers	10^5	centimeters
kilograms	10^3	grams
newtons	10^5	dynes
joules	10^7	ergs

SCIENTIFIC NOTATION AND POWERS OF TEN

Many times in science, especially in astronomy, we are confronted by extremely large or extremely small numbers. For example, the approximate number of stars in our Galaxy is 400 billion, or 400,000,000,000. Numbers of this size make the arithmetical operations of multiplication and division extremely cumbersome. Therefore, it is convenient to use what is known as *scientific notation,* which involves powers of ten. We begin by explaining exponents and their manipulations.

If *n* (known as an exponent) is a positive integer, then a^n is defined as the number 1 multiplied *n* times by *a*; for example for $n = 3$, *a* raised to the third power is

$$a^3 = 1 \cdot a \cdot a \cdot a.$$

By convention any number raised to the zeroth power is equal to 1. If *a* is the number 10, then we may form a table of powers of 10 with their accepted prefixes and symbols (Table A1.7). For example, 1 million light years is 1 mega-light year, and its abbreviation is Mly.

There are rules for multiplication, division, raising to a

TABLE A1.4
Physical Constants

Quantity	Symbol	Value in CGS Units
Velocity of light in vacuum	c	2.998×10^{10} cm/s
Gravitational constant	G	6.667×10^{-8} cm^3/g·s^2
Planck's constant	h	6.626×10^{-27} erg·s
Mass of hydrogen atom	m_H	1.673×10^{-24} g
Mass of proton	m_p	1.6725×10^{-24} g
Mass of electron	m_e	9.109×10^{-28} g

TABLE A1.5
Conversion Factors for Lengths: CGS to Special Units

Multiply	By	To Obtain
centimeters (cm)	1.000×10^{-8}	angstroms (Å)
earth radii ($R_\oplus$)	6.371×10^8	centimeters (cm)
solar radii ($R_\odot$)	6.960×10^{10}	centimeters (cm)
astronomical units (AU)	1.496×10^{13}	centimeters (cm)
astronomical units (AU)	2.149×10^2	solar radii ($R_\odot$)
light years (ly)	9.461×10^{17}	centimeters (cm)
light years (ly)	1.359×10^7	solar radii ($R_\odot$)
light years (ly)	6.324×10^4	astronomical units (AU)
parsecs (pc)	3.086×10^{18}	centimeters (cm)
parsecs (pc)	2.063×10^5	astronomical units (AU)
parsecs (pc)	3.262	light years (ly)

TABLE A1.6
Conversion Factors for Mass, Time, and Energy: CGS to Special Units

Multiply	By	To Obtain
earth masses ($M_\oplus$)	5.977×10^{27}	grams (g)
solar masses ($M_\odot$)	1.989×10^{33}	grams (g)
years (y)	3.156×10^{7}	seconds (s)
Galactic years	2.30×10^{8}	years (yr)
electron volts (eV)	6.242×10^{13}	ergs
solar luminosity ($L_\odot$)	3.82×10^{33}	ergs per second (ergs/s)

TABLE A1.7
Powers of Ten

Word	Number	Power	Prefix	Symbol
trillion	1,000,000,000,000	10^{12}	tera	T
billion	1,000,000,000	10^{9}	giga	G
million	1,000,000	10^{6}	mega	M
thousand	1,000	10^{3}	kilo	K or k
hundred	100	10^{2}	hecto	h
ten	10	10^{1}	deca	da
unit	1	10^{0}		
tenth	0.1	10^{-1}	deci	d
hundreth	0.01	10^{-2}	centi	c
thousandth	0.001	10^{-3}	milli	m
millionth	0.000,001	10^{-6}	micro	μ
billionth	0.000,000,001	10^{-9}	nano	n
trillionth	0.000,000,000,001	10^{-12}	pico	p

power, or extracting a root for numbers expressed in exponent form:

1. Multiplication is accomplished by adding exponents. For example, where the dot denotes multiplication,

$$10^{-3} \cdot 10^{2} \cdot 10^{4} = 10^{-3+2+4} = 10^{3} = 1000.$$

2. Division is accomplished by subtracting exponents. For example,

$$10^{4} \div 10^{2} = 10^{4-2} = 10^{2} = 100.$$

3. Raising a number to a power is accomplished by multiplying exponents. For example,

$$(10^{2})^{5} = 10^{2 \cdot 5} = 10^{10} = 10,000,000,000$$

or

$$(10^{-4})^{2} = 10^{-4 \cdot 2} = 10^{-8} = 0.000,000,01$$

or

$$(2 \times 10^{2})^{3} = 2^{3} \times 10^{2 \cdot 3} = 8 \times 10^{6} = 8,000,000.$$

4. Extracting a root is also accomplished by multiplying exponents. For example,

$$(10^{6})^{\frac{1}{2}} = 10^{6 \cdot \frac{1}{2}} = 10^{3} = 1000$$

or

$$(10^{-12})^{\frac{1}{3}} = 10^{-12 \cdot \frac{1}{3}} = 10^{-4} = 0.0001.$$

With these rules in mind we can now move to define scientific notation and consider some examples of its use:

Scientific notation is the expression of any number as the product of a number between 1 and 10 times a power of 10.

As examples of scientific notation for expressing numbers, consider the following examples:

-   2,380,000,000   may be written as 2.38×10^9
-   86,496   may be written as 8.6496×10^4
-   0.0005492   may be written as 5.492×10^{-4}

From these examples a general rule is apparent for determining the numerical value and algebraic sign of the power of 10 to be used in expressing the number in scientific notation:

The number of places the decimal point is shifted indicates the numerical value of the power of 10; if the shift is to the left, the algebraic sign is positive, and if the shift is to the right, the algebraic sign is negative.

For example, in the number 86,496 above, the decimal is shifted four places to the left so that the power of 10 is 4 and its algebraic sign is positive, or 8.6496×10^4. For the number 0.0005492 above the decimal is shifted four places to the right so that the power of 10 is also 4 but its algebraic sign is negative, or 5.492×10^{-4}.

As mentioned above, scientific notation finds its greatest utility in multiplication or division, such as in Newton's law of gravitation (page 49):

$$F = \frac{Gm_1m_2}{d^2}$$
$$= \frac{(6.667 \times 10^{-8}\ \text{cm}^3/\text{g} \cdot \text{s}^2)(5.977 \times 10^{27}\ \text{g})(9.0 \times 10^4\ \text{g})}{(6.371 \times 10^8\ \text{cm})^2}$$
$$= \frac{(358.6 \times 10^{23})}{(40.59 \times 10^{16})} \frac{\text{cm}^3 \cdot \text{g} \cdot \text{g}}{\text{g} \cdot \text{s}^2 \cdot \text{cm}^2} = \frac{(3.586 \times 10^{25})}{(4.059 \times 10^{17})} \frac{\text{g} \cdot \text{cm}}{\text{s}^2}$$
$$= 8.834 \times 10^7\ \text{dynes}.$$

When numbers expressed in scientific notation are to be added or subtracted, then the following basic rule must be applied:

In addition or subtraction the power of 10 must be the same for all numbers to be added or subtracted, which will also be the appropriate power of 10 of the answer.

For example, let us add the mass of the sun and earth, as we might do in Newton's modified form of Kepler's Third Law (page 53):

$$M_\odot + M_\oplus = (1.989 \times 10^{33}\ \text{g}) + (5.977 \times 10^{27}\ \text{g})$$
$$= (1.989 \times 10^{33}\ \text{g}) + (0.000005977 \times 10^{33}\ \text{g})$$
$$= 1.989005977 \times 10^{33}\ \text{g}$$

It should also be noted that the units for each quantity in the last two examples are manipulated algebraically just as are the numbers. Thus in multiplication and division the units are multiplied and divided in order to obtain resultant units for the answer. And in addition and subtraction numbers may be added or subtracted if they have the same units. Hence an extremely valuable means of verifying the correctness of a series of algebraic operations is to carry out first the operations with the units only. The result must be the correct units for the desired answer; that is, if the desired answer is a force, then the units will not be those of energy.

MATHEMATICAL SYMBOLS, PROPORTIONS, AND RATIOS

Mathematical symbols are a magnificent shorthand for science, allowing scientists to express amazingly complex ideas, concepts, and relations in a reasonably simple universal language. Part of learning to read the language of science is to know the meanings of the symbols listed in Table A1.8.

In proportionalities the important point is not the equality of things but how some quantity, called a *dependent* variable, depends on one or several other quantities, called *independent* variables. To remove the proportionality, we must introduce into the equation a *constant of proportionality* that incorporates the units of measurement for the quantities involved. As an example, if k and G are appropriate constants of proportionality, then

$$b = \frac{k}{d^2} \quad \text{and} \quad F = \frac{Gm_1m_2}{d^2},$$

where G is known as the constant of gravitation.

Often mathematical relations can be formed as simple ratios, such as in measuring the Doppler effect:

$$\frac{\Delta\lambda}{\lambda} = \frac{v}{c},$$

where $\Delta\lambda$ is the change in wavelength, λ the wavelength, v the radial velocity, and c the velocity of light. Thus the equation says that the ratio of the two wavelengths, $\Delta\lambda$ and λ, is the same as the ratio of the two velocities, v and c.

ARC MEASUREMENT, AREAS, AND VOLUMES

A circle on a flat plane can be divided into 360 degrees of arc (360°), with each degree further subdivided into 60 minutes of arc (60′) and each minute of arc into 60 seconds of arc (60″). Thus there are 2.16×10^4′, or 1.296×10^6″, in 360°. Another unit of measure for angle is the radian, there being 2π, or 6.2831852, radians in 360°. Thus 1 radian = 57.29578°, or 1 radian = 206,264.8″.

A useful relation when dealing with circles, such as one of the principal reference circles on the celestial sphere (see Appendix 2), is the length of an arc of the circle. The relation is

TABLE A1.8
Mathematical Symbols

Symbol	Meaning	Examples
$=$	equal	a equals the sum of b and c: $a = b + c$
$\propto$	proportional to	square of the sidereal period is proportional to the cube of the semimajor axis: $P^2 \propto a^3$; or apparent brightness is inversely proportional to distance squared: $b \propto 1/d^2$
$\simeq$	very nearly equal to	$3.998 \simeq 4$
$\approx$	roughly, or approximately, equal to	$3.61 \approx 4$; or $87.5 \approx 100$
$>$	greater than	$a > b$, such as $4 > 3$
$<$	less than	$a < b$, such as $4 < 7$

Arc length =
$$\text{(radius of arc)} \times \text{(subtended angle in radians)},$$

where the units for the arc length are the same as those for the radius of arc. Since there are 2π radians (360°) subtended by the arc of a whole circle, the circumference (C) is

$$C = 2\pi r = 6.28r = \pi d = 3.14d,$$

where r is the radius and d is the diameter ($d = 2r$) of the circle.

In many places in the text we discuss areas and volumes. Areas are defined as products of lengths; that is, lengths squared, while volume is area times length, or length cubed. The area (A) of a circle in a plane is

$$A = \pi r^2 = 3.14r^2 = \frac{\pi d^2}{4} = 0.785d^2.$$

Another useful area for astronomical considerations is the surface area of a sphere. The sphere area (S) is

$$S = 4\pi r^2 = 12.56r^2,$$

while the volume (V) enclosed by the surface is given by

$$V = \tfrac{4}{3}\pi r^3 = 4.19r^3.$$

To locate an astronomical object on the sky, astronomers employ coordinate systems similar in concept to the system for plotting a point in a plane. However, the *celestial sphere,* which is the apparent sphere of the sky, is not a plane: It is the inside of a sphere, and one that rotates relative to our location on the surface of the earth. On the celestial sphere distances are measured in degrees along the arc of a great circle. A *great circle* is the arc formed by passing a plane through the center of a sphere. Thus a great circle divides a sphere into two equal halves. All other circles on a sphere that are not great circles are called *small circles.*

There are several features common to all astronomical coordinate systems. Each has a principal axis, or *polar axis,* about which the system rotates. The points of intersection of this axis and the celestial sphere are the *poles* of the system. Perpendicular to the principal axis is a great circle, which is the principal reference circle along which one coordinate is measured. Finally on the celestial sphere there are secondary reference circles, which are great circles perpendicular to the principal reference circle. There are an infinite number of these secondary reference circles, which meet at the poles of the principal axis, as shown in Figure A2.1.

HORIZON COORDINATE SYSTEM

The principal axis of the system is defined as parallel to the direction of gravity. Extended upward, it intersects the celestial sphere at the point known as the *zenith,* which is directly overhead for an observer (Figure A2.2). The principal reference circle, the *astronomical horizon,* is the great circle marked on the celestial sphere by a plane perpendicular to the zenith-nadir axis and tangent to the earth at the point of the observer. There are four reference points on the astronomical horizon: north, east, south, and west. They are 90° from each other.

If we project onto the celestial sphere the terrestrial meridian of the longitude passing through the observer's position, it defines the great circle called the *celestial meridian.* The meridian passes through the north and south points of the horizon as well as the zenith. It also contains the two points that are the intersection of the earth's axis of rotation extended to the celestial sphere: the *north* and *south celestial poles.* The position of the celestial poles relative to the north and south points of the horizon depends upon the observer's latitude. For an observer in the northern hemisphere the angular distance of the north celestial pole above the north point of the horizon is equal to the observer's latitude.

The coordinates in the horizon system are known as azimuth and altitude. *Azimuth* is the angular distance measured from the north point of the horizon along the as-

tronomical horizon. The secondary reference circles in this system are known as *vertical circles.* Thus the azimuth is measured to the foot of the vertical circle passing through the object of interest. As an example the azimuth of the east point of the horizon is 90°; the south point, 180°; and the west point, 270°. *Altitude* is the angular distance of the object of interest above or below the horizon measured along a vertical circle. The altitude of the zenith, for example, is +90°. In Figure A2.2 the altitude of the star is about 60°, and its azimuth is about 220°.

The major disadvantage of the horizon coordinate system is that it is peculiar to the observer and not at all a general system. Also, since the altitude and the azimuth of an object, such as Jupiter, continually change as the earth rotates, one must know the exact location and time at which an altitude and azimuth are measured in order for them to have any meaning to anyone besides the observer. A more general system is the equatorial coordinate system.

FIGURE A2.1
Basic plan of the various astronomical coordinate systems. Beginning with the principal axis, the principal reference circle is everywhere 90° from the poles, while the infinity of secondary reference circles is perpendicular to the principal reference circle.

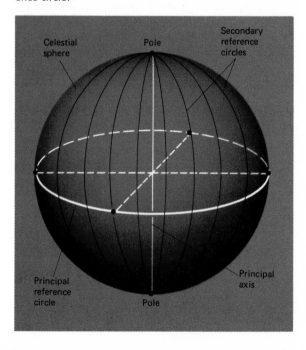

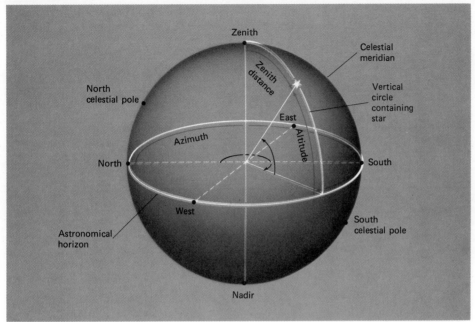

FIGURE A2.2
The horizon system of coordinates. For a star in the southwestern part of the sky the measurement of the azimuth is along the astronomical horizon and the measurement of the altitude is along a vertical circle.

EQUATORIAL COORDINATE SYSTEM

The principal axis of the equatorial coordinate system coincides with the axis of rotation of the earth. Its poles are the north and south celestial poles, as defined in the preceding section. The principal reference circle is the *celestial equator*, while the secondary reference circles are called *hour circles*, as shown in Figure A2.3. This system differs from the horizon system in that the reference circles are on the celestial sphere and thus rotate with it relative to the observer. In the horizon system the astronomical horizon and the vertical circles remain fixed relative to the observer, and the celestial sphere moves relative to them. Figure A2.4 shows the relationship between the equatorial and horizon systems for one particular time.

As the earth rotates from west to east, a star traces a path across the sky called a *diurnal* (*daily*) *circle* from east to west. For an observer located as shown in Figure A2.3 at intermediate geographic latitudes, the sky rotates at an oblique angle, the value depending upon the value of the latitude. The celestial equator passes through the east and west points of the horizon regardless of the geographic latitude of the observer.

The coordinates in the equatorial system are called right ascension and declination. *Declination* is the angular distance of an astronomical object north or south of the celestial equator measured along the hour circle through the object. Thus the declination of the north celestial pole is +90°, while that of the south celestial pole is −90°. In Figure A2.4 the declination of the star is about +60°. The other coordinate is called *right ascension* and is measured along the celestial equator from the point of intersection of the celestial equator with the annual apparent path of the sun on the celestial sphere, the *ecliptic*. This point is called the *vernal equinox* since the sun moves through this point on or about March 21 each year. Right ascension is measured eastward in units of time rather than degrees of arc to the foot of the hour circle passing through the object of interest. Since a 360° rotation by the earth corresponds to 24 hours, then 1 hour equals 15° of arc, or 1° equals 4 minutes, and so on. The star in Figure A2.4 has a right ascension of about 6 hours, or 90°. The angular distance of the hour circle from the observer's celestial meridian is called the *hour angle* of the star. For the example in Figure A2.4 the hour angle shown is about 1.5 hours, or 22.5° of arc.

It is obvious that the right ascension and declination of a celestial object remain fixed as the sky rotates. This makes it possible to catalog astronomical objects by their right ascensions and declinations in the same way in which places on earth are cataloged by their longitudes and latitudes. Because of precession (see page 22), the vernal equinox slides westward along the ecliptic by about 50″ per year. Thus over a number of years a star's right ascension and declination change so that cataloged positions are not accurate indefinitely and must be updated periodically to correct for the effect of precession.

In addition to the horizon and equatorial systems there are two less frequently used systems of astronomical coor-

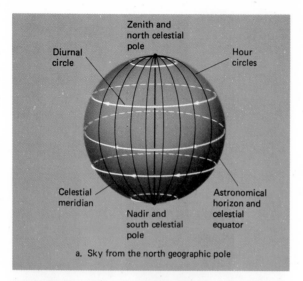

a. Sky from the north geographic pole

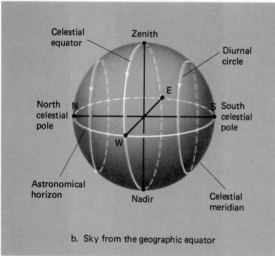

b. Sky from the geographic equator

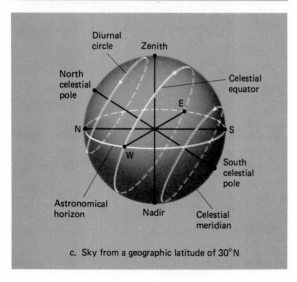

c. Sky from a geographic latitude of 30°N

dinates, the *ecliptic* and the *galactic* systems. The principal reference circle for the ecliptic system is the ecliptic; for the galactic system it is the central plane of the Milky Way. The four systems are summarized in Table A2.1.

STAR MAPS

Although astronomical coordinate systems are necessary in order to make precision observations, a general knowledge of the sky can be obtained with the aid of star charts (see Figures A2.5 to A2.8). The starting point for learning the constellations is Polaris, the North Star, which is located about 1° from the north celestial pole.

Five prominent constellations revolve around Polaris: Ursa Major, the Big Bear; Ursa Minor, the Little Bear; Cepheus, the King; Cassiopeia, the Queen; and Draco, the Dragon. These are called *circumpolar constellations* because they are visible all night and every night to observers located north of about 40° north latitude.

To locate the five circumpolar constellations, refer to Figure A2.5, in which Polaris is in the upper part of the figure and the five constellations are around it. Polaris, which marks the end of the handle of the Little Dipper and the tip of the tail of the Little Bear, locates the constellation of Ursa Minor. This constellation is easy to locate because its clear, bright asterism is oriented so that either dipper can always pour into the other. The two end stars in the bowl of the Big Dipper, Merak and Debhe, are called the "pointers" because a line drawn through them always points to Polaris.

Autumn Constellations

Three beautiful constellations are visible almost directly overhead in the autumn months (see Figure A2.5). They are Cygnus, the Swan, Lyra, the Harp, and Aquila, the Eagle, and they all lie in the Milky Way. The brightest star in each of these constellations—Deneb in Cygnus (the tail of the swan), Vega in Lyra (one corner of the small triangle), and Altair in Aquila (the middle star in the straight line)—form the *summer triangle*, which is almost a right triangle. The asterisms are the Northern Cross for Cygnus, a rectangle and triangle for Lyra, and a straight line of three stars for Aquila.

Winter Constellations

The winter hunting scene, the outstanding feature of the winter sky (Figure A2.6), comprises several constellations; Orion, the Mighty Hunter; his two dogs, Canis Major, the Big Dog, and Canis Minor, the Little Dog; his adversary, Taurus, the Bull; and his prey, Lepus, the Hare. The asterism of Orion is the hourglass. The star Betelgeuse (bright red) marks the right shoulder. The star Rigel marks the left knee. Orion wears a belt of three stars, the top star, Mintaka, lying almost on the celestial equator. The curved row

FIGURE A2.3
Relationship between equatorial and horizon coordinate systems for three different geographical locations.

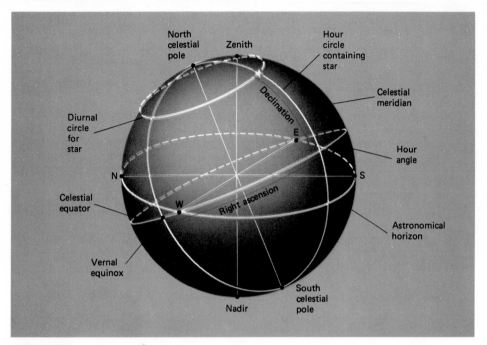

FIGURE A2.4

The equatorial system of coordinates, including its relationship to the horizon system. As the earth rotates, the hour angle of the vernal equinox continually increases, repeating after a 360° rotation of the earth.

of stars in front and to the west of Orion represents the lion, which Orion is holding with his left hand. Below Orion is Lepus, and following the mighty hunter, to the east, are his two dogs. Canis Major has no asterism because the outline of a dog can be easily visualized from the arrangement of the stars. The brightest star in the sky is Sirius, which marks the nose of the Big Dog. The con-

stellation of Canis Minor is difficult to visualize because most of its stars are very dim. Its brightest star, Procyon, together with Sirius and Betelgeuse form the *winter triangle*. Completing the scene is Taurus, a zodiacal constellation, which is in front and to the west of Orion. Its asterism is the open cluster of stars, the Hyades, that marks the head of the bull. Only the front part of the bull's body is

TABLE A2.1
Astronomical Coordinate Systems

Coordinate System	Principal Axis	Principal Reference Circle	Coordinate (units)	Secondary Reference Circles	Coordinate (units)
Horizon	Zenith-nadir	Astronomical horizon	Azimuth (0°–360°)	Vertical circle	Altitude (±0°–90°)
Equatorial	North-south celestial pole	Celestial equator	Right ascension (0ʰ–24ʰ)	Hour circle	Declination (±0°–90°)
Ecliptic	North-south ecliptic pole	Ecliptic	Celestial longitude (0°–360°)	No name	Celestial latitude (±0°–90°)
Galactic	North-south galactic pole	Plane of Milky Way	Galactic longitude (0°–360°)	No name	Galactic latitude (±0°–90°)

visible because the Greeks imagined that the bull was swimming in the Mediterranean Sea. The right eye of the bull is the bright star Aldebaran. The open cluster of the Pleiades is located in the left shoulder of the bull.

Spring Constellations

The prominent constellations in Figure A2.7) are three zodiacal constellations: Cancer, the Crab, Leo, the Lion, and Virgo, the Virgin. The brightest and most beautiful constellation is Leo, and its asterism is a sickle and a triangle. The sickle represents the head of the lion; the triangle, its hindquarters. To the west of the sickle is a hazy spot of light, which is the Praesepe open cluster, also called the "beehive," located between two faint stars. This star group marks the constellation of Cancer, whose stars are very dim. To the east of the triangle in Leo is the constellation of Virgo. The location of this constellation is also marked by the bright star Spica, which represents the sheaf of wheat in Virgo's left hand.

Summer Constellations

There are several prominent summer constellations shown in Figure A2.8: Boötes, the Bear Driver; Corona Borealis, the Northern Crown; Hercules, the Kneeler; Serpens, the Serpent; Ophiuchus, the Serpent Carrier; and the zodiacal constellations of Libra, the Scales, Scorpius, the Scorpion, and Sagittarius, the Archer.

FIGURE A2.5
AUTUMN SKIES

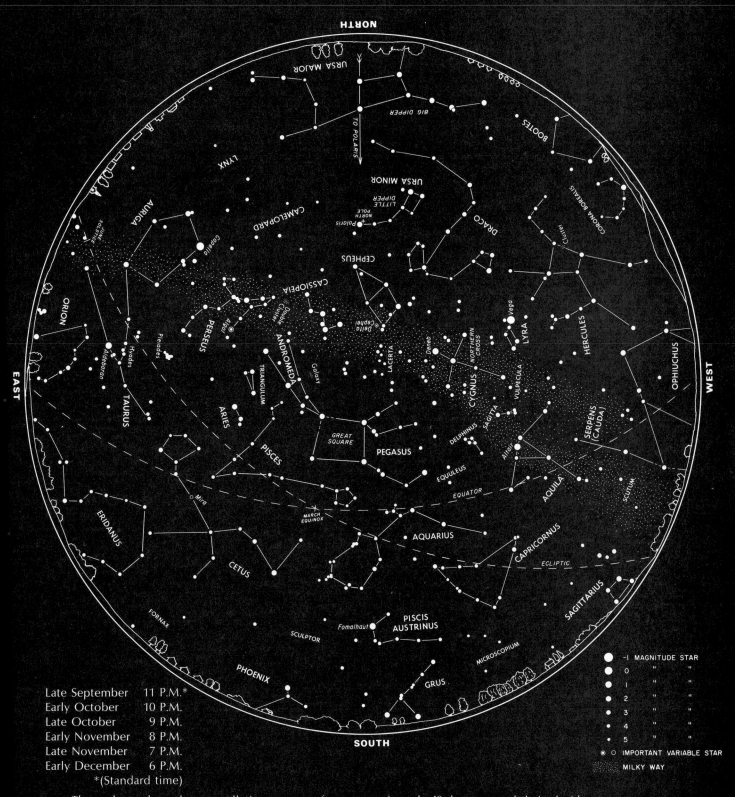

Late September 11 P.M.*
Early October 10 P.M.
Late October 9 P.M.
Early November 8 P.M.
Late November 7 P.M.
Early December 6 P.M.
*(Standard time)

These charts show the constellations as seen from approximately 40 degrees north latitude (the average latitude of the continental United States, around which most of the population is grouped). To best use

CHARTS BY GEORGE LOVI

FIGURE A2.6
WINTER SKIES

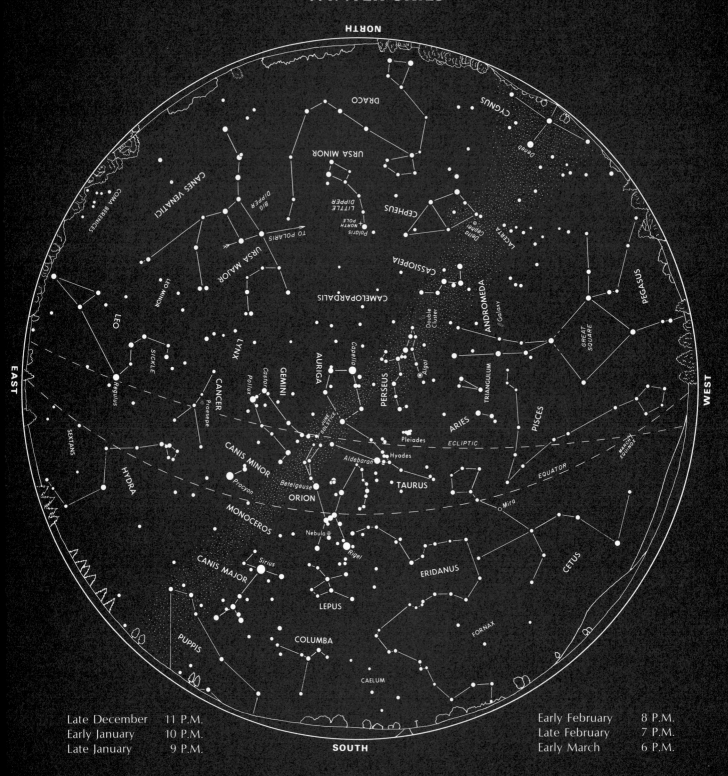

Late December   11 P.M.
Early January   10 P.M.
Late January   9 P.M.

Early February   8 P.M.
Late February   7 P.M.
Early March   6 P.M.

these charts, rotate the chart so that the label for the direction you are facing appears at the bottom. It is not necessary to follow an exact time schedule. Although all constellations that are visible above the horizon at the stated times are indicated, only the more prominent and/or most important constellations have lines connecting their principal stars.

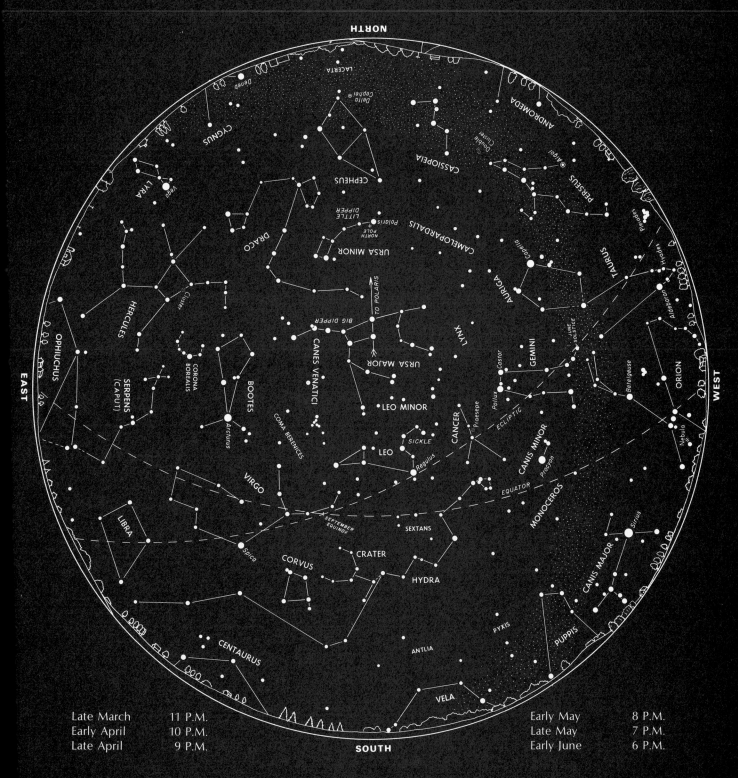

FIGURE A2.7
SPRING SKIES

Late March   11 P.M.
Early April   10 P.M.
Late April   9 P.M.

Early May   8 P.M.
Late May   7 P.M.
Early June   6 P.M.

Two coordinate lines are shown: the celestial equator and the ecliptic. It is important to remember that the moon and the planets stay close to the ecliptic—so that they often appear as strange "stars" within the

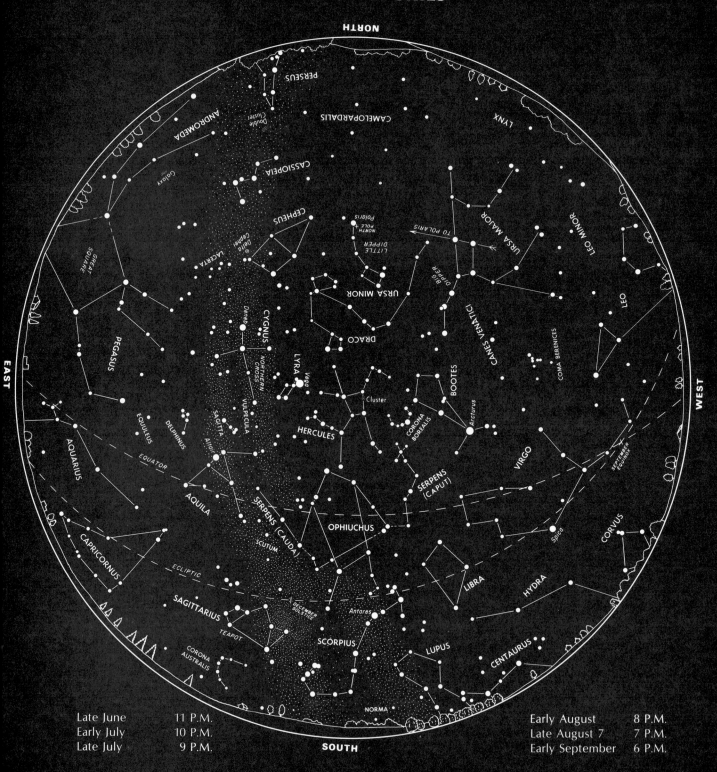

constellations that lie along this circle. Notice also the four season points along the ecliptic—the equinoxes and solstices—which indicate the sun's position at the beginning of each season.

Appendix 3
Time Systems and the Calendar

Time is measured basically by the earth's rotation. A convenient reference position from which to count time is the observer's celestial meridian. Either the vernal equinox (γ) or the sun serves as "hour hand" to denote passage of time since each marker is carried westward by the rotation of the sky.

Sidereal Time

One *sidereal day* is equal to the interval between two successive crossings of the vernal equinox over the observer's meridian. It corresponds to one complete rotation of the earth on its axis. The *local sidereal time* (LST) is given by the local hour angle of the vernal equinox (LHAγ), which is equal to the sum of any object's right ascension and hour angle (see Figure A2.4). It progresses uniformly at the rate of 15° to the hour and can be read from a sidereal clock whose rate is that of the earth's rotation. The clock reads $0^h0^m0^s$ at the instant the vernal equinox crosses the meridian; $1^h0^m0^s$ when the vernal equinox has moved 15° west of the meridian, corresponding to a local hour angle of 1^h; and so on around the sky through 24^h, when a new sidereal day begins.

Solar Time

Apparent solar time is given by the local hour angle of the sun, (LHA$\odot$) + 12^h, to start the solar day at midnight. One apparent solar day is equal to the interval between two consecutive passages of the sun over the observer's meridian. Time measured by the apparent or real sun is slightly variable for two reasons: First, solar apparent annual motion along the ecliptic varies a little because of the earth's orbital eccentricity (recall Kepler's law of areas, page 35); second, movement of the real sun along the ecliptic when projected onto the celestial equator where hour angle is measured varies slightly. A more suitable clock time is one that does not change in an irregular manner from day to day.

This is accomplished by introducing an imaginary sun called the *mean sun*. It moves uniformly eastward along the celestial equator at a daily rate that is equal to the average daily rate of the real sun so that both suns arrive at the vernal equinox simultaneously 1 year later which is where they started together. *Mean solar time* is equal to the local hour angle of the mean sun, (LHA$\overline{\odot}$) + 12^h. Since mean solar time progresses uniformly, clocks can run on this kind of time, which is kept by us in everyday life. The greatest difference between the apparent and mean solar time, which varies during the year, is about $\pm15^m$.

The *tropical year*, the year of the seasons, is equal to 365.2422 mean solar days. It represents the time it takes the sun to complete one revolution around the ecliptic with respect to the vernal equinox. The *sidereal year*, on the other hand, is the period of the sun's revolution with respect to the stars, equivalent to the period of the earth's orbital revolution. The mean solar day is longer than the sidereal day by 3^m56^s of solar time because of the earth's annual motion around the sun; see Figure A3.1.

STANDARD TIME

Obviously mean solar time is different at places separated by differences in longitude. To get around this inconvenience, the *standard time system* was introduced in 1884. There are 24 time zones, each theoretically 15° wide with the zero time zone centered on Greenwich and proceeding east or west of Greenwich. Through the approximate center of each zone runs the standard meridian, whose local mean time is the standard time within the zone. In the United States and Canada the standard meridians are at 60°, 75°, 90°, 105°, 120°, 135°, and 150°. In the continental United States the time zones are eastern standard time (EST), central standard time (CST), mountain standard time (MST), and Pacific standard time (PST), respectively 5^h, 6^h, 7^h, and 8^h behind Greenwich mean time (GMT), also known as universal time (UT). The time changes by 1^h as the traveler crosses a zone boundary. The zones frequently have irregular boundaries to suit local conditions.

The successive time zones run either to the west or to the east of Greenwich until they meet halfway around the world at 180° longitude at the international date line, which passes through the center of the 12^h zone. The half portion of the zone west of the line is 1 day ahead of the other half east of the line even though the standard time is the same in each half. A traveler crossing the date line from Tokyo to San Francisco, for example, gains a day; one traveling in the opposite direction loses a day.

EARLY CALENDARS

The common calendar of ancient peoples was based on the lunar cycle of 29.5 days because changes in the phases of the moon are readily apparent. Since 12 lunar months cannot be contained in a tropical year a whole number of times, people later added an extra month from time to time to bring the seasons back on schedule. Attempts to synchronize the lunar month with the year never proved satisfactory. The Egyptians were the first to base their calendar on a tropical year of $365\frac{1}{4}$ days. Their year consisted of 12 months of 30 days each, with 5 extra days at the end of the year set aside for celebrations. The early Romans employed a complicated lunar calendar of 10 months.

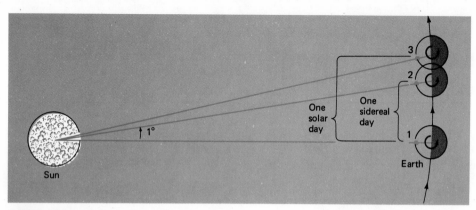

FIGURE A3.1
Why the solar day is longer than the sidereal day. With the observer at position 1, the sun is on the celestial meridian. After a 360° rotation relative to the stars the earth has moved about 1° along its orbit, and the observer will be at position 2, where the sun is 1° east of the celestial meridian. The earth must rotate one additional degree, which is done in approximately 4 minutes, in order to bring the sun to the observer's meridian in position 3.

JULIAN CALENDAR

The imprecise Roman calendar was replaced by one that became known as the *Julian calendar* because it was ordered by Julius Caesar with the advice of the astronomer Sosigenes. The year before its adoption (on January 1, 45 B.C.) had to be made 445 days long in order to bring the seasons back on schedule. The Julian calendar consisted of 365 days with every fourth year a leap year of 366 days.

In A.D. 321 the emperor Constantine introduced the 7-day week and set aside Sunday as the first day of the week to be considered as a Christian day of worship. The Council of Nicaea, convened in A.D. 325, established the rule for Easter Sunday so that it is the first Sunday after the fourteenth day of the moon (around the time of full moon) on or after the date of the vernal equinox.

GREGORIAN CALENDAR

Now the true length of the tropical year is $365^d5^h48^m46^s$, or about 11^m14^s shorter than $365\frac{1}{4}$ days. Hence the longer calendar year results in a discrepancy of 3 days in 400 years. By A.D. 1582 the accumulated difference amounted to 10 days so that the date of the vernal equinox had retreated from March 21 at the time of the Council of Nicaea to March 11. Accordingly Pope Gregory XIII called upon the astronomer Clavius to revise the Julian calendar. It was decreed that 10 days be dropped from the calendar. The day following Thursday, October 4, 1582, thereby became Friday, October 15, 1582. To avoid a future calendar discrepancy, only the century year divisible by 400 was to be a leap year (such as 1600 and 2000). This *Gregorian calendar* was readily adopted in Catholic countries. The Lutherans and Protestants finally made the adoption in 1700. When Great Britain and the American colonies changed to the Gregorian calendar in 1752, September 2 was followed by September 14. Early in this century other countries in Europe and Asia also followed. The present calendar is accurate 1 day in 3300 years.

Appendix 4
Periodic Table of the Elements

The number to the left of the chemical symbol is the atomic number; the number below is the atomic mass. When the atomic mass is not accurately known, the mass of the most stable isotope is given in parentheses.

1	2	3	4	5	6	7	8	9	10	11	12	13	14	15	16	17	18
1 H hydrogen 1.008																	2 He helium 4.003
3 Li lithium 6.94	4 Be beryllium 9.01											5 B boron 10.81	6 C carbon 12.01	7 N nitrogen 14.01	8 O oxygen 16.00	9 F fluorine 19.00	10 Ne neon 20.18
11 Na sodium 22.99	12 Mg magnesium 24.31											13 Al aluminum 26.98	14 Si silicon 28.09	15 P phosphorus 30.97	16 S sulfur 32.06	17 Cl chlorine 35.45	18 Ar argon 39.95
19 K potassium 39.10	20 Ca calcium 40.08	21 Sc scandium 44.96	22 Ti titanium 47.90	23 V vanadium 50.94	24 Cr chromium 52.00	25 Mn manganese 54.94	26 Fe iron 55.85	27 Co cobalt 58.93	28 Ni nickel 58.71	29 Cu copper 63.55	30 Zn zinc 65.37	31 Ga gallium 69.72	32 Ge germanium 72.59	33 As arsenic 74.92	34 Se selenium 78.96	35 Br bromine 79.90	36 Kr krypton 83.80
37 Rb rubidium 85.47	38 Sr strontium 87.62	39 Y yttrium 88.91	40 Zr zirconium 91.22	41 Nb niobium 92.91	42 Mo molybdenum 95.94	43 Tc technetium 98.91	44 Ru ruthenium 101.07	45 Rh rhodium 102.91	46 Pd palladium 106.4	47 Ag silver 107.87	48 Cd cadmium 112.40	49 In indium 114.82	50 Sn tin 118.69	51 Sb antimony 121.75	52 Te tellurium 127.60	53 I iodine 126.90	54 Xe xenon 131.30
55 Cs cesium 132.9	56 Ba barium 137.34	See below 57–71	72 Hf hafnium 178.49	73 Ta tantalum 180.95	74 W tungsten 183.85	75 Re rhenium 186.2	76 Os osmium 190.2	77 Ir iridium 192.22	78 Pt platinum 195.09	79 Au gold 196.97	80 Hg mercury 200.59	81 Tl thallium 204.37	82 Pb lead 207.19	83 Bi bismuth 208.98	84 Po polonium (210)	85 At astatine (210)	86 Rn radon (222)
87 Fr francium (223)	88 Ra radium 226.03	See below 89–103	104 Rf rutherfordium (261)	105 Ha hahnium (262)	106?												

Lanthanide Series (rare earth elements)

57 La lanthanum 138.91	58 Ce cerium 140.12	59 Pr praseodymium 140.91	60 Nd neodymium 144.24	61 Pm promethium (146)	62 Sm samarium 150.35	63 Eu europium 151.96	64 Gd gadolinium 157.25	65 Tb terbium 158.93	66 Dy dysprosium 162.50	67 Ho holmium 164.93	68 Er erbium 167.26	69 Tm thulium 168.93	70 Yb ytterbium 170.04	71 Lu lutetium 174.97

Actinide Series

89 Ac actinium (227)	90 Th thorium 232.04	91 Pa protactinium 230.04	92 U uranium 238.03	93 Np neptunium 237.05	94 Pu plutonium (242)	95 Am americium (242)	96 Cm curium (245)	97 Bk berkelium (248)	98 Cf californium (252)	99 Es einsteinium (253)	100 Fm fermium (257)	101 Md mendelevium (257)	102 No nobelium (255)	103 Lw lawrencium (256)

Legend:
- Light Metals
- Transition Heavy Metals
- Nonmetals
- Inert Gases

(*Credits continued from page iv.*)
Peak National Observatory. *Fig. 5.7:* top left to right, Palomar Observatory, California Institute of Technology; Perkin-Elmer Corporation, OTD Division; Kitt Peak National Observatory; middle left to right, Gary Ladd; European Southern Observatory; Dale Cruikshank; bottom left to right, Anglo-Australian Observatory; Cerro Tololo Inter-American Observatory; Palomar Observatory, California Institute of Technology. *Fig. 5.10:* top left and right, Lawrence Berkeley Laboratory, University of California; bottom, Kitt Peak National Observatory. *Fig. 5.13:* Duncan Chesley, Institute for Astronomy, University of Hawaii. *Fig. 5.15:* NASA. *Fig. 5.17:* courtesy E-Systems, Inc. *Fig. 5.18:* top and center right and bottom, NASA; center left, Harvard-Smithsonian Center for Astrophysics.

Chapter 6

Page 117: NASA. *Fig. 6.1:* adapted from a diagram by B. Lovell. *Page 121:* from a poem by Diane Ackerman. Copyright © 1976, Harvard Magazine, Inc. Reprinted by permission. *Fig. 6.6:* photos, Lowell Observatory. *Fig. 6.7:* photo, Palomar Observatory, California Institute of Technology. *Fig. 6.8:* photos, Lick Observatory. *Fig. 6.10:* W. M. Sinton, Institute for Astronomy, University of Hawaii. *Fig. 6.12:* photo, U.S. Naval Observatory. *Fig. 6.13:* all NASA.

Chapter 7

Page 140: NASA. *Page 149:* from Christopher Morley, "The Hubbub of the Universe" in *Translations from the Chinese* (New York: Doubleday, Page & Company, 1927). Reprinted by permission of Harper & Row. *Fig. 7.8:* prepared from a *National Geographic* map. *Fig. 7.9:* courtesy John McHone, Arizona State University. *Fig. 7.13:* Jack Finch/Aurora Borealis Photo. *Fig. 7.14:* NASA. *Fig. 7.15:* Lick Observatory. *Fig. 7.16:* top left to right, Lick Observatory; Palomar Observatory, California Institute of Technology; middle left to right, Lick Observatory; NASA; bottom, NASA. *Fig. 7.17:* both NSSDC/NASA. *Figs. 7.18 and 7.19:* NASA. *Fig. 7.20:* Donald Davis and Donald Wilhelm, U.S. Geological Survey, Menlo Park, CA. *Fig. 7.21:* from *Sky and Telescope*, 53:3 (March 1977), p. 166. Redrawn by permission.

Chapter 8

Page 169: NASA. *Figs. 8.1–8.3:* NSSDC/NASA. *Figs. 8.4 and 8.5:* ARC/NASA. *Fig. 8.6:* top to bottom, NSSDC/NASA; NSSDC/NASA; James W. Head, Brown University; NSSDC/NASA. *Page 176:* Yerkes Observatory. *Figs. 8.7–8.10:* NASA. *Fig. 8.11:* NSSDC/NASA. *Figs. 8.13–8.15:* Jet Propulsion Laboratory, NASA. *Fig. 8.16:* Andrew Chaikin. *Fig. 8.17:* Meteor Crater Enterprises.

Chapter 9

Page 197: NASA. *Fig. 9.1:* adapted from Jet Propulsion Laboratory diagrams, courtesy *Sky and Telescope*. *Fig. 9.2:* both NASA. *Fig. 9.4:* both Jet Propulsion Laboratory, NASA. *Fig. 9.5:* left, Jet Propulsion Laboratory, NASA; right, NASA. *Fig. 9.6:* NASA. *Figs. 9.7 and 9.9:* Jet Propulsion Laboratory, NASA. *Figs. 9.10–9.13:* NASA. *Fig. 9.14:* all NASA. *Fig. 9.15:* all NASA. *Fig. 9.17:* adapted from a diagram by NASA. *Fig. 9.19:* right, Lowell Observatory; below, Mount Wilson and Las Campanas Observatories, Carnegie Institution of Washington.

Chapter 10

Fig.10.2: reproduced by permission of the Smithsonian Institution Press from the *Smithsonian Institution Annual Report, 1949,* "The Origin of the Earth," by Thornton Page: Plate 1. Washington,

D.C.: Government Printing Office, 1950. *Page 226:* from Arthur Guiterman, "Ode to the Amoeba," in *Gaily the Troubador,* © E. P. Dutton, 1936. Renewed 1964 by Vida Lindo Guiterman. Reprinted by permission of Louise H. Sclove. *Fig. 10.7:* Stella Snead/Bruce Coleman, Inc. *Fig. 10.13:* J. W. Schopf and Elso S. Barghoorn. *Fig. 10.14:* photo by John Fields, The Trustees, The Australian Museum.

Chapter 11

Page 247: By permission, Harvard College Observatory, Cambridge, MA. *Fig. 11.1:* top left to right (photos), all Mount Wilson and Las Campanas Observatories, Carnegie Institution of Washington; bottom left to right, Princeton University Observatory; High Altitude Observatory; upper right, Mount Wilson and Las Campanas Observatories, Carnegie Institution of Washington; lower right, Lick Observatory. *Fig. 11.6:* Mount Wilson and Las Campanas Observatories, Carnegie Institution of Washington. *Fig. 11.7:* Dr. Robert Leighton, California Institute of Technology. *Fig. 11.10:* both Kitt Peak National Observatory. *Fig. 11.12:* Dr. Mitsuo Kanno, Hida Observatory. *Fig. 11.13:* all Mount Wilson and Las Campanas Observatories, Carnegie Institution of Washington. *Fig. 11.14:* Sacramento Peak Observatory, Association of Universities for Research in Astronomy, Inc. *Fig. 11.16:* left, NASA; middle and right, Big Bear Solar Observatory. *Fig. 11.17:* courtesy of Harvard College Observatory/NASA. *Fig. 11.19:* Lockheed Solar Observatory. *Fig. 11.20:* From "The Case of the Missing Sunspots," by John Eddy, *Scientific American,* May 1977, © Scientific American, Inc. All rights reserved.

Chapter 12

Page 277: Mount Wilson and Las Campanas Observatories, Carnegie Institution of Washington. *Page 279:* Reprinted with permission of Charles Scribner's Sons from "Octaves XI," *The Collected Poems of Edwin Arlington Robinson* (New York: Macmillan Co., 1935, 1937). *Fig. 12.2:* Dr. Martha Liller, Harvard College Observatory. *Figs. 12.4, 12.5, and 12.8:* from *An Atlas of Low-Dispersion Grating Stellar Spectra,* Kitt Peak National Observatory. *Fig. 12.6a:* After L. H. Aller. *Fig. 12.13:* adapted from *Astronomy: The Cosmic Journey,* Second Edition, by William K. Hartmann. © 1982 by Wadsworth, Inc. Reprinted by permission of Wadsworth Publishing Company, Belmont, CA 94002. *Fig. 12.18:* Lick Observatory. *Fig. 12.19:* European Southern Observatories. *Fig. 12.20:* Palomar Observatory, California Institute of Technology.

Chapter 13

Page 305: plate selected by Dr. D. J. MacConnell, University of Michigan Observatory photograph. *Page 308:* courtesy Yerkes Observatory. *Fig. 13.6:* after H. L. Johnson. *Fig. 13.7:* after A. Sandage and H. L. Johnson. *Page 319:* courtesy Yerkes Observatory.

Chapter 14

Page 330: Palomar Observatory, California Institute of Technology. *Fig. 14.1:* photo, Mount Wilson and Las Campanas Observatories, Carnegie Institution of Washington. *Page 333:* Bell Laboratories. *Fig. 14.3:* adapted from *Astronomy: The Cosmic Journey,* Second Edition, by William K. Hartmann. © 1982 by Wadsworth, Inc. Reprinted by permission of Wadsworth Publishing Company, Belmont, CA 94002. *Fig. 14.6:* Palomar Sky Survey. *Page 340:* from "Desert Places" from *The Poetry of Robert Frost,* edited by Edward Connery Lathem. Copyright 1936 by Robert Frost. Copyright © 1964 by Lesley Frost Ballantine. Copyright © 1969 by Holt, Rinehart and Winston. Reprinted by permission of Holt, Rinehart & Winston, Publishers. *Fig. 14.8:* Kitt Peak National Observatory. *Fig. 14.9:* Kitt

Peak National Observatory photo, Mathew Schneps and Edward Wright. *Fig. 14.10:* Mount Wilson and Las Campanas Observatories, Carnegie Institution of Washington. *Figs. 14.11 and 14.12:* Palomar Observatory, California Institute of Technology. *Page 344:* from Robinson Jeffers, "Nova." Copyright 1937 and renewed 1960 by Robinson Jeffers. Reprinted from *The Selected Poetry of Robinson Jeffers,* by Robinson Jeffers, by permission of Random House, Inc. *Fig. 14.13:* Palomar Observatory, California Institute of Technology. *Fig. 14.14:* courtesy of Bart J. Bok, University of Arizona. *Fig. 14.15:* redrawn by permission of The University of Chicago Press from *Astrophysical Journal,* vol. 141, p. 993. All rights reserved. © 1965 by The American Astronomical Society. *Fig. 14.17:* From "The Birth of Massive Stars," by M. Zeilik, *Scientific American,* April 1978. Scientific American, Inc. All rights reserved. Background photo, Kitt Peak National Observatory. *Fig. 14.18:* based on a diagram by Merle Walker, Lick Observatory.

Chapter 15

Page 353: painting by Lois Cohen, Griffith Observatory. *Fig. 15.1:* after Icko Iben, Jr. *Page 356:* from Robinson Jeffers, "Margrave." Copyright, 1932 and renewed 1960 by Robinson Jeffers. Reprinted from *The Selected Poetry of Robinson Jeffers,* by Robinson Jeffers, by permission of Random House, Inc. *Fig. 15.5:* after Icko Iben, Jr. © 1971 *Astronomical Society of the Pacific,* vol. 83, p. 697. Redrawn by permission. *Fig. 15.7:* after B. Paczynski and C. R. O'Dell. *Fig. 15.13:* J. Wampler and J. C. Miller, Lick Observatory. *Figs. 15.18 and 15.20:* from William Kaufmann, *The Cosmic Frontiers of General Relativity,* pp. 176, 228. Copyright © 1977 by Little, Brown and Company (Inc.). Redrawn by permission.

Chapter 16

Page 382: National Gallery, London. *Fig. 16.1:* Lund Observatory. *Fig. 16.2:* Mount Wilson and Las Campanas Observatories, Carnegie Institution of Washington. *Fig. 16.4:* redrawn from the *Astrophysical Journal;* from a map prepared by Dennis Downes, Alan Maxwell, and M. L. Meeks. *Fig. 16.10:* adapted from *Astronomy: The Cosmic Journey,* Second Edition, by William K. Hartmann. © 1982 by Wadsworth, Inc. Reprinted by permission of Wadsworth Publishing Company, Belmont, CA 94002. *Fig. 16.11:* courtesy of G. Westerhout. *Fig. 16.14:* both Max Planck Institut für Radioastronomie, Bonn. *Fig. 16.17:* left, Naval Research Laboratories; right, Goddard Space Flight Center. *Fig. 16.18:* top, adapted from John Archibald Wheeler, "The Universe as Home for Man," *American Scientist,* vol. 62 (1974), p. 684. *Fig. 16.19:* top, both NASA; bottom, Harvard-Smithsonian Center for Astrophysics.

Chapter 17

Page 412: Palomar Observatory, California Institute of Technology; inset, © Dr. Eric B. Jensen, Kitt Peak National Observatory, 36-inch telescope. *Fig. 17.1:* Lick Observatory. *Page 414:* Mount Wilson and Las Campanas Observatories, Carnegie Institution of Washington. *Fig. 17.3:* all Palomar Observatory, California Institute of Technology. *Fig. 17.4:* Dr. Francesco Bertola. *Figs. 17.5 and 17.6:* Mount Wilson and Las Campanas Observatories, Carnegie Institution of Washington. *Fig. 17.7:* Kitt Peak National Observatory. *Fig. 17.9:* European Southern Observatory. *Fig. 17.10:* Cerro Tololo Inter-American Observatory. *Fig. 17.11:* Dr. B. A. Vorontsov-Valyaminov, Sternberg Astronomical Institute, Moscow University. *Fig. 17.12:* photo from Hale Observatories; drawings redrawn by permission of the publisher from "Violent Tides between Galaxies" by A. Toomre and J. Toomre, *Scientific American,* December 1973. © Scientific American, Inc. All rights reserved. *Fig. 17.14:* Mount Wilson and Las Campanas Observatories, Carnegie Institution of Washington. *Fig. 17.15:* © 1978 by the International Astronomical

Union. *Fig. 17.16:* adapted from *Astronomy: The Cosmic Journey,* Second Edition, by William K. Hartmann. © 1982 by Wadsworth, Inc. Reprinted by permission of Wadsworth Publishing Company, Belmont, CA 94002. *Fig. 17.17:* top left to right, Palomar Observatory, California Institute of Technology; two middle, courtesy of Leon van Speybroeck, Harvard-Smithsonian Center for Astrophysics; Darrel Emerson, Institut du Radio Astronomie Millimetrique, Universitaire de Grenoble; middle, courtesy of Leon van Speybroeck, Harvard-Smithsonian Center for Astrophysics; Max Planck Institut für Radioastronomie, Bonn; bottom, B. Milliard. *Fig. 17.18:* Palomar Observatory, California Institute of Technology.

Chapter 18

Page 440: courtesy Allen Sandage, Mount Wilson and Las Campanas Observatories, Carnegie Institution of Washington. *Fig. 18.2:* Dr. W. W. Morgan. *Fig. 18.4:* top left to right, National Radio Astronomy Observatory, operated by Associated Universities, Inc. under contract with the National Science Foundation; Palomar Observatory, California Institute of Technology; National Radio Astronomy Observatory; middle, Palomar Observatory, California Institute of Technology; inset, Lick Observatory; bottom, Palomar Observatory, California Institute of Technology; far right, by permission of the publisher from Wellington, Miley, and van der Laan, *Nature,* vol. 244 (1974), p. 502. Redrawn by permission. Photo by Palomar Sky Survey. *Fig. 18.5:* all Palomar Observatory, California Institute of Technology. *Fig. 18.6:* Lick Observatory. *Fig. 18.7:* photo from European Southern Observatory 3.6-meter telescope by P. A. Wehinger, S. Wyckoff, and T. Gehren. *Fig. 18.8:* Dr. Stephen C. Unwin, California Institute of Technology. *Page 451:* From George Gamow, *A Star Called the Sun.* Copyright © 1964 by George Gamow. Reprinted by permission of Viking Penguin Inc. *Fig. 18.9:* Courtesy Dr. G. Chincarini, Palomar Observatory. *Fig. 18.10:* Dr. Alan Stockton, Institute for Astronomy, University of Hawaii. *Fig. 18.11:* Kitt Peak National Observatory; inset, Palomar Observatory, California Institute of Technology. *Fig. 18.12.* courtesy of Christine Jones, Harvard-Smithsonian Center for Astrophysics. *Fig. 18.13:* Kitt Peak National Observatory. *Fig. 18.14:* Palomar Observatory, California Institute of Technology. *Fig. 18.15:* adapted courtesy of K. Lang et al. and *The Astrophysical Journal,* published by the University of Chicago Press; © 1979 the American Astronomical Society. *Fig. 18.16: Sky and Telescope* diagram by Robert Hess, courtesy R. Brent Tully. *Fig. 18.18:* adapted courtesy of S. A. Gregory and L. A. Thompson and *The Astrophysical Journal,* published by the University of Chicago Press; © 1979 The American Astronomical Society. *Fig. 18.19:* adapted courtesy of H. J. Rood and *The Astrophysical Journal,* published by the University of Chicago Press; © 1979 The American Astronomical Society.

Chapter 19

Page 466: University Library, Cambridge, England. *Fig. 19.7:* courtesy Halton Arp, Mount Wilson and Las Campanas Observatories.

Chapter 20

Page 492: courtesy of the Boston Museum of Science and permission of Harvard College Observatory, Cambridge, MA. *Fig. 20.2:* top and middle, courtesy Museum of Modern Art Film Library; bottom, Dr. Alan Hynek, Center for U.F.O. Studies. *Fig. 20.3:* Reprinted by permission of the publisher from "The Search for Extraterrestrial Intelligence," by Carl Sagan and Frank Drake, *Scientific American,* May 1975. © Scientific American, Inc. All rights reserved. *Figs. 20.4–20.6:* NASA. *Fig. 20.8:* Sherman J. Larson, Center for U.F.O. Studies. *Fig. 20.9:* Norman Hammond, *Ancient Mayan Civilization,* Rutgers University Press, 1982, from a rubbing by Merle Greene Robertson. *Fig. 20.10:* NASA.

Index

Numbers in *italics* refer to tables and figures.

The Elements

Chemical Symbol	Name	Atomic Number	Atomic Weight	Number of Stable Isotopes	Solar Abundance[a] (H = 1×10^{12})
Ac	Actinium	89	227	0	
Al	Aluminum	13	27.0	1	3.3×10^6
Am	Americium	95	242	0	
Sb	Antimony	51	121.8	2	1.0×10^1
A	Argon	18	40.0	3	1.0×10^6
As	Arsenic	33	74.9	1	
At	Astatine	85	210	0	
Ba	Barium	56	137.3	7	1.2×10^2
Bk	Berkelium	97	248	0	
Be	Beryllium	4	9.0	1	1.4×10^1
Bi	Bismuth	83	209.0	1	$< 7.9 \times 10^1$
B	Boron	5	10.8	2	1.3×10^2
Br	Bromine	35	79.9	2	
Cd	Cadmium	48	112.4	8	7.1×10^1
Ca	Calcium	20	40.1	6	2.2×10^6
Cf	Californium	98	252	0	
C	Carbon	6	12.0	2	4.2×10^8
Ce	Cerium	58	140.1	4	3.5×10^1
Cs	Cesium	55	132.9	1	$< 7.9 \times 10^1$
Cl	Chlorine	17	35.5	2	3.2×10^5
Cr	Chromium	24	52.0	4	5.1×10^5
Co	Cobalt	27	58.9	1	7.9×10^4
Cu	Copper	29	63.6	2	1.1×10^4
Cm	Curium	96	245	0	
Dy	Dysprosium	66	162.5	7	1.1×10^1
Es	Einsteinium	99	253	0	
Er	Erbium	68	167.3	6	5.8
Eu	Europium	63	152.0	2	5.0
Fm	Fermium	100	257	0	
F	Fluorine	9	19.0	1	3.6×10^4
Fr	Francium	87	223	0	
Gd	Gadolinium	64	157.3	7	1.3×10^1
Ga	Gallium	31	69.7	2	6.3×10^2
Ge	Germanium	32	72.6	5	3.2×10^3
Au	Gold	79	197.0	1	5.6
Hf	Hafnium	72	178.5	6	6.3
Ha	Hahnium	105	262	0	
He	Helium	2	4.0	2	6.3×10^{10}
Ho	Holmium	67	164.9	1	
H	Hydrogen	1	1.0	2	1.0×10^{12}
In	Indium	49	114.8	2	4.5×10^1
I	Iodine	53	126.9	1	
Ir	Iridium	77	192.2	2	7.1
Fe	Iron	26	55.9	4	3.2×10^7
Kr	Krypton	36	83.8	6	
La	Lanthanum	57	138.9	2	1.3×10^1
Lr	Lawrencium	103	256	0	
Pb	Lead	82	207.2	4	8.5×10^1
Li	Lithium	3	6.9	2	1.0×10^2
Lu	Lutetium	71	175.0	2	5.8
Mg	Magnesium	12	24.3	3	4.0×10^7
Mn	Manganese	25	54.9	1	2.6×10^5
Md	Mendelevium	101	257	0	
Hg	Mercury	80	200.6	7	$< 1.3 \times 10^2$
Mo	Molybdenum	42	95.9	7	1.4×10^2